PRINCIPLES OF
ANALYTICAL ELECTRON
MICROSCOPY

PRINCIPLES OF ANALYTICAL ELECTRON MICROSCOPY

Edited by

David C. Joy

AT&T Bell Laboratories
Murray Hill, New Jersey

Alton D. Romig, Jr.

Sandia National Laboratories
Albuquerque, New Mexico

and

Joseph I. Goldstein

Lehigh University
Bethlehem, Pennsylvania

PLENUM PRESS • NEW YORK AND LONDON

Library of Congress Cataloging in Publication Data

Principles of analytical electron microscopy.

Includes bibliographies and index.
1. Electron microscopy. I. Joy, David C., 1943– . II. Romig, Alton D. III. Goldstein, Joseph, 1939–
TA417.23.P75 1986 502'.8'25 86-16877
ISBN 0-306-42387-1

A Division of Plenum Publishing Corporation
233 Spring Street, New York, N.Y. 10013

Printed in the United States of America

PREFACE

Since the publication in 1979 of *Introduction to Analytical Electron Microscopy* (ed. J. J. Hren, J. I. Goldstein, and D. C. Joy; Plenum Press), analytical electron microscopy has continued to evolve and mature both as a topic for fundamental scientific investigation and as a tool for inorganic and organic materials characterization. Significant strides have been made in our understanding of image formation, electron diffraction, and beam/specimen interactions, both in terms of the "physics of the processes" and their practical implementation in modern instruments. It is the intent of the editors and authors of the current text, *Principles of Analytical Electron Microscopy*, to bring together, in one concise and readily accessible volume, these recent advances in the subject.

The text begins with a thorough discussion of fundamentals to lay a foundation for today's state-of-the-art microscopy. All currently important areas in analytical electron microscopy—including electron optics, electron beam/specimen interactions, image formation, x-ray microanalysis, energy-loss spectroscopy, electron diffraction and specimen effects—have been given thorough attention. To increase the utility of the volume to a broader cross section of the scientific community, the book's approach is, in general, more descriptive than mathematical. In some areas, however, mathematical concepts are dealt with in depth, increasing the appeal to those seeking a more rigorous treatment of the subject.

Although previous experience with conventional scanning and/or transmission electron microscopy would be extremely valuable to the reader, the text assumes no prior knowledge and therefore presents all of the material necessary to help the uninitiated reader understand the subject. Because of the extensive differences between this book and *Introduction to Analytical Electron Microscopy*, the current volume is far more than a second edition. *Principles of Analytical Electron Microscopy* easily stands alone as a complete treatment of the topic. For those who already use the first text, *Principles of Analytical Electron Microscopy* is an excellent complementary volume that will bring the reader up to date with recent developments in the field.

The text has been organized so that it can be used for a graduate course in analytical electron microscopy. It makes extensive use of figures and contains a complete bibliography at the conclusion of each chapter. Although the book was written by a number of experts in the field, every attempt was made to structure and organize each chapter identically. As such, the volume is structured as a true textbook. The volume can also be used as an individual learning aid for readers wishing to extend their own areas of expertise since the text has been compartmentalized into discrete topical chapters.

This preface would be incomplete if we did not acknowledge those who participated directly or indirectly in our efforts. The editors thank the many organizations and individuals who made *Principles of Analytical Electron Microscopy* possible. Without their support and assistance, the project would have never been completed. The Microbeam Analysis Society (MAS) and Electron Microscopy Society of America (EMSA) must be acknowledged for their initial sponsorship, which was essential in the earliest stages of this project. J. I. Goldstein expresses his gratitude for research support from the Materials Science Program of the National Aeronautics and Space Administration and from the Earth Sciences Division of the National Science Foundation. We all appreciate the encouragement and support of AT&T Bell Laboratories. D. C. Joy specifically acknowledges the support of AT&T Bell Laboratories management: L. C. Kimmerling, Manager, Materials Physics Research Department; G. Y. Chin, Director, Materials Research Laboratory; W. P. Schlichter, Executive Director, Materials Science and Engineering Division; and A. A. Penzias, Vice President, Research.

Finally, but most importantly, we all express our greatest appreciation to Sandia National Laboratories, operated by AT&T Technologies, Inc., for the United States Department of Energy under Contract Number DE-AC04-76DP00789. A. D. Romig, Jr., specifically acknowledges the support of Sandia Laboratories management: W. B. Jones, Supervisor, Physical Metallurgy Division; M. J. Davis, Manager, Metallurgy Department; R. L. Schwoebel, Director, Materials and Process Sciences; and W. F. Brinkman, Vice President, Research. It is through the generosity of Sandia National Laboratories that the text could be cast into its final form.

Our highest praise must go to Joanne Pendall, our Sandia Laboratories technical editor, who skillfully transformed the authors' rough drafts into an immaculate and professionally finished product. Without her hard work and dedicated efforts, the entire project would have never reached completion. We also acknowledge the support of the entire technical writing group at Sandia: K. J. Willis, Supervisor, Publication Services Division; D. Robertson, Manager, Technical Information Department; and H. M. Willis, Director, Information Services. The contributions of D. L. Humphreys, graphic art support; W. D. Servis, technical library; and A. B. Pritchard, text processing, are sincerely appreciated. Very special thanks go to our compositors, Emma Johnson, Tonimarie Stronach, and Steven Ulibarri.

D. C. Joy, Bell Laboratories
A. D. Romig, Jr., Sandia National Laboratories
J. I. Goldstein, Lehigh University

CONTENTS

CHAPTER 3 PRINCIPLES OF IMAGE FORMATION *J. M. Cowley*

CHAPTER 1

ELECTRON BEAM–SPECIMEN INTERACTIONS IN THE ANALYTICAL ELECTRON MICROSCOPE

D. E. Newbury

Center for Analytical Chemistry
National Bureau of Standards, Gaithersburg, Maryland

I. INTRODUCTION

Energetic electrons interact with the atoms and electrons of a specimen in a wide variety of ways, many of which can be used to obtain information when using the analytical electron microscope (AEM). It is the purpose of this chapter to describe the properties of these interactions so as to provide a basis for discussions of imaging and microanalysis in subsequent chapters. Only a cursory examination of the complex subject of electron interactions can be presented. A catalog of equations that describe the various forms of electron scattering will be provided, together with examples of the application of these equations to specific calculations. This "user's guide" to electron interactions will be accompanied by references to more complete treatments, which enable the interested reader to find additional detail, e.g., Bethe and Ashkin (1953). The separate topic of diffraction contrast will be treated at length in Chapter 9.

II. SCATTERING

The interaction of the beam electrons with the ionic cores of the atoms and loosely bound electrons takes place through various mechanisms of electron scattering, in which the direction and/or energy of the beam electron are changed, with the possibility of the energy transfer to the specimen and the consequent emission of some form of secondary radiation. Two general types of scattering are recognized:

- Elastic scattering—The direction of the electron trajectory is altered, but the energy remains essentially constant.
- Inelastic scattering—The magnitude of the electron velocity is altered, and the kinetic energy ($E = mv^2/2$, where m is the electron mass and v is the velocity) is reduced. Energy is transferred to the atoms of the sample.

Scattering processes are quantified by means of the cross section, σ, which is the probability that a process will occur. The cross section is defined as

$$\sigma = N/n_t n_i \qquad \text{events/e}^-/(\text{atom/cm}^2) \tag{1}$$

where N is the number of events of a certain type (elastic scattering events, inner shell ionizations, etc.) per unit volume (events/cm^3), n_t is the number of target sites per unit volume (atoms/cm^3), and n_i is the number of incident particles per unit area (electron/cm^2). Although cross sections are usually thought of as having dimensions of area (cm^2) and give the effective "size" of an atom, it should be recognized that the dimensionless quantities present in the definition of Eq. (1) are important in properly employing the cross section in a calculation. Thus, the complete dimensions of Eq. (1) are events/electron/(atom/cm^2).

The cross section is often more conveniently used by transforming it into a mean free path, λ, which is the mean distance the electron must travel through the specimen to undergo an average of one event of a particular type. The cross section can be converted into a mean free path by means of a dimensional argument

$$\sigma \frac{\text{events}}{\text{electron(atom/cm}^2)} \times N_0 \frac{\text{atoms}}{\text{mole}} \times \frac{1}{A} \frac{\text{moles}}{\text{g}} \times \rho \frac{\text{g}}{\text{cm}^3} = \frac{1}{\lambda} \frac{\text{events}}{\text{cm}}$$

or

$$\lambda = A/\sigma N_0 \rho \qquad \text{cm/event} \tag{2a}$$

where N_0 is Avogadro's number, A is the atomic weight, and ρ is the density. The mean free path for a given type of event, i, is obtained by substituting the appropriate cross section, σ_i, in Eq. (2a). If several different processes, a, b, c, etc., can occur, the total mean free path, λ_t, is found by calculating the mean free paths for the individual processes, λ_i, and combining them according to the equation

$$\frac{1}{\lambda_t} = \sum_i \frac{1}{\lambda_i} . \tag{2b}$$

Alternatively, we can calculate the probability that an event will take place, P (events/e^-). From the argument in Eq. (2a), the probability, P, for a given process is described by

$$P\,(\text{events}/e^-) = \frac{\sigma N_o \rho t}{A} \tag{3}$$

where t is the total specimen thickness.

III. ELASTIC SCATTERING

A. Elastic Scattering Cross Sections

As an energetic beam electron passes near the nucleus of an atom, the electron can be scattered elastically by the coulombic field of the nucleus, a process known as nuclear or Rutherford scattering. According to Wentzel (1927) and Mott and Massey (1965), the differential cross section for Rutherford scattering corrected for relativistic effects and screening of the nucleus by inner shell electrons is

$$\frac{d\sigma}{d\Omega} = \frac{Z^2e^4}{16\,E^2}\,|\sin^2(\theta/2) + (\theta_o^2/4)|^{-2}\,|1 - \beta^2\sin^2(\theta/2) + \pi\alpha\beta\,[\sin(\theta/2) - \sin^2(\theta/2)]| \tag{4a}$$

where the element of solid angle $d\Omega = 2\pi\sin\theta\,d\theta$, θ is the scattering angle measured relative to the incident electron trajectory, $0 \le \theta \le \pi$, Z is the atomic number, E is the beam energy expressed in keV units in all equations, θ_o is the screening parameter, and $\beta = v/c$. β can be conveniently calculated from

$$\beta = |\,1 - [1 + (E/511)]^{-2}\,|^{1/2} \tag{4b}$$

where 511 keV is the rest mass of the electron. The screening parameter has been given by Cosslett and Thomas (1964) as $\theta_o = 0.1167\,Z^{1/3}/E^{1/2}$ (radians). The factor, α, in Eq. (4a) (McKinley and Feshbach, 1948) has been given by Mott and Massey (1965) as $\alpha = Z/137$ for light elements. Tabulated values for α appropriate to heavier elements have been given by McKinley and Feshbach (1948). Because the exact cross section given by Eq. (4a) is not reducible to an analytic form due to the term, α, it is often convenient for purposes of estimating cross sections to ignore this factor. The uncertainty that this introduces has been calculated for several elements by Reimer and Krefting (1976), and a figure for germanium taken from their work is shown in Figure 1(a). In general, the magnitude of the deviation of the relativistic Rutherford cross section from the exact Mott cross section is ~20% for Ge at 100 keV, with the deviation varying as a function of angle. The deviation becomes larger for heavier elements and lower beam energies, as shown for gold in Figure 1(b). Reimer and Krefting (1976) have emphasized the importance of using the exact Mott cross section for accurate calculations, especially in the Monte Carlo electron trajectory simulation.

The differential Rutherford cross section can be expressed in the following terms (Bethe and Ashkin, 1953)

$$\sigma_R(\theta) = \frac{e^4 Z^2}{16(4\pi\epsilon_o E)^2}\;\frac{d\Omega}{\left[\sin^2\left(\frac{\theta}{2}\right) + \left(\frac{\theta_o^2}{4}\right)\right]^2} \tag{5a}$$

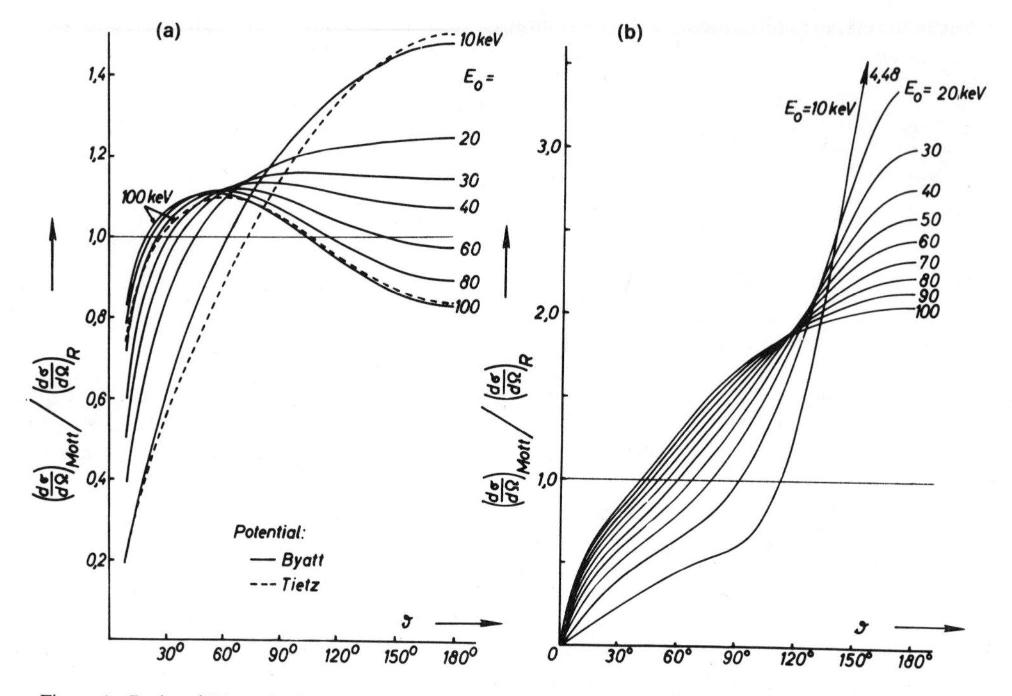

Figure 1 Ratio of Mott elastic scattering cross section to Rutherford elastic scattering cross section (from Reimer and Krefting, 1976). (a) Germanium; (b) Gold.

where $\theta_o^2/4$ is the screening parameter given by $\theta_o^2/4 = 3.4 \times 10^{-3}\ Z^{2/3}/E = \delta$ (E in keV) and ϵ_o is the dielectric constant. This equation can be corrected for relativistic effects by substituting the Bohr radius,

$$a_o = \frac{\epsilon_o\, h^2}{\pi m_o e^2} = 5.29 \times 10^{-9}\ \text{cm} \tag{5b}$$

where h is Planck's constant and m_o is the electron rest mass, and the relativistically corrected electron wavelength,

$$\lambda_R = \frac{h}{\sqrt{2m_oE\left(1 + \dfrac{E}{2m_oc^2}\right)}} = \frac{3.87 \times 10^{-9}}{E^{1/2}\,(1 + 9.79 \times 10^{-4}E)^{1/2}}\ (\text{cm}) \tag{5c}$$

into Eq. (5a) to yield

$$\sigma_R(\theta) = \frac{Z^2\,\lambda_R^4}{64\pi^4 a_o^2}\ \frac{d\Omega}{\left[\sin^2\!\left(\dfrac{\theta}{2}\right) + \dfrac{\theta_o^2}{4}\right]^2} . \tag{5d}$$

The total relativistic screened elastic cross section, σ_e, can be found by integrating Eq. (5d) with the relation $d\Omega = 2\pi \sin\theta d\theta$ over the range $0 \leq \theta \leq \pi$. Following the procedure given by Henoc and Maurice (1976) for this integration

$$\sigma_e = \frac{Z^2 \lambda_R^4}{16\pi^3 a_o^2} \frac{1}{\delta(\delta + 1)} \quad (6)$$

$$= \frac{7.20 \times 10^{13} Z^2 \lambda_R^4}{\delta(\delta + 1)}$$

where $\delta = \theta_o^2/4$.

Another useful form of the elastic scattering cross section is obtained by integrating the cross section of Eq. (5a) over a portion of the angular range from $\theta_1 \leq \theta \leq \pi$. This yields the cross section for elastic events that exceed a scattering angle, θ_1

$$\sigma\,(>\theta_1) = \int_{\theta_1}^{\pi} d\sigma\,(\theta) = 1.62 \times 10^{-20} \frac{Z^2}{E^2} \cot^2(\theta_1/2) \quad (7)$$

which has dimensions of (events $>\theta_1$)/e^-/(atom/cm^2). Plots of calculations made with Eq. (7) for several elements and beam energies are shown in Figures 2(a) and 2(b). The probability of scattering into angles $>\theta_1$ decreases rapidly as θ_1 increases.

B. Elastic Scattering Angles

The probability, $P(\theta)$, for scattering into a particular angular range (0 to θ) can be found from the integral

$$P(\theta) = \int_{\Omega} \frac{\sigma(\theta)}{\sigma_e} d\Omega = \int_{o}^{\theta} \frac{2\pi \sin\theta\, \sigma(\theta)\, d\theta}{\sigma_e} \quad (8)$$

where $\sigma(\theta)$ is the differential cross section (Eq. (5a)), σ_e is the total elastic scattering cross section (Eq. (6)), and $d\Omega = 2\pi \sin\theta d\theta$. Upon substitution, $P(\theta)$ becomes

$$P(\theta) = \int_{o}^{\theta} \frac{\delta(1 + \delta)}{\pi} \frac{2\pi \sin\theta}{[\sin^2(\theta/2) + \delta\,]^2} d\theta \quad (9)$$

$$P(\theta) = (1 + \delta)\,\{1 - [2\delta/(1 - \cos\theta + 2\delta)]\}\ . \quad (10)$$

By making calculations for increasing angles and taking the difference between successive calculations, Eq. (10) can be used to calculate the scattering probability for various angular ranges, as shown for a particular target in Table I. This calculation indicates that for a target of intermediate atomic number (copper) at a typical AEM beam energy (100 keV), the most probable scattering angle is on the order of 2° to 3°. A similar calculation for a high-atomic-number target (gold) in Table I reveals a similar value for the most probable scattering angle, but the peak probability is not as high, and larger scattering angles are more probable in gold than in copper.

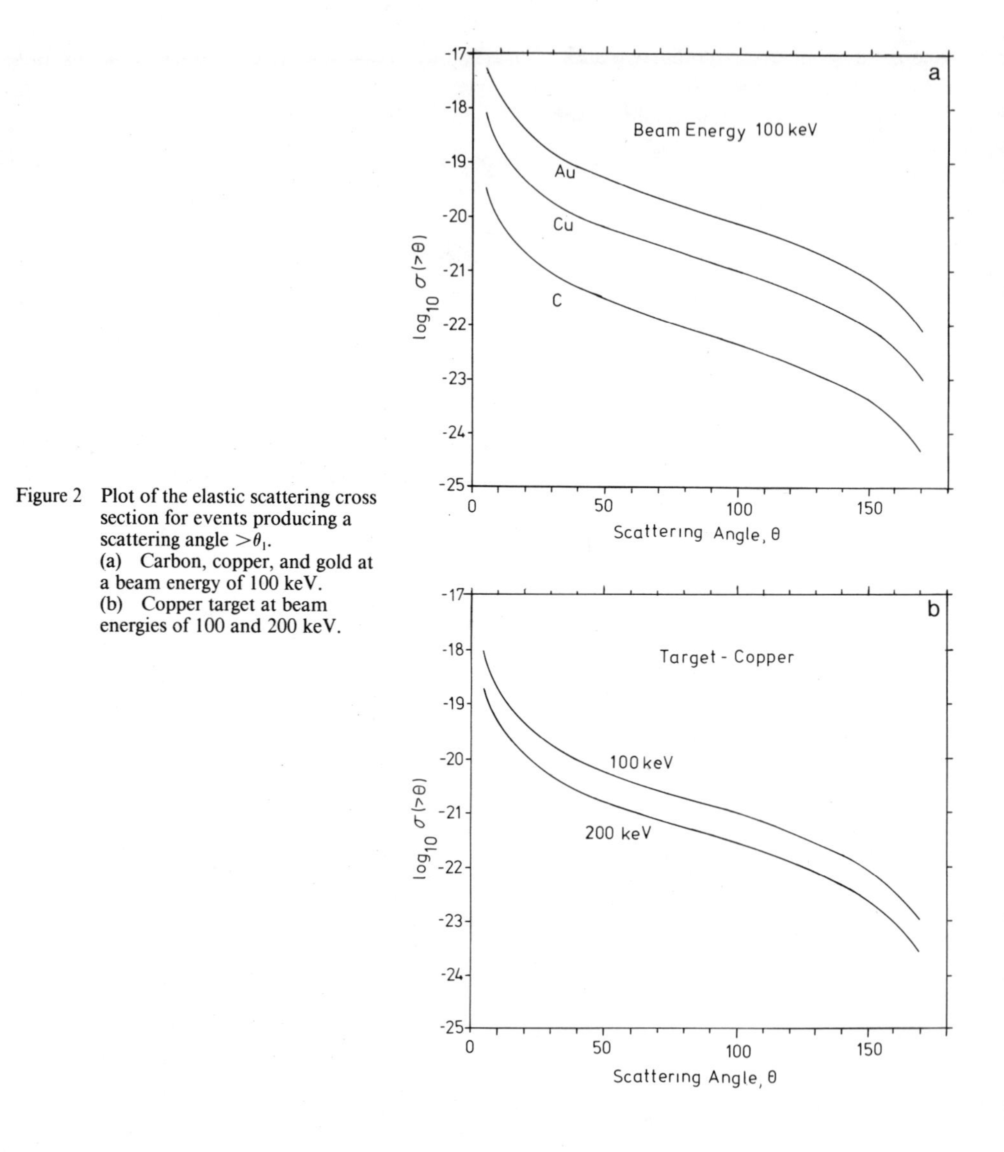

Figure 2 Plot of the elastic scattering cross section for events producing a scattering angle $>\theta_1$.
(a) Carbon, copper, and gold at a beam energy of 100 keV.
(b) Copper target at beam energies of 100 and 200 keV.

C. Elastic Mean Free Path

1. Single Scattering

An important parameter that is useful in assessing the condition of scattering is the elastic mean free path, which is obtained by substituting the total elastic scattering cross section of Eq. (6) into Eq. (2) for the mean free path:

$$\lambda_e = A/\sigma_e N_o \rho \qquad (\text{cm}) \ . \tag{11}$$

TABLE I

Probability* of Elastic Scattering Into Various Angles. Beam Energy, 100 keV.

Scattering Angle, θ (°)	Copper $P(\Delta\theta)$	Gold $P(\Delta\theta)$
1	0.190	0.108
2	0.294	0.219
3	0.195	0.195
4	0.111	0.138
5	0.065	0.092
6	0.039	0.062
7	0.026	0.042
8	0.018	0.030
9	0.012	0.022
10	0.009	0.016
20†	0.030	0.056
30	0.0065	0.011
40	0.0022	0.0040
50	0.0010	0.0018
90†	0.0011	0.0023
180	0.0002	0.0006

*Normalized so that $\Sigma P_\theta = 1$.
†Note change in angle interval for calculation.

Values of the total elastic mean free path calculated from Eq. (11) are presented in Table II. The total elastic mean free path is a measure of the thickness needed for the average electron to undergo one elastic scattering act while passing through the foil. If the foil thickness is increased over that given for λ_e from Eq. (11), the number of scattering events that the average electron undergoes exceeds unity, and the scattering regime is known as plural elastic scattering. An important consequence of this change is the modification of the angular distribution of the scattered electrons from that given by Eq. (10) to a more complex function. Finally, when the number of scattering events per electron exceeds ~25, the scattering regime becomes that of multiple scattering. For foil thicknesses of interest in AEM (e.g., 50 to 200 nm), Table II shows that plural elastic scattering will frequently exist, and multiple elastic scattering may occur for thick sections of high-atomic-number materials. This condition is particularly evident in foils composed of intermediate- and high-atomic-number elements.

2. Multiple Scattering

In the multiple scattering regime, the angular distribution of the transmitted beam is described by a Gaussian equation, which gives the probability of an electron emerging from the foil between θ and $\theta + d\theta$

$$P(\theta)\, d\theta = \frac{2}{\langle\theta_m^2\rangle}\, \theta \exp\left| -\theta^2/\langle\theta_m^2\rangle \right|\, d\theta \tag{12}$$

where $\langle\theta_m^2\rangle$ is the mean square angular deflection (radians). The mean square angular deflection is given by Bethe and Ashkin (1953)

$$\langle\theta_m^2\rangle = \left|3.93 \times 10^4 \frac{Z(Z+1)\rho t}{A E^2}\right| \log \left|\frac{4\pi Z^{4/3} N_o \rho t}{A} \left(\frac{h}{2\pi m v}\right)^2\right| \text{ (radians)} . \quad (13)$$

For a copper target 1 μm thick and for a beam energy of 100 keV, $\theta_m = 0.45$ rad or $\sim 26°$. The variation in $\langle\theta_m^2\rangle$ is proportional to the thickness, t, neglecting the relatively slow variation of the log(t) term, and hence $\theta_m \propto t^{1/2}$ for multiple scattering.

3. Plural Scattering

Plural scattering does not lend itself so readily to the development of analytic solutions such as those for multiple scattering. The theory of Moliere (1948) can be applied to plural scattering to yield tractable equations (Bethe and Ashkin, 1953; Knop and Paul, 1966). The angular distribution function for the probability of scattering into an angular interval, $d\theta$, at an angle, θ, is given by

$$P(\theta)\,\theta\, d\theta = \tau\, d\tau\ |\ 2 \exp(-\tau^2) + B^{-1} f^{(1)}(\tau) + B^{-2} f^{(2)}(\tau) + \ldots | \quad (14)$$

where

$$\tau = \frac{\theta}{\theta_1 B^{1/2}} . \quad (15)$$

θ_1 is given by the expression

$$\theta_1 = \left|\frac{3.93 \times 10^4 Z(Z+1)\rho t}{A E^2}\right|^{1/2} . \quad (16)$$

The parameter B has been tabulated by Moliere as a function of $|\theta_1^2/\theta_a^2|$, and by linear regression, a simple formula accurately describes these tabularized values

$$B = 2.52 \log_{10} |\theta_1^2/\theta_a^2| + 1.211 . \quad (17)$$

TABLE II

Mean Free Paths for Elastic Scattering in Various Targets (calculated with the relativistic Rutherford formula).

Beam Energy (keV)	80	100	150	200
Target (nm)				
Carbon	160	200	300	400
Silicon	95	120	180	240
Copper	17	21	32	42
Gold	8	11	15	22

The parameter $|\theta_1^2/\theta_a^2|$ is given by

$$|\theta_1^2/\theta_a^2| = 7800 \frac{(Z+1)\,Z^{1/3}\,\rho t}{\beta^2\,A\,(1+3.35\,\gamma^2)}, \quad \gamma = \frac{2\pi Z e^2}{hv}. \tag{18}$$

The functions $f^{(n)}$ and (τ) in Eq. (14) are complex integrals that were tabularized in Moliere's paper; this table is reproduced in Table III.

Equations (14) to (18) allow the plural scattering angular distribution to be calculated. Bethe and Ashkin point out that these formulas are limited in application to angles for which $\theta \simeq \sin\theta$, and that plural scattering into large angles must be treated with an even more complex approach.

D. Applications of Elastic Scattering Calculations

1. Beam Broadening

As an application of elastic scattering theory to a practical problem in analytical electron microscopy, we shall consider the development of an analytic solution for beam broadening proposed by Goldstein *et al.* (1977). These authors used the following argument to derive an expression appropriate to the single scattering regime.

TABLE III

Values of $f^{(n)}(\tau)$ from Moliere (1948).

τ	$f^{(1)}(\tau)$	$f^{(2)}(\tau)$
0.0	0.8456	2.49
0.2	0.700	2.07
0.4	0.343	1.05
0.6	−0.073	−0.003
0.8	−0.396	−0.606
1.0	−0.528	−0.636
1.2	−0.477	−0.305
1.4	−0.318	0.052
1.6	−0.147	0.243
1.8	0.0	0.238
2.0	0.080	0.131
2.2	0.106	0.020
2.4	0.101	−0.046
2.6	0.82	−0.064
2.8	0.062	−0.055
3.0	0.045	−0.036
3.2	0.033	−0.019
3.5	0.0206	0.0052
4.0	0.0105	0.0011
5.0	3.82×10^{-3}	8.36×10^{-4}
6.0	1.74×10^{-3}	3.45×10^{-4}
7.0	9.1×10^{-4}	1.57×10^{-4}

The situation to be modeled is illustrated in Figure 3, where the scattering geometry is simplified to the condition that each electron is assumed to undergo one scattering act at the center of the foil of thickness t. The problem is to take an appropriate angular distribution for the scattered electrons and convert that distribution into an equivalent spatial distribution that directly describes the broadening of the beam as a function of depth. The broadening is arbitrarily defined as the diameter of the base of the cone, which contains 90% of the scattered electrons. We thus need to find the scattering angle equal to the semiangle of the cone, $\phi/2$. Consider the mean free path to achieve a scattering angle $>\theta$:

$$\lambda = A/N_o \rho\sigma (>\theta) . \tag{19}$$

In this case, the cross section for scattering into an angle $>\theta$ has been given in Eq. (7).

$$\lambda = A/\left| N_o \rho 1.62 \times 10^{-20} \frac{Z^2}{E^2} \cot^2(\theta/2) \right| \tag{20}$$

or

$$\lambda = A\, E^2/\, |9.75 \times 10^3 \rho Z^2 \cot^2(\theta/2)| . \tag{21}$$

In traveling a mean free path given by Eq. (19), *all* of the electrons would be expected to undergo scattering through an angle $>\theta$. We wish to achieve a condition such that only 10% of the electrons are scattered outside the cone and 90% remain inside. Thus, if we consider the fraction that undergoes scattering [f $(>\theta)$] to be a simple linear function of t/λ, at least for small values of t/λ, then

$$f(>\theta) = \frac{t}{\lambda} = \frac{9.75 \times 10^3 Z^2 \rho t \cot^2(\theta/2)}{A\, E^2} . \tag{22}$$

From the geometry shown in Figure 3, the scattering angle is related to the foil and cone dimensions by the relation for small angles

$$\theta \simeq \tan\theta = \frac{(b/2)}{(t/2)} = \frac{b}{t} . \tag{23}$$

Moreover, for small angles

$$\tan(\theta/2) \simeq \theta/2 \text{ and } \cot(\theta/2) \simeq 2/\theta . \tag{24}$$

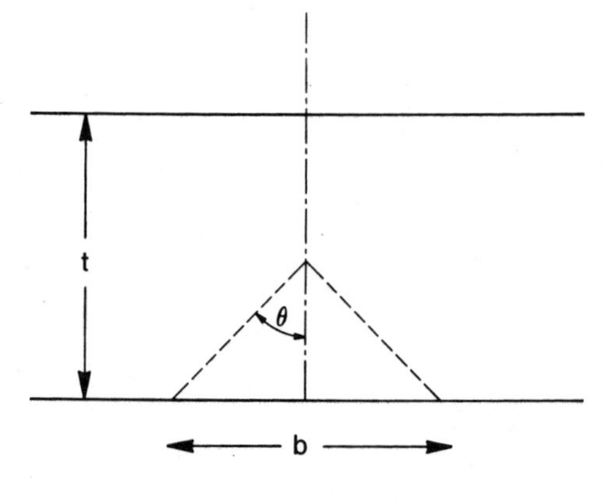

Figure 3 Thin foil scattering geometry assumed in the development of the analytic model for beam broadening by Goldstein *et al.* (1977).

Combining these conditions

$$f(>\theta) = 0.1 = \frac{9.75 \times 10^3 Z^2 \rho t (2/\theta)^2}{A E^2} = \frac{3.9 \times 10^4 Z^2 \rho t^3}{b^2 A E^2} \tag{25}$$

$$b = 6.25 \times 10^2 (\rho/A)^{1/2} (Z/E) t^{3/2} \text{ (cm)} \tag{26a}$$

where b and t are in cm, ρ is in g/cm^3, A is in g/mole, E is in keV. Equation (26) suggests that the broadening increases as the 3/2 power of thickness, linearly with Z, and inversely with E. Experimental data suggest that Eq. (26a) describes beam broadening reasonably well in the single scattering regime (Jones and Lorretto, 1981).

Reed (1982) has published a new derivation of Eq. (26a) in which the artificial condition of scattering at the foil center is replaced by scattering throughout the foil thickness, which requires an integration of the incremental scattering in layers of thickness, dt, over thickness t. The result of this derivation yields an equation with an identical dependence on parameters as Eq. (26a) but with a slightly increased constant

$$k = 7.21 \times 10^2 . \tag{26b}$$

2. Backscattering

Backscattering is defined as the deflection of the electron by the scattering process through an angle >90° relative to the incident direction. After scattering, the electron propagates back through the specimen surface through which it entered. Strictly, this definition of backscattering refers to scattering in a single event and is distinguished from the situation in solid targets where multiple scattering provides enough cumulative small-angle scattering for the beam electron to escape through the entrance surface as a "backscattered" electron.

It is important to consider the process of backscattering under conditions appropriate to analytical electron microscopy because:

(a) As a possible advantage, backscattered electrons provide a useful imaging signal that is atomic-number dependent.
(b) As a possible disadvantage, backscattered electrons are scattered out of the specimen with virtually all of their incident energy and will travel essentially undeflected from the specimen to strike other detectors, such as those for x-ray or cathodoluminescence photons, possibly introducing spurious features in a spectrum.

If we consider a specimen set normal to the beam, then scattering events >90° will result in backscattering. The cross section for the process is given by Eq. (7), with $\theta_1 = 90°$. This cross section can be converted into a probability by converting events/e$^-$/(atom/cm^2) into events/e$^-$ by multiplying by atom/cm^2

$$P_i(>\theta_1) = \text{probability} = \sigma(>\theta_1) N_i t \tag{27a}$$

where

N_i = number of atoms per unit volume
t = thickness.

$$N_i = \rho N_o / A \tag{27b}$$

where

ρ = density (g/cm^3)
N_o = Avogadro's number (atoms/mole)
A = atomic weight (g/mole).

For a 100-nm-thick foil of gold and a beam energy of 100 keV, $P_i = 0.006$, whereas for aluminum under these conditions, $P_i = 0.00017$. Thus, the contrast available goes as the ratio of $\rho Z^2/A$. However, the absolute level of the total signal is quite small, being ~0.6% of the incident-beam current for the gold foil. Considering the low beam currents used in the scanning transmission electron microscopy (STEM) mode with conventional thermionic sources, e.g., 0.1 to 1 nA, the total available backscattered current will only be ~1 pA, indicating that detectors with large solid angles of collection will be needed for any useful backscattered electron imaging. On the other hand, when x-ray spectrometry is of interest, this relatively low backscattering electron flux may be a significant source of interference since x-ray generation is an inefficient process compared to elastic scattering.

IV. INELASTIC SCATTERING

Inelastic scattering processes result from interactions of the beam electrons with the electrons of the atom. Various types of inelastic scattering occur, depending on whether the electrons of the specimen are excited singly or collectively. The energy transferred to the sample can be emitted in a form that carries useful information to the microscopist/analyst, such as secondary electrons, Auger electrons, x-rays, etc. The beam electron loses specific amounts of energy in these inelastic events.

If a particular process is associated with a characteristic energy of excitation, then measurement of the energy spectrum of the electrons transmitted through the specimen (the technique of electron energy-loss spectrometry) can reveal characteristic energy losses useful in identifying that process.

A. Single Electron Excitations

1. Low-Energy Secondary Electrons

Secondary electrons, that is, specimen electrons ejected from the sample by the primary (beam) electrons, are generated as a result of energy transfer from the beam electron to loosely bound conduction-band electrons. The energy transferred to the conduction-band electron is relatively small, ~1 to ~50 eV. The cross section for this process has been given by Streitwolf (1959), per conduction-band electron per unit volume

$$\sigma(E_{SE}) = e^4\, k_F^3/[3\pi\, E(E_{SE} - E_F)^2] \qquad (28a)$$

where

k_F = the wave vector magnitude ($k = 1/\lambda$) corresponding to the Fermi energy (E_F)
E_{SE} = the secondary electron energy
E = the beam energy.

Equation (28a) can be integrated over a range of secondary electron energy to give a total cross section. Since the cross section is undefined at $E_{SE} = E_F$, the lower limit of integration is arbitrarily set at $E_{SE} = E_F + 1$ eV. The upper limit of integration is set, also arbitrarily, at $E_{SE} = 50$ eV, which provides an energy range that covers the practical extent of secondary electron energies

$$\sigma_{SE} = \frac{e^4 k_F^3 A}{3\pi E n_c \rho N_o} \left[\frac{1}{(E_F - E_{SE})} \right]_{E_F + 0.001 \text{ keV}}^{0.050 \text{ keV}} \tag{28b}$$

where n_c is the number of conduction-band electrons per atom, and the factor $(A/\rho N_o)$ places the cross section on a per-atom basis.

Because of the low energy of the secondary electrons, their range in the solid is ~5 nm. Thus, even in a thin foil, only a small fraction of the secondaries generated have a significant probability of escape. Since the surface region of the sample is frequently oxidized or contaminated, the secondary electron signal may not be directly related to the "real" sample. However, the low kinetic energy of the secondaries restricts their lateral movement as well, and since the secondary electron component generated in solid specimens by backscattering electrons does not occur significantly in thin foils at high beam energies, the secondary electron signal is mostly determined by events in the immediate region of the beam impact.

2. Fast Secondary Electrons

Although most numerous secondary electrons are those of low energy, i.e., $E_{SE} <$ 50 eV, collisions are possible between the beam electron and the atomic electron in which a large energy transfer occurs. These so-called fast secondary electrons, although small in number relative to the low-energy secondaries, are of interest in analytical electron microscopy because of their greater range resulting from their higher initial energy. Murata *et al.* (1981) have demonstrated the importance of fast secondary electrons in defining spatial resolution in the exposure of electron beam resists. Joy *et al.* (1982) have studied the influence of fast secondary electrons on the spatial resolution of analysis in the AEM.

Fast secondary electrons can be generated with energies up to that of the incident electron. Since the primary and secondary electrons cannot be distinguished after the collision, the cross sections for the two electrons are added, and the maximum possible energy loss is restricted to $\Delta E \leq 0.5E_o$. The differential relativistic cross section for the process, as given by Moller (1931), is

$$\frac{d\sigma}{d\epsilon} = \frac{\pi e^4}{E^2} \left| \frac{1}{\epsilon^2} + \frac{1}{(1-\epsilon)^2} + \left(\frac{E'}{E'+1} \right)^2 - \frac{2E'+1}{(E'+1)^2 \epsilon(1-\epsilon)} \right| \tag{29}$$

where $\epsilon = \Delta E/E_o \leq 0.5$ and $E' = E_o/511$ keV. The scattering angle of the primary electron after the collision relative to its incident direction is given by

$$\sin^2 \theta = 2\epsilon/(2 + E' - E'\epsilon) . \tag{30a}$$

The scattering angle of the secondary electron relative to the incident primary is given by

$$\sin^2 \phi = 2(1 - \epsilon)/(2 + E'\epsilon) . \tag{30b}$$

For an incident-beam energy of 100 keV and fast secondary electron energies in the range 1 to 10 keV, Eq. (30b) predicts that the scattering angles relative to the incident-beam direction are in the range 84° ($\Delta E = 1$ keV) to 70° ($\Delta E = 10$ keV). For a beam incident normally on a thin foil, the fast secondary electrons tend to travel laterally into the foil, transferring energy from the beam radially outward and into the foil, degrading the spatial resolution.

3. Inner Shell Ionization

The beam electron can interact with a tightly bound inner-shell electron and eject it from the atom. The minimum energy required to remove the electron from the atom is called the critical excitation (ionization) energy, E_c. The cross section for inner-shell ionization can be described by a formula derived by Bethe (1930), which is an approximation to a series expansion

$$\sigma = \frac{\pi e^4 \, b_S \, n_S}{\left(\frac{m_o v^2}{2}\right) E_c} \log \left[c_S \left(\frac{m_o \, v^2}{2}\right) / \, E_c \right] \tag{31}$$

where

n_S = number of electrons in the shell or subshell
m_o = rest mass
v = velocity
e = charge of the electron

and b_S and c_S are constants appropriate to the shell or subshell. At low beam energies ($E < 50$ keV), the term $(m_o v^2/2)$ can, with little error, be set equal to the kinetic energy of the electron ($E = e \times V$), where e is the electron charge and V is the accelerating voltage. The Bethe ionization equation can then be rewritten as

$$\sigma = \frac{\pi e^4 \, b_S \, n_S}{E \, E_c} \log \left(c_S \, E/E_c \right) \; . \tag{32a}$$

Equation (32a) can also be expressed in terms of the overvoltage, $U = E/E_c$, and further substituting for the constant, πe^4, yields

$$\sigma = \frac{6.51 \times 10^{-20} \, b_S \, n_S}{U \, E_c^2} \log \left(c_S \, U \right) \tag{32b}$$

where the dimensions are ionizations/e^- (atom/cm^2) when E is in keV.

At higher beam energies, relativistic effects become more important, and the mass of the energetic electron deviates significantly from the rest mass. Bethe and Fermi (1932) and Williams (1933) have described a relativistic modification of Eq. (31)

$$\sigma = \frac{\pi e^4 \, b_S \, n_S}{\left(\frac{m_o v^2}{2}\right) E_c} \left\{ \log \left[c_S \left(\frac{m_o v^2}{2}\right) / E_c \right] - \log \left(1 - \beta^2\right) - \beta^2 \right\} \tag{33}$$

where $\beta = v/c$. The term $(m_o v^2/2)$ in Eq. (33) can be calculated by substituting $(m_o v^2/2) = (m_o c^2 \beta^2/2)$ where $m_o c^2 = 511$ keV and β is given by Eq. (4b).

The choice of the constants, b_S and c_S, to be substituted in Eq. (31) and (33) has been a subject of considerable discussion, which has been reviewed in detail by Powell (1976a, b). While Bethe (1930) has defined the terms physically, the b_S and c_S values are more typically taken as parameters that are adjusted to fit specific experimentally determined cross-section data. Fow low beam energies ($E < 20$ keV) and the overvoltage range $4 \leq U \leq 25$, Powell (1976a, b) recommends for K-shells, $b_K = 0.9$ and $c_K = 0.65$. Specifically, these constants were determined from fits to data for low-atomic-number elements (C, N, O, and Ne) at low, nonrelativistic beam energies. For higher-atomic-number targets and higher beam energies, substantially different values of b_K and c_K are found. For example, $b_K = 1.05$ and $c_K = 0.51$ for nickel in the overvoltage range $5.5 \leq U \leq 22$ (Pockman *et al.*, 1947).

Another approach to the values of the constants involves fixing c_S at unity, which has the advantage of ensuring that the cross-section expression produces positive values down to the ionization edge, and then fitting b_S. The resulting b_S values have been taken as parameters with an atomic number dependence (Brown, 1974)

$$c_K = 1 \qquad b_K = 0.52 + 0.0029\,Z \tag{34a}$$

$$c_{L_{23}} = 1 \qquad b_{L_{23}} = 0.44 + 0.0020\,Z \tag{34b}$$

Finally, when the relativistic Eq. (33) is used, the incorporation of $(m_o c^2 \beta^2/2) = (m_o v^2/2)$ has an important numerical consequence. Expressed in keV units, $(m_o c^2 \beta^2/2)$ deviates sharply from the electron kinetic energy, $E = eV$. For example, $(m_o c^2 \beta^2/2) = 76.8$ keV when $E = 100$ keV. Thus, if c_S is taken as unity, substitution of $(m_o c^2 \beta^2/2)$ will produce values of the logarithmic term in Eq. (33) that are negative for a substantial energy range above the edge. To produce a positive result for the cross section in the energy range above the edge, c_S and b_S must be adjusted appropriately for the incorporation of $(m_o c^2 \beta^2/2)$.

Numerous other formulas exist to describe the ionization cross section (Inokuti, 1971). As an example of an alternative relativistic cross section, Kolbenstvedt (1967) described the following equation

$$\sigma = 10^{-24} \frac{0.275\,(E' + 1)^2}{E'_c\, E'\,(E' + 2)} \left| \log\,[1.19E'\,(E' + 2)/E'_c] - E'\,(E' + 2)/(E' + 1)^2 \right|$$
$$+ \frac{0.99\,(E' + 1)^2}{E'_c\, E'\,(E' + 2)} \left| 1 - (E'_c/E')\,\{1 - [E'^2/2\,(E' + 1)^2] + [(2E' + 1)\log\,(E'/E_c)/(E' + 1)^2]\}\right| \tag{35}$$

where E' and E'_c are expressed in terms of the electron rest energy ($m_o c^2 = 511$ keV) so that $E' = E/511$ and $E'_c = E_c/511$ where E and E_c have been previously defined.

A comparison of the ionization cross section for nickel calculated from these formulas and experimental data of Pockman *et al.* (1945) is presented in Table IV. In this instance, the cross sections produce similar results, with the best fit given by the Bethe-Fermi relativistic cross section and the empirical fits of Pockman *et al.* and Powell. In fact, the change in the constants b and c in the fits by Powell and by Pockman *et al.* does not significantly alter the quality of the fit, the change in the constant within the log term being offset by the change in the linear term. Further experimental data in the 50- to 300-keV energy range is needed to adequately test the cross-section calculations.

TABLE IV

Cross Section for Inner-Shell Ionization for Nickel (K-shell).

Beam Energy (keV)	Experiment	Calculations				
	Pockman *et al.* (1945)	Pockman *et al.* (1947)	Powell (1976a)	Bethe-Fermi (1932)	Kolbenstvedt (1967)	Brown (1974)
50	3.8E-22	3.7E-22	3.8E-22	3.7E-22	4.4E-22	3.4E-22
75	3.2E-22	3.3E-22	3.3E-22	3.2E-22	3.7E-22	2.8E-22
100	2.8E-22	3.0E-22	2.9E-22	2.8E-22	3.2E-22	2.3E-22
150	2.2E-22	2.4E-22	2.3E-22	2.4E-22	2.7E-22	1.8E-22
200	1.9E-22	2.0E-22	1.9E-22	2.1E-22	2.4E-22	1.5E-22

Dimensions: ionizations/e^-/(atom/cm^2)

B. Interactions With Many Electrons

1. Bremsstrahlung

A beam electron that passes through the Coulomb field of an atom can undergo deceleration, which decreases the magnitude of its velocity and its kinetic energy. The lost kinetic energy is emitted as a photon of electromagnetic radiation by the beam electron (not by the medium as in Cerenkov radiation). This "bremsstrahlung" or "braking radiation" forms a continuous distribution of photon energies from zero up to the incident energy. The x-ray continuum is the principal component of the background of the x-ray spectrum from a thin foil, provided that the extraneous artifact sources of radiation (scattered electrons, remotely excited x-rays, etc.) are excluded from the detector.

The production of bremsstrahlung is directly related to the direction of the velocity of the beam electron. In thin samples where the majority of electron trajectories are not altered significantly by elastic scattering, the anisotropy of the bremsstrahlung can be observed because the directional nature of the bremsstrahlung contributions produced by the individual electrons is not randomized.

Several cross sections for the production of bremsstrahlung by energetic electrons are available in the literature. The simplest of these is based upon the energy dependence obtained by Kramers (1923) for solid specimens. This formula neglects the directional anisotropy of the bremsstrahlung and is most useful for purposes of estimating the total bremsstrahlung intensity at a particular beam energy, E, and bremsstrahlung energy, E_b

$$I = \frac{1.43 \times 10^{-21} Z^2}{4\pi} \frac{(E - E_b)}{E E_b} . \tag{36}$$

Note that σ in the previous cross-section equations has been replaced by I in the equations for bremsstrahlung to reflect a change in units to photon energy/unit energy interval/e^-/(atom/cm^2)/sr.

Kirkpatrick and Wiedmann (1945) derived algebraic equations that yield results approximating those derived from more rigorous calculations of the Sommerfeld theory. Strictly, these equations do not contain a relativistic correction and are only applicable to low beam energies (10 keV or less) and low-atomic-number targets since screening is neglected (Koch and Motz, 1959). The equations of Kirkpatrick and Wiedmann do take account explicitly of the anisotropy of the bremsstrahlung production with electron direction. Despite the limitations noted above, the Kirkpatrick-Wiedmann equations have been frequently used with reasonable success in the range 10 to 100 keV (Brown *et al.*, 1975). The Kirkpatrick-Wiedmann expressions consider the anisotropy of the bremsstrahlung components for an electron propagating normal (along the x-axis) to a plane of atoms in the y-z plane. In the following equations, note that transformations have been applied from electrostatic units to keV and in units of I from erg/sr/unit frequency interval/(atom/cm^2) to keV/keV energy interval/sr/(atom/cm^2) following Statham (1976).

$$I_x = \frac{0.2998Z^2}{E} \left| 0.252 + a\,[(E_b/E) - 0.135] - b\,[(E_b/E) - 0.135]^2 \right| \times 1.51 \times 10^{-24} \quad (37a)$$

where

$$a = 1.47B - 0.507A - 0.833$$

$$b = 1.70B - 1.09A - 0.627$$

$$A = \exp(-0.223E/0.2998Z^2) - \exp(-57E/0.2998Z^2)$$

$$B = \exp(-0.0828E/0.2998Z^2) - \exp(-84.9E/0.2998Z^2)$$

$$I_y = I_z = \frac{0.2998Z^2}{E} \left| -j + \frac{k}{(E_b/E) + h} \right| \times 1.51 \times 10^{-24} \quad (37b)$$

where

$$h = (-0.214y_1 + 1.21y_2 - y_3)/(1.43y_1 - 2.43y_2 + y_3)$$

$$j = (1 + 2h)\,y_2 - 2\,(1 + h)y_3$$

$$k = (1 + h)(y_3 + j)$$

$$y_1 = 0.220\,[1 - 0.390\exp(-26.9E/0.2998Z^2)]$$

$$y_2 = 0.067 + 0.023/[(E/0.2998Z^2) + 0.75]$$

$$y_3 = -0.00259 + 0.00776/[(E/0.2998Z^2) + 0.116]$$

At a particular angle, θ, in the x-z plane, the intensity is given by

$$I_\theta = I_x \sin^2\theta + I_y + I_z \cos^2\theta \quad (38)$$

and integration of I over 4π sr gives the total intensity

$$I_T = \frac{8\pi}{3}(I_x + I_y + I_z)\ . \quad (39)$$

A more rigorous equation for the cross section for bremsstrahlung generated by electrons in the 50- to 500-keV range is given by Koch and Motz (1959)

$$\frac{d\sigma}{dE_b d\Omega} = \frac{7.95 \times 10^{-26}}{8\pi\, 137} \frac{1}{E_b} \frac{p}{p_o} \left| \frac{8\sin^2\theta\,(2T_o^2 + 1)}{p_o^2\Delta_o^4} - \frac{2\,(5T^2 + 2TT_o + 3)}{p_o^2\Delta_o^2} \right.$$

$$- \frac{2\,(p_o^2 - E_b^2)}{Q^2\Delta_o^2} + \frac{4T}{p_o^2\Delta_o} + \frac{L}{pp_o}\left[\frac{4T_o\sin^2\theta\,(3E_b - p_o^2\,T)}{p_o^2\Delta_o^4}\right.$$

$$\left. + \frac{4T_o^2\,(T_o^2 + T^2)}{p_o^2\Delta_o^2} + \frac{2 - 2\,(7T_o^2 - 3TT_o + T_o^2)}{p_o^2\Delta_o^2} + \frac{2E_b\,(T_o^2 + TT_o - 1)}{p_o^2\Delta_o}\right]$$

$$\left. - \frac{4\epsilon_1}{p\Delta_o} + \frac{\epsilon_2}{pQ}\left[\frac{4}{\Delta_o^2} - \frac{6E_b}{\Delta_o} - \frac{2E_b\,(p_o^2 - E_b^2)}{Q^2\Delta_o}\right]\right| \tag{40}$$

where

$$L = \log\left|\frac{TT_o - 1 + pp_o}{TT_o - 1 - pp_o}\right|$$

$$\Delta_o = T_o - p_o\cos\theta$$

$$\epsilon_1 = \log\left|\frac{T + p}{T - p}\right|$$

$$\epsilon_2 = \log\left|\frac{Q + p}{Q - p}\right|$$

$$Q^2 = p_o^2 + E_b^2 - 2p_oE_b\cos\theta$$

$$p = [E_f\,(E_f + 2)\,]^{1/2}$$

$$p_o = [E_o\,(E_o + 2)\,]^{1/2}$$

$$T = E_f + 1$$

$T_o = E_o + 1$ (T is the total electron energy, $E + m_oc^2$)

$$E_f = E_o - E_b$$

where

E_o = incident energy
E_b = photon energy
E_f = final electron energy

E_o, E_f, E_b, T, and T_o are expressed in terms of m_oc^2 units (511 keV).

Koch and Motz (1959) state that this cross section tends to underestimate the true value by a factor of 2, at most, in the energy range 100 to 500 keV. Moreover, as a Born-approximation calculation, the conditions $(2\pi Z/137\beta_f)$ and $(2\pi Z/137\beta_o) \ll 1$ where $\beta_f = v_f/c$ and $\beta = v_o/c$ must hold. At 100 keV, the quantity $(2\pi/137\beta)$ is 0.084, which indicates that the formula is strictly limited to low-Z targets and low bremsstrahlung energies ($v_f \cong v_o$). These equations are discussed extensively in a subsequent chapter.

2. Plasmon Scattering

The Coulomb field of the beam electron can perturb electrons of the solid at relatively long range as it passes through the foil. The beam electron can excite oscillations, called plasmons, in the conduction electron "gas" that exists in a metallic sample with loosely bound outer-shell electrons. The cross section for plasmon excitation is given by Ferrel (1956), per conduction-band electron per unit volume

$$\frac{d\sigma(\theta)}{d\Omega} = \frac{1}{2\pi a_o} \frac{\theta_p}{\theta^2 + \theta_p^2} \tag{41}$$

where

a_o = Bohr radius (5.29×10^{-9} cm)
$\theta_p = \Delta E_p/2E$.

Plasmon scattering results in an energy transfer, ΔE_p, in the range 3 to 30 eV, depending on the target. Hence, scattering angle $\theta_p = \Delta E_p/2E$ for an aluminum target ($\Delta E_p = 15$ eV) is of the order of $\theta_p = 1.5 \times 10^{-4}$ rad for E = 100 keV. Plasmon scattering is so sharply peaked forward that the total plasmon cross section, σ_p, can be found by setting $d\Omega = 2\pi \sin\theta d\theta \cong 2\pi\theta d\theta$:

$$\sigma_p = \int d\sigma(\theta) = \frac{\theta_p}{2\pi a_o} \int_0^{\theta_1} \frac{2\pi\theta d\theta}{\theta^2 + \theta_p^2} . \tag{42a}$$

Taking the upper integration limit as $\theta_1 = 0.175$ rad, where $\theta \approx \sin\theta$, and incorporating the factor $(n_c A/\rho N_o)$ to put the cross section on a per-atom/cm^2 basis, gives

$$\sigma_p = \frac{n_c A\, \theta_p}{2N_o \rho a_o} [\ln(\theta_p^2 + 0.175^2) - \ln(\theta_p^2)] . \tag{42b}$$

In plasmon scattering, beam electrons lose energy in quantized units that are detected as strong features in electron energy-loss spectra. The wavelengths of the plasmons excited in the target are of ~100-atom spacings, so that the electron loss spectral information derived from plasmon scattering is not well localized (Hirsch *et al.*, 1965).

3. Thermal Diffuse Scattering

Because of thermal vibrations, the atoms in a crystal oscillate about their nominal lattice sites. The beam electron passing through the crystal can interact with these crystal oscillations, creating or annihilating phonons (Hirsch *et al.*, 1965). The energy change associated with this scattering is very small (~0.02 eV). The angular deflection imposed on the beam electron is significant and is observed in electron diffraction patterns as a diffuse background between the Bragg diffraction spots, which are less intense as a result of thermal scattering. Thermal diffuse scattering increases as the temperature of the crystal increases.

V. CONTINUOUS ENERGY LOSS APPROXIMATION

The discrete mechanisms of inelastic scattering can be combined into a treatment that approximates energy loss as a continuous process, yielding a relationship for the rate of energy loss, dE, with distance traveled in the specimen, dS (Bethe, 1933)

$$\frac{dE}{dS}\frac{keV}{cm} = -2\pi e^4 N_o \left|\frac{pZ}{EA}\right| \ln \left|\frac{1.166E}{J}\right| \tag{43}$$

where J is the mean ionization potential, given by Berger and Seltzer (1964)

$$J = (9.76Z + 58.5Z^{-0.19}) \times 10^{-3}\ (keV)\ . \tag{44}$$

Equation (43) is useful for estimating energy losses through various foil thicknesses. However, it should be recognized that in the AEM, electrons passing through foils of practical thickness usually undergo only single or plural inelastic scattering so that the energy loss given by Eq. (43) is only approximate. In reality, as shown by the electron energy-loss spectrum, electrons are transmitted through a thin foil, with energies corresponding to zero loss as well as all possible discrete inelastic processes.

VI. COMPARISON OF CROSS SECTIONS

The elastic scattering cross section and the cross sections for several of the inelastic scattering processes are plotted as a function of electron energy for an aluminum target in Figure 4. The probability for inelastic scattering exceeds the probability for elastic scattering. Among the inelastic processes, plasmon scattering is the most probable process that significantly alters the energy of the beam electron.

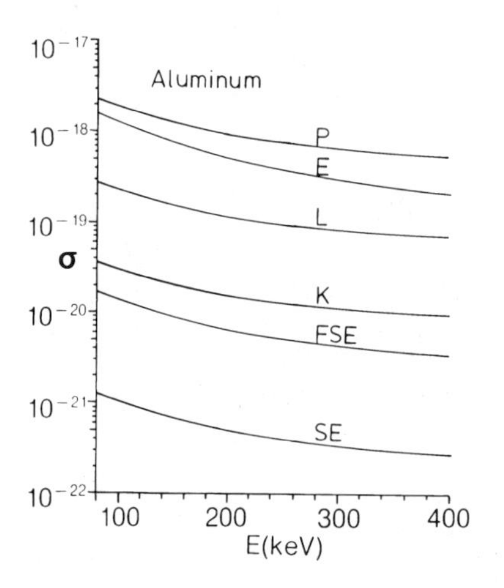

Legend:

P = Plasmon
E = Elastic
L = L-shell ionization
K = K-shell ionization
FSE = Fast secondary electrons, $E_{SE} > 50$ eV
SE = Slow secondary electrons, $E_{SE} < 50$ eV

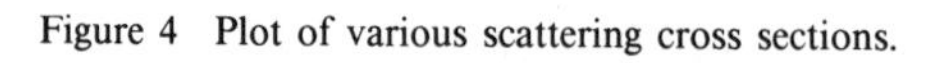

Figure 4 Plot of various scattering cross sections.

VII. SIMULATION OF INTERACTIONS

Frequently, it is important to carry out quantitative calculations of the interaction of beam electrons with a target in order to interpret data obtained with the AEM: image contrast, diffraction pattern details, and x-ray and electron energy-loss analytical spectra. Two different approaches are used for such calculations based on the dual nature of the electron-specimen interaction: wave-nature and particle-nature. For the calculation of diffraction patterns and the contrast seen in images of crystalline materials, the wave-nature of the electron is considered. The incident electron wave interacts with the potential field of the assemblage of charged particles (ionic cores, "free" electrons) in the sample; this interaction results in modifications to the phase of the electron wave while its amplitude is modified, due to inelastic scattering. This major topic of image formation will be considered in a subsequent chapter.

The alternative treatment of the interaction, in which the electrons are considered as discrete particles scattering from discrete target atoms over all possible angles, is useful for the calculation of such phenomena as beam spreading in the target, backscattering, and effects at interfaces. The discrete particle treatment of electron-specimen interaction is embodied in the technique of Monte Carlo electron trajectory simulation. Detailed treatments of the Monte Carlo technique as applied to AEM are available in the literature (Newbury and Myklebust, 1979, 1981; Kyser, 1979). A brief outline will be given here to illustrate the principles of the technique.

In the Monte Carlo electron trajectory simulation, the interaction of the beam electron with the target is calculated in a stepwise fashion, a single element of which is illustrated in Figure 5. The electron undergoes an interaction at location P_N that causes it to be deviated by an angle θ from its previous path. It travels a distance, S, along the new path, where S is the step length of the calculation, until it undergoes another scattering event at point P_{N+1}. It is common practice in Monte Carlo simulations to consider that only elastic scattering produces significant angular deviations, so that scattering angle θ in Figure 5 is derived from the angular distribution for elastic scattering, Eq. (9). Since the elastic scattering angle can take on any value from 0° to 180°, specific values are chosen through the use of a random number (hence, the name Monte Carlo). A linearly distributed random number, R, where $0 \leq R \leq 1$ is substituted for $P(\theta)$ in Eq. 10, to yield a scattering angle selection formula

$$\cos \theta = 1 - [2\delta R/(1 + \delta - R)] \ . \tag{45}$$

Figure 5 Fundamental calculation step in a Monte Carlo electron trajectory simulation, showing elastic scattering angle θ, step length S, and azimuthal scattering angle γ.

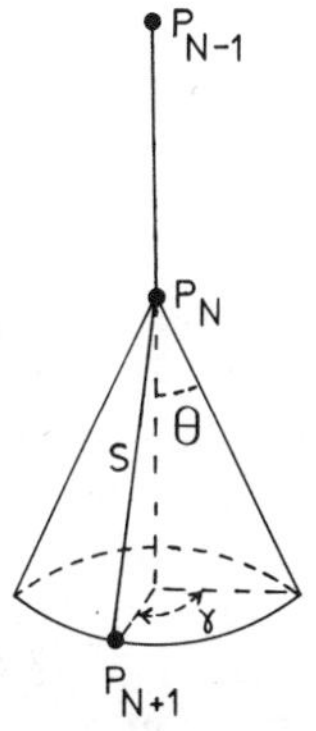

When a large number of choices of θ are made through a sequence of random numbers, the distribution is that given by Eq. (9). The azimuthal scattering angle in the base of the cone in Figure 5 can take on any value in the range 0° to 360°, so that this angle, γ, is chosen by the equation, with another choice of R, $0 \leq R \leq 1$

$$\gamma = R \times 360° \ . \tag{46}$$

The step length, S, of the calculation is proportional to the total elastic mean free path, λ, given by Eq. (11). Since the real path is distributed about the mean, a distribution is applied to S by an equation of the form

$$S = -\lambda \log_e R \tag{47}$$

where R is another linearly distributed random number. From the scattering angles and the step length, the x, y, z coordinates of P_{N+1} can be calculated.

Although inelastic scattering has been neglected so far as angular deviations are concerned, energy loss due to such scattering is considered by means of the Bethe continuous energy loss relation, Eq. (43). Thus, for a particular increment in path length along the trajectory, energy loss ΔE is approximated by

$$\Delta E = S \times (dE/dS) \ . \tag{48}$$

Because the path length segments are approximately equal to the elastic mean free path, the change in energy along an individual segment is small, so that the dependence of dE/dS upon E may be neglected.

Since the energy of the electron and the distance it travels are continuously known, individual inelastic scattering processes, such as inner-shell ionization or fast secondary electron production, can be calculated. The production of x-rays, for example, along a path length segment, S, is calculated according to the relation

$$I_S \ (\text{x-rays}/e^-) = \sigma_I N_o \rho \omega S/A \tag{49}$$

where

σ_I = cross section for inner-shell ionization
N_o = Avogadro's number
ρ = density
ω = fluorescence yield (x-rays/ionization)
A = atomic weight

The great power of the Monte Carlo technique is derived from the continuous calculation of electron energy, direction of flight, and position. Complex boundary conditions, such as target shape and changes in composition at an interface, can be taken into account. Transmission and backscattering of the electron can be calculated, and since the direction of flight and energy of the exciting electron are known, the influence of the electron on a detector can be calculated. The production of a signal of interest can be calculated as a function of position in the foil. The details of calculational techniques for the partitioning of steps to generate a histogram distribution have been described by Newbury and Myklebust (1984).

The principal disadvantage of the Monte Carlo technique arises from the necessity of using random numbers to select scattering parameters from their possible ranges. Since a different random number sequence generates a different trajectory, a statistically valid number of trajectories must be calculated. In general, the number of trajectories necessary will depend on the precision required, with the relative standard deviation (RSD) of the calculation given by

$$\mathrm{RSD} = \bar{n}^{-1/2} \tag{50}$$

where $\bar{n}$ is the mean number of trajectories contributing to the parameter of interest, e.g., backscattering. A second disadvantage of the Monte Carlo technique arises from the need to assume that the specimen is amorphous. In principle, a crystallographic array of atoms could be considered, but this has so far proven intractable in practical calculations.

Examples of Monte Carlo simulations of electron trajectories in thin foils of several materials and at three different beam energies are given in Figure 6. The utility of Monte Carlo simulations to convey a strong graphical representation of the electron-beam-specimen interaction volume is illustrated by these plots. For quantitative calculations, the parameters of interest are calculated in numerical form. An example of such a calculation for the x-ray signal observed as a function of position across an interface between two phases of different compositions is shown in Figure 7, taken from the work of Romig, Newbury, and Myklebust (1982). The quality of agreement between experiment and Monte Carlo calculation is satisfactory.

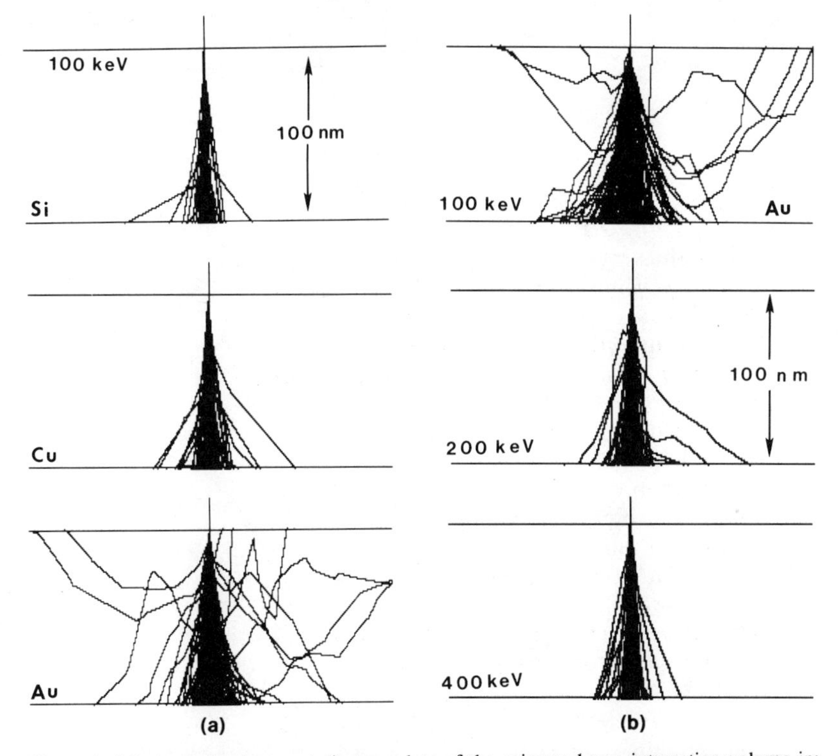

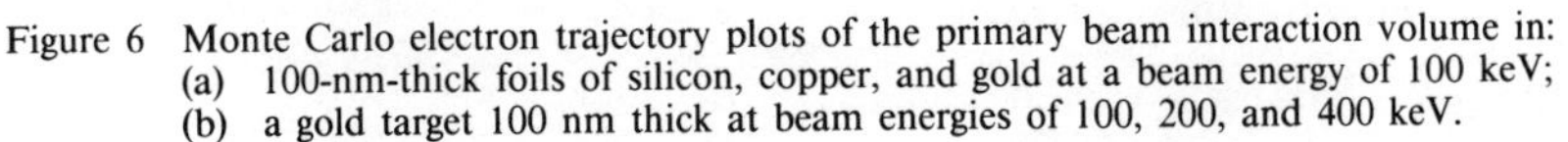
Figure 6 Monte Carlo electron trajectory plots of the primary beam interaction volume in: (a) 100-nm-thick foils of silicon, copper, and gold at a beam energy of 100 keV; (b) a gold target 100 nm thick at beam energies of 100, 200, and 400 keV.

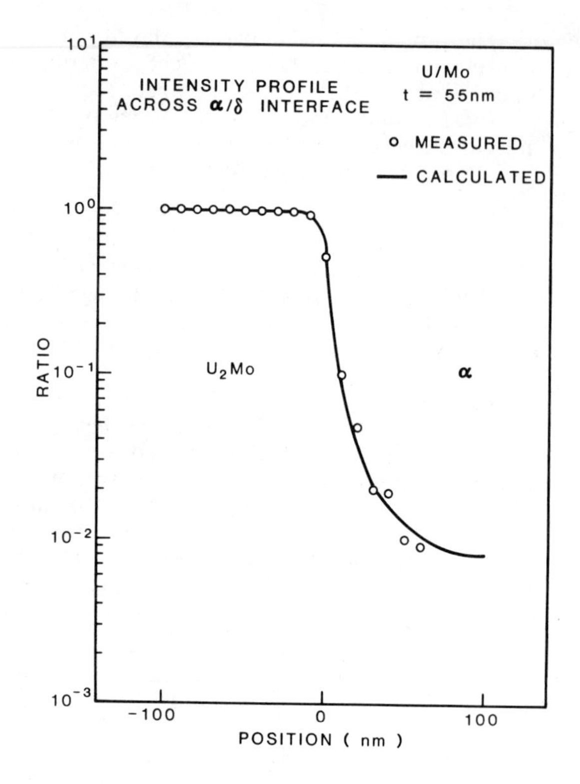

Figure 7 Monte Carlo calculations and experimental measurements of x-ray signal from solute element in a uranium matrix as a function of position across an interphase boundary (from Romig et al., 1982).

TABLE OF CHAPTER VARIABLES

a_o	Bohr radius
A	Atomic weight
b	Beam broadening
b_s	Constant in ionization cross section
B	Parameter in plural elastic scattering expression
c	Speed of light
c_s	Constant in ionization cross section
e	Electron charge
E	Kinetic energy (keV)
E_b	Energy of a bremsstrahlung photon
E_c	Critical ionization energy
E_f	Final beam electron energy
E_F	Fermi energy
E_o	Incident beam energy
E_{SE}	Secondary electron energy
E'	$E_o/511$ (keV)
h	Planck's constant
I	Bremsstrahlung intensity

I_x, I_y, I_z	Bremsstrahlung intensity resolved along coordinate axes (beam electron parallel to x-axis)
J	Mean ionization potential
k	Constant in beam broadening equation
k_F	Wave vector corresponding to Fermi energy
L	Term in Koch-Motz bremsstrahlung cross section
m	Mass
m_o	Rest mass of electron
n_c	Number of conduction-band electrons per atom
n_i	Number of incident particles per unit area
n_s	Number of electrons in shell or subshell
n_t	Number of target sites per unit volume
N	Events
N_i	Number of atoms per unit volume
N_o	Avogadro's number
p, p_o	Terms in Koch-Motz bremsstrahlung cross section
P	Probability
Q	Term in Koch-Motz bremsstrahlung cross section
R	Random number
t	Thickness
T, T_o	Total electron energy, $E + M_oc^2$
U	Overvoltage, E/E_c
v	Velocity
Z	Atomic number
α	Z/137 for light elements
β	Relativistic parameter, v/c
γ	$2\pi\ Ze^2/hv$
δ	Screening parameter, $\theta_o^2/4$
ΔE	Energy loss of primary electron
Δ_o	Term in Koch-Motz bremsstrahlung cross section
ϵ	$\Delta E/E_o$
ϵ_o	Dielectric constant
ϵ_1, ϵ_2	Terms in Koch-Motz bremsstrahlung cross section
θ	Scattering angle of the primary (beam) electron
θ_a, θ_1	Characteristic angles in plural elastic scattering expression
θ_m	Mean angular deflection
θ_o	Screening parameter
θ_p	Plasmon loss scattering angle
λ	Mean free path
λ_t	Total mean free path
ρ	Density
σ	Cross section
σ_R	Rutherford cross section
τ	$\theta/\theta_1\beta^{1/2}$
ϕ	Scattering angle of secondary electron
Ω	Solid angle

REFERENCES

Berger, M. J., and Seltzer, S. M. (1964), Natl. Res. Council Pub. 1133, Washington, DC, p. 205.

Bethe, H. A. (1930), Ann. der Phys. (Leipzig) *5*, 325.

——— (1933), Handb. Phys. (Berlin) *24*, 273.

Bethe, H. A., and Ashkin, J. (1953), in "Experimental Nuclear Physics" (E. Segre, ed.), Wiley, New York, vol. 1, p. 166.

Bethe, H. A., and Fermi, E. (1932), Zeit. f. Physik *77*, 296.

Brown, D. B. (1974), in "Handbook of Spectroscopy" (J. W. Robinson, ed.), CRC Press, Cleveland, vol. 1, p. 248.

Brown, D. B., Gilfrich, J. V., and Peckerar, M. C. (1975), J. Appl. Phys. *46*, 4537.

Cosslett, V. E., and Thomas, R. N. (1964), Br. J. Appl. Phys. *16*, 883.

Ferrel, C. R. (1956), Phys. Rev. *101*, 554.

Goldstein, J. I., Costley, J. L., Lorimer, G. W., and Reed, S. J. B. (1977), Proc. 10th Annl. SEM Symp. (O. Johari, ed.), SEM Inc., Chicago, vol. 1, p. 315.

Henoc, J., and Maurice, F. (1976), in "Use of Monte Carlo Calculations," Natl. Bur. of Stds. Special Pub. 460, Washington, DC.

Hirsch, P. B., Harvie, A., Nicholson, R. B., Pashley, D. W., and Whelan, M. J. (1965), in "Electron Microscopy of Thin Crystals," Butterworth's, London.

Inokuti, M. (1971), Rev. Mod. Phys. *43*, 297.

Jones, I. P., and Loretto, M. H. (1981), J. Micros. *124*, 3.

Joy, D. C., Newbury, D. E., and Myklebust, R. L. (1982), J. Micros. *128*, 1.

Kirkpatrick, P., and Wiedmann, L. (1945), Phys. Rev. *67*, 321.

Knop, G., and Paul, W. (1966), in "Alpha, Beta and Gamma Spectroscopy" (K. Siegbahn, ed.), North-Holland, Amsterdam, vol. 1, p. 1.

Koch, H. W., and Motz, J. W. (1959), Rev. Mod. Phys. *31*, 920.

Kolbenstvedt, H. (1967), J. Appl. Phys. *38*, 4785.

Kramers, M. A. (1923), Phil. Mag. *46*, 836.

Kyser, D. F. (1979), in "Introduction to Analytical Electron Microscopy" (J. Hren, J. Goldstein, and D. C. Joy, ed.), Plenum Press, New York, p. 199.

McKinley, W. A., and Feshbach, H. (1948), Phys. Rev. *74*, 1759.

Moliere, G. (1948), Z. Naturforsch. *3A*, 78.

Moller, C. (1931), Z. Phys. *70*, 786.

Mott, N. F., and Massey, H. W. W. (1965), in "The Theory of Atomic Collision," Oxford Univ. Press, England.

Murata, K., Kyser, D. F., and Ting, C. M. (1981), J. Appl. Phys. *7*, 4396.

Newbury, D. E., and Myklebust, R. L. (1979), Ultramicroscopy *3*, 391.

——— (1981), in "Analytical Electron Microscopy, 1981" (R. H. Geiss, ed.) San Francisco Press, San Francisco.

——— (1984), in "Electron Beam Specimen Interactions for Microscopy, Microanalysis and Lithography" (D. F. Kyser, ed.), SEM Inc., Chicago, p. 153.

Pockman, L. T., Webster, D. L., Kirkpatrick, P., and Harworth, K. (1945), Phys. Rev. *67*, 153.

——— (1947), Phys. Rev. *71*, 330.

Powell, C. J. (1976a), Rev. Mod. Phys. *48*, 33.

——— (1976b), in "Use of Monte Carlo Calculations," Natl. Bur. of Stds. Special Pub. 460, Washington, DC, p. 97.

Quinn, J. J. (1962), Phys. Rev. *126*, 1453.

Reed, S. J. B. (1982), Ultramicroscopy *7*, 405.

Reimer, L., and Krefting, E. R. (1976), in "Use of Monte Carlo Calculations," Natl. Bur. of Stds. Special Pub. 460, Washington, DC, p. 45.

Romig, A. D., Jr., Newbury, D. E., and Myklebust, R. L. (1982), in "Microbeam Analysis, 1982" (K. Heinrich, ed.), San Francisco Press, San Francisco, p. 88.

Statham, P. J. (1976), X-ray Spectr. *5*, 154.

Streitwolf, H. W. (1959), Ann. der Phys. (Leipzig) *3*, 183.

Wentzel, G. (1927), Zeit. f. Physik *40*, 590.

Williams, E. J. (1933), Proc. Roy. Soc. (London) *A139*, 163.

——— (1960), Phys. Rev. *58*, 292.

CHAPTER 2

INTRODUCTORY ELECTRON OPTICS

R. H. Geiss

IBM Research Laboratory, San Jose, California

A. D. Romig, Jr.

Sandia National Laboratories, Albuquerque, New Mexico

I. INTRODUCTION

The purpose of this chapter is to present an introductory, nonmathematical background in electron optics. The level is geared to the user of an electron microscope who is interested in understanding the qualitative features of the electron optical column but does not want to design a microscope. Because of this, mathematical statements are made and not derived in general and, in fact, mathematics is kept to a minimum with little more than a basic background required.

Although the discussion will center on magnetic lenses, as they are the most popular, an introduction to the terminology, laws, and techniques will be given with light optical principles. Electrostatic lenses will be discussed briefly, especially as the electron gun is an electrostatic lens. Although aberrations are discussed in some detail, it is important to remember that a good approximate picture of the image formation process may be easily derived ignoring aberrations and adhering only to Gaussian optics. Subsequent chapters in this book will describe applications to image formation. For those interested in more complete and/or mathematical treatments, a list of texts on electron optics is included.

II. GEOMETRIC OPTICS

A. Refraction

The fundamental law of geometrical optics as it pertains to light optics is Snell's law. This law describes the refraction of a light wave at the interface between two media with differing indices of refraction, n_i. It is written

$$\frac{\sin\theta_1}{\sin\theta_2} = \frac{n_2}{n_1} \tag{1}$$

where θ_1 and θ_2 are the angles of incidence and refraction with respect to the interface normal. The physical construction is diagrammed in Figure 1.

The refractive index, n_i, of a substance is simply the ratio of the velocity (speed) of light in a vacuum to the velocity of light in the substance, $n = c/v_\ell$. In vacuum the speed of light, $c \simeq 3 \times 10^8$ m/s, while in liquid or solid media such as oil or glass, light travels more slowly. Hence, the refractive index of a substance is always >1, and may be as large as 2.5. More commonly, n lies in the range of 1.3 to 1.7. The index of refraction of some common materials is given in Table I.

TABLE I

Index of Refraction of Some Common Materials for Light with $\lambda = 589.3$ nm (yellow light from sodium flame).

Material	Index of Refraction
Diamond	2.42
Crown Glass	1.52
Fused Quartz	1.46
Water (at 20°C)	1.333
CO_2	1.00045
Air	1.00029
Vacuum	1.00000

From Snell's law it then follows that a light wave passing from air into a glass lens will be bent closer to the interface normal on entering and vice versa on exiting. This effect is illustrated for three lens configurations in Figure 2.

The convergent or divergent properties of a lens are described with respect to incident light rays parallel to an axis passing through the center of the lens about which the lens is rotationally symmetric. It follows that lenses with greater curvature or larger refractive index will deflect light more. Such properties are associated with the "strength" of a lens. As will be discussed later, the concept of refractive index may be applied to electron optics in the case of both electrostatic and magnetic lenses.

B. Cardinal Elements

Consider a bundle of light rays parallel to the axis of rotational symmetry incident on a converging lens as shown in Figure 3. The rays are converged by the action of the lens and pass through the point, F_i, lying on the axis. This point is called the *image focal point* of the lens and is one of the cardinal elements. All rays parallel to the lens axis will pass through this point.

The intersection of the incident parallel rays with the extended exit rays passing through F_i forms a plane perpendicular to the lens axis, intersecting this axis at O.

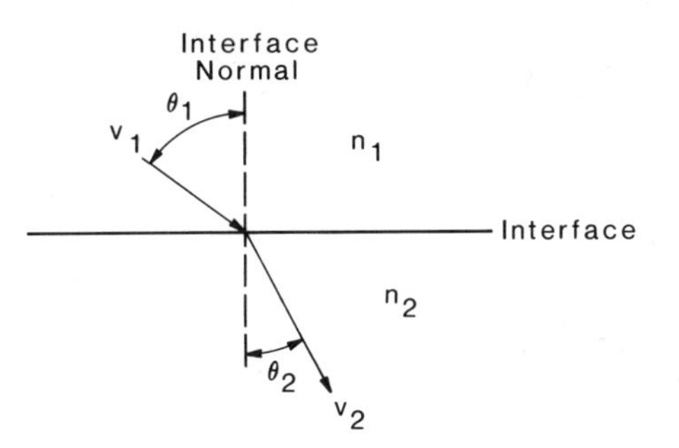

Figure 1 Geometrical construction showing Snell's law.

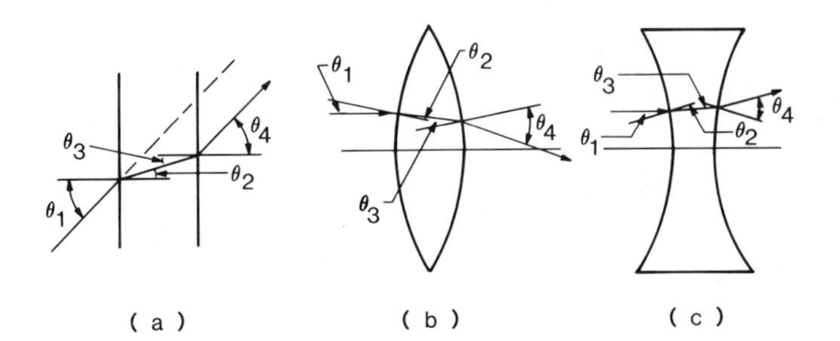

Figure 2 Light wave path through various types of lens. (a) A plane parallel-sided lens simply shifts the incident light waves; $\theta_1 = \theta_4 > \theta_2 = \theta_3$. (b) A convex, or positive, lens acts to converge the light waves; $\theta_1 > \theta_2$, $\theta_4 > \theta_3$. (c) A concave, or negative, lens acts to diverge the incident light waves; $\theta_1 > \theta_2$, $\theta_4 > \theta_3$.

This plane is called the image principal plane and the point at O is the *image principal point*. The image principal point is another cardinal element of the lens.

A third, and most familiar, cardinal element is the length described by $\overline{OF_i}$. This is the *image focal length*, f_i, and is the distance from the image principal point to the image focal point.

A similar construction for a bundle of parallel rays incident from the opposite side of the lens would define the *object focal point*, F_o; the *object principal point*, $O^{1'}$, and the *object focal length*, $O^{1'}F_o = f_o$. In the case drawn in Figure 3, the image principal point coincides with the object principal point, $O = O^{1'}$. This occurs for what is called a thin lens. Conversely, a more common construction especially in electron lenses is the thick lens, where the image and object principal planes do not coincide. The geometrical construction for a thick lens is shown in Figure 4.

C. Real and Virtual Images

If an object is placed in front of a converging lens in a plane intersecting the axis at a point, A, beyond the object focal point, F_o, a real image will be formed after the lens in a plane intersecting the axis at point B beyond the image focal point, F_i. This is illustrated in Figure 5.

By convention, rays are always considered to travel from left to right or top to bottom on any diagram. The object is always in the space to the left or above the lens and is described as being "in front of" or "before" the lens. The image space is described as "after" or "behind" the lens. The coordinate origin is the center of the lens, with "plus" to the right and above the horizontal axis describing the lens

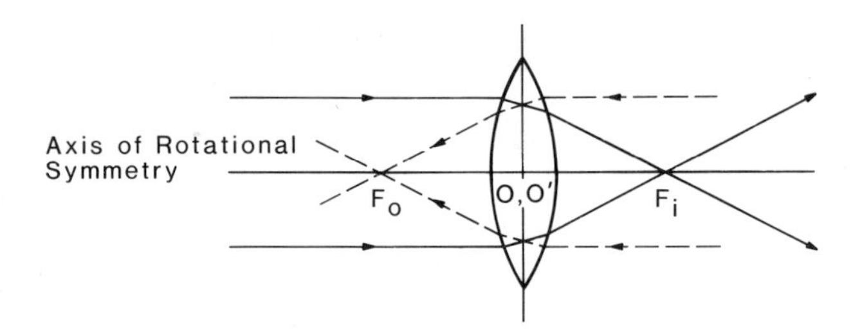

Figure 3 Geometrical construction used to determine image cardinal elements.

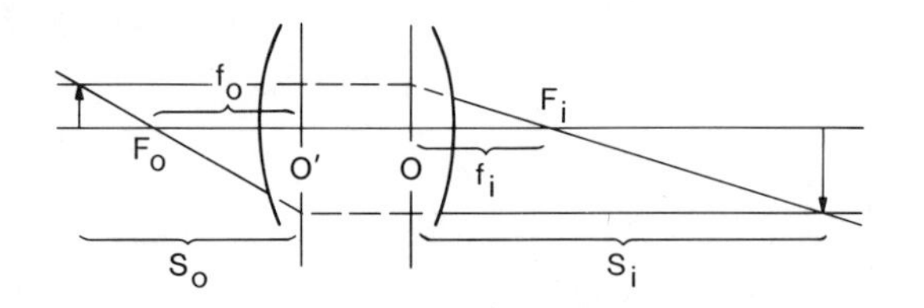

Figure 4 Cardinal elements for a thick lens showing image and object focal points (F_i and F_o), principal points (O and O′), and focal lengths (f and f_o), respectively. Note the principal plane is determined by the intersection of the extended incident and existing rays.

symmetry as in Figures 3, 4, and 5. With the graphical approach, the size of the lens does not matter as far as locating the image is concerned. It is only necessary to define the principal planes and allow them to extend as far as necessary to intersect the appropriate rays. This is demonstrated in Figure 5 by the path taken by "Ray 2." Obviously, even though the lens may not be large enough to let the diagrammed rays go through, in practice other rays will pass through and form the image. A third ray, the ray which passes through the center of the lens, is often convenient to use in diagramming the object-image relationship, especially when determining *virtual images.*

If the object is placed between the object focal point, F_o, and principal point, O', as shown in Figure 6, a different kind of image will be formed. That is, if a screen were placed in the image plane at B, no real image would be observed, but an apparent, or virtual, image is formed. Virtual image formation is actually used in some electron microscopes by the intermediate lens in the formation of low-magnification images.

Virtual images are formed by diverging lenses since only diverging rays appear to come from them. The formation of a virtual image by a diverging lens is shown in Figure 7. Since all electron lenses of interest here are converging lenses, the diverging lens will not be discussed further.

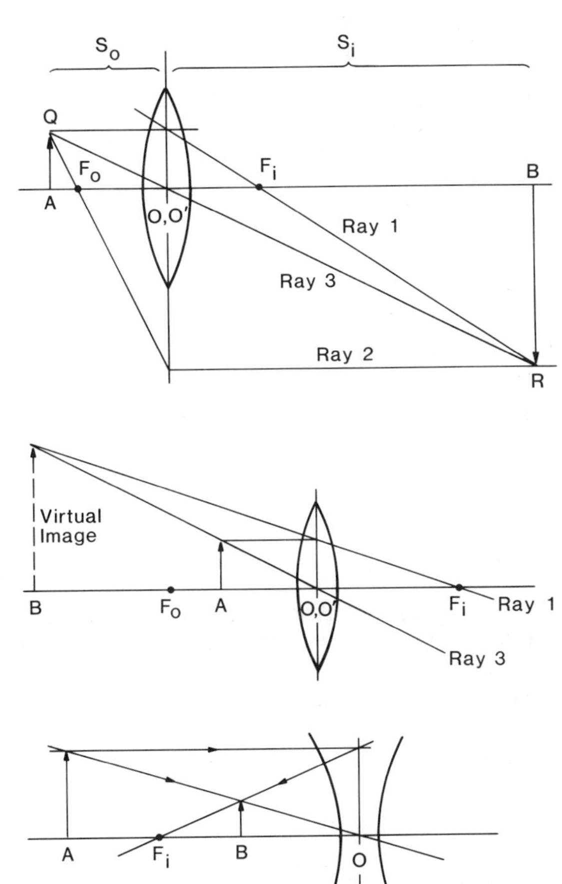

Figure 5 Geometric construction showing the formation of a real image by a thin lens. Ray 1 leaves from Q parallel to the axis and is refracted at the image principal plane through the image focal point, F_i. Ray 2 passes through the object focal point, F_o, and is refracted parallel to the axis at the object principal plane. Ray 3 passes undeviated through the center of the lens.

Figure 6 Formation of a virtual image by a converging lens. It is required that the object, A, be located between F_o and O' ($=O$). The image is formed in the plane derived by the intersection of Ray 1 and Ray 3 (previously described in Figure 5, no Ray 2 is possible here).

Figure 7 Formation of a virtual image at B from an object located at A. The image plane at B is determined by the intersection of the exit rays.

D. Lens Equations

A few important relationships for lenses can be expressed in simple mathematical terms. These relationships are true for light and electron optical lenses. Consider the refractive indices of the media in front of and behind a lens to be n_o and n_i, respectively; then

$$\frac{n_o}{n_i} = \frac{f_o}{f_i} \tag{2}$$

where f_o and f_i correspond to the object and image focal lengths. It can also be shown that

$$\frac{n_o}{S_o} + \frac{n_i}{S_i} = \frac{n_o}{f_o} = \frac{n_i}{f_i} \tag{3}$$

where S_o and S_i are given in Figures 4 and 5. Here, S_o is the distance, $O^{1'}A$, between the object principal point and the object; and S_i is the distance, OB, between the image principal point and the image. Since in most instances $n_o = n_i$, we have $f_o = f_i = f$, and the lens equation becomes

$$\frac{1}{S_o} + \frac{1}{S_i} = \frac{1}{f} \, . \tag{4}$$

The magnification, M, of the object is given by

$$M = \frac{Y_i}{Y_o} = -\frac{S_i}{S_o} \tag{5}$$

where Y_i and Y_o are the height of the image and the object, respectively. Sign convention here allows that, for real image formation, the image is inverted. Thus if Y_o is positive, Y_i will be negative, whereas both S_i and S_o are positive numbers. In the more general case where $f_o = f_i$, the expression for transverse magnification becomes

$$M = \frac{Y_i}{Y_o} = \frac{f_o}{S_o - f_o} = \frac{S_i - f_i}{f_i} = -\frac{S_i f_o}{S_o f_i} \, . \tag{6}$$

E. Paraxial Rays

Thus far all the discussion of image formation has assumed an idealized case, that is, a lens with no aberrations that maps the object in a point-to-point correspondence onto the image. More specifically, second and higher orders of inclination of the illumination with respect to the optical axis have been ignored. Mathematically, if the angle of inclination is γ, then we may replace the $\tan\gamma$ by its sine or its arc, i.e., in the series expansion of $\sin\gamma$

$$\sin\gamma = \gamma - \frac{\gamma^3}{3!} + \frac{\gamma^5}{5!} - \ldots \tag{7}$$

the higher order terms may be neglected and the approximation $\sin\gamma \simeq \gamma$ used. (Note: γ is in radians, but $\sin\gamma = \sin\theta$ when θ is in degrees.) Rays intercepting an object point through a range of γ satisfying this approximation are called *paraxial rays*, and the image formed by these rays is called the *Gaussian image*. The study of optics neglecting the second and higher order terms is often referred to as: the ideal case, paraxial ray case, stigmatic imaging, first order theory, or Gaussian imaging. Image formation by a bundle of incident paraxial rays is shown in Figure 8. Here all the paraxial rays emerging from any point in the object at A will again pass through a single point in the image at B. That is, cach point in the image is *conjugate* to a point in the object. For example, P is conjugate to P_o, A is conjugate to B, and the image plane is said to be conjugate to the object plane.

III. ELECTROSTATIC LENSES

A. Refraction

The foregoing discussion applied in particular to light optics and glass lenses but can be applied equally as well to electron optics. The concepts of image formation and cardinal elements are identical in electron and light optics. And the basic laws of refraction (Snell's law) and rectilinear propagation under constant refractive index are the same. The only problem is to translate these basic concepts from light optics into electron optics.

An electron of charge, $-e$, experiences a force, $\underline{F} = -e\underline{E}$, when exposed to an electrostatic field, $\underline{E}$. (The designation $\underline{E}$ means the field strength is a vector, that is, has magnitude and direction.) More conveniently, the field strength is described in terms of the change of scalar potential, ϕ_i over incremental distance, $\underline{s}$, that is, $\underline{E} = -\delta\phi/\delta\underline{s}$ where $\delta\underline{s}$ is in the direction of $\underline{E}$ and may be written in cartesian coordinates $\delta\underline{s}^2 = \delta x^2 + \delta y^2 + \delta z^2$. Thus, the work done by the force when an electron moves through the electrostatic field from point A to B is given by

$$W = \sum_A^B \underline{F} \times \delta\underline{s} = e\sum_A^B \frac{\delta\phi}{\delta\underline{s}} \times \delta\underline{s} = e\phi_B - e\phi_A \tag{8}$$

so the result is independent of the path but depends only on the potential at the end points. Conservation of energy also requires that this change in potential energy of the electron in going from point A to B is equal to the change in kinetic energy, i.e.,

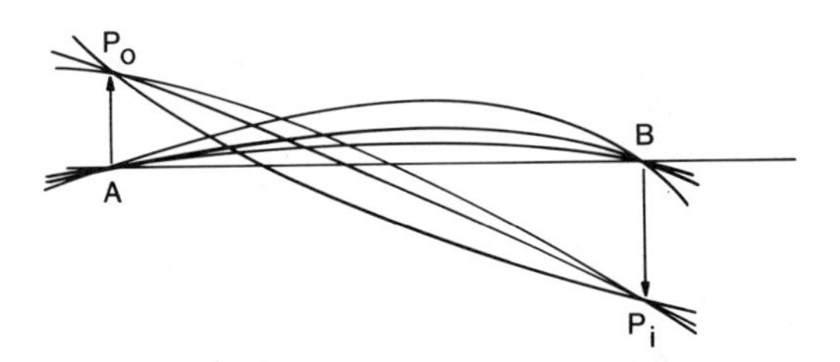

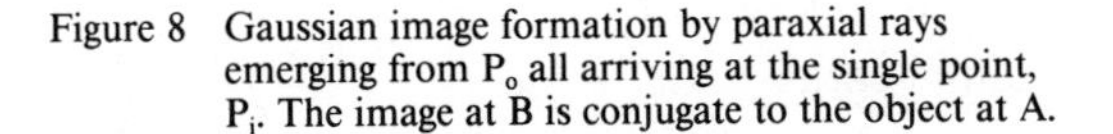
Figure 8 Gaussian image formation by paraxial rays emerging from P_o all arriving at the single point, P_i. The image at B is conjugate to the object at A.

$$e\phi_B - e\phi_A = \frac{1}{2}mv_B^2 - \frac{1}{2}mv_A^2 \tag{9}$$

where m is the mass of the electron and v, the scalar component of the velocity. If point A is at zero potential with the electron at rest ($v_A = 0$), then, for $\phi \geq 0$,

$$\frac{1}{2}mv^2 = e\phi_i \tag{10a}$$

thus

$$v = \left(\frac{2e}{m}\phi_i\right)^{1/2} \tag{10b}$$

assuming that a relativistic correction for the electron mass is not required. In electron (particle) physics, the index of refraction, n, is directly proportional to the velocity. (The inverse relationship was shown for light waves.) Thus

$$n \propto v \propto \sqrt{\phi_i} \tag{11}$$

and the electron equivalent of Snell's law, known as Bethe's law of refraction, becomes

$$\frac{\sin\theta_1}{\sin\theta_2} = \frac{n_2}{n_1} = \frac{v_2}{v_1} = \sqrt{\frac{\phi_2}{\phi_1}} \tag{12}$$

where ϕ_i is the potential expressed in the units of volts. Bethe's law of refraction shows that as the refractive index is increased (i.e., the potential increased), the electrons will be accelerated toward the higher potential.

B. Action of Electrostatic Lenses

As has just been shown, an electron moving from one region of electrostatic potential to another experiences refraction in a manner similar to a light wave encountering a glass lens. However, the change in potential, or refractive index, in an electrostatic lens is continuous in space compared to the abrupt change that occurs at the interface of a glass lens. It thus becomes necessary to determine the potential distribution within the electrostatic lens in order to trace the trajectories of the electrons. Unfortunately, the mathematics are very complex, and complete analytical solutions are found for only the most simple electrode configurations. Instead, experimental methods are usually employed to determine the equipotentials. The path of an electron moving through a simple potential distribution can be approximated by imposing a discontinuous change in refractive index as each equipotential line is encountered. This would be similar to having a group of glass lenses back to back with continuously increasing or decreasing curvature. Consider the case of two coaxial cylinders of the same diameter at different potentials as shown in Figure 9. The equipotential lines are indicated by the dotted curves, and the proportionate change in potential across the lens is indicated at a few increments. The electrostatic field would be normal to these equipotential lines.

An electron entering the lens from the left, at the lower potential, will be refracted toward the axis at each convex equipotential and away from the axis at each concave equipotential as it proceeds past the lens center to the right. The opposite will be true if the electron were to enter the lens from the right at the higher potential and proceed toward the lower potential. Here the convex equipotential lines diverge the electron, and concave equipotential lines converge the electron. The net effect for any electrostatic lens, however, will be to act as a converging lens since an electron is accelerated by the longitudinal (parallel to the axis) component of the potential gradient as it proceeds toward a higher potential—or decelerated if it is proceeding toward a lower potential. Consequently, the electron always spends more time in regions where there is a convergent effect than in regions where there is a divergent effect.

The cardinal elements of an electrostatic lens are defined in the manner discussed for a glass lens and are shown in Figure 10 for a cylinder lens and Figure 11 for a unipotential lens.

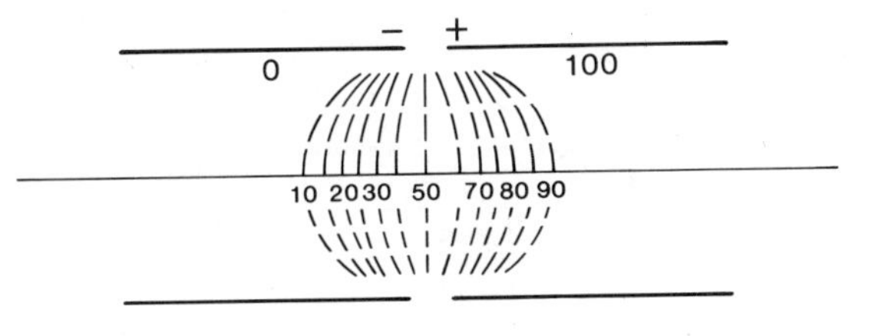

Figure 9 Schematic of coaxial electrostatic lens cylinders showing equipotential lines.

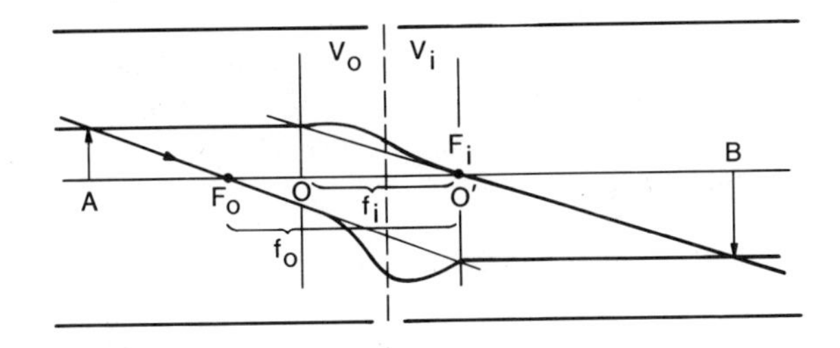

Figure 10 Schematic of an electrostatic cylinder lens with $V_o > V_i$, showing cardinal elements of the lens as previously defined.

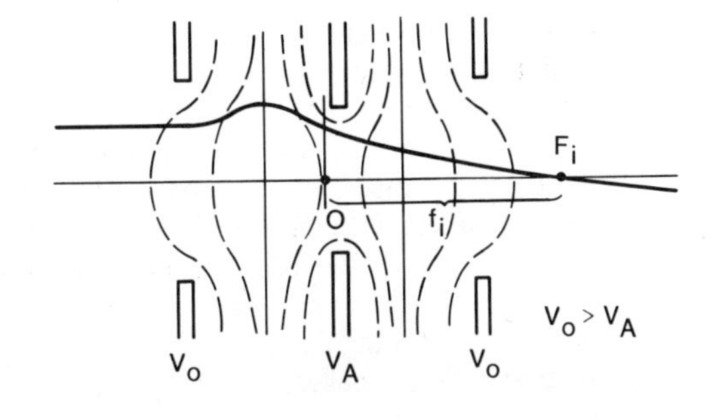

Figure 11 Unipotential or Einsel lens, showing electron trajectory and image cardinal elements. This lens has a symmetrical potential gradient; thus, the electron trajectory through the lens is independent of the direction of incidence.

Notice from the figures that the lens action is different, depending on the direction of the incident electron with respect to the direction of the potential gradient. As can be seen, the image and object focal lengths are different. The relevant lens equations from which one can obtain the focal lengths and image magnification are

$$\frac{f_o}{S_o} + \frac{f_i}{S_i} = 1 \tag{13a}$$

$$\frac{Y_i}{Y_o} = -\frac{f_o}{f_i}\frac{S_i}{S_o} \tag{13b}$$

and

$$\frac{f_i}{f_o} = -\sqrt{\frac{\phi_i}{\phi_o}} = -\sqrt{\frac{V_i}{V_o}}\ . \tag{13c}$$

This last expression is the Lagrange-Helmholtz relation and is derivable from the relation between refractive index and potential shown previously.

C. Types of Electrostatic Lenses

1. Single-Aperture Lens

Probably the simplest type of electrostatic lens is the single-aperture lens, which consists of a single aperture separating two regions of different but uniform fields.

2. Cylinder Lens

Another straightforward lens is that shown in Figures 9 and 10, consisting of two coaxial cylinders of equal diameter. In these lenses the potentials in object and image space are constant and unequal. Cylinder lenses are used in cathode-ray tubes of various kinds.

3. Unipotential or Einsel Lens

This lens is the most important electrostatic lens used in electron microscopy aside from the electron gun, which is an electrostatic accelerating lens. Electron guns will be discussed later. An Einsel lens generally consists of three circular apertures, with the potential on the outer two electrodes the same, usually at the same potential as the anode of the gun. The central electrode is insulated from the outer electrodes and is held at a different potential, often the same as the filament, which is at a high negative voltage. An example of an Einsel lens is shown in Figure 11.

Electrostatic lenses are not frequently used in electron microscopes because of the hazards associated with very high voltages and the fact that even the best electrostatic lenses are inferior to the best magnetic lenses. Electrostatic lenses have, however, been used for electron diffraction cameras since the electrons do not spiral about the optical axis as they do in magnetic lenses, and therefore the diffraction pattern does not rotate with respect to the image.

IV. MAGNETIC LENSES

A. Action of a Homogeneous Field

The action of a homogeneous magnetic field on an electron is quite different from that of an electrostatic field on an electron or the action of optical lenses on light. In the latter two cases, the velocity of the electron or light wave was changed in the direction of propagation by the action of the refractive medium. When an electron encounters a magnetic field, its speed is unchanged while its direction is normal to both the field direction, $\underline{B}$, and the electron velocity, $\underline{v}$. In mathematical terms the electron experiences a force given by

$$\underline{F} = -e\,(\underline{v} \times \underline{B}) \tag{14}$$

where the x indicates the vector or cross product between $\underline{v}$ and $\underline{B}$. Compared with the force from an electrostatic field, $\underline{E}$ (given previously as $\underline{F} = -e\underline{E}$, which acts in the direction of the field), the magnitude of the magnetic force is

$$F = B\,e\,v\,\sin\theta \tag{15}$$

where θ is the angle between $\underline{v}$ and $\underline{B}$. Physically, the action can be described by the "right-hand rule," where the thumb, first, and second fingers of the right hand are held at right angles to each other, approximating a rectangular coordinate system as shown in Figure 12. The direction of the Field is along the direction of the Forefinger; the electron velocity, or Speed, in the direction of the Second finger; and the resultant force on the electron, or Thrust, in the direction of the Thumb.

If the field magnetic were uniform and infinite, the electron would follow a circular path with radius

$$R = \frac{m_o v}{eB} = \frac{1}{B}\left[\frac{2m_o V_r}{e}\right] \tag{16}$$

where $v = (2e/m_o\,V)^{1/2}$ as shown before. V_r is the relativistically corrected accelerating voltage given by

$$V_r \cong V_o\,(1 - 10^{-6}\,V_o)$$

and m_o is the rest mass of the electron $= 9.1 \times 10^{-31}$ kg.

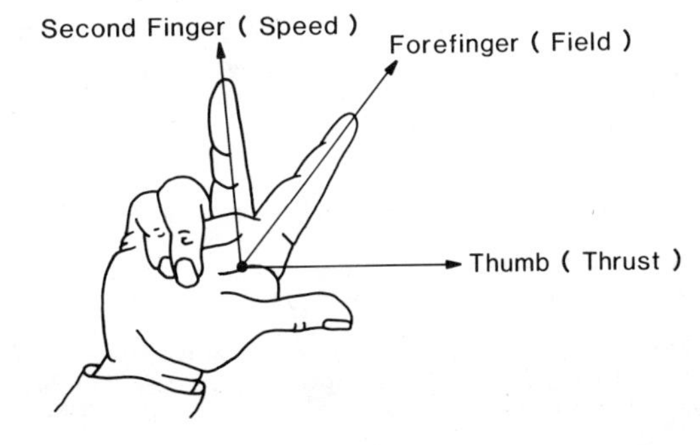

Figure 12 Right-hand coordinate system using thumb, first, and second fingers on the right hand.

If the electron were injected into this homogeneous magnetic field at some angle, θ, to the axis of the field, the velocity could be resolved into two components: one, a longitudinal component, $\underline{v}_\ell$, along the field axis and other, $\underline{v}_t$, a transverse component perpendicular to the field. As $\underline{v}_\ell$ is parallel to $\underline{B}$, $\underline{v}_\ell \times \underline{B} = 0$, and the longitudinal component will be unchanged. The transverse component will describe a circle of radius given by Eq. (16) perpendicular to the field direction. The resultant path of the electron will therefore be a helix of fixed radius having an axis parallel to the field.

B. Action of an Inhomogeneous Field

In all lens configurations encountered in an electron microscope, the magnetic field distribution is inhomogeneous, that is, varying in either magnitude or direction in space. Since almost all magnetic lenses are rotationally symmetric, it is convenient to describe the field in a cylindrical coordinate system—r, θ, and Z—as shown in Figure 13, referenced to a rectangular coordinate system.

The simplest form of a magnetic lens is a single loop of wire forming a circular conductor of radius R, centered on Z_0 and carrying a current, I, as shown in Figure 14(a). The magnetic flux density at any point, Z, on the axis is given by

$$B(z) = \frac{\mu_o I R^2}{2 (R^2 + Z^2)^{3/2}} \tag{17a}$$

which reduces to

$$B(Z = 0) = \frac{\mu_o I}{2 R} \tag{17b}$$

at the center of the loop.

If there were N closely wound turns of radius R, instead of just a single turn, the flux density at Z = 0 would be approximate

$$B (Z = 0) = \frac{\mu_o NI}{2 R} (Wb/m^2) \tag{18}$$

(MKS units are used throughout; thus, $\mu_o = 4\pi \times 10^{-7}$ Wb/A $\times$ m, B(z) is in Wb/m^2, I is in amperes, and all dimensions are in meters.)

A complete magnetic lens is made if the coil is surrounded by a yoke and polepieces made of ferromagnetic materials, such as iron, as shown in Figure 14(b). The magnetic field produced in the gap of this lens is the sum of two components: (i) the flux density produced by the coil, B_c, and (ii) the field produced by the ferromagnetic polepieces, B_{pp}. Usually, $B_{pp} \gg B_c$, so that the maximum field in the gap is limited by the saturation magnetization of the polepieces material to values on the order of 2.0 to 2.5 Wb/m^2 (20 to 25 kG).

A reminder on magnetic units and relationships:
(i) $\underline{B}$ is the magnetic induction or flux density and in the MKS system has units of Wb/m^2 or Teslas.
(ii) $\underline{H}$ is the magnetic field intensity and has units amperes/meter in the MKS system.
(iii) In all systems $\underline{B} = \mu\underline{H}$ where μ is the magnetic permeability. In vacuum $\mu = \mu_o = 4\pi \times 10^{-7}$ Wb/A $\times$ m in the MKS units.
(iv) For the more familiar CGS system, B is in gauss, H is in oersteds, and $\mu_o = 1$ in vacuum. Thus, in a vacuum, $\underline{B} = \underline{H}$ and the two are often used interchangeably.
(v) The conversion units from MKS to CGS are 1 Wb/m^2 (= 1 Tesla) = 10^4 G.

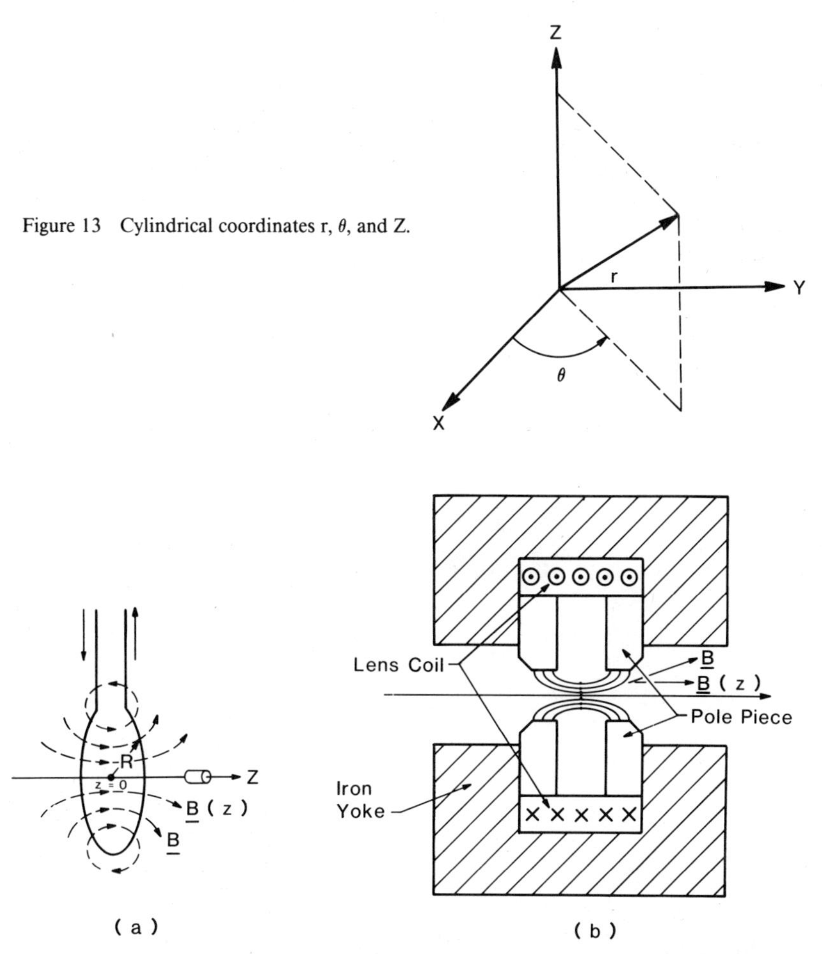

Figure 13 Cylindrical coordinates r, θ, and Z.

Figure 14 Magnetic field distribution in (a) a single turn of wire and; (b) a magnetic lens with polepieces.

The flux density distribution in the gap may be represented by a longitudinal component, $\underline{B}(z)$, and a radial component, $\underline{B}(r)$. (Note: $\underline{B}(z)$ denotes the magnetic flux density distribution along the z-axis only, e.g., $\underline{B}(z) = \underline{B}(r = 0, z)$.) The variation in these components along the lens axis, z, is depicted in Figure 15 for a lens with polepieces having bore D and gap S.

The action of a magnetic lens on an electron may be easily described in terms of these two components of the magnetic induction. If an electron with velocity parallel to the z-axis enters the field, it initially encounters only the radial component, $\underline{B}(r)$. Applying the right-hand rule, one finds the force on the electron will cause it to assume a trajectory in the θ direction. This follows from $F = -e\ \underline{v}_z \times \underline{B}_r = \underline{F}_\theta$. Note that if there were no radial component of the magnetic field in the lens, the field

A word on semantics: As all magnetic lenses considered are axially symmetric, the description of a symmetrical, or asymmetrical, magnetic lens refers to whether or not the bores of the two polepieces are equal. Also, since a lens is axially symmetric, $\underline{B}(\theta) = 0$.

would be homogeneous and the electron with velocity in the z-direction would be completely unaffected by the magnetic lens. The radial component of the field causes the electron to change direction such that its velocity is now normal to $\underline{B}_z$ and it now experiences a force $\underline{F} = e\, \underline{v}_\theta \times \underline{B}_z = \underline{F}_r$, which is radially directed toward the z-axis. The combined components of the field thus cause the electron to spiral toward the z-axis until it crosses it at some point and then to spiral back out again as depicted in Figure 16. In other words, a bundle of parallel electron rays will be focused to a point. This focusing action is caused by the $\underline{B}(z)$ component of the induction; thus, $\underline{B}(z)$ is the component most frequently discussed. It is also the component that is most easily measured.

C. Paraxial Ray Equations

The radial force on the electrons, $\underline{F}_r$, can be calculated from the right-hand rule to be

$$\underline{F}_r = -e\, \underline{v}_\theta \times \underline{B}(z) \tag{19}$$

and after some mathematical substitution and rearranging becomes

$$\underline{F}_r = -\left[\frac{e^2}{4m}\right] B^2(z)\underline{r} \ . \tag{20}$$

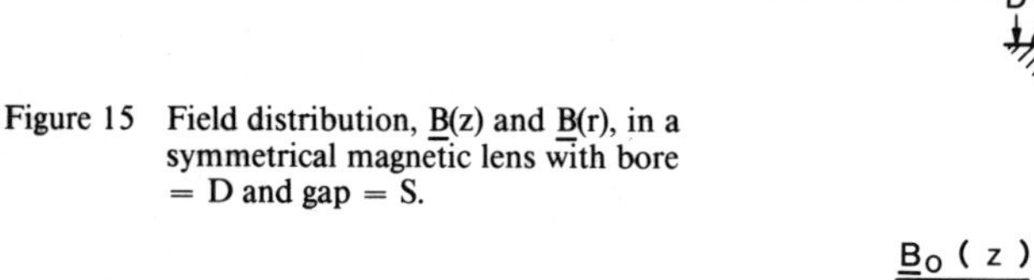

Figure 15 Field distribution, $\underline{B}(z)$ and $\underline{B}(r)$, in a symmetrical magnetic lens with bore = D and gap = S.

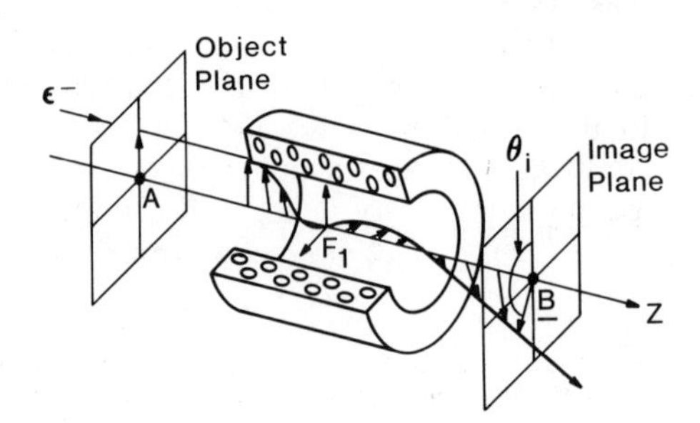

Figure 16 Image formation with a magnetic lens, depicting electron rotation about the z-axis.

This equation states that the radial force drives the electrons toward the z-axis (because of the negative sign in front of the expression) and is directly proportional to the distance, r, of the electron from the axis, which is the principle of a focusing lens. The equation also shows that all magnetic lenses are converging lenses because the equation is independent of the sign of B(z).

Since force is equal to mass times acceleration, $\underline{F} = m \times \underline{a}$, and the acceleration is defined as the second derivative of the spatial coordinate, $\underline{r}$, with respect to time (e.g., $\underline{a} = d^2\underline{r}/dt^2$), one obtains

$$\underline{F}_r = m\underline{a}_r = m\frac{d^2\underline{r}}{dt^2} = -\left[\frac{e^2}{4m}\right]B^2\underline{r} \qquad (21a)$$

or, rearranging

$$\frac{d^2\underline{r}}{dt^2} + \frac{e^2}{4m^2}B^2\underline{r} = 0 \ . \qquad (21b)$$

This gives the equation of motion of an electron in a plane normal to the z-axis. It can also be shown that

$$\frac{dz}{dt} = \left[\frac{2eV}{m}\right]^{1/2} \qquad (22)$$

along the z-axis. Combining these two equations under the assumption of paraxial rays and that r is not too far from the z-axis gives the "paraxial-ray equation," which describes the motion of an electron in a rotating meridional plane

$$\frac{d^2\underline{r}}{dz^2} + \frac{e}{8m_oVr}B^2(z)\underline{r} = 0 \ . \qquad (23a)$$

A meridional plane is determined by the original location of the electron in field free space and the z-axis. The function, $d^2\underline{r}/dz^2$, may be thought of as the change in slope of the electron path with respect to the z-axis. Rewriting the equation as

$$\frac{d^2\underline{r}}{dz^2} = -\frac{e}{8m_o}\frac{B^2(z)}{V_r}\underline{r} \qquad (23b)$$

one can see that the greater $B^2(z)/V_r$, the faster the electron is bent toward the axis, and vice versa.

The rotation of the electron about the z-axis as it proceeds in the z-direction may be written as

$$\frac{d\theta}{dz} + \left[\frac{e}{8m_oV_r}\right]^{1/2}\underline{B}(z) = 0 \ . \qquad (24)$$

This equation may be integrated to give the total rotation, $\Delta\theta$, of the electron as it moves from point z_0 to z_1:

$$\Delta\theta = -\int_{z_0}^{z_1}\left[\frac{e}{8m_0Vr}\right]^{1/2}\underline{B}(z) \times dz \ . \qquad (25)$$

The notation, $\int_{z_0}^{z_1} B(z)dz$, means the sum of all the increments, B(z)dz, from $z = z_0$ to $z = z_1$.

The important point is that $\Delta\theta$ is a functional of B(z), so that if B(z) is of the opposite sign, $\Delta\theta$ will be in the opposite direction. This property is often used in the design of electron microscopes by incorporating sequential lenses having fields of opposite sign to nullify the image rotation.

D. Bell-Shaped Fields

An analytical solution to the paraxial ray equation is obtained only if the axial induction distribution, B(z), can be expressed. Following the form of the equation shown previously, describing the induction distribution around a single loop of wire, the induction distribution

$$\underline{B}(z) = \frac{\underline{B}\,(z = 0)}{1 + \left[\frac{z}{a}\right]^2} \tag{26}$$

known as Glaser's bell-shaped field (Glaser, 1956), is found to be a good approximation to the field distribution in the gap of a symmetrical magnetic lens. In the equation, a is half the full width at half maximum (FWHM) of the induction distribution, B(z), at B(z)/2, and $B(z = 0)$ is the maximum flux density along the axis at $z = 0$.

Substituting this bell-shaped field distribution into the paraxial ray equation yields

$$\frac{d^2\underline{r}}{dz^2} + \frac{eB_o^{\,2}}{8m_oVr}\left\{\frac{\underline{r}}{\left[1 + \left(\frac{z}{a}\right)^2\right]^2}\right\} = 0\ . \tag{27a}$$

The conventional method for finding solutions to such an equation is to make substitutions for the variables, r and z, which puts the equation into a form whose solution is recognizable. For example, if one makes the substitutions $x = z/a = \cot\phi$ and $y = r/a$, the equation becomes

$$\frac{d^2y}{d\phi^2} + 2\cot\phi\,\frac{dy}{d\phi} + ky^2 = 0 \tag{27b}$$

where $k^2 = (eB_o^{\,2}/8m_oV)a^2$ is called a lens parameter. One solution to this equation is

$$y = \frac{1}{\omega}(\sin\omega\phi/\sin\phi) \tag{28}$$

where $\omega^2 = k^2 + 1$. It is apparent that y is simply a reduced variable relating to the radial distance, r, of the electron from the z-axis and $\cot\phi$ is a function converting linear distance into angular measure as shown in Figure 17. "a" is a straight line parallel to the z-axis. The slope of y at any point is given by

$$y' = \frac{dy}{dz} = \frac{dy}{d\phi} = \frac{d\phi}{dz} = \frac{1}{\omega a}(\sin\omega\phi\ \ \omega s\phi - \omega\sin\phi\ \cos\omega\phi)\ . \tag{29}$$

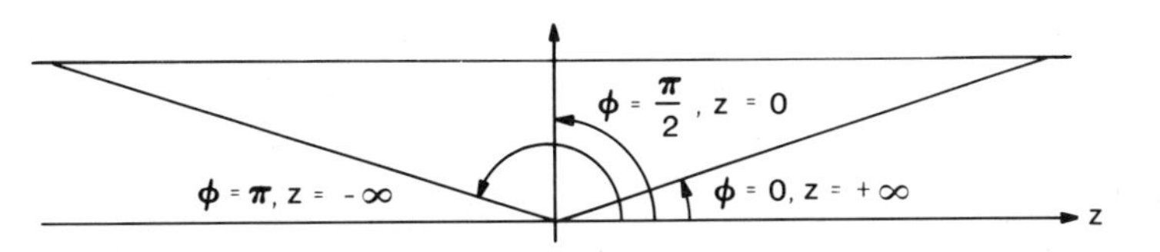

Figure 17 Shows relationship between ϕ and z.

E. Lens Excitation Parameters ω and k^2

It is instructive to determine under what conditions an incident beam of electrons parallel to the axis exits the field still parallel to the z-axis, e.g., to calculate under what conditions $y' = 0$ for $\phi = 0$ and π.

Calculation of y' for $\phi = 0$ ($z = +\infty$) shows $y' = 0$ identically, which is considered to be the incident beam. It remains to calculate under what conditions $y' = 0$ for $\phi = \pi$ ($z = -\infty$). As is easily seen, this yields $y' = 1/\omega a \sin\omega\pi = 0$, which requires $\sin\omega\pi = 0$. From trigonometry this condition is satisfied when ω = integer, e.g., 0, 1, 2, 3 . . . .

For a beam to enter and exit the lens parallel to the axis requires that it either be unaffected by the field or cross the axis. Actually, the beam may cross the axis a number of times as shown below. Remembering $\omega^2 = k^2 + 1$ where $k^2 = (eB_o^2/8m_oV_r)\, a^2$, one can tabulate the number of axial crossings of the electron trajectory for various values of ω and thus k^2. This is done in Table II. Obviously, the cases $\omega = 0$, 1 are uninteresting. The case $\omega = 2$, $k^2 = 3$, is the condition under which the beam crosses the z-axis one time which, by symmetry arguments, must occur in the center of the lens. This operating condition is called the *telefocus condition.* It may be observed in the image mode of a transmission electron microscope (TEM) if the objective lens has sufficient "excitation," first-condenser lens (C1) strong, and second-condenser lens (C2) off. Almost all scanning transmission electron/transmission electron microscopes (STEM/TEM) have a sufficiently strong objective lens for this operation, but only a few TEM microscopes do. And most conventional objective lenses operate under the condition $k^2 < 3$.

TABLE II

Relationship Between ω and k^2 and the Number of Beam Crossovers of the Z-axis.

ω	ω^2	k^2	Comment
0	0	−1	meaningless
1	1	0	no field, beam is undeviated
2	4	3	beam crosses z-axis one time
3	9	8	beam crosses z-axis two times
4	16	15	beam crosses z-axis three times

The case $\omega = 3$, $k^2 = 8$ provides for two axial crossovers of the electron beam, each at the same distance from the lens center. Some of the present day S(TEM) microscopes are using lenses with this excitation and locating the specimen near the first crossover.

The case $\omega = 4$, $k^2 = 15$ provides three crossovers of the z-axis. To achieve this would require such a high magnetic field, assuming reasonable focal properties are maintained, that such lenses have not been realized for commercial use. The table could be continued for $\omega = 5, 6, 7, \ldots$, etc., with an increase in the number of crossovers, but a discussion of these would be purely academic.

For the noninteger values of ω, the beam crosses the z-axis but does not exit parallel to it. In the conventional lens, $k^2 < 3$, so the beam crosses the optical axis once. For $3 < k^2 < 8$, the beam crosses the axis twice, and either may be used as a focal point for image formation, etc. As almost all TEM and STEM objective lenses operate at lens excitations such that $k^2 < 3$, this discussion will be restricted to lenses of this type. Some typical electron trajectories are drawn in Figure 18 for various values of k^2.

F. Cardinal Elements of Magnetic Lenses

From the discussion of lens crossovers for various values of lens excitation, it is easily seen that the crossover for all but the weakest lens will occur while the electron is still under the influence of the magnetic field. This is different than previously discussed for electrostatic lenses, where the focal point is usually outside the potential field of the lens. A typical electron trajectory in a magnetic lens operated such that $k^2 < 3$ (but not too weak) is shown in Figure 19.

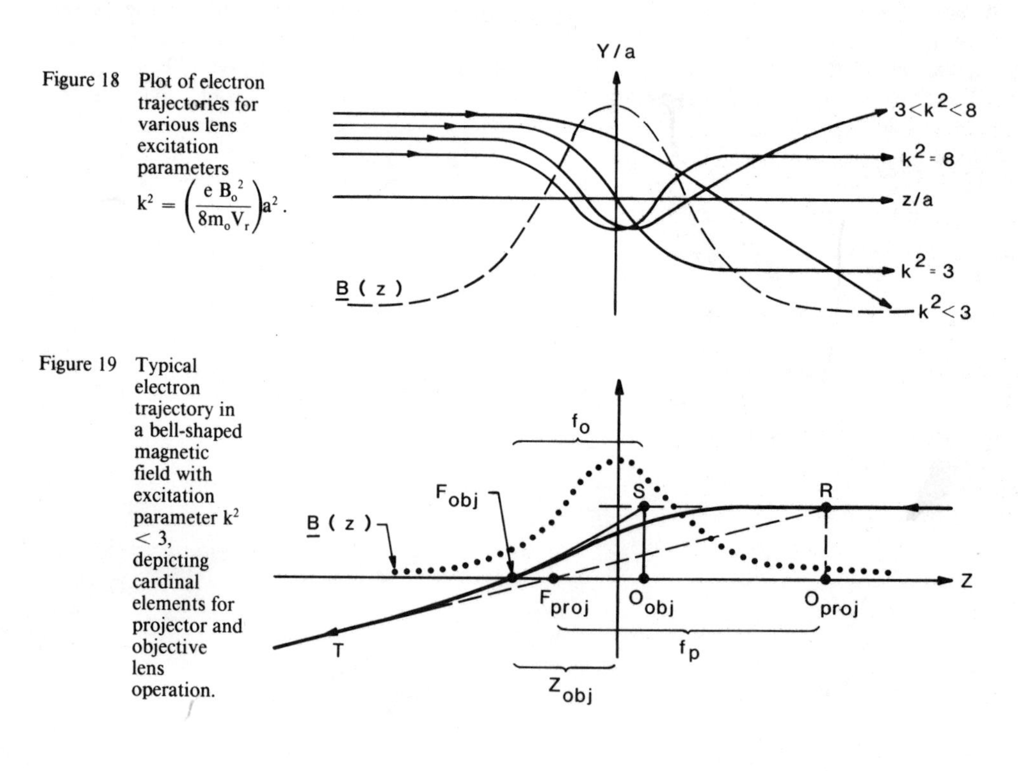

Figure 18 Plot of electron trajectories for various lens excitation parameters $k^2 = \left(\frac{e B_o^2}{8 m_o V_r}\right) a^2$.

Figure 19 Typical electron trajectory in a bell-shaped magnetic field with excitation parameter $k^2 < 3$, depicting cardinal elements for projector and objective lens operation.

As one can see in Figure 19, there are two focal lengths, focal points, and principal planes defined for a single trajectory through the lens. Since the trajectory intersects the z-axis while still in the magnetic field, the path will continue to curve after passing the axis such that the asymptote, RT, to the trajectory at a large distance from the lens will not intersect the z-axis at the crossover point. The intersection of this asymptote and the extended incident beam defines the principal plane at O_{proj}, and the distance ($F_{proj} - O_{proj}$) defines the projector focal length, fp. These cardinal elements are those used to define the properties for lenses used as magnification lenses, condenser lenses, etc., and are defined as the general class of projector lenses. The object for this lens configuration is outside or just inside the magnetic field.

If the object is placed well into the magnetic field, as is the case in many modern TEM objective lenses, a different set of cardinal elements is required to describe the lens behavior. These cardinal elements are determined by the intersection of the beam trajectory with the z-axis defining the focal point, F_{obj}. The intersection, s, of a tangent to F_{obj} and the extension of the incident beam direction define the principal plane at O_{obj}, and the distance ($O_{obj}\ F_{obj}$) defines the objective focal length, f_o.

Quantitatively, the cardinal elements can be described in terms of the solution to the paraxial-ray equation. The projector focal length is given by

$$f_p = \frac{1}{y'} = + \frac{a\omega}{\sin(\pi\omega)} = a\omega \csc(\pi\omega) \tag{30}$$

where y' is evaluated at $z = \infty$ (or $\phi = \pi$). Obviously, a projector focal length is undefined when the exit beam is parallel to the axis, e.g., when $\omega = 2, 3, 4$.

The objective focal point is determined by the zero of the function $y = (\omega a)(\sin\omega\phi/\sin\phi)$, for example, where the electron beam crosses the axis. This occurs when $\sin(\omega\phi) = 0$ or $\omega\phi = n\pi$, where n is any integer. The location of this focal point with respect to the lens center is obtained from evaluating $z/a = \cot(\phi)$ at $\phi = n\pi/\omega$. Addressing the case $n = 1$ only, one obtains $Z_{obj} = a\cot(\pi/\omega)$ as the distance of the objective focal point from the center of the lens and $Z_{im} = -a\cot(\pi/\omega)$ as the corresponding distance of the image focal point from the center of the lens. Z_{obj} and Z_{im} are often called the midfocal points.

In a TEM objective lens, therefore, the diffraction pattern appears in the back focal plane (BFP) of the lens located at Z_{im}, and the objective aperture is positioned there. The specimen is usually situated in front of the objective midfocal point, Z_{obj}.

The objective focal length, f_o, as determined by the length $F_{obj}\ O_{obj}$, is

$$\frac{1}{f_o} = -\frac{1}{y}\frac{dy}{dy}\bigg|_{z=0} = -\left(\frac{1}{a}\right)\sin\left(\frac{\pi}{\omega}\right) \tag{31a}$$

for the case $n = 1$, or

$$f_o = -a\csc\left(\frac{\pi}{\omega}\right). \tag{31b}$$

Similarly, the image focal length is given by $f_i = a\csc(\pi/\omega)$. The minimum value of the focal length is obtained when $\sin(\pi\omega) = 1$ at $\omega = 2$, corresponding to the telefocal lens excitation $k^2 = 3$. At $k^2 = 3$, $f_o = -a$, and $f_i = a$ with $Z_{obj} = -Z_{im} = 0$, the midfocal points are located at the lens center $Z = 0$. For this unique operating condition, the specimen is placed at the center of the lens, $Z = 0$, and the BFP is located at $Z = a$.

The location of the principal points with respect to the lens center is given by

$$Z_{pi} = -a \cot\left(\frac{\pi}{2\omega}\right) \tag{32a}$$

and

$$Z_{po} = a \cot\left(\frac{\pi}{2\omega}\right) \tag{32b}$$

for the image and object midprincipal points, respectively.

In Table III, normalized values of these cardinal elements have been calculated for various lens excitation parameters, k^2, with the relationships derived above.

Defining $\ell_i = s_i - f_i$ and $\ell_o = s_o - f_o$ where s_i and s_o were defined in Figures 4 and 5, one can show that

$$\ell_o \ell_i = f_o f_i \tag{33}$$

which is the electron optical equivalent to Newton's formulas in light optics. ℓ_i represents the distance between the image and the image focal point; ℓ_o represents the corresponding distance for the object. In the usual case where $f_o = -f_i$, then $\ell_o \ell_i = -f^2$. Newton's formula demonstrates that an image will be formed in real space independent of the object's being immersed in the lens field.

It can also be shown that the magnification is given by

$$M = -\frac{f_o}{\ell_o} = -\frac{\ell_i}{f_i}\,. \tag{34}$$

TABLE III

Values of the Cardinal Elements with Respect to Lens Center and the Focal Length (normalized to the half width of the magnetic field, a) as Functions of k^2. The image cardinal elements are determined assuming the incident beam is coming from $Z = -\infty$.

k^2	Midprincipal Point $Z_{po}/a = -Z_{pi}/a$	Midfocal Point $Z_{im}/a = -Z_{obj}/a$	Focal Length f/a
1	0.496	0.761	1.257
1.5	0.651	0.442	1.093
2	0.782	0.248	1.020
2.5	0.897	0.109	1.006
3	1.000	0.000	1.000
4	1.181	0.167	1.014
5	1.340	0.297	1.043
6	1.482	0.403	1.079

The rotation of the image with respect to the object is found to be

$$\Delta\theta = \frac{nk_\pi}{(k^2 + 1)^{1/2}} = \frac{nk_\pi}{\omega} . \tag{35}$$

Since $n \leq (k^2 + 1)^{1/2}$, the most frequently encountered case will be with $n = 1$; therefore, $\Delta\theta = k_\pi/\omega$. For example, if $k^2 = 1$, $\omega^2 = 2$, and $\Delta\theta \cong \pi/\sqrt{2} \cong 127°$ rotation.

G. Objective Lenses

The excitation of objective lenses found in most of the commercial TEMs is $k^2 < 3$. Thus, the specimen is positioned above the lens, and the focal point is below lens center. The electron ray path for such a lens was shown in Figure 19. Since the specimen lies in front of the lens center, the prefield of the lens is not strong enough to image a plane conjugate to the specimen in real space before the lens. Hence, the plane conjugate to the specimen is virtual and lies behind the lens, and as a result, the illumination conditions at the specimen are defined by the second condenser lens of the microscope. This means that to obtain the nearly parallel illumination needed for highest resolution, C2 must be strongly overfocused, resulting in a large probe. Small probes may be obtained by forming the crossover of the C2 lens at the specimen plane, but only under conditions which result in a very convergent beam.

Spherical aberration, which shall be discussed soon, essentially determines the resolving power of a lens. In the "normal" lens ($k^2 < 3$) discussed here, it has been found that the spherical aberration coefficient is somewhat greater than the focal length of the lens and therefore limits the ultimate resolution for a fixed geometry of lens polepieces. Resolution can be improved, however, by immersing the specimen further into the field. In particular, two types of lenses with specimens deeply immersed have been studied.

If the specimen is positioned at the exact center of the lens, the excitation is given by $k^2 = 3$. This corresponds to the telefocal condition previously discussed with parallel incident and exit rays. A lens in this configuration is known as the single-field *condenser objective lens.* In this lens, developed by Riecke (1962), the first half of the magnetic field lies in front of the specimen and acts as a short focal length condenser, while the second half of the field lies behind the specimen and acts as an objective lens; thus the term condenser-objective. The prefield of this lens is sufficiently strong to image the plane conjugate to the specimen in real space before the lens, usually at the C2 aperture position. With a special alignment procedure developed by Riecke (1962), it is thus possible to form a demagnified image of the C2 aperture at the specimen plane. This provides a small probe with nearly parallel illumination.

By immersing the specimen even further into the lens field where $3 < k^2 < 8$, the electron trajectory will now have two crossovers. Such a lens is called a *second zone lens.* The specimen may be placed near the second crossover, but its position is not as critical as in the condenser objective lens of Riecke. Again, the prefield of this lens is sufficiently strong to image the plane conjugate to the object in real space in front of the lens; however, the location of this conjugate plane is within the gap of the lens, so it is not readily accessible, and the procedure to obtain parallel illumination at the specimen closely resembles that used in a "normal" objective lens although smaller probes may be used. Second zone lenses have not been used frequently in microscopes because of the large excitation required to produce the magnetic field. They do provide a distinct advantage in microscopes combining STEM with TEM when the specimen is placed at the first crossover of the lens. Operating the lens under strong excitation provides a small convergent probe for STEM, and a reduction in the excitation allows normal TEM operation.

Frequently, the lens excitation, k_1^2, is given as the ratio of the square of the number of ampere-turns, $(NI)^2$, to the relativistic accelerating voltage, V_r, that is

$$k_1^2 = \beta_c (NI)^2/V_r \,. \tag{36}$$

The proportionality constant, β_c, is related to the lens geometry and field and is given by

$$\beta_c = \frac{e\mu_o^2}{32_m}\left[\frac{B_o}{B_p}\right]^2\left[\frac{D}{S}\right]^2 = 0.0087\left[\frac{B_o}{B_p}\right]^2\left[\frac{D}{S}\right]^2 \tag{37}$$

where B_o is the maximum field in the gap, $B = \mu_o NI/S$ is the magnetic flux density that would exist in the gap if $S \gg D$, and D and S are the polepiece bore and gap dimensions, respectively. Hence

$$k_1^2 = 0.0087\left[\frac{B_o}{B_p}\right]^2\left[\frac{D}{S}\right]^2\frac{(NI)^2}{V_r}\,. \tag{38}$$

The lens excitation, k_1^2, used here is that described by Liebmann *et al.* (1949) in their extensive discussion of lens properties and is related to the lens excitation, k^2, used throughout this chapter by $k_1^2 = (R_\ell/a)^2\, k^2$; k^2 is the parameter used by Glaser and most modern-day authors; R_ℓ is the bore radius, $R_\ell = D/2$. Some computed values of β_c, B_o/B_p, and R_ℓ/a are given for a few typical values of S/D in Table IV.

The optical parameters of objective and projector lenses have been calculated under a variety of geometrical conditions, S/D, and excitation conditions, k^2 of k_1^2. These data are available in many of the references given, and the serious reader is advised to look there.

V. LENS ABERRATIONS AND DEFECTS

A. Spherical Aberration

Thus far, paraxial illumination and Gaussian imaging have been assumed. Consequently, there is a conjugate point-to-point correspondence between the object and the image. Such is not the case in real optical systems, and performance is usually

TABLE IV

Values of β_c, B_o/B_p, and R_ℓ/a as f(S/D).

S/D	β_c	B_o/B_p	R_ℓ/a
1	0.0064	0.860	0.9099
2	0.0021	0.987	0.4975
3	0.0010	0.999	0.3332

determined by the minimization of aberrations. Figure 20 illustrates one of the aberrations.

The image of object point P will be formed at point Q in the Gaussian image plane for all paraxial rays, P A Q. However, if the rays are not paraxial, that is, if sin $\alpha \neq \alpha$, they will be bent more at the periphery of the lens. The result is that the image point, Q, will be displaced a distance, Δr, to Q′, as shown in the figure for the nonparaxial ray, P B Q′. Point P will thus have an apparent radius, $\Delta r/M$, where M is the image magnification. This effect is called spherical aberration and is one of the principal factors limiting the resolution of a TEM.

The addition of a third-order term, $\alpha^3/3!$, is usually sufficient to allow the approximation $\sin \alpha \cong \alpha - \alpha^3/3!$ for magnetic lenses. The image distortion, Δr, is related to α^3 through $\Delta r = C_s\alpha^3$, where C_s is defined as the coefficient of spherical aberration. Compared to the objective focal length, $C_s > f$ for k^2 small and $C_s < f$ for $k^2 \geq 3$ approaching a constant value for large k^2.

B. Pincushion, Barrel, and Spiral Distortion

When spherical aberration is present in a lens, e.g., $\Delta r = C_s\alpha^3$, the image magnification will vary in proportion to the cube of the distance of the image point from the axis. The result is a distorted image. There are two kinds of distortion (pincushion and barrel) that are rather common in TEMs, especially at low magnification, and a third kind (spiral) that is not frequently seen. Pincusion distortion occurs when the magnification of the image increases with the distance of the image point from the center, as shown in Figure 21(b). Barrel distortion is the opposite of pincushion and occurs when the magnification decreases with distance from the center, as shown in Figure 21(c). Spiral distortion occurs when the angular rotation of an image point depends on the distance of the point from the axis. The images are sigmoidal-shaped, as shown in Figure 21(d).

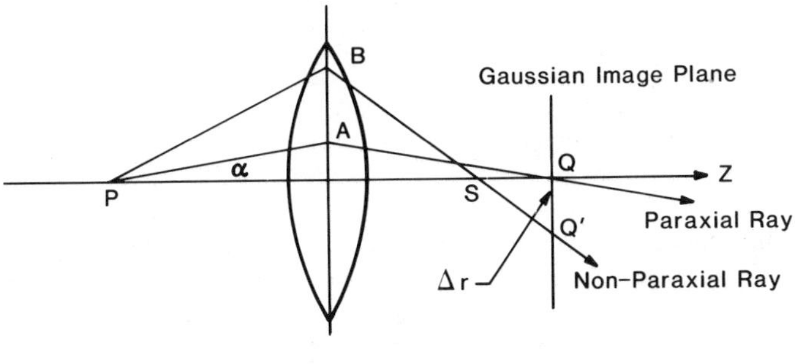

Figure 20 Paraxial ray, P A Q, and nonparaxial ray, P B Q_1, illustrating spherical aberration in the Gaussian image plane.

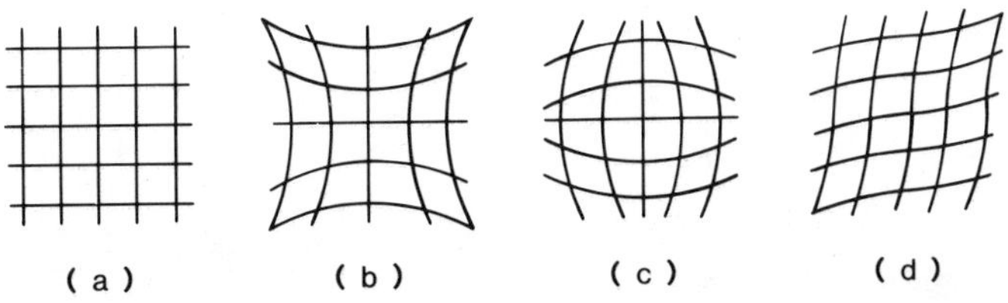

Figure 21 Distortion in a magnetic lens: (a) object; (b) pincushion distortion; (c) barrel distortion; (d) spiral distortion.

C. Astigmatism

Even with the high technology of today, production of a magnetic lens with perfect axial symmetry has not been possible. The result is that the image plane for objects lying in one direction will be different from the image plane for objects lying in another direction. This means, for example, that the " – " in an H will be focused in one plane and the "| |" in another. Hence, there is no sharp image plane—only a plane of least confusion between the two sharply focused images. Lenses with this effect are said to be astigmatic; however, devices called stigmators used in all modern electron microscopes can completely compensate for this lens imperfection. Stigmators are usually used in both the objective lens and the illumination system and sometimes in the magnification system of lenses.

D. Chromatic Aberration

Because of refraction by the electric or magnetic fields of a lens, electrons leaving a point, P, with different velocities will be brought to a focus at different points, Q and Q′, in accordance with their velocity, as shown in Figure 22; e.g., the greater the velocity, the longer the focal length.

As in the case with spherical aberration, the result is that point P will be imaged as a disc of radius $\Delta r/M$. By geometrical arguments, it can be shown that

$$\Delta r = C_c \alpha \frac{\Delta V}{V} \tag{39}$$

where C_c is defined as the coefficient of chromatic aberration and α is the angle of divergence. Its numerical values are similar to C_s and are usually on the order of f (a few mm). Differences in electron velocity or electron potential, V, may come from many sources. Among the most prevalent are fluctuations of the accelerating voltage and the inelastic scattering within the specimen. Chromatic aberration usually limits the resolution in thick specimens.

E. Boersch Effect

Another, although minor, source of the spread in electron velocities in the beam comes from the Boersch effect (Boersch, 1954). Laws of physics provide that particles having the same charge repel; thus, the electrons may interact with each other, creating transverse velocity components and thereby changing the longitudinal components. This effect is noticeable only when the electrons are at a very high current density crossover, such as at the gun. For a heated tungsten filament, the Boersch effect gives rise to an overall energy spread of the beam of 1 to 2 eV, while for a cold field emitter, the energy spread is only a few tenths of an eV.

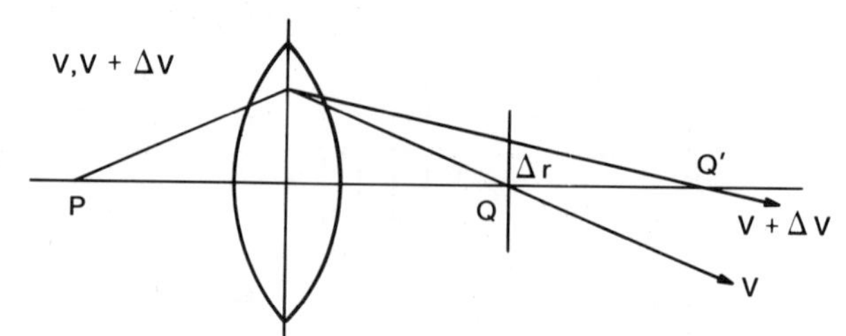

Figure 22 Electrons leaving point P with different velocities (because of potentials V and V + ΔV) will be brought to a focus at Q and Q′, respectively. This effect is called chromatic aberration.

VI. SPECIAL MAGNETIC LENSES

There are so many different designs of magnetic lenses that it is impossible to discuss them all here. However, a brief description of a few of the most interesting designs will be given.

A. Quadrapole and Octapole Lenses

Quadrapole lenses are formed by arranging electrodes of alternate potential or magnetic poles of opposite polarity, as shown in Figure 23, in a plane normal to the electron trajectory.

Because the field lines are normal to the electron velocity, much stronger focusing action may be achieved with much less power expenditure. Since the action of the lens is to form a line image from a point source, however, a single quadrapole is not useful as an imaging lens. Crossing quadrapoles of opposite polarity will yield point focus, but other properties are unattractive. Octapoles are similar to quadrapoles but have eight electrodes or poles of opposite sign. By combining octapoles and quadrapoles, it is possible to design a lens system free of spherical aberration, but mechanical alignment factors prohibit their practical use.

B. Pancake and Snorkel Lenses

In the quest to attain greater resolution, the requirements of low aberration lenses have dictated polepieces with extremely small bore and gap. The problem with this approach is that the gap becomes increasingly small, and difficulties in specimen manipulation become significant. Thus, practical limits are established on S and D depending on the degree of specimen manipulation required. Mulvey (1974) has proposed two lenses of unconventional design that may offer new hope for overcoming these problems.

The first is the "pancake" lens, Figure 24, which consists of a partly shielded, flat or helical coil, producing a very high magnetic field in front of the lens. Calculations indicate that the axial flux density distribution, B(z), can have an appreciably lower C than that of a comparable standard polepiece lens. Because of the physical geometry of the lens, there is ample space above the lens at the field maxima for specimen manipulation.

The second is the "snorkel" lens, which consists of a central magnetic core and outer shroud with an enclosed coil that can be excited to a high current density. The lens, as shown in Figure 25, may be thought of as a modified pancake lens—modified by the addition of the central iron core in the general shape of a conical snout with a hole in the center.

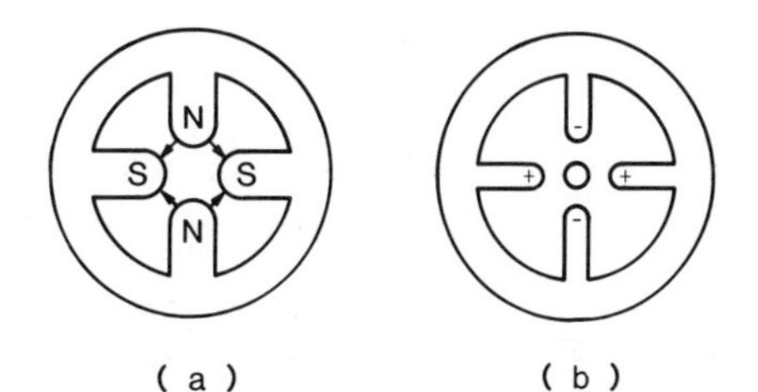

Figure 23 Schematic of (a) magnetic and (b) electrostatic quadrapole lenses.

The field distribution in this lens is such that the electrons may be focused in front of the lens, and thus the specimen may be located there, allowing easy access for study. Munro (1974) has calculated the field distribution in the snorkel lens and found that, to a good approximation, snorkel lenses may be regarded as "half-pancake" lenses. This analysis assumed a solid iron core as might be used for an SEM rather than one with a hole in it for the TEM as pictured here.

VII. PRISM OPTICS

A. Magnetic Sectors

As previously discussed, the magnetic force on a moving electron is described by the expression

$$\underline{F}_o = e\underline{v} \times \underline{B} \tag{40}$$

and the electron path will be normal to both $\underline{v}$ and $\underline{B}$. In a uniform magnetic field, $\underline{B}$, the electron will describe a circular path of radius r, which depends on the velocity or energy of the incident electron.

$$\underline{r} = \frac{m_o\underline{V}}{e\underline{B}} = \frac{1}{\underline{B}}\left[\frac{2m_o}{e} \times V_r\right]^{1/2} . \tag{41}$$

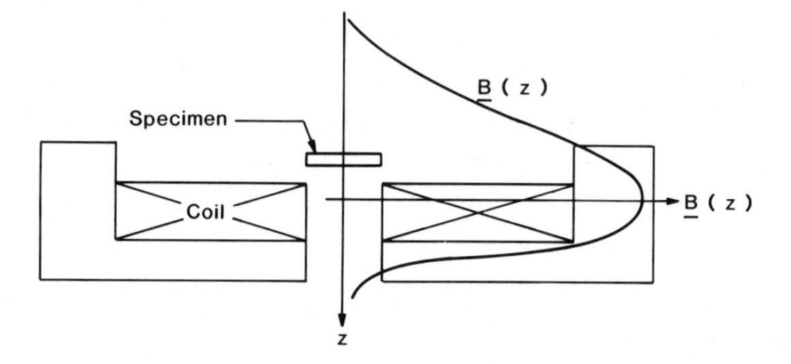

Figure 24 The geometry and magnetic field distribution of a "pancake" lens.

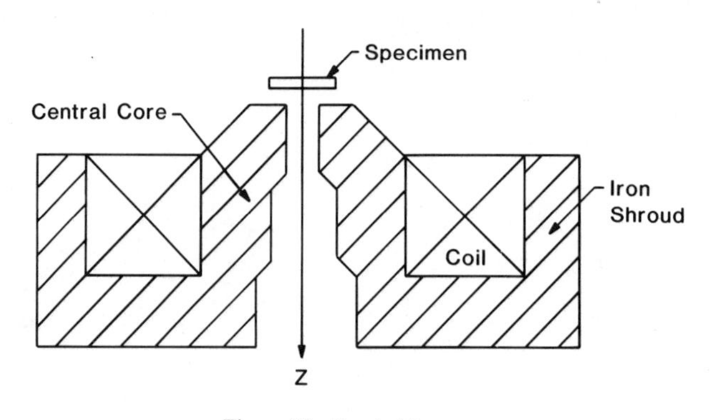

Figure 25 Snorkel lens.

If the energy of the electron is changed by an amount, ΔV, the radius of the electron trajectory will be correspondingly changed to

$$r \pm \Delta r = \frac{1}{\underline{B}}\left[\frac{2m_o}{e}\right]^{1/2} \times (V_r \pm \Delta V_r)^{1/2} . \tag{42}$$

Considcr a magnet consisting of two identical poles in the form of a circular sector of angle ϕ with mean radius r_o and having a homogeneous magnetic field, $B_o(R)$, normal to the sector along each circle R of radius r, as shown in Figure 26. This is the ideal case and is sufficient for these purposes.

In this simple case, the sector may be considered as equivalent to a thin lens with coincident object and image principal points at O, focal points at F_o and F, respectively, and corresponding focal lengths F_o = O F_o and f = O F. An object at P will thus be imaged at Q. It can be shown that points P, C, and Q are collinear and that

$$\phi_p + \phi_q + \phi = \pi . \tag{43}$$

Once P has been selected and the sector geometry known, only PC has to be determined to find Q. The linear magnification is given by −CQ/CP. In the case where CQ = P, the linear magnification is −1, resulting in "stigmatic-operation." Also

$$p = q = r_o \cot(\phi/2) . \tag{44}$$

When a beam of electrons passes through a specimen, it is expected that some, or all, of the electrons will lose a portion of their incident energy through one or more of the many mechanisms of inelastic scattering that can take place in the specimen. Thus, a monoenergetic beam incident on a specimen will probably exit with a spectrum of energy. (The details of this topic, electron energy-loss in solids, will be discussed in later chapters.) It is often important to study the energy-loss spectrum and magnetic sectors as described here since they are being used widely for this purpose today.

Recalling that the radius of the electron trajectory depends on the electron energy, it is easy to see that a beam of electrons containing a spectrum of energies (velocities) at point P in Figure 26 will be dispersed along a line through point Q. Generally, it is preferred to place a small slit at Q and record with an electron-sensitive detector only those electrons passing through the slit. Since the radius, r, is a function of v/B, the ratio of the electron velocity (energy) to the applied magnetic field (if v/B is kept constant) will remain the same, and by changing the magnetic field, electrons having different energy will selectively be passed through the slit at Q.

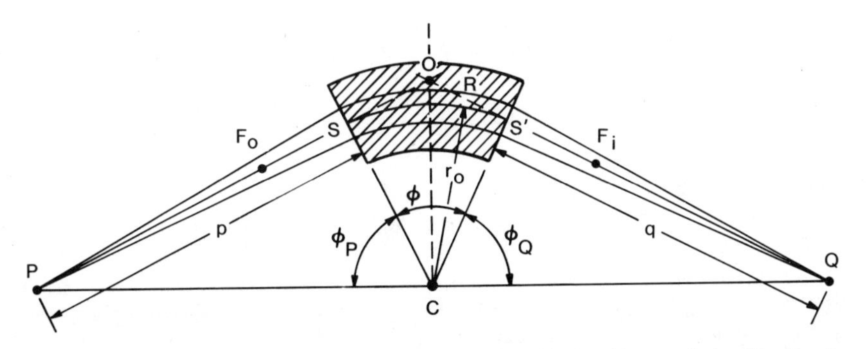

Figure 26 Diagram of magnetic sector of angle ϕ, showing object and image relationships P and Q, respectively.

B. Electrostatic Sectors

In a manner analogous to the magnetic field, an electron entering a uniform electric field, $\underline{E}_o$, will be deflected into a circle of radius, $\underline{r}_o$, where

$$\underline{r}_o = \frac{mV^2}{e\underline{E}_o} = \frac{2V_o}{\underline{E}_o}\,. \tag{45}$$

All the quantities have been described before. The major difference between electrostatic and magnetic fields is that the force on the electron by an electrostatic field is in the direction of the field (in the negative sense in the case of an electron), that is, $\underline{F}_e = -e\underline{E}$, rather than perpendicular to the field as is the case for a magnetic field.

One can qualitatively describe electrostatic sectors in much the same way as magnetic sectors. The optics are a bit more complicated but can still be discussed in terms of cardinal elements. Because of this and the fact that electrostatic sectors are not often used in electron microscopes, they will not be discussed here. One major problem in their use is worth mentioning, though. For an electric field to have any measurable effect on deflecting an electron with an accelerating voltage, V_o, it is essential that the potential of the field be $\sim V_o$. For keV electrons, this then requires that the potential across the sector would have $\sim$100 keV, so that one encounters problems of high voltage supply and breakdowns, etc.

C. Wien Filter

Despite the problem associated with high voltage, there is one popular use for the electrostatic field in electron velocity analyses: By crossing magnetic and electric fields, it is possible to create a situation in which electrons with particular values of energy will be unaffected by the fields, while other electrons with all other energies will be affected.

For example, if a homogeneous magnetic field, $\underline{B}$, acts over a distance, ℓ, electrons entering the field will be deflected through angle α at right angles to electron velocity $\underline{v}$ and field $\underline{B}$, given by

$$\sin\alpha = \frac{\ell}{R} = \frac{\ell \underline{B} e}{m\underline{v}} \tag{46}$$

as shown in Figure 27(a). Similarly, an electron incident on a region of length, ℓ, containing an electric field, $\underline{E}$, will be deflected through an angle, β, given by

$$\sin\beta = \frac{\ell}{R} = \frac{\ell e\underline{E}}{m\underline{v}} \tag{47}$$

as seen in Figure 27(b).

If the magnetic and electric fields are applied at right angles to each other (crossed) as shown in Figure 27(c), the magnetic field will deflect the electron to the left, and the electric field will deflect it to the right. If the fields are properly adjusted, the respective deflections will cancel, and certain electrons will pass through undeflected. This requires $\sin\alpha = \sin\beta$, that is

$$\frac{\ell Be}{mv} = \frac{\ell eE}{mv} \tag{48}$$

which reduces to $v = E/B$. Thus, for a particular set of field values, $\underline{B}$ and $\underline{E}$, only electrons with velocities $v = E/B$ will pass through the crossed fields undeflected. By appropriately placing entrance and exit slits, a device so constructed acts as an electron velocity filter. Such a device is called a Wien filter. Various models have been built, some obtaining very high resolution, i.e., on the order of a few meV.

VIII. OPTICS OF THE ELECTRON MICROSCOPE

A. Introduction

The optical components of an electron microscope may be conveniently broken into three components according to function. The illumination composed of the electron gun and two condenser lenses, provides the electrons to "illuminate the specimen." The objective lens, which has been discussed in detail, is the heart of any electron microscope and forms the image of the specimen, usually with a magnification of 50 to 100 times. The magnification system consists of three or four projector lenses, often identified as diffraction, intermediate, and projector lenses, and provides the final magnification of the image. The total magnification in some transmission electron microscopes may exceed one million times, with each lens contributing about 20 times in a three-lens system.

In SEM and dedicated STEM microscopes, there is no need for the magnification system, and such microscopes will have only one or two condenser lenses and an objective lens. Magnification in scanning microscopes is inversely related to the extent of the area scanned; e.g., a large area scan gives low magnification, and a small area scan gives high magnification.

Before discussing the operation of the complete electron microscope, it is necessary to discuss the illumination system in some detail.

A key element of the illumination system is the electron gun. It is the electron gun that ultimately determines the electron-beam parameters, including beam spot size, d_p, and beam current, i_b. The beam current controls the image contrast and x-ray signal intensity. The spot size determines the image resolution and may influence the x-ray spatial resolution. Unfortunately, these parameters are coupled to one another. Since the beam current density,

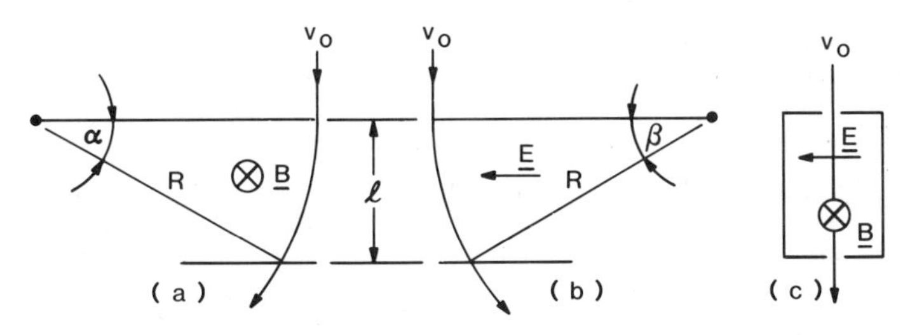

Figure 27 Electron paths through electrostatic and magnetic fields. (a) Deflection by a magnetic field, $\underline{B}$; (b) deflection by an electric field, $\underline{E}$; (c) crossed fields yielding no deflection for $v = E/B$.

$$J_b = \frac{i_b}{\pi\left[\frac{d_p}{2}\right]^2} \tag{49}$$

is relatively constant for a given type of electron gun (more precisely, electron source brightness β is fixed for a given electron source), large values of i_b correspond to large values of d_p, and small values of d_p correspond to small values of i_b. High rates of x-ray generation and high levels of image contrast require large values of i_b, which results in a large d_p. Improved x-ray spatial resolution and improved image resolution require minimizing d_p, which reduces i_b. The ideal situation would be one in which i_b is maximized and d_p is minimized.

Three types of electron guns are currently available: (1) thermionic emission W-filament, (2) heated LaB_6, and (3) cold field emission. These electron sources are listed in order of increasing J_c (and therefore larger i_b for a given d_p), increasing cost, and increasingly stringent vacuum requirements. The W-filament is by far the most common electron source in use today, although many of the newest generation TEM/STEM instruments do use LaB_6 sources. Cold field emission sources are found only on a few dedicated STEMs. It is appropriate now to consider briefly these types of electron guns (adapted from Goldstein *et al.*, 1981).

Both the W-hairpin and LaB_6 guns emit electrons by the process of thermionic emission. In this process, at sufficiently high temperatures, a certain percentage of the electrons become sufficiently energetic to overcome the work function, E_w, of the cathode material and escape the source. The emission current density, J_c, obtained from the filament of the electron gun by the thermionic emission, is expressed by the Richardson law,

$$J_c = A_c T^2 \exp(-E_w/kT) \text{ A/cm}^2 \tag{50}$$

where A_c (A/cm^{2}K^2) is a constant that is a function of the material, k is Boltzmann's constant, and T(K) is the emission temperature.

The filament or cathode has a V-shaped tip that is ~5 to 100 μm in radius. The filament materials that are used, W or LaB_6, have high values of A and low values of work function E_w. Specific filament types are discussed in the following sections. The filament is heated directly or indirectly with a filament supply and is maintained at a high negative voltage (50 – 300 kV) during operation. At the operating filament temperature, the emitted electrons leave the V-shaped tip and are accelerated to ground (anode) by the potential between the cathode and anode. The configuration for a typical electron gun is shown in Figure 28.

Surrounding the filament is a grid cap or Wehnelt cylinder with a circular aperture centered at the filament apex. The grid cap is biased negatively between 0 and 2500 V with respect to the cathode. The effect of the electric field formed in such a gun configuration, the filament, Wehnelt cylinder, and anode, causes the emitted electrons from the filament to converge to a crossover of dimension d_o. Figure 28 also shows the equipotential field or voltage lines which are produced between the filament grid cap and the anode. The constant field lines are plotted with respect to the filament (cathode) voltage and vary between 0 at the filament to a negative potential (up to −2500 V) at the grid cap, and to the large positive potential (up to 200 kV) at the anode or anode plate.

The emitted electrons are accelerated through this voltage field and attempt to follow the maximum voltage gradient, which is perpendicular to the field lines. Where

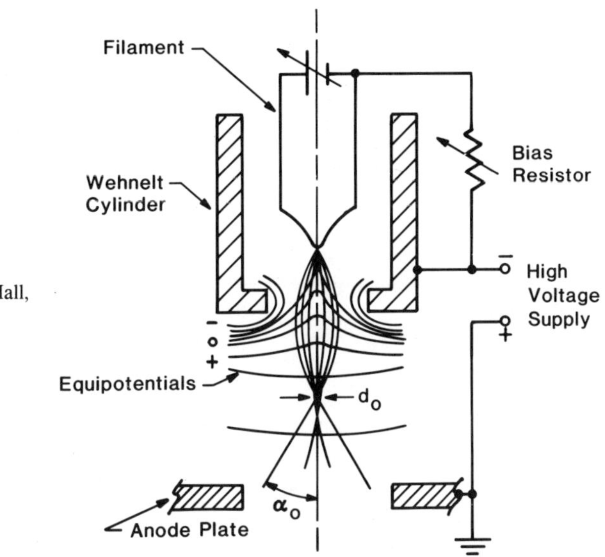

Figure 28 Configuration of self-biased electron gun (adapted from Hall, 1953).

the constant field lines represent a negative potential, however, the electrons are repelled. Figure 28 shows the paths of electrons through the field or constant voltage lines. Note the focusing action as the electrons approach and are repelled from the negatively biased grid cap. By use of the grid cap, the electrons are focused to a crossover of dimension d_o and divergence angle a_o below the Wehnelt cylinder. The negative bias on the grid cap and its placement with respect to the tip of the filament control the focusing action. The intensity distribution of the electrons at crossover is usually assumed to be Gaussian. The condenser and probe-forming lenses produce a demagnified image of this crossover to yield the final electron probe.

The current density in the electron beam at crossover represents the current that could be concentrated into a focused spot on the specimen if no aberrations were present in the electron lenses. This current density, J_b (A/cm²), is the maximum intensity of electrons in the electron beam at crossover and can be defined as

$$J_b = \frac{i_b}{\pi(d_o/2)^2} \tag{51}$$

where i_b represents the total beam or emission current measured from the filament.

It is desirable, in practice, to obtain the maximum current density in the final image. Since the maximum usable divergence angle of the focused electron beam is fixed by the aberrations of the final lens in the imaging system, the most important performance parameter of the electron gun is the current density per unit solid angle. This is called the electron beam brightness, β, and is defined as

$$\beta = \frac{\text{current}}{(\text{area})(\text{solid angle})} = \frac{4i}{\pi^2 d^2 \alpha^2} \ \text{A/cm}^2\text{sr} . \tag{52}$$

The steradian is defined as the solid angle subtended at the center of a sphere of unit radius by a unit area on the surface of the sphere; it is dimensionless. For the electron gun, $i = i_b$, $\alpha = \alpha_o$, and $d = d_o$. The electron beam brightness remains constant throughout the electron optical column even as i, d, and α change. As shown by Langmuir (1937), the brightness has a maximum value given by

$$\beta = \frac{J_c e E_o}{\pi kT} \text{ A/cm}^2\text{sr} \tag{53}$$

for high voltages, where J_c is the current density at the cathode surface, E_o is the accelerating voltage, e is the electronic charge, and k is Boltzmann's constant. The brightness can be calculated as $\beta = 11600\ J_c E_o/\pi T$, where the units are A/cm^2 for J_c and V for E_o. The current density can then be rewritten as

$$J_b = \pi \beta \alpha_o^2 \tag{54}$$

and the maximum current density is

$$J_b = J_c \frac{eE_o\alpha_o^2}{kT} . \tag{55}$$

The theoretical current density per unit solid angle (brightness) for a given gun configuration can be approached in practice provided an optimum bias voltage is applied between cathode and grid cap.

B. Tungsten Hairpin Cathode

The tungsten cathode is a wire filament ~0.01 cm in diameter, bent in the shape of a hairpin, with a V-shaped tip that is ~100 μm in radius. The cathode is heated directly as the filament current, i_f, from the filament supply is passed through it. For tungsten, at a typical operating temperature of 2700 K, J_c is equal to 1.75 A/cm^2 as calculated from the Richardson expression (Eq. (50)), where $A_c = 60$ A/cm^{2}K^2 and $E_w = 4.5$ eV. At normal operating temperatures, the emitted electrons leave the V-shaped tip from an emission area of ~100 × 150 μm. The filament wire diameter decreases with time because of tungsten evaporation. Therefore, the filament current necessary to reach the operating temperature and to obtain filament saturation decreases with the age of the filament. Filament life also decreases with increasing temperature. At an emission current of 1.75 A/cm^2, filament life should average 40 to 80 hours in a reasonably good vacuum. Although raising the operating temperature has the advantage of increasing J_c, this would be accomplished only at the loss of filament life.

For a tungsten filament operated at 2700 K and cathode current density J_c of 1.75 A/cm^2, the brightness at 100 kV, as calculated from Eq. (52), is ~25 × 10^4 A/(cm^{2}sr). If the operating temperature is increased from 2700 to 3000 K, the emission current and brightness can be raised from $J_c = 1.75$ A/cm^2 and $\beta = 25 \times 10^4$ A/(cm^{2}sr) at 100 kV to $J_c = 14.2$ A/cm^2 and $\beta = 15 \times 10^5$ A/(cm^{2}sr). However, although the brightness increases by over a factor of 7, the filament life decreases to an unacceptable low of 1 hour, owing to W evaporation. Although it is of increasing interest to obtain cathodes of higher brightness, the W filament has served the TEM community well over the last 30 years. The W filament is reliable, its properties are well understood, and it is relatively inexpensive. Therefore, for the majority of problems where high-brightness guns are not a necessity, the W filament will continue to play an important role.

C. The Lanthanum Hexaboride (LaB_6) Cathode

As new techniques in scanning microscopy are developed, the need for sources of a higher brightness than the tungsten hairpin filament described in the previous section becomes more evident. From the Richardson expression, Eq. (50), it can be seen that the cathode current density, and hence the brightness, can be increased by lowering work function E_w or by increasing the value of constant A_c. A considerable amount of work has therefore gone into searching for cathode materials that would have a lower value of E_w, a higher value of A_c, or both. The lower work function is usually of most significance because, at the operating temperature of 2700 K, each 0.1 eV reduction in E_w will increase J_c by $\sim$1.5 times.

The most important potential cathode material with a low value of E_w so far developed has been lanthanum hexaboride, which was first investigated by Lafferty (1951). Lanthanum hexaboride, LaB_6, is a compound in which the lanthanum atoms are contained within the lattice formed by the boron atoms. When the material is heated, the lanthanum can diffuse freely through the open boron lattice to replenish material evaporated from the surface. It is this action of the lanthanum that gives LaB_6 its low work function. Other rare earth hexaborides (praseodymium, neodymium, cerium, etc.) have a similar property and low work functions, but they have been less extensively investigated.

The measured work function (Swanson and Dickinson, 1976) for polycrystalline LaB_6 is $\sim$2.4 eV, with A_c typically 40 A/cm^2K^2. This means that a current density equal to that produced by the conventional tungsten filament is available at an operating temperature of only $\sim$1500 K and that at 2000 K nearly 100 A/cm^2 would be expected. This ability to produce useful cathode current densities at relatively low temperatures is important for two reasons. First, the rate of evaporation will be low, so that a long operating life can be expected compared to the relatively short life of a tungsten hairpin. Second, from the Langmuir relation (Eq. (53)), it is apparent that for two sources, both having the same cathode current density and accelerating voltage but operated at 1500 and 3000 K, respectively, the source at 1500 K will have twice the brightness of the otherwise equivalent source at 3000 K. Substantial advantages therefore appear to be offered by LaB_6 over tungsten as an emitter material.

There are, however, compensating disadvantages that have slowed the acceptance of LaB_6. The material is very chemically reactive when hot and readily forms compounds with all elements except carbon and rhenium. When this occurs, the cathode is "poisoned" and ceases to be an efficient emitter. This reactivity also means that the cathode can be operated only in a good vacuum since an oxide, characteristically purple in color, forms in any pressure above 100 μPa. This oxide also impairs the performance. Finally, LaB_6 is available commercially only as a fine-grain (5-μm-diameter) powder; therefore, considerable processing is needed to form a useful cathode assembly.

The first practical LaB_6 gun was developed by Broers (1969) and is shown schematically in Figure 29. The cathode was made from powdered LaB_6, which was hot pressed and sintered to form a rod $\sim$1 mm square in cross section and 1.6 cm long. One end of the rod was milled to a fine point, with a radius of only a few micrometers. The other end was held in an oil-cooled heat sink. In this way the sharp emitting end of the cathode could be heated while maintaining the other end of the rod at a low temperature, where its reactivity would not be a problem. The heat was applied from a coil of tungsten wire wrapped around, but not touching, the rod. A high current was passed through this coil, which was also held at a negative potential

Figure 29 Gun configuration for LaB_6 cathode (from Broers, 1969).

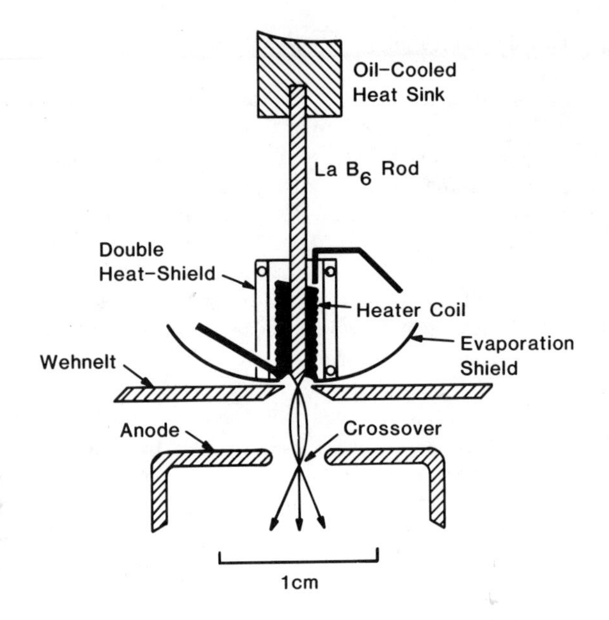

relative to the cathode rod. Heating then occurred through a combination of radiation and electron bombardment. To prevent lanthanum from being evaporated onto other surfaces around the gun, evaporation shields were placed around the heater. The temperature at the tip of the rod was in the range 1700 to 2100 K. In operation this gun was able to produce cathode current densities of $\sim$65 A/cm^2 at a temperature of 1850 K, with a lifetime of several hundred hours. At 100 kV, this corresponds to a brightness of $\sim 10^7$ A/(cm^{2}sr), which is at least five times greater than that of a tungsten filament operating at its maximum temperature.

The diameter of the end of the rod is such that the emitting area often consists of only one grain of LaB_6. During operation this grain is gradually eroded by evaporation, and an adjacent grain becomes the emitter. Because the work function of LaB_6 is a function of the crystallographic orientation of the emitting face, these random changes in the tip cause periodic fluctuations in the output of the gun.

Another effect can also change the emission from the tip. Because of the small radius of the emitting cathode, and in the presence of a high bias voltage, a very high electric field gradient exists locally. This field, which may be 10^6 V/cm or more, will modify the potential barrier seen by an electron trying to leave the cathode by reducing the effective value of work function E_w. The amount by which E_w is reduced will obviously depend on the strength of the field outside the cathode, but this "Schottky" effect can typically reduce E_w by 0.1 eV or more (Broers, 1975), leading to increased emission from the tip. Increased emission is desirable, but if the tip geometry changes—as it will if the crystallite at the point is evaporated away—the field at the cathode will change, thus altering the effective work function and hence the emission.

Several manufacturers have recently made available the LaB_6 gun for use on their newest generation TEM/STEMs. The necessity for vacuums of better than 100 μPa in the gun region requires improved gun pumping. This, plus the greater difficulty in fabricating LaB_6 filaments, leads to an increased cost for the LaB_6 gun. Nevertheless, the increased brightness of the LaB_6 emitter yields a significantly smaller probe size at the same beam current, or a larger beam current at the same probe size, than a tungsten emitter. This improvement in performance will clearly justify the increased cost in many applications.

D. Field-Emission Gun (FEG)

The electron sources described so far have relied on the use of high temperature to enable a fraction of the free electrons in the cathode material to overcome the E_w barrier and leave. There is, however, another way to generate electrons that is free from some of the disadvantages of thermionic emission; this is the process of field emission. In field emission, the cathode is in the form of a rod with a very sharp point at one end (typically ~100-nm diameter or less). When the cathode is held at a negative potential relative to the anode, the electric field at the tip is so strong ($>10^7$ V/cm) that the potential barrier discussed above becomes very narrow as well as reduced in height. As a result, electrons can "tunnel" directly through the barrier and leave the cathode without needing any thermal energy to lift them over the barrier (Gomer, 1961). A cathode current density between 1000 and 10^6 A/cm^2 is obtained in this way, giving an effective brightness many hundreds of times higher than that of a thermionic source at the same operating voltage, even though the field emitter is at room temperature.

The usual cathode material is tungsten because the field at the tip is so high that the cathode experiences a very large mechanical stress, which only very strong materials can withstand without failing. However, other substances, such as carbon fibers, have been used with some success. Because E_w is a function of the crystal orientation of the surface through which the electrons leave, the cathode must be a single crystal of a chosen orientation (usually <111> axial direction) to obtain the lowest E_w and hence the highest emission. The expected E_w of the cathode is only obtained on a clean material, that is, when no foreign atoms of any kind are on the surface. Even a single atom on the surface will increase the work function and lower the emission. In a vacuum of 10 μPa, a monolayer of atoms will form in less than 10 seconds; therefore, it is clear that field-emission demands a very good vacuum if stable emission is to be obtained. Ideally, the field-emission tip is used in a vacuum of 10 nPa or better. Even in that condition, however, a few gas molecules will land on the tip from time to time and cause fluctuations in the emission current. Eventually the whole tip will be covered, and the output will become very unstable. The cathode must then be cleaned by rapidly heating it to a high temperature (2300 K) for a few seconds. Alternatively, the tip can be kept warm (1100 to 1300 K), immediately reevaporating most of the impinging molecules. In this case, acceptably stable emission is maintained even in a vacuum of 100 nPa or so.

The effective source or crossover size, d_o, of a field emitter is only ~10 nm as compared with 10 μm for LaB_6 and 50 μm for a tungsten hairpin. No further demagnifying lenses are therefore needed to produce an electron probe suitable for high-resolution applications. However, unless the gun, and the lenses following, are designed to minimize electron optical aberrations, most of the benefits of the high brightness of the source, with its small source size, will be lost.

Table V compares several characteristics of the various types of electron guns.

TABLE V

Electron Gun Characteristics at 100 keV.

Characteristic	W-Hairpin	LaB_6	Cold Field Emission
Filament Temperature (K)	2700	2000	300
J_c (A/cm^2)	1.75	100	10^4
d_p (μm)	30	5	0.5
i_b (μA)	100	200	20
α_c (rad)	10^{-2}	10^{-2}	10^{-4}
β ($A/cm^2 \times sr$)	2.5×10^5	1×10^7	2×10^8
Filament Life (hr)	30	500	1000
Vacuum (torr)	10^{-5}	10^{-7}	10^{-10}

E. Condenser Lens System

The purpose of the condenser lens system is to deliver electrons from the gun crossover to the specimen under the multitude of conditions needed by the microscopist. For conventional transmission microscopy, focusing a high-intensity, reduced image of the gun crossover is required at the specimen plane at one extreme and at infinity at the other extreme. This provides nearly parallel illumination of considerably reduced intensity at the specimen. For scanning electron microscopy, the singular requirement is for a small-diameter, high-intensity probe at the specimen.

In principle, a single condenser lens situated about halfway between the gun and the specimen would be adequate for TEM, but not for STEM unless a FEG (field-emission gun) was used. With a single condenser lens the smallest spot at the specimen would be approximately the same size as the gun crossover since the magnification of the lens would be approximately unity. Modern microscopes universally incorporate two condenser lenses, with one (the second condenser, C2) placed approximately halfway between the specimen and the gun, and the other (C1) halfway between the gun and C2. The demagnifying factor of the combined lenses is usually sufficient to produce submicrometer probes at the specimen. Since C1 is located closer to the gun than a single condenser lens would be, a larger number of electrons are collected; thus, a higher current may be focused into the probe.

With STEM microscopes, the specimen is immersed in the magnetic field of the objective lens to the extent that this prefield acts as another condenser lens, C3. For STEM operation, C1 and C2 are adjusted to provide a small beam of almost parallel electrons to the prefield of the objective lens which, in turn, focuses the beam to probes as small as 1.5 nm in diameter on the specimen, or smaller for a FEG.

Ray diagrams illustrating these different situations are drawn in Figure 30. As shown, the angular aperture of the illumination is defined by angle α_i. When the gun crossover is not focused at the specimen plane, α_i is defined as shown in Figure 30(c). The angular aperture of the illumination is a measure of the parallelism of the incident probe. The effective angle of the source is defined by α in Figure 30(c). This is the angle of the undeviated ray from the crossover with respect to the optical axis; it is useful for calculating the crossover image at the crossover of C1. Subsequent angles can be defined at each lens crossover as the electrons proceed down the column; of particular significance is angle α_i, previously defined.

The angular aperture of illumination, α_i, is determined by the combined strengths of C1 and C2. That is, to have small α_i, it is essential to operate C1 with a short focal length (strong). C2 is then adjusted to have its crossover above, at, or below the specimen plane depending on the minimum illumination required to view the specimen. For a fixed C1, the relationship between C2 excitation and angular aperture follows the general trend shown in Figure 31.

The intensity of illumination is proportional to the square of the angular aperture of illumination, α_i^2. Consequently, the illumination at the specimen drops off very rapidly as the second condenser excitation is changed from the crossover condition

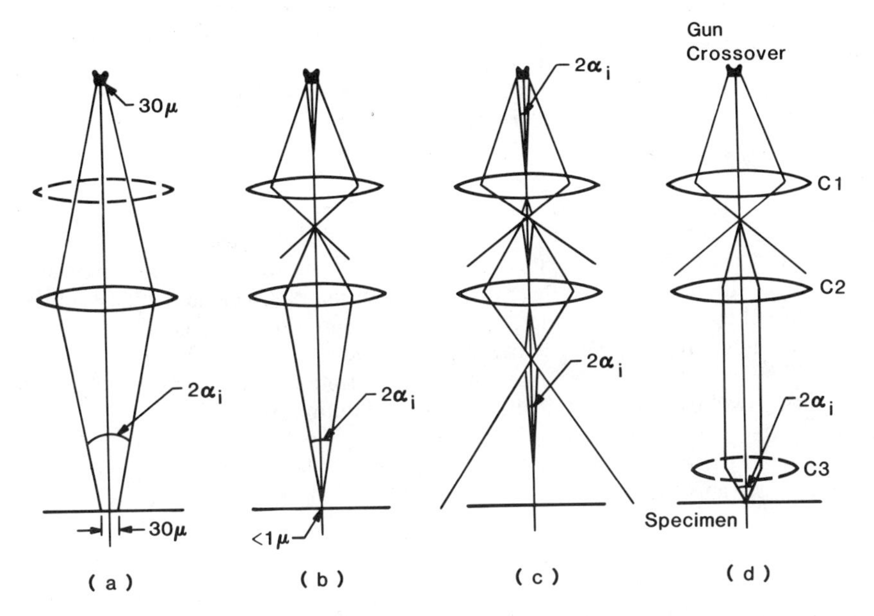

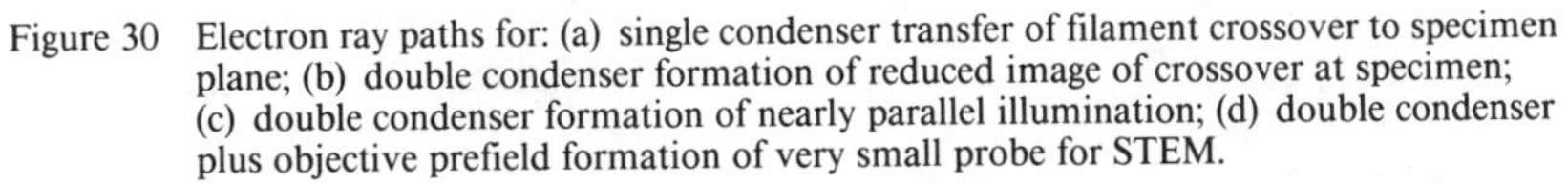
Figure 30 Electron ray paths for: (a) single condenser transfer of filament crossover to specimen plane; (b) double condenser formation of reduced image of crossover at specimen; (c) double condenser formation of nearly parallel illumination; (d) double condenser plus objective prefield formation of very small probe for STEM.

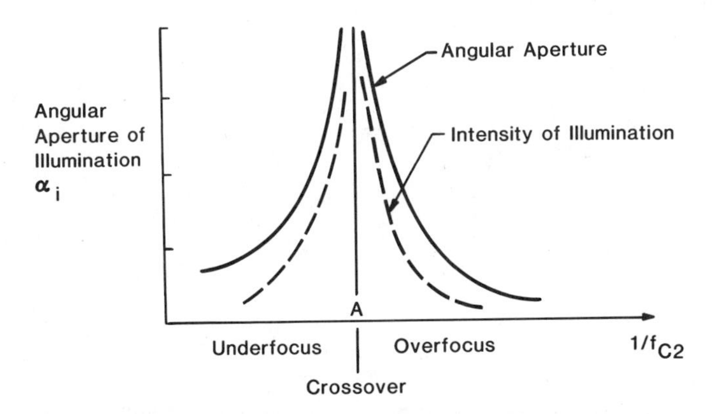

Figure 31 Relationship between the strength of C2 (1/f) and α_i as indicated by the axial position of C2 crossover with respect to the specimen plane.

(point A in Figure 31). From the figure, it can be seen that the smallest α_i are obtainable with C2 overfocused, e.g., with the crossover above the specimen.

Recently, Cliff and Kenway (1982) have shown that spherical aberration in the probe-forming (condenser) lens system in the STEM mode can have serious consequences with respect to both image resolution and x-ray spatial resolution. (The reader is referred to subsequent chapters concerning image formation and x-ray analysis for a more detailed discussion of these topics.) Traditionally, spherical aberration in the probe-forming lens has been assumed negligible, and the electron probe is considered to have a Gaussian spatial distribution. However, as illustrated in Figure 32, spherical aberration can cause significant non-Gaussian tails to form. This will increase the area of electron impingement and have a detrimental effect on resolution. This effect can be minimized with a smaller C1 aperture, but it will also result in corresponding loss in electron-beam current.

F. Coherence

The term "coherence" refers to the range of phase differences in the illuminating beam as it approaches the specimen. If the electrons come from a single point source, all the waves in the incident beam are in phase with one another, and the illumination is said to be coherent. On the other hand, if the source of electrons is so large that there is no phase relationship between the incident waves, the illumination is incoherent. The situation at the filament in an electron microscope is somewhere between these two extremes, and the incident illumination is defined as partially coherent.

Consider two waves coming from a source of width d, passing through separate slits S_1 and S_2 in an otherwise opaque screen. If $d = 0$, the waves are coherent, and

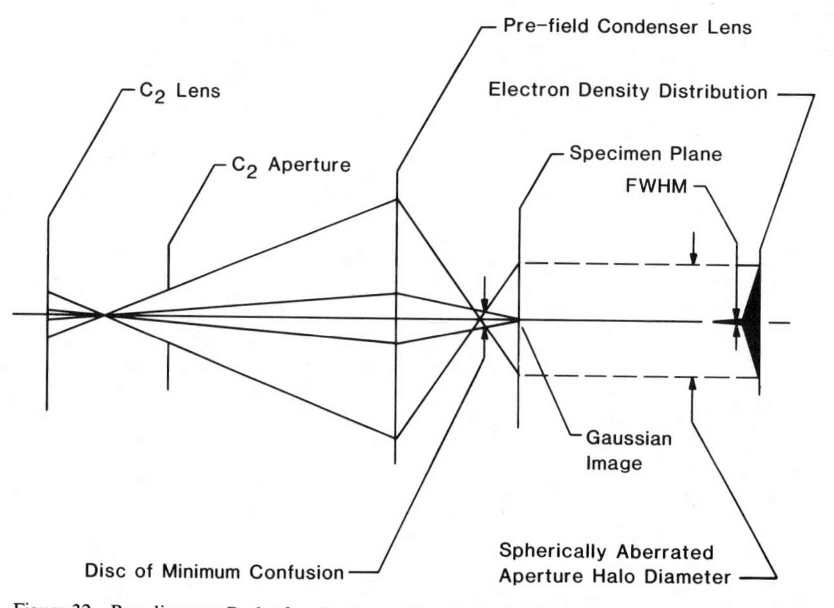

Figure 32 Ray diagram: Probe-forming lens with spherical aberration (from Cliff and Kenway, 1982).

the intensity, I_w, at some plane after the slits is given by

$$I_w = I_1 + I_2 + I_{12}^* . \tag{56}$$

I_1 and I_2 represent the intensity distribution in the individual waves, and I_{12}^* is the intensity resulting from interference between the two waves. As d gets larger, I_{12}^* becomes smaller until at some point d is so large that $I_{12}^* = 0$ and the two waves are said to be incoherent.

In electron microscopy, coherence is related to the angular aperture of the illumination, α_i, and to the dimensions of the region on the specimen over which the illumination appears coherent. This is an extremely important concept when imaging objects that give rise to interference effects. The length, a_ℓ, over which the illumination is said to be coherent is given by

$$a_\ell = \frac{\lambda}{2\alpha_i} . \tag{57}$$

Typical illumination conditions for crossover at 100 keV have $\alpha_i = 10^{-3}$ rad; thus $a = \lambda/2\alpha_i = 0.037 \times 10^{-10}/2 \times 10^{-3} = 18 \times 10^{-10}$ m (= 1.8 nm), and the illumination is not very coherent. With C2 crossover considerably overfocused, it is reasonable to have $\alpha_i = 2.5 \times 10^{-5}$ rad, giving coherence length a = 74 nm at the specimen. This illumination condition would allow interference effects such as Fresnel fringes, magnetic domains, or lattice images to be obtained.

G. Magnification Lens System

The properties of the individual lenses that make up the magnification lens system have been previously described. All the lenses are designed as projector lenses, and their most important role is to magnify the image of the specimen formed by the objective lens. Usually, the collective lens action in the magnification process has been programmed into the microscope, so that the operator has little control over the individual lenses. The total magnification of the final image is the product of the individual lens magnifications, e.g., in a four-lens system (including the objective lens)

$$M_{total} = M_{obj} \times M_{diff} \times M_{int} \times M_{proj} . \tag{58}$$

In normal operation the excitation, and therefore the magnification of the objective lens, is almost constant since it is used for providing the primary image of an object located in a prescribed plane. For the same reason, the excitation of the projector lens is usually fixed since the location of the final image is fixed at the viewing screen. Consequently, almost the whole range of magnification is obtained by varying only the diffraction and intermediate lenses. At very low magnification, say 50 times, a slightly different optical situation is required. Here, because of the intrinsic high magnification of the objective lens, it is either turned off or operated at a reduced excitation, with the diffraction lens now effectively acting as a long-focal-length objective lens. Ray diagrams showing low and high magnification ranges and standard diffraction pattern formation are shown in Figure 33.

In obtaining diffraction patterns, the object plane for the diffraction lens is adjusted to be the back focal plane of the objective lens. Recall from previous discussion that the diffraction pattern of the specimen is at the back focal plane of the objective lens. This plane intersects the axis at the objective focal point at the location of the objective aperture. The selected area aperture is positioned at the image plane

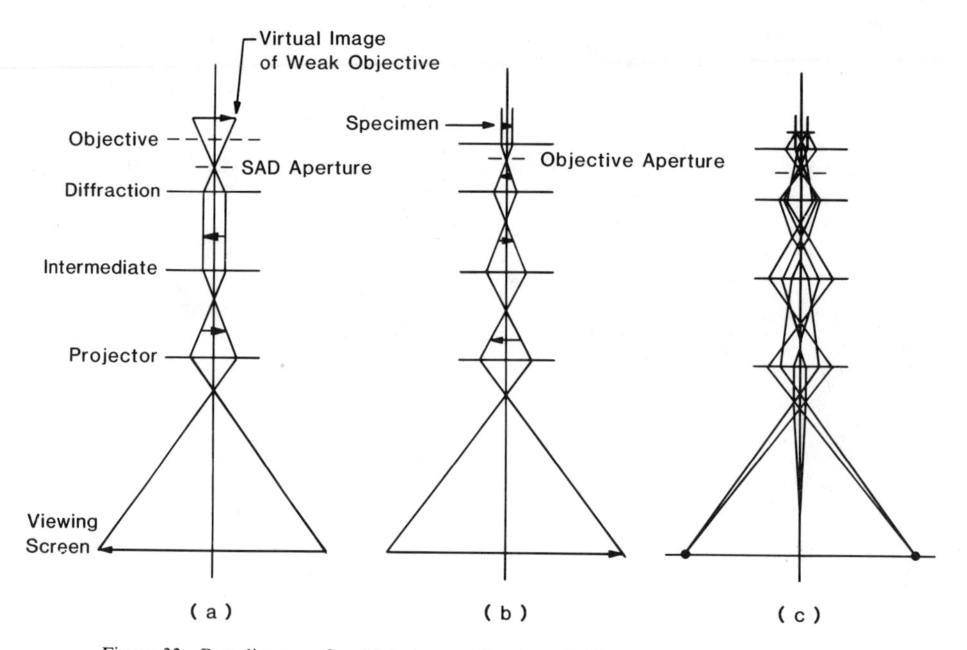

Figure 33 Ray diagrams for: (a) low magnification; (b) high magnification; (c) diffraction.

of the objective lens, which is the object plane of the diffraction lens in Figure 33(b), normal magnification operation. In terms of paraxial imaging, the final viewing screen is conjugate to the objective image plane under normal magnification operation and conjugate to the back focal plane (sometimes called the image focal plane) of the objective lens in diffraction imaging. By suitable alteration of the diffraction and intermediate lens excitation, it is possible to have the final viewing screen conjugate to any plane in the optical column.

In both dedicated STEMs and in the STEM operating mode of modern TEM/STEMs, the postspecimen lens system is not used for image magnification. Rather, the primary function of this lens system is to match selected portions of the characteristic distribution of scattered electrons leaving the specimen to defectors having a fixed angular acceptance or physical size. This system is important for two applications, microdiffraction (see later chapter by Cowley) and electron energy-loss spectroscopy (EELS) (see later chapters by Joy). The important consideration with respect to electron diffraction is that these lenses allow one to operate over a wide range of camera lengths.

In the EELS application, the function of this lens system is to allow a desired range of scattering angles to be matched to an electron spectrometer whose physical acceptance angle is limited by aberrations. These lenses angularly compress the diffracted electron beams for acceptance into the spectrometer. The operational theory for these lenses is not different from that for any other magnetic electron lens, like those discussed previously. These lenses also suffer from the same aberrations. It is worth noting, however, that chromatic aberrations are typically more severe because energy-loss interactions between the incident electron beam and the specimen may impart a significant energy distribution to the electrons. Several configurations of electron lenses and detectors have been used. For additional detail on this subject, the reader is referred, for example, to the work of Egerton (1981) or Buggy *et al.* (1981).

IX. COMPARISON OF CTEM AND STEM OPTICS

The optical systems of the CTEM have been described in detail here. Using Figures 29 and 32, one should be able to follow the electron path down the column for the standard modes of microscope operation. For the sake of completeness, an electron ray diagram for a complete microscope in the high-magnification mode is drawn in Figure 34.

With the CTEM, the best image quality will be obtained by operating C1 as strongly as possible while still providing the necessary illumination at the object plane with C2 defocused.

All previous discussion pertains to CTEM with an objective lens excitation $k^2 \neq 3$ and usually with $k^2 < 3$. For the unique excitation $k^2 = 3$, the operation of the condenser lenses to control illumination conditions is very restricted. Again, C1 is operated as strongly as possible to form the smallest effective source, but C2 is now fixed to the value of excitation that focuses its crossover to the front focal plane of the objective lens. This is the plane before the objective lens which intersects the optical axis at F_o and is conjugate to the objective back focal plane (BFP). Since it is

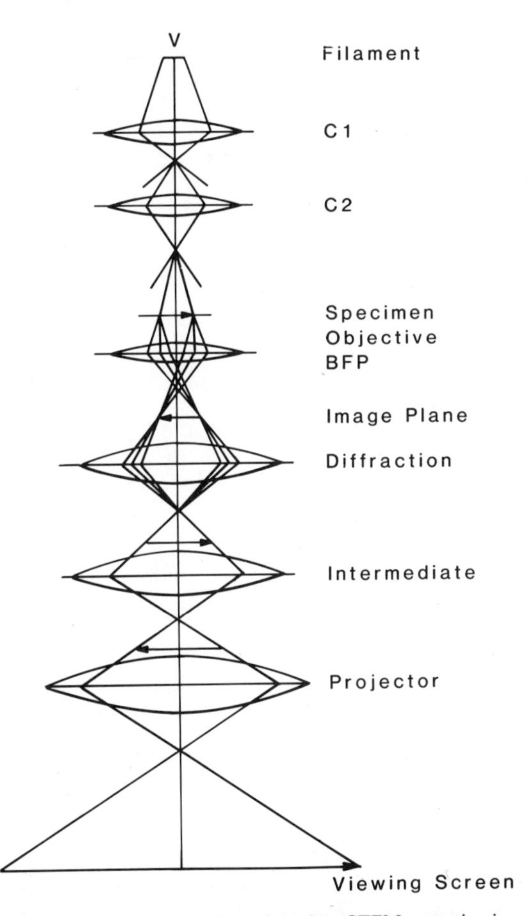

Figure 34 Ray diagram for a complete CTEM operating in the high-magnification mode.

conjugate to the BFP, the front focal plane may be viewed on the final screen by observing the BFP in diffraction. With properly adjusted illuminating conditions, the probe of diameter d at the specimen will be defined by the demagnified image of the C2 aperture of diameter D_{C2} (which is conjugate to the specimen). The demagnification factor is obtained from $M_d = \ell_o/f_o$, where ℓ_o is the distance from the C2 aperture to F_o, and f_o is the focal length of the objective lens; thus, $d = \ell/M\ D_{C2}$. The electron optical path from the gun to the objective lens is shown in Figure 35 for this case when $k^2 = 3$. Riecke (1962a) has described an alignment procedure for this optical configuration and has used it to obtain small-area, 10-nm-diameter diffraction patterns having an angular resolution of $\sim 10^{-5}$ rad.

The illumination requirements of the STEM microscope require as small a probe as possible, usually 0.5 – 2 nm, with sufficient beam current to yield a usable image.

As discussed previously, this small probe is obtained by immersing the specimen in the field of a highly excited objective lens to the extent that the prefield of this lens will form the small probe from the nearly parallel incident illumination. If the specimen is placed in the center of a properly energized symmetrical lens, $k^2 = 3$, the lens operates in the telefocal condition described before. That is, all bundles of parallel rays passing through F_o will be focused on the specimen, regardless of their inclination to the optical axis. Thus, a set of scan coils adjusted such that the pivot point of the coils is at F_o will provide the necessary incident probe conditions. To obtain the parallel illumination, it is convenient to turn off C2 and use a small C2 aperture to define the angular aperture of the probe.

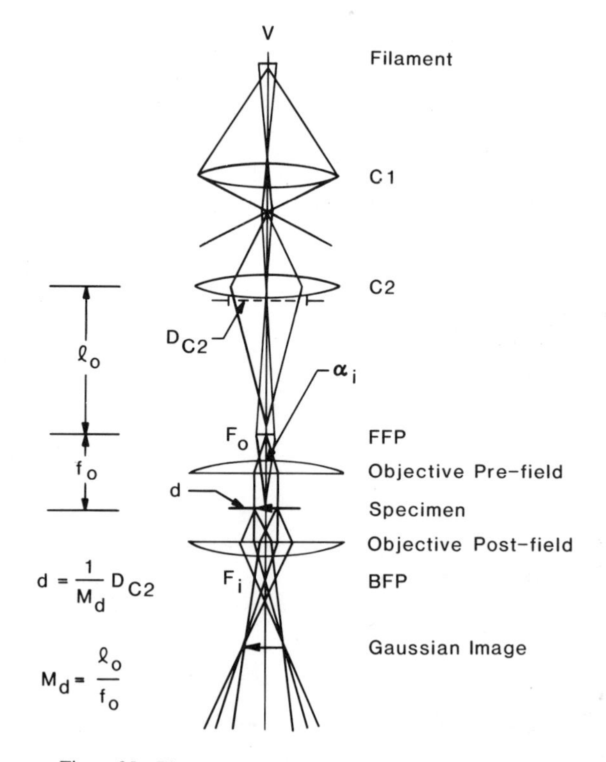

Figure 35 Electron ray paths for the case $k^2 = 3$ and C2 focused to F_o.

Since F_o is conjugate to F_i, the incident beam after passing through the specimen will pivot about F_i, which defines the BFP of the objective lens. A detector placed at F_i (as is the case in a dedicated STEM) or anywhere in the optical path after the specimen where F_i is imaged (the case for STEM/TEM) will gather the required signal to form the image. In STEM/TEM microscopes, this is achieved by placing a detector near the final viewing screen and adjusting the magnification lens system to focus the diffraction pattern on the viewing screen. A schematic electron ray path for the optical condition is drawn in Figure 36(a). The optical requirements are very similar, with a very strongly excited objective lens, $3 < k^2 < 8$; thus, only the objective lens is drawn in Figure 36(b).

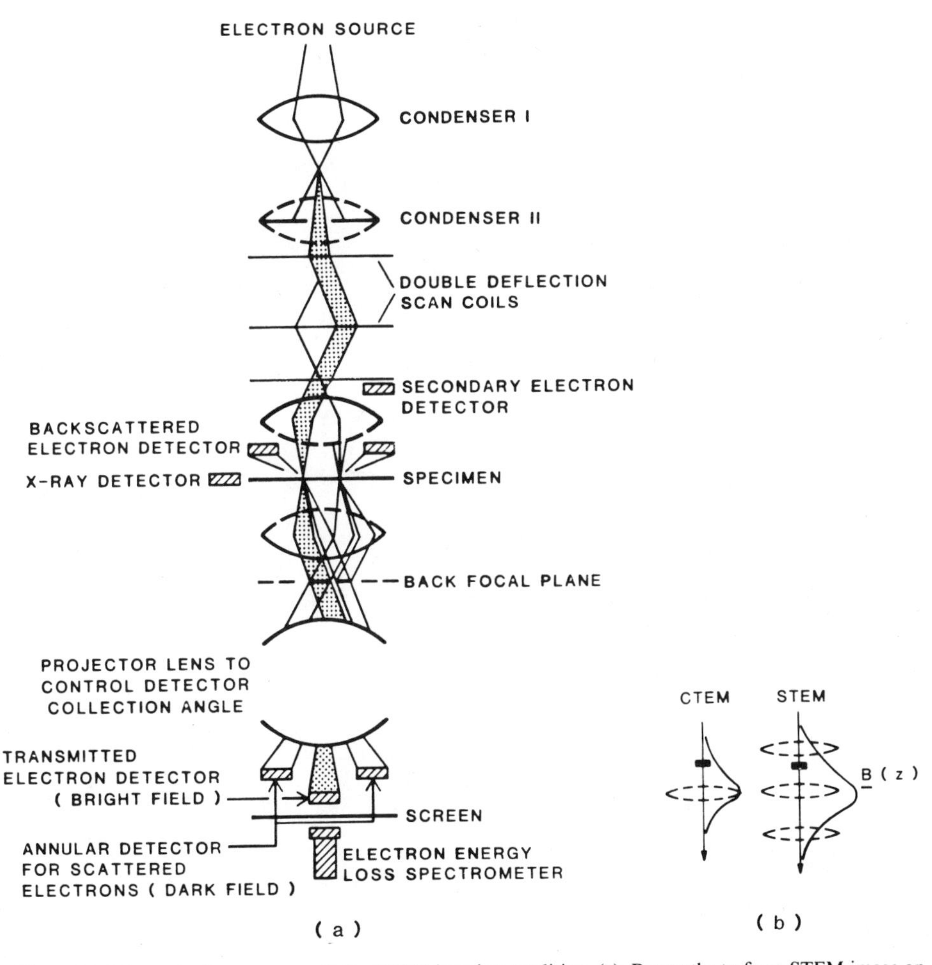

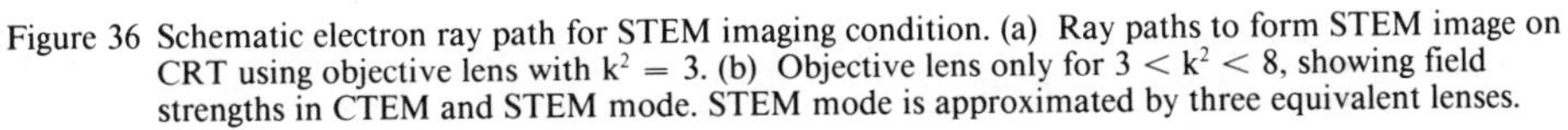
Figure 36 Schematic electron ray path for STEM imaging condition. (a) Ray paths to form STEM image on CRT using objective lens with $k^2 = 3$. (b) Objective lens only for $3 < k^2 < 8$, showing field strengths in CTEM and STEM mode. STEM mode is approximated by three equivalent lenses.

The angular aperture of the incident probe, α_i, assuming nearly parallel illumination from the condenser lens system with a C2 aperture of diameter D_{C2}, is given as

$$\alpha_i + D_{C2}/2f_o \ . \tag{59}$$

For most STEM microscopes, this angle is on the order of 5×10^{-3} rad compared to normal TEM imaging with $\alpha_i < 10^{-3}$ rad.

X. CONCLUSION

The basic principles of TEM and STEM microscope operation have been outlined. With the simple concepts of geometrical optics it is, in theory, possible to approximate the path of any electron in the microscope column. Drawing a few ray diagrams for the standard operational conditions will undoubtedly help the operator visualize the predicament of the electron anywhere in the column.

TABLE OF CHAPTER VARIABLES

a	Acceleration
a_r	Radial component of accleration in cylindrical coordinates
a_θ	Angular component of acceleration in cylindrical coordinates
a_ℓ	Coherent illumination length
A_c	Material constant for thermionic emission
B	Magnetic flux density ($\underline{B}$ = vector)
$\underline{B}_r$	Radial component of $\underline{B}$
$\underline{B}_z$	Axial component of $\underline{B}$
c	Speed of light, 3×10^8 m/s
C_c	Coefficient of chromatic aberration
C_s	Coefficient of spherical aberration
d_o	Diameter of electron beam at initial crossover
d_p	Electron probe diameter
D_{C2}	Diameter of condenser aperture
e	Charge on an electron 1.6×10^{-19} Coulombs
E	Electrostatic field ($\underline{E}$ = vector)
E_w	Work function of filament material
f_i	Image focal length
f_o	Object focal length
f_p	Projector focal length
F	Force ($\underline{F}$ = vector)
F_i	Image focal point
F_o	Object focal point
$\underline{F}_r$	Radial component of force
$\underline{F}_0$	Angular component of force
i_b	Electron beam current
I	Current passed through magnetic windings
I_w	Intensity of an electron wave

J_b	Electron beam current density
J_c	Emission current density
k	Lens excitation parameter
k	Boltzmann's constant, 8.62×10^{-5} eV/K
ℓ	Path length of electron in field $\underline{B}$
ℓ_i	Image-to-image focal point distance
ℓ_o	Object-to-object focal point distance
m	Mass of an electron
m_o	Rest mass
M	Magnification
n_i	Indices of refraction
N	Turns in a magnetic coil
r_o	Radius of path followed by electron with energy V in electrostatic field E_o
R	Radius of path followed by electron with velocity $\underline{v}$ in field $\underline{B}$
R_ℓ	Lens bore radius
s	Incremental distance
S_i	Image to lens distance
S_o	Object to lens distance
t	Time
T	Absolute temperature
v_ℓ	Velocity of light in a medium
v, v_i, v_o	Electron velocities ($\underline{v}$ = vector)
$\underline{v}_\theta$	Angular component of velocity
$\underline{v}_z$	Axial component of velocity
V	Accelerating voltage
V_r	Relativistically corrected voltage
W	Work done on electron moving through field $\underline{E}$
y	Reduced variable that relates radial distance of electron from z-axis to lens excitation parameters
Y_i	Image height
Y_o	Object height
Z_{im}	Image midfocal point
Z_{obj}	Object midfocal point
Z_{pi}	Image midprincipal point
Z_{po}	Object midprincipal point
α	Electron beam angle of divergence
β	Brightness of electron source
β_c	Lens proportionality constant
γ	Illumination angle of inclination
Δr	Radius of image disc due to lens aberrations
θ	Angle between $\underline{F}$ and $\underline{B}$
θ_i	Angles of refraction
μ_o	Magnetic permeability of free space, $4\pi \times 10^{-7}$ Wb/A/m
ϕ	Angle of circular sector
$\phi_{i, A, B, 1, 2}$	Scalar electrostatic potentials
ω	Angular constant for lens excitation ($\omega^2 = k^2 + 1$)

ACKNOWLEDGMENTS

This work was performed at Sandia National Laboratories and supported by the US Department of Energy under contract number DE-AC04-76DP00789.

The thorough manuscript review by M. J. Cieslak and F. G. Yost is sincerely appreciated.

RECOMMENDED READING AND REFERENCE MATERIAL

The following is a list of books on electron optics and related topics listed in the author's order of preference, with some comments on each.

Goldstein, J. I., Newbury, D. E., Echlin, P, Joy, D. C., Fiori, C. E., and Lifshin, E. (1981), "Scanning Electron Microscopy and X-ray Microanalysis," Plenum Press, New York.

Electron optics in the SEM and electron microprobe. X-ray analysis of bulk specimens.

Hawkes, P. W. (1972), "Electron Optics and Electron Microscopy," Taylor and Francis Ltd., London.

Excellent book describing elements of electron optics and application to microscopes. Scanning microscopy and image interpretation included.

Grivet, P. (1972), "Electron Optics," 2nd ed., Pergamon Press, Oxford.

An excellent monograph dealing with the principles of electrostatic and magnetic lenses and the applications of electron optics to various electron instruments. Available as a paperback in two volumes.

ElKareh, A. B., and ElKareh, J. C. J. (1970), "Electron Beams, Lenses, and Optics," 2 vol., Acad. Press, New York.

Extensive treatment with many tables for calculation of lens parameters, including excellent treatment of aberrations. Recommended for designers.

Kempler, O., and Barnett, M. E. (1971), "Electron Optics," Cambridge Univ. Press, London.

In the same vein as Grivet, recommended for further reading by those with good background in physics and mathematics.

Hall, C. E. (1966), "Introduction to Electron Microscopy," McGraw-Hill, New York.

Fundamentals of light and electron optics and limited applications. Good beginning book.

Septier, A. (ed.) (1967), "Focusing of Charged Particles," 2 vol., Acad. Press, New York.

Volumes cover the whole field of particle optics, including electrons and ions. High-intensity beams, particle accelerators, and prism optics are well covered.

Oatley, C. W. (1972), "The Scanning Electron Microscope," Cambridge Univ. Press, London.

Good introduction to electron optics of SEM. Especially good discussion of probe formation.

Bauer, R., and Cosslett, V. E. (ed.) (1966-present), "Advances in Optical and Electron Microscopy," Acad. Press, London.

Series of books with research and review articles by esteemed authors covering a variety of topics relating to electron optics and microscopy.

Marton, L. (1950-present), "Advances in Electronics and Electron Physics," Acad. Press, London.

Series of books with review articles on a wide range of topics, many relating to electron optics.

Listed alphabetically are some rather complete articles relating to electron optics, with brief comments.

Agar, A. W., Alderson, R. H., and Chescoe, D. (1974), in "Practical Methods in Electron Microscopy" (Glauert, ed.), North-Holland, Amsterdam.

Good introduction to the electron microscope, with discussions pertaining to general principles of use and operation.

Glaser, W. (1956), Handb. Phys. *33*, 123.
Thorough discussion of theoretical electron optics. Discussion of bell-shaped field. In German.

Mulvey, T., and Wallington, M. J. (1973), Reports Prog. Phys. *36*, 347.
Extensive article concerning electron lenses and their design. Many curves and tables. Extensive references.

Siegel, B. (1975), Proc. of SEM, pt. III (O. Johari, ed.), IITRI, Chicago, p. 647.
Tutorial on electron optics, with special reference to the SEM.

Listed alphabetically are technical references either referred to in this chapter or generally recommended to those with advanced backgrounds.

Boersch, H. (1954), Z. Phys. *139*, 115.

Broers, A. N. (1969), J. Phys. E. *2*, 273.

——— (1975), Proc. of SEM, pt. III (O. Johari, ed.), IITRI, Chicago, p. 662.

Buggy, T. W., Craven, A. J., and Ferrier, R. P. (1982), "Analytical Electron Microscopy" (R. H. Geiss, ed.), San Francisco Press, San Francisco.

Cliff, G., and Kenway, P. B. (1982), "Analytical Electron Microscopy" (R. H. Geiss, ed.), San Francisco Press, San Francisco, p. 107.

Egerton, R. F. (1981), "Analytical Electron Microscopy" (R. H. Geiss, ed.), San Francisco Press, San Francisco, p. 235.

Francken, J. C., and Heeres, A. (1974), Optik *37*, 483.

Gomer, R. W. (1961), "Field Emission and Field Ionization," Harvard Univ. Press, Cambridge, Massachusetts.

Hall, C. E. (1953), "Introduction to Electron Microscopy," McGraw-Hill, New York.

Kammenga, W., Verster, J. L., and Frenchen, J. C. (1968), Optik *28*, 442.

Koike, H. (1978), JEOL News *16E*, 26.

Lafferty, J. M. (1951), J. Appl. Phys. *22*, 299.

Langmuir, D. B. (1937), Proc. IRE *25*, 977.

Liebmann, G., *et al.* (1949), Proc. Phys. Soc. B. *62*, 753.

——— (1950a), Br. J. Appl. Phys. *1*, 92.

——— (1950b), Phil. Mag. *41*, 1143.

——— (1951), Proc. Phys. Soc. B. *64*, 972.

Liebmann, G., *et al.* (1952), Proc. Phys. Soc. B. *65*, 188.

——— (1953), Proc. Phys. Soc. B. *66*, 448.

——— (1955a), Proc. Phys. Soc. B. *68*, 682.

——— (1955b), Proc. Phys. Soc. B. *68*, 757.

Liebmann, G., and Grad, E. M. (1951), Proc. Phys. Soc. B. *64*, 956.

Mulvey, T. (1974), Proc. of SEM, pt. I (O. Johari, ed.), IITRI, Chicago, p. 43.

Mulvey, T., and Newman, C. D. (1973), "Scanning E. M.: Systems and Applications," Inst. of Physics, London, p. 16.

Munro, E. (1974), Proc. of SEM, pt. I (O. Johari, ed.), IITRI, Chicago, p. 35.

Munro, E., and Well, D. C. (1976), Proc. of SEM, pt. I (O. Johari, ed.), IITRI, Chicago, p. 27.

Riecke, W. D. (1962a), Optik *19*, 81.

——— (1962b), Optik *19*, 169.

——— (1972), Proc. 5th Euro. Cong. on E. M., Inst. of Physics, London, p. 98.

Riecke, W. D., and Ruska, E. (1966), Proc. 6th Intl. Conf. on E. M. (R. Uyeda, ed.), Maruzen Press, Tokyo, vol. 1, p. 19.

Ruska, E. (1965), Optik *22*, 319.

Suzuki, S., Akashi, K., and Tochigi, H. (1968), Proc. 26th EMSA Mtg., Claitor's Pub. Div., Baton Rouge, p. 320.

Swanson, C. W., and Dickinson, T. (1976), Appl. Phys. Lett. *28*, 578.

CHAPTER 3

PRINCIPLES OF IMAGE FORMATION

J. M. Cowley

Department of Physics
Arizona State University, Tempe, Arizona

I. INTRODUCTION

Until about twenty years ago, a rather simple set of assumptions was sufficient for the interpretation of electron microscope images in all but a few isolated instances, and even today a great deal of electron microscopy can be understood on the same simple basis.

For the microscopy of thin films of near-amorphous materials with resolution no better than ~10 Å, including most work on biological samples, it is sufficient to assume that the image intensity depends directly on the amount of scattering matter present, with the darker areas of positive prints corresponding to thicker regions or heavier atoms in the specimen. The complications of phase-contrast imaging are avoided when a small objective aperture angle is used to enhance the image contrast. The effects of radiation damage of the specimen by the incident electron beam usually limit the useful resolution to 10 to 20 Å or worse, so that there is no incentive to make use of the 3- to 4-Å resolution available with modern electron microscopes or to become involved with the complications of image interpretation that attend the use of these levels of resolution.

For materials scientists, a different set of assumptions, based on the simplest approximations to dynamical diffraction theory, have usually been sufficient to deal with the imaging of the crystalline samples normally used. For resolutions no better than ~10 Å, with crystal thicknesses 100 to 1000 Å for 100-keV electrons, and with a reasonably careful selection of crystal orientations, the two-beam approximation is sufficient for a good qualitative interpretation of images of metals, semiconductors, and other materials having small unit cell dimensions. The image intensity varies strongly and almost sinusoidally with crystal thickness and varies rapidly with changes of crystal orientation around the angles for which the incident-beam direction is close to the Bragg angle for prominent crystal lattice planes. The strain fields around dislocations or other extended crystal defects give characteristic patterns of contrast that are well documented and usually easily recognized.

Increasingly, however, these relatively simple assumptions are proving to be insufficient. Most electron microscopes now sold offer resolutions in the range of 2.5 to 5 Å, and many can show image detail on a 1-Å scale. Unless a microscopist either knowingly or carelessly allows the microscope performance to be grossly degraded, image detail will constantly appear that cannot be interpreted in terms of the familiar concepts. Also, increasingly, microscopists are learning that if they take advantage of the enormous increase in information made available by the greatly improved microscope performance, the electron microscope becomes a vastly more powerful tool for research.

A further technical development, which likewise requires a reorientation in the thinking of the practical electron microscopist, is the introduction on a commercial basis of new imaging modes. In particular, scanning transmission electron microscopy (STEM) is becoming commonplace. It is provided both in the dedicated STEM instruments and also as an essential part of the analytical electron microscopy (AEM) capabilities of the modern TEM/STEM instruments. While, as we will show, there are formal analogies that suggest TEM and STEM imaging processes are basically similar, there are practical instrumental factors that require a very different approach to image formation and analysis in the two cases.

With these developments in mind, we summarize in this review an approach to the principles of image formation that allows an understanding of the image contrast features seen under high-resolution conditions or with unconventional electron optics. For this purpose, we must go back to basic ideas of electron scattering and electron optics. Fortunately, for electron microscopy it is possible to make a number of simplifying assumptions, such as the small-angle scattering approximation, that allow us to avoid a great deal of complication. Even so, it is impossible to provide anything like an adequate treatment of the subject within a relatively brief review. We will attempt only a rough outline of the essential ideas, with references to sources where more complete treatments may be found.

Many electron microscopists will prefer nonmathematical descriptions of the ideas involved. The mathematical statements, however, are such compact, powerful, and explicit descriptions of the ideas that they should be included. Therefore, as compromise, each topic will be discussed first in a loose verbal fashion, and this will usually be followed by a distinct mathematical statement that can be ignored without loss of continuity or concept.

For a review of electron microscope imaging on the older, simpler assumptions, the reader is referred to such books as those by Agar *et al.* (1974) or Wischnitzer (1981) for biological electron microscopy, or by Hirsch *et al.* (1965, 1977), Bowen and Hall (1975), or Thomas and Goringe (1979) for materials science. The best overall review of the theory appropriate for high-resolution electron microscopy is the book of Spence (1981). Relevant reviews have been provided in conference reports such as those edited by Kihlborg (1979) and Eyring (1982). The ideas of physical optics on which the current approach is based are well summarized in books such as Nussbaum and Phillips (1976) or Born and Wolf (1975). The mathematical sections of this review follow the treatment of Cowley (1981), which should be consulted, along with Spence (1981), for a more complete treatment of most of the topics.

A. CTEM and STEM

The essential parts of the two main types of TEM are illustrated in Figure 1. In the more familiar, so-called "conventional" transmission electron microscopy (CTEM), a source of electrons (B) (an electron gun) is used to illuminate the specimen. The transmitted electrons, diverging through a relatively large angle (commonly $\sim 10^{-2}$ rad), are focused by the objective lens to form an image (A). A series of two or more lenses provide further stages of magnification before the image is recorded. One or more condenser lenses may be placed between the source and the specimen to control the convergence, coherence, and intensity of the beam falling on the specimen. It is the objective lens, however, that is most important in determining the resolution and contrast of the images. A more complete description of the physical nature of the electron optical components, the focusing properties of electromagnetic lenses, and the hardware of electron microscopes is provided in a subsequent chapter.

For scanning transmission electron microscopy (STEM), the essential components of the system can be considered as the reverse of those for CTEM. The objective lens is used to form the image of a small electron source at the specimen position. Of the electrons transmitted through the specimen, some are chosen by a small detector to give the image signal. To form a two-dimensional image, which is displayed on a cathode ray tube (CRT), deflector coils or plates are installed in the vicinity of the objective lens to scan the incident electron beam over the specimen. The same scanning currents, which are applied to these deflector coils, are applied with suitable amplification to the deflector coils of the display CRT. The voltage generated in the detector in the STEM instrument is amplified and used to modulate the intensity of the beam in the CRT. The magnification of the image produced in this way is then given simply by the ratio of the amplitudes of the beam deflections in the two cases.

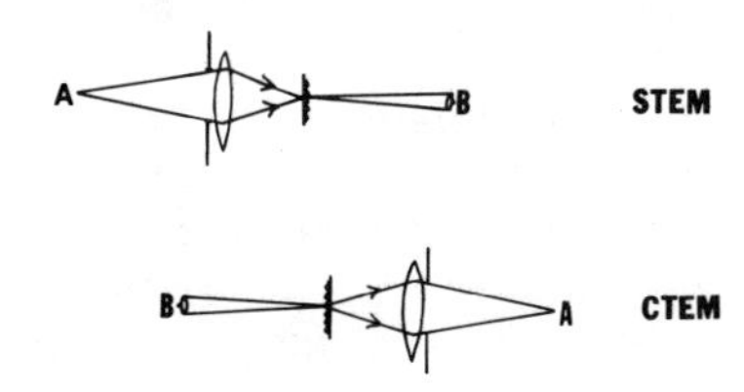

Figure 1 Reciprocal relationship between essential elements for STEM and CTEM instruments.

As an oversimplified picture, the image contrast can be considered to be given by the variation in the number of electrons transmitted through the specimen and emerging within a given small range of directions. The image resolution may then be considered as being determined by the diameter of the incident electron beam at the specimen position.

It was this simple picture of STEM which led many early workers to the impression that this imaging process could be described in terms of an incoherent imaging picture, in which the intensity distribution at the specimen was calculated by considering the geometric optics image of the source, broadened by diffraction at the objective aperture and the lens aberrations. Then the intensities transmitted directly through the specimen and the intensities scattered out of the incident beam could be calculated from the scattering cross sections of the atoms; thus, the intensity measured by the detector could be determined. On this basis, it was concluded, quite falsely, that the STEM imaging process was essentially different from that of CTEM. For CTEM the usual assumption is that of coherent imaging. The incident beam is approximated by a plane wave. The scattered amplitude is calculated for each direction; these wave amplitudes are combined to form the image amplitude and hence the image intensity.

This simple, incoherent imaging picture can be used without serious error for many cases of practical interest in both forms of microscopy. But since, in the case of each technique, we will be concerned with the coherent imaging phenomena that appear under high-resolution conditions and with the diffraction effects given by crystalline samples, we start from the coherent imaging picture, which is valid for both cases.

The relation between the two imaging modes is provided by the reciprocity principle. This relates to the idealized case of point sources and point detectors but may be extended, within limits, to describe actual situations (Cowley, 1969). The principle may be stated as follows: The wave amplitude at point B in a system due to a point source at point A is the same as the amplitude that would be produced at A by a point source at B. This holds, provided that the transmission through the system involves only elastic scattering processes and only scalar fields (i.e., no magnetic fields for electrons). It may hold for electrons in the presence of magnetic fields if the directions of the magnetic fields are reversed when the direction of the electron wave propagation is reversed. If there are inelastic scattering processes involved, the relationship can be applied to intensities but not to the amplitudes of the electron waves, provided that the change of energy of the electrons is not detectable under the experimental conditions involved (Doyle and Turner, 1968). Within these restrictions, we see that the relationship between CTEM and STEM image intensities can be established by reference to Figure 1. In these diagrams, we have drawn geometric ray paths as a matter of convenience only. It must be emphasized that we are concerned with wave amplitudes and only with initial and final amplitudes, not with the paths or directions of electron beams through the system.

For STEM we consider a point source at A giving an image intensity at a point in the detector plane at B. This image intensity will be identical with that in CTEM for point A in the image when electrons are emitted from point source B, provided that the lenses, lens apertures, and other dimensions are the same. The extension to deal with practical sources and detectors of finite size comes from the assumption that, for a finite source, all points emit independently, so that the observed intensity is the sum of the intensities from all source points considered separately. Likewise, most detectors are incoherent in that the measured intensity is the sum of the intensities for all points of the detector aperture. Thus, the effect of a finite source size in CTEM will be exactly the same as the effect of a finite detector size in STEM, provided that the CTEM source may be assumed to be completely incoherent (which is not true in some cases when a field-emission gun is used). The effect of having a finite source size for STEM will be the same as integrating the intensity over a finite image area in CTEM. This will obviously result in a loss of resolution if this image area in CTEM is larger than the scale of the resolved detail produced in the image plane and detectable with a high-resolution recording medium such as an ideal photographic plate.

As a consequence of the reciprocity relationship, it can be confidently predicted that, for identical electron optical components and equivalent sources and detectors, the resolution and contrast of STEM images will be the same as for CTEM images. In fact, the whole range of phase contrast and amplitude contrast effects, Fraunhofer and Fresnel diffraction effects, and bright-field and dark-field imaging behavior, familiar in CTEM, have been reproduced in STEM.

The reciprocity-based relationship does not imply, of course, that the various types of observation are made with equal ease by the two methods. Many instrumental factors influence the results, making some types of experimental result very easy by using one method but difficult or virtually impossible by using the other. CTEM and STEM in practice show very few areas of overlap in their preferred modes of operation and must be considered as complementary rather than competitive techniques.

B. STEM and CTEM in Practice

Diagrams suggesting the principal operating modes of STEM and CTEM instruments, for imaging and diffraction, are given in Figure 2.

Some of the finer points of difference between these two forms of microscopy will emerge from our later treatment of imaging theory, but it is useful at this stage to summarize the more obvious points of difference arising from instrumental factors.

For some situations, especially for very thick specimens, the reciprocity relationship cannot be applied because the image intensities are appreciably affected by the energy losses associated with inelastic scattering. Most CTEM instruments have no provision for energy filtering of the electrons transmitted through the specimen. For thick specimens, the number of electrons that have lost energy by inelastic scattering processes may be comparable with, or greater than, the number that have lost no appreciable amount of energy. For the consequent spread of electron energies of several hundred electron volts, the effects of the chromatic aberrations of the imaging lenses can be to degrade the image resolution and contrast considerably. In a STEM instrument used without an electron energy analysis, all the electrons transmitted by the specimen are recorded equally, irrespective of their energy losses, so that the corresponding loss of resolution or contrast does not occur.

For most STEM instruments, an energy-loss spectrometer is provided. This can be used to filter out all but the no-loss electrons, or else electrons having a given energy loss. Then the image contrast must be different again. Only rarely are completely equivalent energy-filtering devices used for STEM and CTEM that provide a useful reciprocity relationship for such imaging of thick specimens.

The most important practical differences between CTEM and STEM arise from the fact that the means for detecting and recording the images are different. In CTEM a static two-dimensional image is formed, and the intensity distribution is recorded by the simultaneous integration of the dose at all points in a two-dimensional detector, usually a photographic emulsion. The STEM image information is detected for one image point at a time. The detector, usually a phosphor-photomultiplier combination, provides an electrical image signal in serial form for display on a cathode ray tube or for recording in any analogue or digital form. A very important practical limitation for STEM is that, in order to get the good signal-to-noise ratio needed for high-quality imaging, the number of electrons scattered from each picture element within a short period must be large (10^4 or more). For high-resolution imaging, this number of electrons must be concentrated within a very small probe size. The requirement is, therefore, for a very-high-intensity electron source, such as can only be approached by use of a field-emission gun.

An analysis of the limitations of STEM imaging by these factors has been given, for example, by Strojnik (1973). One is usually faced with a compromise. With a given source brightness, it is possible to achieve a desired signal-to-noise ratio for a given resolution either by decreasing the number of picture elements or by increasing the

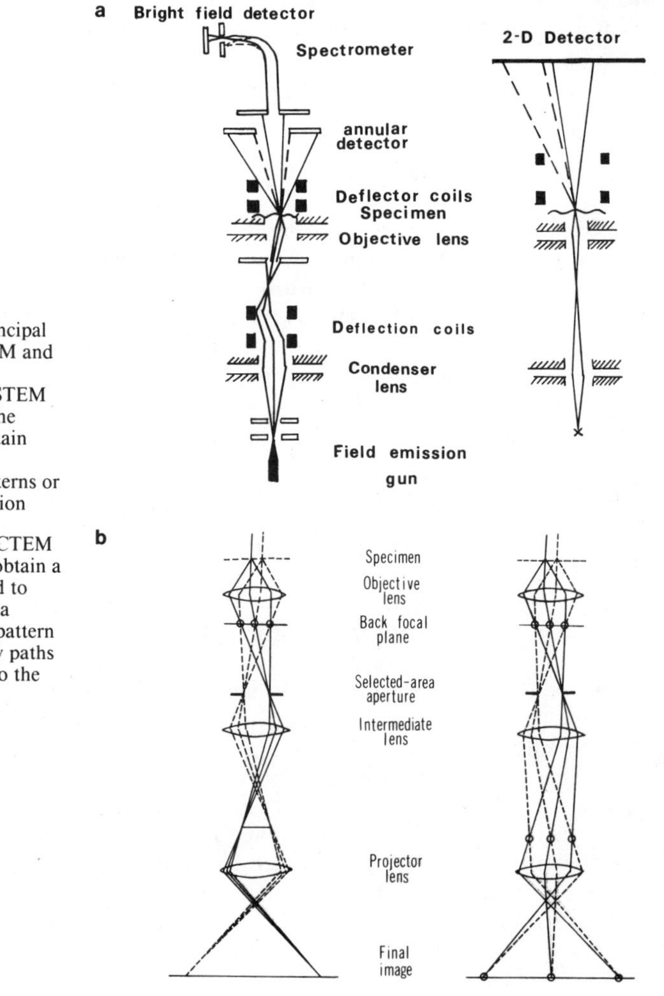

Figure 2 Diagrams suggesting principal operating modes of STEM and CTEM instruments.
(a) Ray diagrams for a STEM instrument used in the imaging mode to obtain convergent-beam microdiffraction patterns or selected-area diffraction patterns.
(b) Ray diagrams for a CTEM instrument used to obtain a magnified image and to obtain a selected-area electron diffraction pattern showing only the ray paths from the specimen to the viewing screen.

exposure time. The latter alternative is often precluded by inconvenience and by the difficulties of maintaining long-term stability of the electron optics and of the specimen. Usually, therefore, STEM suffers because the picture quality may be relatively poor and the picture area may be small.

On the other hand, the serial nature of the STEM image data has allowed enormous flexibility for on-line image evaluation and image processing or for recording the image in an analogue or digital form for subsequent analysis or image processing. On-line, the image signal may be filtered to reduce unwanted noise or interference, or its contrast may be enhanced by the subtraction of a constant background. With the recent improvements in the capabilities of electronic circuitry and the availability of fast minicomputers, it is becoming feasible to perform on-line image analysis, using Fourier transform or autocorrelation processes to measure defocus and lens aberrations and to detect periodic components or other suspected features of the image.

Current developments in electronics are making it feasible to extend these methods of image processing and image analysis to images produced at TV rates. Then the same possibilities become available in both CTEM and STEM instruments. For CTEM a chosen part of the image may be recorded with a TV camera, giving a signal that may be fed into the image processing system. The capability of CTEM to provide integration over time of a two-dimensional image in order to improve the signal-to-noise ratio is matched in STEM by the use of TV-rate digital image storage and accumulation devices. Future developments along these lines will no doubt be rapid and will provide the much-needed means for obtaining quantitative image data under well-defined imaging conditions with accurate values of the relevant instrumental parameters.

A remaining and important advantage of STEM is that, by use of multiple detector systems, it is possible to obtain a number of different image signals at the same time. These image signals may be combined—by addition, subtraction, multiplication, or division or any combination of these processes—to give images designed to emphasize a particular aspect of the image information. Quite early in the development of STEM, Crewe and colleagues (Langmore *et al.,* 1973) used two images. One was produced by using an annular detector that collected all electrons scattered outside of the main, directly transmitted electron beam; the other was produced by using an energy filter on the transmitted beam to collect those electrons that had lost energy through small-angle, inelastic scattering processes. The ratio of these two image signals gives, to a first approximation, a signal that is dependent on the atomic number of the scattering atoms. This is very appropriate for the detection of a small number of heavy atoms supported on films of low atomic number (Isaacson *et al.,* 1977). Various proposals have been made for combining bright-field signals obtained from multiple detectors within the central, transmitted electron beam (see the review by Burge and Van Toorn, 1980). Cowley and Jap (1976) have discussed the use of multiple detectors within the diffraction patterns produced outside of the central beam, and equipment for realizing these possibilities has been described by Strahm and Butler (1981).

C. Analytical Electron Microscopy (AEM)

Currently, AEM provides the basis for the design of most new commercial electron microscopes and is an increasingly powerful research tool for many areas of solid state science. It arose from the realization that the techniques of microdiffraction and microanalysis using energy-loss spectroscopy or energy-dispersive spectroscopy (ELS or EDS), all of which have been developed separately in specialized instruments, could be combined with the imaging capabilities of the electron microscope to provide an important expansion of the information that could be obtained about small specimen areas. An essential component of each of the added techniques is that a useful signal can be produced when an electron beam of small diameter strikes an identifiable small area of a specimen. The signal may be detected with the beam held stationary on the specimen, as is usually the case with microdiffraction, or may be recorded as a function of time as the beam is moved. In each case, it is clear that the STEM imaging system is more closely allied with the requirements of the added techniques. It would seem, then, that the ideal AEM should be a dedicated STEM instrument. In practice, there are good arguments for using the dedicated STEM instrument for some purposes and the TEM/STEM instruments for other purposes. In the case of microdiffraction, for example, a comparison made by Chan *et al.* (1981) shows that, in spite of theoretical equivalence, each type of instrument has important advantages for particular types of application.

The various aspects of AEM will be covered in detail in the remainder of this book. Because both TEM and STEM are, in theory and practice, integral components of AEM, we will proceed in this chapter to develop the basic ideas of imaging in the way that allows an understanding of both imaging modes.

II. DIFFRACTION AND IMAGING

As a basis for understanding the imaging with electrons, we first review the fundamental principles of the two main steps of the process: the interaction of electrons with matter, and the formation of an image by a lens system.

The term "diffraction" is often used here in preference to the more general term "scattering" because it implies the importance of coherent interference effects. Our description of electron microscope imaging is based on the initial assumption of a coherent incident plane (or spherical) electron wave falling on the specimen, giving rise to scattered waves that are brought together and interfere to form the image. In electron microscopy, much more than in optical microscopy, we are conscious of the existence of definable diffracted waves because we can readily see and use the pattern of diffracted beams, the diffraction pattern formed in the back focal plane of the objective lens (Figure 2(b)).

Electrons are scattered much more strongly by matter than x-rays or visible light. A single atom can scatter enough electrons to allow its detection in an electron microscope. Monomolecular layers can give strong diffraction effects with electrons, but thicknesses of a micrometer or more are needed for comparable relative diffraction intensities for x-rays.

For even very large assemblies of atoms, the amplitude of scattered x-rays is much less than the amplitude of the incident beam. This provides the basis for the common, useful, "kinematical" or "single-scattering" approximation, which is justifiable because, if the amplitude of single scattering is very small, the amplitude of doubly scattered radiation will be negligible. Then the amplitude of the scattered or diffracted x-rays is given as a function of the scattering angle by a simple mathematical operation, the Fourier transform, applied to the electron density distribution in the sample. When this angular distribution is observed very far from a scattering atom (or group of atoms scattering independently because there is no significant correlation in their positions), it is described in terms of an atomic scattering factor, characteristic of the type of atom and listed conveniently in such tables as the *International Tables for X-Ray Crystallography* (Vol. III and IV). When diffraction by an assembly of atoms is considered, one merely adds together the scattered amplitudes for all atoms, with phase factors depending on their relative positions. For crystals, as is well known, the regularity of the atom arrangement leads to a reinforcement of the scattered amplitude in a regularly spaced set of strong, diffracted beams. The process of working back from the observed intensities of these beams, through the Fourier transform relationship to the relative atomic positions, is the foundation of x-ray crystal structure analysis.

This kinematical scattering approximation is so elegant and so relatively simple, as compared with multiple-scattering theory, that it is used for electrons whenever it seems even halfway reasonable (and often when it does not). But for electrons, the scattering by atoms is very strong. There can be appreciable multiple scattering within a single, moderately heavy atom, and for very heavy atoms, double and triple scattering can rarely be neglected.

An important factor in this respect is that, because electrons have short wavelengths (0.037 Å for 100-keV electrons) relative to atomic dimensions, the scattering angles are small ($\sim$10 mrad). When electrons enter a crystal in the direction of one of its principal axes, the electrons travel along rows of atoms, and electrons scattered by one atom can be rescattered by one of the subsequent atoms in the row. For crystals, this mutliple scattering is very important. The diffraction effects are strongly influenced, so that for electron diffraction by crystals, one must usually look beyond the kinematical approximation and deal with the complications of the dynamical (or coherent multiple-scattering) theory.

Fine detail, at an atomic level, can be seen in electron microscopes only because the scattering of electrons by atoms is very strong. Because it is strong, however, severe complications can arise for all but the ideal cases of very thin specimens containing only light atoms.

The Physical Optics Analogy

It was Boersch (1947) who first stated clearly the alternative view that, instead of considering electrons as particles being bounced off atoms, one could consider an electron wave being transmitted through the potential field of the charged particles in a sample, and the main effect of the potential field is to change the phase of the electron wave. Atoms or assemblies of atoms then constitute phase objects for electron waves, as thin pieces of glass or unstained biological sections do for light waves. In the usual, idealized picture, a plane wave enters the sample. The wave at its exit surface has a distribution of relative phase values depending on the variations of the potential field it has traversed, plus relatively small changes of amplitude corresponding to the loss of electrons by inelastic scattering processes. The angular distribution of electron wave intensity (the number density of electrons detected) at a large distance from the object is then given by squaring the magnitude of the wave amplitude in the Fraunhofer diffraction pattern, given by a Fourier transform of the exit complex wave amplitude distribution. This diffraction pattern intensity distribution will be the same as is given by the single-scattering kinematical theory only if the phase changes of the wave in the specimen are small. Large phase changes correspond to the presence of strong multiple scattering.

On this basis, we can build up a consistent theory of electron diffraction and of electron microscope image contrast using direct analogies with the concepts of elementary physical optics, including Fraunhofer diffraction, Fresnel diffraction, phase contrast imaging, and the wave optical formulation of the theory of lens action, aberrations, image contrast, and resolution.

III. DIFFRACTION PATTERNS

If a plane parallel beam of radiation strikes a specimen, the angular distribution of emergent radiation (as seen from a distance that is large compared with the specimen dimensions) is the Fraunhofer diffraction pattern. It is convenient to consider the intensity distribution as a function, not of the scattering angle, ϕ, but of the parameter, $u = 2\lambda^{-1} \sin(\phi/2)$.

For electrons, wavelength λ is small, angles ϕ are small, and in the diffraction pattern on any plane of observation, distances between features are closely proportional to differences of u values (Figure 3(a)). In the two dimensions of the plane of observation, coordinates x,y are proportional to parameters u,v derived from the components of diffraction angles ϕ_x, ϕ_y.

For single atoms, or for random arrays of many atoms, the diffraction pattern intensities are proportional to the square of the atomic scattering factor f(u) and fall off smoothly with scattering angle (Figure 3(b)). Any systematic correlation between atom positions is reflected in a modulation of scattered electron intensity distribution. For most biological and nonbiological materials considered to be "amorphous," the only correlation of atom positions is that resulting from interatomic bonds of closely prescribed lengths. This gives a modulation of the diffraction intensities, with a periodicity roughly proportional to the reciprocal of the bond lengths. Since there is no preferred direction for the bonds, however, these modulations are smeared out into diffuse, circular halos, and the percentage modulation of the intensities is usually quite low (Figure 3(c)).

The extreme case of correlated atom positions is the strictly periodic arrangement of atoms in a perfect crystal. To the periodicity a of the repetition in the crystal, there corresponds the set of regularly spaced, sharp diffraction spots, with separations proportional to $1/a$ (Figure 3(d)). For a very thin crystal lying nearly perpendicular to the incident beam and having periodicities a and b in two perpendicular directions, the diffraction pattern will be a regular cross-grating of sharp spots, with separations proportional to $1/a$ and $1/b$ in the two directions (see Figure 11(b)).

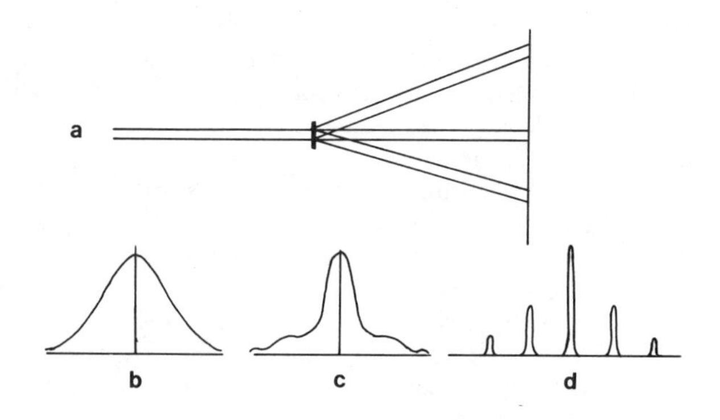

Figure 3 Fraunhofer diffraction. (a) Formation of a diffraction pattern by transmission through a thin specimen. Intensity distribution in diffraction pattern for (b) a single atom, (c) an amorphous material, and (d) a thin, single crystal.

The perfectly periodic two-dimensional case is an idealization that is seldom realistic. If the lateral dimensions of the crystal are small, the diffraction spots will be smeared out by an amount inversely proportional to the crystal dimensions. If the crystal is distorted or bent, the spots will be spread into arcs or the intensities will be changed. If there are crystal defects, if the atoms are disordered on the lattice sites, or if the atoms have thermal vibrations, there will be diffuse scattering in the background between the spots. Usually, of course, the crystal is three-dimensional, and the dimension in the beam direction can have a strong effect on the presence or absence of diffraction spots, can make the diffraction pattern intensities highly sensitive to small crystal tilts, and can give rise to a multitude of complicated dynamical diffraction effects.

Mathematical Formulation

The kinematical approximation is equivalent to the approximation of excluding all but the first term in the Born series. Starting from the integral formulation of the Schrodinger equation for a plane incident wave of wave vector k_o, the wave function at point r for the scattered wave having wave vector k is

$$\psi(r) = \psi^0(r) + \frac{\mu}{4\pi} \int \frac{\exp[-ik(r - r')]}{r - r'} \phi(r')\psi(r')dr'. \tag{1}$$

If we assume that we can replace $\psi(r')$ in the integral by the incident plane wave, $\psi^0(r) = \exp[-ik_o \cdot r']$, and consider the solution at a large distance, $r = R$, this can be written

$$\psi(r) = \exp[-ik_oR] + (\mu/R)\exp[-ik_oR] \times g(u) \tag{2}$$

where

$$\begin{aligned} g(u) &= \int \phi(r)\exp[-i(k - k_o) \cdot r]\,dr \\ &= \int \phi(r)\exp[2\pi i u \cdot r]\,dr \end{aligned} \tag{3}$$

and

$$|u| = 2\lambda^{-1} \sin \phi/2$$

i.e., the scattering amplitude, g(u), is given by the Fourier transform of the specimen potential distribution. This applies for any distribution of scattering matter, as well as for a single atom when the scattered amplitude is the atomic scattering amplitude, f(u).

For any assembly of atoms at positions r_i, one can write

$$\phi(r) = \Sigma_i \phi_i(r) * \delta(r - r_i) \tag{4}$$

where the $*$ sign denotes a convolution, so that

$$g(u) = \Sigma_i f_i \exp[2\pi i u \cdot r] \tag{5}$$

and the intensity distribution is proportional to

$$I(u) = |g(u)|^2 = \Sigma_i \Sigma_j f_i f_j \exp[2\pi i u \cdot (r_i - r_j)]. \tag{6}$$

For the special case of a periodic object

$$\phi(\mathrm{r}) = \Sigma_h F_h \exp[2\pi i h \cdot r] \tag{7}$$

where h represents the vector to a reciprocal lattice point denoted by the set of indices h,k,l, and the diffraction pattern amplitudes are given by the intersection of the Ewald sphere $[u = 2\pi(k-k_o)]$ with the distribution

$$g(u) = \Sigma_h F_h \delta(u - h) \tag{8}$$

which is the set of weighted reciprocal lattice points.

In the physical optics formulation, for incident wave $\psi_o(xy)$ and object transmission function q(xy), the wave at the exit face of specimen $\psi_e(xy)$ is $\psi_o(xy) \times q(xy)$, and the Fraunhofer diffraction pattern is given by the two-dimensional Fourier transform

$$\Psi(uv) = \int\int \psi_e(xy) \exp[2\pi i(ux + vy)]dx\,dy. \tag{9}$$

For a thin object, the transmission function represents the change of phase of the electron wave on traversing the potential field of the specimen

$$q(xy) = \exp[-i\sigma\phi(xy)] \tag{10}$$

where $\phi(xy) = \int\phi(r)dz$ and $\sigma = \pi/\lambda E$, where E is the accelerating voltage. The assumption that $\sigma\phi(xy) \ll 1$ then gives the equivalent of the single-scattering approximation.

The wave at any finite distance R from the specimen may be calculated by using the Fresnel diffraction formula

$$\psi(xy) = \psi_e(xy) * p(xy) \tag{11}$$

where p(xy) is the propagation function, given in the usual small-angle approximation by

$$p(xy) \simeq (i/R\lambda) \exp[-ik(x^2 + y^2)/2R]. \tag{12}$$

For thicker specimens, when Eq. (10) does not apply, one can calculate diffraction effects by dividing the specimen into thin slices and applying alternately the phase change for each slice, Eq. (10), and the propagation, Eq. (12), to the next slice. This is the basis for the dynamical diffraction formulation of Cowley and Moodie (Cowley, 1981). Alternatively, one can go back to the Schroedinger equation in the differential form and find solutions for the wave in a crystal, subject to suitable boundary conditions, as in Bethe's original dynamical theory (Hirsch *et al.,* 1965).

IV. THE ABBE THEORY: CTEM IMAGING

One of the most important properties of a lens is that it forms a Fraunhofer diffraction pattern of an object at a finite distance. Using for convenience a ray diagram to indicate a wave-optical process, we suggest in Figure 4(a) that an ideal lens brings parallel radiation to a point focus at the back focal plane. If a specimen is placed close to the lens, all radiation scattered through the same angle ϕ by different parts of the specimen will be brought to a focus at another point in the back focal plane, separated by a distance proportional to ϕ (in the small-angle approximation) from the central spot. The intensity distribution on the back focal plane is thus the Fraunhofer diffraction pattern, suitably scaled.

The more usual function of a lens is to form the image of a specimen. Then radiation scattered from any one point of the object is brought together at one point of the image, as in Figure 4(b). At the same time, however, the Fraunhofer diffraction pattern is formed at the back focal plane, as in Figure 4(a). The imaging process may then be described as the formation of a diffraction pattern of the object in the back focal plane (a Fourier transform operation) plus the recombination of the diffracted beams to form the image (a second Fourier transform operation).

For a perfect, ideal lens, the reconstruction of the object transmitted wave in the image plane would be exact since all radiation leaving the object is brought back with exactly the right phase relationships to form the image. For a real lens, some of the diffracted radiation is stopped when it falls outside the lens aperture. The phase relationships are upset by lens aberrations. These changes can be considered to take place on the back focal plane, where the wave function is multiplied by a "transfer function" that changes its amplitude and phase. Correspondingly, on the image plane, the complex amplitude distribution of the wave function can be considered as smeared out or convoluted by a smearing function that limits the resolution and affects the contrast.

It would be convenient if we could make an equivalent statement for image intensities, i.e., that the effect of lens apertures and aberrations is to smear out the intensity distribution by means of a smearing function that can be related (by a Fourier transform operation) to a contrast transfer function characterizing the lens. This is the usual practice for light optics, but it does not apply in general for electron microscopy. The difference is that the usual imaging with light is incoherent; each point of the object emits or scatters light independently, with no phase relationship to the light from neighboring points. In electron microscopy, the conditions usually approach those for coherent imaging, with a definite phase relationship between the waves transmitted through neighboring parts of the specimen. It is only under very special circumstances (which fortunately occur quite often) that the relatively simple ideas developed for light optics can be used as reasonable approximations for the electron case.

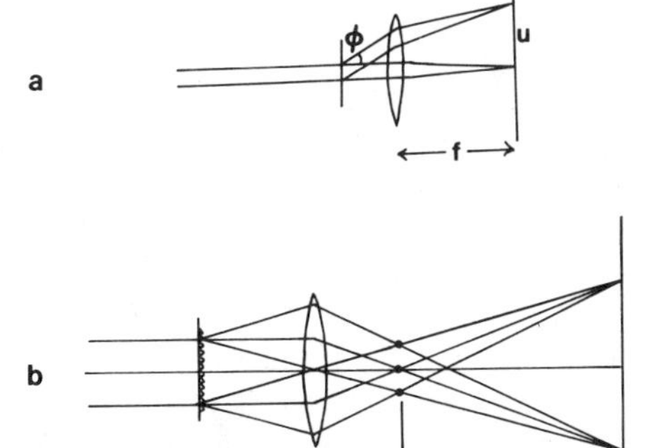

Figure 4 Ray diagrams of lens action. (a) Use of a lens to produce a focused Fraunhofer diffraction pattern in the back focal plane. (b) Ray diagram used to illustrate the wave-optical Abbe theory of imaging.

For the coherent, electron optical case, the transfer function that modifies the amplitudes of the diffraction pattern in the back focal plane can be written in terms of the coordinates u,v as

$$T(uv) = A(uv) \exp[i\chi(uv)] \tag{13}$$

where the aperture function, A(uv), is zero outside the aperture and unity within it. The phase factor $\chi(uv)$ is usually written as including only the effects of defocus Δf and the spherical aberration constant C_s since it is assumed that astigmatism has been corrected and other aberrations have negligible effect. Then

$$\chi(uv) = \pi\Delta f\lambda(u^2 + v^2) + \frac{1}{2}\pi C_s\lambda^3(u^2 + v^2)^2. \tag{14}$$

The corresponding smearing functions, given by Fourier transform of Eq. (13), are in general complicated and have complicated effects on the image intensities except in the special cases we will discuss later.

A. Incident-Beam Convergence

So far we have considered only the ideal case, that the specimen is illuminated by a plane parallel electron beam. In practice, the incident beam has a small convergence, and this may have important effects on the diffraction pattern and on high-resolution images.

Two extreme cases can be considered. First, it may be appropriate to assume that the electrons come from a finite, incoherent source, with each point of the source emitting electrons independently (Figure 5(a)). This is a useful assumption for the usual hot-filament electron gun.

Then for each point in the source, the center point of the diffraction pattern (and the whole diffraction pattern intensity distribution) will be shifted laterally—and also modified in the case of relatively thick specimens. The main effect on the image will result from the different effects of the transfer function, Eq. (13), when applied to the differently shifted diffraction patterns. In the extreme case that the incident-beam convergence angle is much greater than the objective aperture angle, we approach the incoherent imaging situation common in light optics, and all interference effects (including the production of out-of-focus phase contrast) disappear.

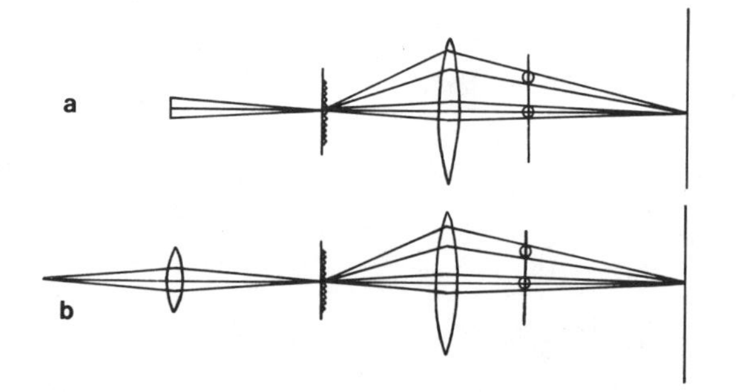

Figure 5 Illumination of the specimen in a microscope with a convergent beam for the case of (a) a finite incoherent source and (b) a point source with a lens used to give coherent radiation with the same convergence angle.

The other extreme case of convergent illumination is that occurring when the incidence convergent wave is coherent, e.g., when it is formed by focusing a point source on the specimen (Figure 5(b)). This situation is approached when a field-emission gun is used. Then it is the amplitude distribution in the diffraction pattern that is smeared out, and there will be complicated additional interference effects within the pattern (Spence and Cowley, 1978). Provided that the convergence angle is not too great, the effects on the image will be small, except in the case of images taken far out of focus.

B. Chromatic Aberration

The focal lengths of electron lenses depend on the electron energies and also on the currents in the lens windings. A variation in the accelerating voltage (ΔE), or a variation in the lens current (ΔI), during the recording time will have the effect of smearing the intensity distribution in the image. The intensities for different electron energies are added incoherently. The loss of resolution is given by

$$\Delta = C_c \alpha \left| \frac{\Delta E}{E} + \frac{2\Delta I}{I} \right| \tag{15}$$

where C_c is the chromatic aberration contrast and α is the objective aperture angle.

In addition, a change of electron energy gives a change of electron wavelength $\lambda \propto E^{-1/2}$, and this in turn affects the phase factor, Eq. (14).

There are important chromatic aberration effects associated with energy loss of electrons that suffer inelastic scattering. These will be considered in the subsequent sections.

C. Mathematical Formulation

For an incident wave amplitude, $\psi_0(xy)$, and an object transmission function, $q(xy)$, the diffraction pattern amplitude on the back focal plane of an ideal lens is

$$\Psi_e(uv) = Q(uv) * \Psi_0(uv) \tag{16}$$

where capital letters are used for the Fourier transforms of the corresponding real space functions. With the transfer function

$$T(uv) = A(uv) \exp[i\chi(uv)] \tag{17}$$

the image amplitude is given by a second Fourier transform as

$$\psi(xy) = \psi_e(xy) * t(xy) \tag{18}$$

where the spread function $t(xy)$ is given by

$$\begin{aligned} t(xy) &= c(xy) + i\, s(xy) \\ &= \mathcal{F}A(uv) \cos[\chi(uv)] + \mathcal{F}A(uv) \sin[\chi(uv)]. \end{aligned} \tag{19}$$

The symbol $\mathcal{F}$ represents the Fourier transform operation. In Eq. (18), the magnification factor $(-R/R_0)$ has been ignored. For an isolated point source, the intensity of the image is $|t(xy)|^2$, which is the spread function for incoherent imaging.

The corresponding contrast transfer function for incoherent imaging is then given by Fourier transform as $T(uv) * T^*(-u,-v)$.

For a plane incident wave from a direction (u_1,v_1), the diffraction plane amplitude, Eq. (16), becomes

$$[\Psi_c(uv) * \delta(u - u_1, v - v_1)]\, A(uv) \exp[i\chi(uv)] \tag{20}$$

and the image, Eq. (18), becomes

$$\psi_c(xy) \exp[2\pi i(u_1x + v_1y)] * t(xy).$$

For a coherent source, the image intensity from a finite source is thus

$$|\iint S(u_1v_1)\, [\psi_c(xy) \exp 2\pi i(u_1x + v_1y) * t(xy)]\, du_1dv_1|^2$$

and for an incoherent source it is

$$\iint |S(u_1v_1)|^2 \times |\psi_c(xy) \exp[2\pi i(u_1x + v_1y)] * t(xy)|^2\, du_1dv_1\,. \tag{21}$$

Here $S(u_1v_1)$ is the source function.

In the case of a thin object for which the effect on the incident wave can be represented by a two-dimensional transmission function, the effect of beam convergence can thus be expressed as in Eq. (20) by a convolution in the diffraction plane. For thicker objects, the formulation is much more complicated.

The effect of chromatic aberration cannot be represented by a convolution in the diffraction plane. It is given by summing the image intensity distributions for all wavelengths and focal length variations.

D. Inelastic Scattering

Other chapters in this volume will deal with the inelastic scattering processes in detail. Here we briefly summarize the effects on the image formation.

Electrons may be scattered by phonons, i.e., by the waves of thermal vibration of atoms in the specimen. The energy losses suffered by the incident electrons are so small ($\sim 10^{-2}$ eV) that no appreciable chromatic aberration effect is introduced. For an ideal lens of infinite aperture, all thermally scattered electrons would be imaged along with the elastically scattered electrons, and the effect on the image would be negligible. For a lens with a finite aperture, the effect of thermal scattering is not significant for amorphous specimens. For crystalline specimens, however, the elastic scattering is often concentrated into a few sharp diffraction spots, whereas the thermal scattering is diffusely spread over the background. The objective lens aperture is used to select the incident beam spot or one of the diffracted beam spots. This may cut off much of the thermal diffuse scattering, and the loss of these electrons can be attributed to an absorption function, included in the calculations of image contrast by adding an out-of-phase term or by adding a small imaginary part to make the effective crystal potential into a complex function.

The other important means by which incident electrons can lose energy is by excitation of the bound or nearly-free electrons of the sample into higher energy states. This may involve collective excitations, as when a plasmon oscillation is generated, or else single-electron excitations. The excitations most relevant for imaging involve energy losses ranging from a few eV to ~ 30 eV and scattering angles that are much smaller than the usual elastic scattering angles. For light-atom materials, the number of electrons inelastically scattered in this way may exceed the number elastically scattered.

The electrons that have lost energy will be imaged by the microscope lenses in exactly the same way as the elastically scattered electrons except that, because of the change of energy, the image will be defocused. The consequent smearing of the image will usually be ~10 Å for thin specimens and so will not be important for low-resolution or medium-resolution electron microscopy. For considerably higher resolution imaging, the inelastically scattered electrons will produce a slowly varying background to the image detail (on a scale of 5 Å or less) produced by elastically scattered electrons.

Even if the electrons inelastically scattered from a thin specimen are separated out by use of an energy filter and brought to the best focus, the image they produce does not have good resolution. This follows because the inelastic scattering process is not localized. The interaction with the incident electrons is through long-range Coulomb forces; consequently, the position of the scattering event cannot be determined precisely, and the image resolution cannot be better than ~10 Å for energy losses of $< \sim 200$ eV.

For thicker specimens, the inelastically scattered electrons can be scattered again, elastically, and give a good, high-resolution image if properly focused. The situation is almost the same as for an incident beam, of slightly reduced energy and slightly greater beam divergence, undergoing only elastic scattering.

V. STEM IMAGING

For many purposes, STEM imaging can be understood by analogy with the well-established ideas of CTEM imaging, using the principle of reciprocity. For many STEM modes, however, there is no convenient CTEM analogue, and it is better to discuss image formation and calculate intensities without reference to CTEM.

Figure 6 illustrates the STEM system. The beam incident on the specimen has a convergence defined by the objective aperture. For every position of the incident beam, a diffraction pattern is produced on the detector plane. This is a convergent-beam diffraction (CBED) pattern, in which the central beam and each crystal diffraction spot will form a circular disc that will be of uniform intensity for a very thin crystal but may contain complicated intensity modulations for thicker crystals (Figure 7).

The intensity distribution in the CBED pattern contains a wealth of information concerning the atomic arrangements within the small regions illuminated by the beam. The interpretation and use of the "microdiffraction" patterns obtained when the incident beam is held stationary on the specimen is the subject of a subsequent chapter (see also Cowley, 1978a). For the coherent convergent incident beam produced with a field-emission gun, CBED intensities may be strongly modified by interference effects (Spence and Cowley, 1978).

If the detector aperture is small and is placed in the middle of the central spot of the CBED pattern (Figure 6), the imaging conditions will be those for the usual CTEM bright-field imaging mode. As the detector diameter is increased, the signal strength is increased. The bright-field phase-contrast term for a thin specimen passes through a maximum value and then decreases to zero as the collector aperture size approaches the objective aperture size. For strongly scattering thin objects, the second-order bright-field "amplitude contrast" term increases continually with collector aperture size. The collector aperture size and defocus giving the best contrast and resolution can be determined for any type of specimen by use of appropriate calculations (Cowley and Au, 1978).

The most common STEM mode is the dark-field mode introduced by Crewe and colleagues (Crewe and Wall, 1970), in which an annular detector collects nearly all the electrons scattered outside of the central beam. This mode of dark-field imaging is considerably more efficient than the usual dark-field CTEM modes (there is a larger image signal for a given incident-beam intensity); it is therefore valuable for the study of radiation-sensitive specimens. Images may be interpreted under many circumstances as being given by incoherent imaging, with an image intensity proportional to the amount of scattering matter present. This convenient approximation breaks down for crystalline specimens when the amount of scattering depends strongly on crystal structure, orientation, and thickness. Also, it fails near the high-resolution limits when interference effects can modify the amount of scattered radiation that is undetected because it falls within the central beam (Cowley, 1976).

The annular-detector dark-field mode has been used with great effect for detecting individual heavy atoms and their movements (Isaacson *et al.*, 1977) and also for quantitatively measuring the mass of biological macromolecules (Lamvik and Langmore, 1977).

When the annular detector is used to collect all scattered radiation, however, the information contained with the diffraction pattern is lost. In principle, using a two-dimensional detector array to record the complete CBED pattern for each image point could add considerably to the information obtained about the specimen. Structural details smaller than the resolution limit could be deduced or recognized by means of

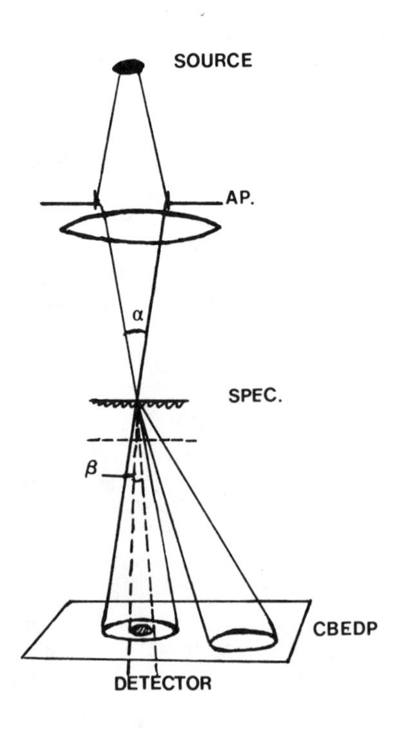

Figure 6 Formation of a convergent-beam diffraction pattern in the detector plane of a STEM instrument showing the position of the bright-field detector aperture.

structure analysis or pattern recognition techniques (Cowley and Jap, 1976). More immediate and practical means for using some of this CBED information include special detectors and masks to form image signals by separating out particular features of the pattern (diffraction spots, diffuse background, Kikuchi lines). Equipment designed to allow the exploitation of these possibilities is now in operation (Cowley and Au, 1978; Cowley, 1980).

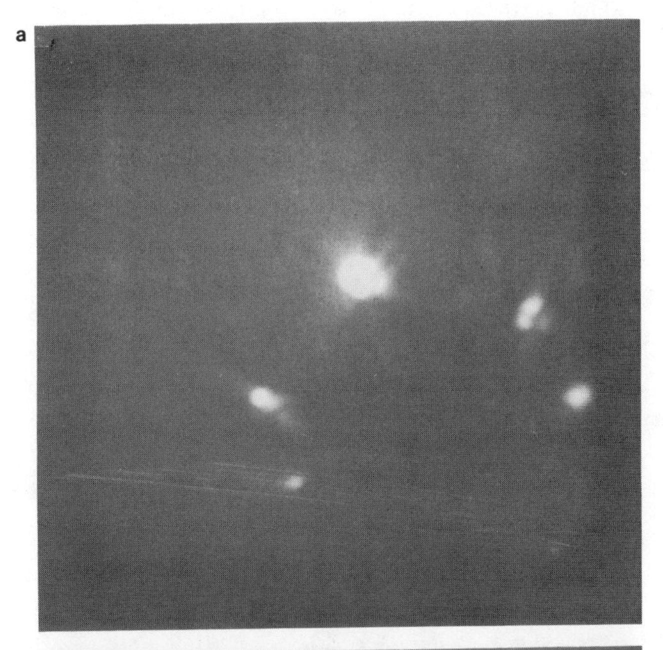

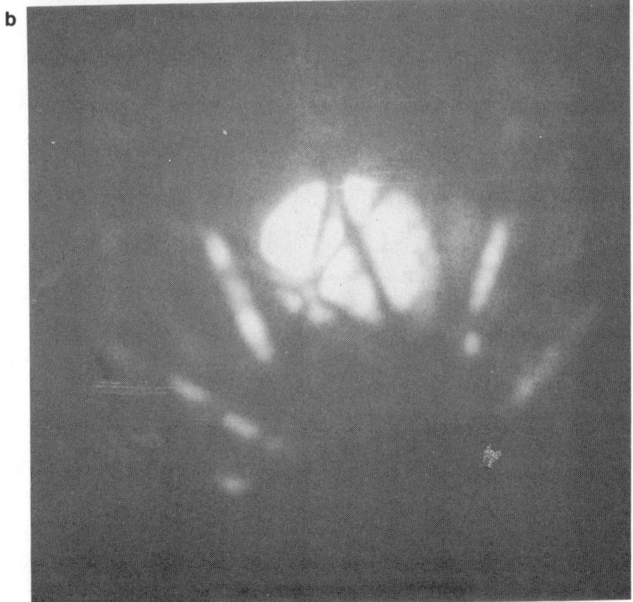

Figure 7 Convergent-beam electron diffraction patterns from a small region (<30 Å diameter) of a MgO crystal for objective aperture sizes (a) 20 μm and (b) 100 μm.

A further means for dark-field imaging in STEM is provided by the convenient addition of an electron energy filter to separate the elastically and inelastically scattered electrons or to pick those electrons from any part of the CBED pattern that have lost any specified amount of energy. The resolution of such images may be restricted by the limited localization of the inelastic scattering process, but the method has important possibilities in providing means for locating particular types of atom or particular types of interatomic bonds within specimens (see subsequent chapters on ELS). The versatility of the STEM detection system allows the inelastic dark-field image signals to be combined with the signals from other dark-field or bright-field detectors in order to emphasize particular features of the specimen structure.

Mathematical Description

For a point source, the wave incident of the STEM specimen is given by Fourier transform of the objective lens transfer function

$$\psi_o(r) = \mathcal{F}\{A(u) \exp[i\chi(u)]\} = c(r) + i\, s(r) \tag{22}$$

where r and u are two-dimensional vectors. For a thin specimen with transmission function $q(r - R)$, where R is the translation of the specimen relative to the incident beam (or vice versa), the wave amplitude on the detector plane is

$$\Psi_R(u) = [Q(u) \exp(2\pi i u \cdot R)] * A(u) \exp[i\chi(u)] \tag{23}$$

and the intensity distribution is $I_R(u) = |\Psi_R(u)|^2$. The image signal corresponding to beam position R is then

$$J(R) = \int I_R(u)\, D(u)\, du \tag{24}$$

where D(u) is the function representing the detector aperture.

In the limiting case of a very small axial detector, D(u) is replaced by $\delta(u)$, and Eq. (24) becomes

$$J(R) = |q(R) * t(R)|^2 \tag{25}$$

which is equivalent to Eq. (18) and applies for plane-wave illumination in CTEM.

VI. THIN, WEAKLY SCATTERING SPECIMENS

For a very thin object, we may assume (neglecting inelastic scattering) that the only effect on an incident plane wave is to change its phase by an amount proportional to the projection of the potential distribution in the beam direction. Writing this in mathematical shorthand, if the potential distribution in the specimen is $\phi(xyz)$, the phase change of the electron wave is proportional to

$$\phi(xy) = \int \phi(xyz) dz\ . \tag{26}$$

If the value of the projected potential $\phi(xy)$ is sufficiently small, it can readily be shown that, for bright-field CTEM with a small source or bright-field STEM with a small detector, the image intensity can be written as

$$I(xy) = 1 + 2\sigma\phi(xy) * s(xy) \tag{27}$$

where σ is the interaction constant. The $*$ sign represents a convolution or smearing operation.

This means that image contrast, given by the deviation of the intensity from unity, is described directly as the projected potential smeared out by the spread function or smearing function s(xy), which describes the effects of the defocus and lens aberrations in determining the resolution. Provided that the spread function is a clean, sharp peak, the image will show a well-shaped circular spot for each maximum in the projected potential, i.e., for each atom or group of atoms. The smearing function depends on the defocus of the objective lens and on its aberrations. For the Scherzer optimum defocus (Scherzer, 1949), which depends on the electron energy and the spherical aberration constant, the smearing function is a sharp, narrow, negative peak (Figure 8(a)), so that the image shows a small dark spot for each atom. This spot is the clearest, sharpest peak attainable with a given objective lens. For 100-keV electrons and $C_s = 2$ mm, for example, the optimum defocus is ~950 Å underfocus, and the corresponding width of the spread function limits the resolution to ~3.5 Å.

The imaging conditions are usually discussed in terms of the modification of the wave amplitude on the back focal plane of the objective lens. The equivalent of Eq. (27) is that the diffraction pattern amplitude is multiplied by $A(uv)\sin[\chi(uv)]$, where $A(uv)$ is the aperture function and $\chi(uv)$ is the phase factor given in Eq. (14). For there to be a good representation of the object in the image, the function $\sin\chi$ should be as close to unity over as large a region of the diffraction pattern as possible so that the diffracted beams will have the correct relative amplitudes when they recombine to form the image. For the optimum defocus, Figure 8(a) shows that the $\sin\chi$ function is small for low-angle scattering but is near unity for a wide range of scattering angles

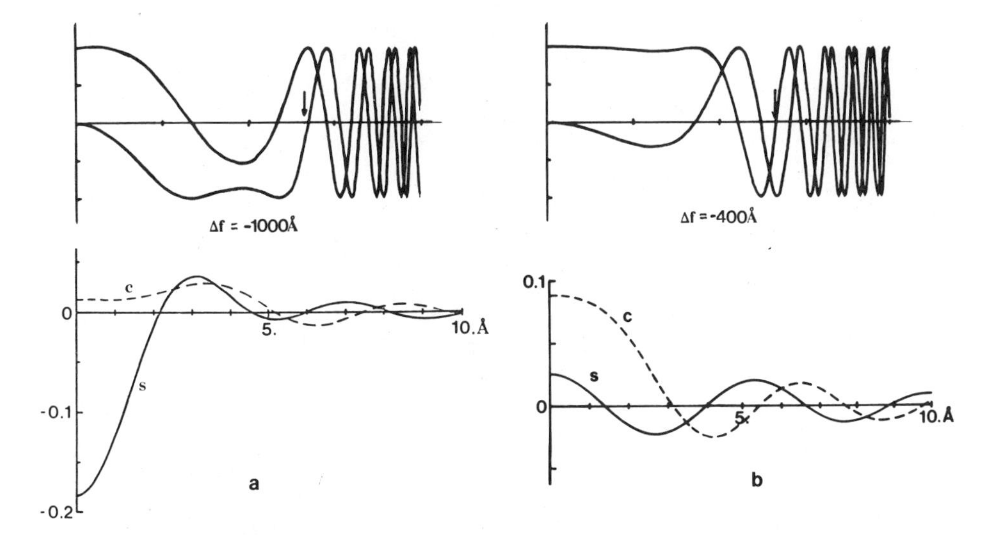

Figure 8 Real and imaginary parts of the transfer function [$\cos\chi(u)$ and $\sin\chi(u)$] and the corresponding spread functions c(r) and s(r) for defocus (a) − 1000 Å and (b) − 400 Å, for 100 keV, $C_s = 2$ mm, aperture radius $u = 0.265$ Å^{-1} (arrows).

(corresponding to spacings in the object of ~20 to 3.5 Å) before it goes into wild oscillations at high angles. This suggests that slow variations of potential will not be seen in the image, 3.5- to ~20-Å detail will be well represented, and finer detail will be represented in a very confused manner.

This mode of imaging is thus of very little use for biologists who wish to see detail on the scale of 10 Å or greater with good contrast. Their needs are discussed in the following section.

It is common practice to insert an objective aperture to eliminate all the high-angle region past the first broad maximum of the $\sin\chi$ function so that all scattered waves contributing to the image will do so in the correct phase. Then the image will be interpretable in the simple intuitive manner in that, to a good approximation, it can be assumed that dark regions of the picture correspond to concentrations of atoms, with a near-linear relationship between image intensity and the value of the smeared projected potential.

It is only for weakly scattering objects that the well-known assumptions for dark-field STEM imaging with an annular detector are useful. For such specimens, the image intensity can be assumed, as a first approximation, to depend on the total amount of scattering by the atoms illuminated by the incident beam and so is proportional to $\phi^2(xy)$. This is smeared out by a spread function sharper than that for bright-field imaging. Hence, the resolution of dark-field STEM images is often better than for bright-field STEM or CTEM by a factor of ~1.5 (Figure 12). It must be noted that the image contrast is then related to the square of the projected potential distribution and that this could introduce some complications. For example, a small hole in a sample (a negative peak of projected potential) could give the same contrast as an atom or group of atoms.

A. The Weak-Scattering Approximation in Practice

The restrictions to very thin, weakly scattering objects [$\sigma\phi(xy) \ll 1$] are often not so severe as might be assumed from initial calculations, especially for light-atom materials. It is always possible to subtract a constant or slowly varying contribution to the projected potential since a constant phase change has no effect on the imaging (except to make a very small change in the amount of defocus).

For an amorphous specimen (for example, a layer of evaporated carbon 100 Å thick), the phase change due to the average potential ϕ_0 may be 1 or 2 rad. Since there are many atoms in this thickness, overlapping at random, the projected potential, however, may not differ greatly from ϕ_0, and the relative phase changes may be only a fraction of a radian (Figure 9(a)). Then the weak-phase object approximation, Eq. (27), holds quite well, and this simple approximation is more likely to fail because of the neglect of three-dimensional scattering effects than because of large phase changes.

On the other hand, a single, very heavy atom (Figure 9(b)) may have a maximum of projected potential no greater than for the amorphous light-atom film (Figure 9(a)), but the deviation from the average projected potential is much greater, and the error in the use of the weak-phase object approximation may be large.

Similarly large deviations from the average occur in the projections of the potential for crystals when the incident beam is parallel to the rows or planes of atoms. If the same atoms as in the amorphous film (Figure 9(a)) were rebuilt into a crystal, the projected potential could be as in Figure 9(c), and the weak-scattering approximation could fail badly.

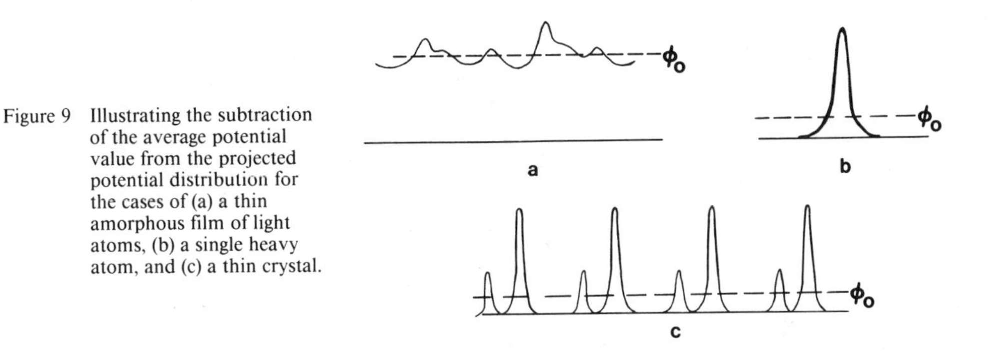

Figure 9 Illustrating the subtraction of the average potential value from the projected potential distribution for the cases of (a) a thin amorphous film of light atoms, (b) a single heavy atom, and (c) a thin crystal.

B. Beam Convergence and Chromatic Aberration

For the case of weakly scattering objects, the consideration of these complicating factors is greatly simplified. It has been shown by Frank (1973), and by Anstis and O'Keefe (1976) for the particular case of crystals, that the effects of beam divergence and chromatic aberration can be included by applying an "envelope function" that multiplies the transfer function and reduces the contribution of the outer, high-angle parts of the diffraction pattern. For most high-resolution imaging, it is one or the other of these envelope functions, rather than the ideal transfer function, that effectively limits the resolution.

For current 100-keV electron microscopes used with the incident beam focused on the specimen to give the high intensity needed for high-magnification imaging, the incident-beam convergence may limit the resolution to 4 Å or more. For the 1-MeV microscopes that have been used for high-resolution imaging with apparent point-to-point resolution approaching 2 Å (Horiuchi *et al.,* 1976), the limiting factor seems to be the chromatic aberration effect due to lens current and high-voltage instabilities.

The results of calculations by O'Keefe, illustrating the forms for these envelope functions and the consequent limitations on resolution, are shown in Figure 10.

The use of an envelope function to take account of convergence is simple and convenient but is strictly applicable only for weakly scattering objects. Also, it is applicable only for small angles of illumination in CTEM or small collector angles in STEM. Calculations for STEM (Cowley and Au, 1978) have shown that, if the objective aperture size is correctly chosen, an increase in collector aperture angle can give a slight improvement in resolution as the signal strength increases to a maximum. Then for relatively large collector angles, approaching the objective aperture angle, there may be conditions of reversed contrast with an appreciably better resolution. The influence of CTEM illumination angle or of STEM collector size is, however, a rather complicated function of the defocus, the objective aperture size, the spherical aberration constant, and the strength of the scattering by the specimen.

With care, the resolution limitations resulting from beam convergence, chromatic aberration, and such other factors as mechanical vibrations or stray electrical or magnetic fields may be reduced. It is possible, for example, to achieve a very small beam convergence with adequate intensity for high-magnification images by using a very bright source, such as a field-emission gun. The resolution limitations represented by the envelope functions will then be less important. If a sufficiently large, or no, objective aperture is used, diffracted beams from the outer parts of the diffraction pattern (where the transfer function is oscillating rapidly) will contribute to the image.

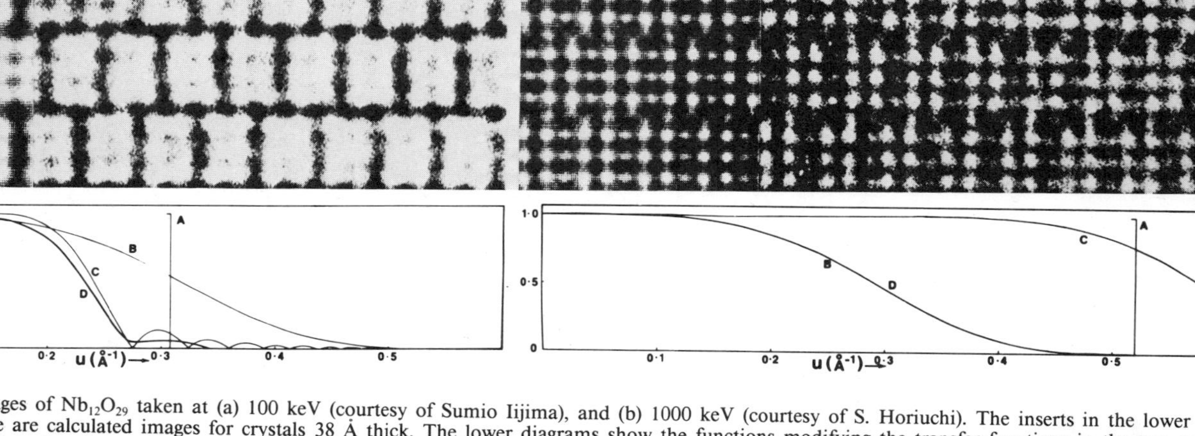

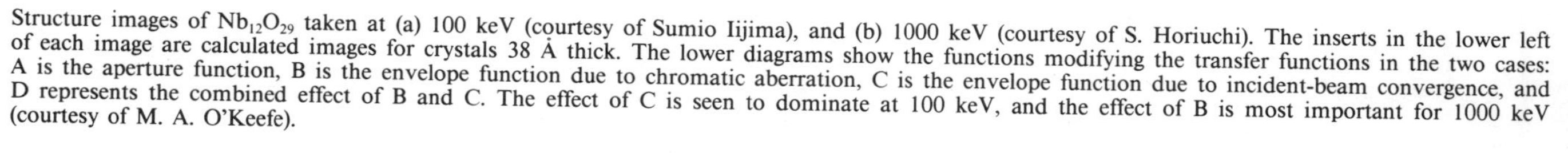

Figure 10 Structure images of $Nb_{12}O_{29}$ taken at (a) 100 keV (courtesy of Sumio Iijima), and (b) 1000 keV (courtesy of S. Horiuchi). The inserts in the lower left of each image are calculated images for crystals 38 Å thick. The lower diagrams show the functions modifying the transfer functions in the two cases: A is the aperture function, B is the envelope function due to chromatic aberration, C is the envelope function due to incident-beam convergence, and D represents the combined effect of B and C. The effect of C is seen to dominate at 100 keV, and the effect of B is most important for 1000 keV (courtesy of M. A. O'Keefe).

This will produce detail in the image on a very fine scale, but since the diffracted beams are recombined with widely different phase changes, this fine detail will not be directly interpretable in terms of specimen structure. Obviously some information about the specimen structure is contained in this detail. To extract the information, however, it will be necessary to use indirect image processing methods or else to seek agreements between observed image intensities and intensities calculated on the basis of postulated models.

Most proposed image processing techniques depend on the idea that the difficulties caused by ambiguities of phase and loss of information around the zero points of the transfer function can be overcome by using two or more images obtained with different amounts of defocus (Saxton, 1978). Whereas some progress has been made along these lines, few clear indications of improved resolution have been obtained. The alternative of calculating images from models of the specimen structure is realistic only for crystals.

C. Mathematical Formulation

For a weak-phase object, we assume $\sigma\phi(xy) \ll 1$, and Eq. (10) becomes

$$q(xy) = \exp[-\ i\sigma\phi(xy)] = 1 - i\sigma\phi(xy)\ .$$

For parallel incident radiation, the wave function in the back focal plane of a lens modified by the transfer function is

$$\Phi(uv) = [\delta(uv) - i\sigma\Phi(uv)] \times A(uv) \exp[i\chi(uv)] \tag{28}$$

and on the image plane the wave function is

$$\psi(xy) = [1 - i\sigma\phi(xy)] * [c(xy) + i\ s(xy)] \tag{29}$$

where c(xy) and s(xy) are the Fourier transforms of A(uv) cos $\chi(uv)$ and A(uv) sin $\chi(uv)$. Neglecting terms of second order in $\sigma\phi$, the image intensity becomes

$$I(xy) = 1 + 2\sigma\phi(xy) * s(xy) \tag{30}$$

so that we may consider the image to be produced by a system with contrast transfer function A sinχ.

For dark-field imaging in CTEM with a central beam stop that removes only the forward-scattered beam, the δ-function and $\phi(0,0)$, from Eq. (28), we are left with only second-order terms and obtain

$$I_{DF}(xy) = [\sigma\phi'(xy) * c(xy)]^2 + [\sigma\phi'(xy) * s(xy)]^2 \tag{31}$$

where $\phi'(xy) = \phi(xy) - \overline{\phi}$, and the average potential $\overline{\phi}$ is equal to $\Phi(0,0)$.

For the case of STEM with a small axial detector, Eq. (25) becomes

$$J(R) = |[1 - i\sigma\phi(R)] * [c(R) + i\ s(R)]|^2 \tag{32}$$

$$\simeq 1 + 2\sigma\phi(R) * s(R)$$

which is identical with Eq. (30).

When the detector angle is not much smaller than the objective aperture angle, the bright-field image is given by replacing the spread function s(R) by $t_1(R)$ where

$$t_1(R) = s(R)[d(R) * c(R)] - c(R)[d(R) * s(R)] \tag{33}$$

where d(R) is the Fourier transform of the collector aperture function D(u) of Eq. (24). The form of $t_1(R)$ and its dependence on defocus and collector aperture size has been explored by Cowley and Au (1978). For the CTEM case, Eq. (21) reduces to the same form, representing the effect of beam convergence from a finite source. For a small collector angle in STEM, or a small finite convergence angle in CTEM, the function d(R) will be a broad, slowly varying peak, and d(R)s(R) = 0

since

$$\int s(R)dR = 0.$$

Then

$$t_1(R) \simeq s(R)[d(R) * c(R)]$$

or the effective transfer function is

$$A(u) \sin[\chi(u)] * \{D(u) \times A \cos[\chi(u)]\}. \tag{34}$$

Convolution with this narrow function will have the effect of smearing out the transfer function and reducing the oscillations by a greater amount as the period of the oscillations becomes smaller. That is, the effect will be to multiply transfer functions such as those of Figure 8(a) by a rapidly decreasing envelope function.

If it is assumed that a dark-field STEM image is obtained by using an annular detector that collects all the scattered radiation (neglecting the loss of scattered radiation that falls within the central beam disc), the expression of Eq. (23) with Q(uv) = $i\sigma\phi(uv)$ gives

$$J_{DF}(R) = \sigma^2\phi^2(R) * [c^2(R) + s^2(R)] \tag{35}$$

which is equivalent to the incoherent imaging of a self-luminous object having an intensity distribution $\sigma^2\phi^2(R)$. The result of Eq. (35) is, of course, different from Eq. (31). The two dark-field imaging modes are not equivalent.

The approximation of incoherent imaging fails significantly for detailed imaging of dimensions near the resolution limit, i.e., when the oscillations of $\Phi(u)$ are of size comparable to that of the hole in the annular detector.

VII. THIN, STRONGLY SCATTERING SPECIMENS

As suggested, the simplifying approximation of Eq. (27) can fail even for a single heavy atom. For thin crystals, and for amorphous materials containing other than light atoms, we must use a better approximation. For sufficiently thin specimens and for sufficiently high voltages (thicknesses of <50 to 100 Å for 100 keV, or 200 to 500 Å for 1 MeV, depending on the resolution being considered and the required accuracy), we can use the approximation that the transmission function of the specimen involves a phase change proportional to the projected potential. The phase change, even when referred to an average equal to the average projected potential, may be several radians. We must then consider not only the first-order term in $\sigma\phi(xy)$, as in Eq. (27), but also higher-order terms. For the case of parallel incident radiation, Eq. (27) is replaced by

$$I(xy) = 1 + 2 \sin\sigma\phi(xy) * s(xy) - 2[1 - \cos\sigma\phi(xy)] * c(xy) \tag{36}$$

or, taking only second-order terms into account,

$$I(xy) = 1 + 2\sigma\phi(xy) * s(xy) - \sigma^2\phi^2(xy) * c(xy) . \tag{37}$$

This means that there are two components to the image intensity. The first, as before, is given by the projected potential smeared out by the smearing function s(xy), which is equivalent to multiplying the diffraction pattern amplitudes by $A(uv)\sin[\chi(uv)]$. The new term is given by the square of the projected potential (or, more accurately, by square of the positive or negative deviation from the average projected potential) smeared out by the smearing function c(xy). Use of this smearing function is equivalent to multiplying a component of the diffraction pattern by the function $A(uv)\cos[\chi(uv)]$ illustrated in Figure 8(a).

The two components of the image signal, arising from the ϕ and ϕ^2 terms, are sometimes referred to as phase contrast and amplitude contrast terms, respectively. We prefer to call them first- and second-order phase contrast terms because they both arise in the imaging of a pure phase object.

For the optimum defocus, Figure 8(b) shows that c(r) is relatively broad and featureless, whereas the function s(r) has a sharp, negative peak. This means that, for high-resolution detail, the $\sin\sigma\phi$ term in Eq. (36) will give good contrast, but the $\cos\sigma\phi$ term will be smeared out into the background.

The situation for medium- or low-resolution imaging is best seen from the curves of Figure 8(a) or (b). The part of the diffraction pattern corresponding to $\sin\sigma\phi$ is multiplied by $\sin\chi$, which is close to zero for the small scattering angles that correspond to slow variations in projected potential. The portion corresponding to $(1 - \cos\sigma\phi)$ is multiplied by $\cos\chi$, which is close to unity for small scattering angles; hence, this will dominate the image contrast.

This situation is familiar in biological electron microscopy. To obtain good contrast in the useful range of resolution (details on a scale >10 to 20 Å), it is advisable to use a heavy-atom stain that will give a strong $(1 - \cos\sigma\phi)$ signal. This signal is then strongly imaged because of the relatively large values of the $\cos\chi$ function. A small objective aperture size is used to remove the high-angle scattered radiation, which adds only a confused background to the image and reduces the image contrast. Detailed calculations on models of stained biological objects have confirmed this interpretation (Cowley and Bridges, 1979). Alternatively, one can go far out of focus to bring the larger parts of the $\sin\chi$ function closer to the origin to give appreciable contrast for the larger-scale image detail. This is the preferred mode when unstained specimens with small $\sigma\phi$ values are used.

The case of thin crystals is distinctive in that it involves more quantitative consideration of more specialized diffraction patterns; it will thus be treated separately.

Mathematical Formulation

Provided that we may use the phase-object approximation, the transmission function of the object is written

$$q(r) = \exp[-i\sigma\phi(r)] \tag{38}$$

and the image intensity for a parallel beam incident is

$$I(r) = |\exp[-i\sigma\phi(r)] * [c(r) + i\, s(r)]|^2. \tag{39}$$

Extending this to the case of a finite incoherent source in CTEM or to a finite collector aperture (with aperture function D(u)) in STEM, we find that, when we neglect the amount of "scattered" radiation included in the collector aperture for bright-field imaging, the image intensity can be written

$$I_{BF}(r) = D_o + 2 \sin \sigma\phi(r) * t_1(r) - 2[1 - \cos \sigma\phi(r)] * t_2(r) \tag{40}$$

where D_o is the integral over D(u), $t_1(r)$ is given by Eq. (34) and

$$t_2(r) = s(r)[d(r) * s(r)] + c(r)[d(r) * c(r)]. \tag{41}$$

The form of the functions $t_1(r)$ and $t_2(r)$ under various conditions of defocus and aperture size has been calculated by Cowley and Au (1978).

For a very small collector aperture size, Eq. (40) reduces to Eq. (37). If the collector aperture is increased until it is equal to the objective aperture size,

$$t_1(r) = 0,$$

$$t_2(r) = c^2(r) + s^2(r).$$

Then the bright-field signal has the form

$$I_{BF}(r) \simeq D_o\{1 - \sigma^2\phi^2(r) * [c^2(r) + s^2(r)]\}$$

$$= D_o [1 - I_{DF}(r)] \tag{42}$$

where $I_{DF}(r)$ is, in this case, the dark-field signal calculated on the assumption that all scattered radiation, including that contained in the central beam spot, is detected to form the signal.

VIII. THIN, PERIODIC OBJECTS: CRYSTALS

For very thin specimens, there is no difference between the phase contrast imaging of periodic and nonperiodic objects, provided that we limit ourselves to the Scherzer optimum defocus and use an objective aperture to remove the higher-angle parts of the diffraction pattern beyond the flat part of the transfer function. The projected potential, or the sine and cosine of the scaled projected potential, will be imaged according to Eq. (27) or (36). Thus, for thin crystals viewed down one of the unit cell axes, the image is periodic with the periodicity of the unit cell projection and shows the distribution of the atoms within the unit cell in projection, within the limitations of the point-to-point resolution of the microscope. With present-day electron microscopes, direct structure analysis of a crystal is possible by direct visualization of atoms in this way for a wide range of materials for which the heavier atoms are separated in projection by distances of 2 to 3 Å or more (Kihlborg, 1979). Since the imaging does not depend on the periodicity, crystal defects may be imaged with the same clarity as the perfect crystal structure, provided that the defects do not have a three-dimensional structure, which gives an unduly complicated two-dimensional projection (Figure 11). This provides a unique opportunity to study the configurations of atoms in individual imperfections of the structure and is rapidly broadening our understanding of the nature of the defects in many types of materials (Cowley, 1978b).

The imaging of crystals, on the other hand, for the first time allows quantitative correlations of image intensities with known specimen structures. Such a possibility can enormously expand the power and range of applications of electron microscopy. To achieve this, however, it is necessary to refine both the experimental techniques and the interpretive methods.

Crystals must be aligned with an accuracy of a small fraction of a degree in two directions by use of a tilting stage. The crystal thickness must be determined with reasonable precision. The amount of defocus, the aberration constants, and the aperture size of the objective lens must be accurately known.

On the other hand, it is rarely sufficient to use a simple approximation such as Eq. (36) to calculate image intensities. This phase-object approximation has been shown to give significant errors for crystal thicknesses of ~20 Å for 100-keV electrons. Reliable calculations for image interpretation must involve the use of three-dimensional, many-beam dynamical diffraction theory, with either the matrix formulation of Bethe's original dynamical theory of electron diffraction or, more usually, the multislice formulation of Cowley and Moodie's dynamical theory (Cowley, 1981). In the latter, the crystal is subdivided into a number of very thin slices perpendicular to the beam. Each slice acts as a thin phase object. Between slices, the electron wave propagates according to the usual laws of Fresnel diffraction. Standard computer programs for these operations are now available.

Taking into account practical experimental parameters such as incident-beam convergence and chromatic aberration, the extent to which agreement can be obtained between observed image intensities and intensities calculated by these methods is demonstrated in Figure 10. Computer programs can be extended to deal with cases of defects in crystals by using the assumption of periodic continuation. That is, the image of a single defect is assumed to be exactly the same as that of a defect in a periodic array of well-separated defects that forms a superlattice with a large unit cell.

This method for the study of crystal structures and crystal defects has recently been extended to a wide variety of inorganic compounds and minerals. Improved resolution, approaching 2 Å, has been achieved by use of the high-resolution, high-voltage microscopes now in operation (Horiuchi *et al.*, 1976; Uyeda *et al.*, 1978-9).

Relatively little of this type of crystal structure imaging has been done by STEM. Those STEM instruments that include the necessary capabilities for observing diffraction pattern and orientation of specimens usually have relatively poor image resolution because the objective lens polepiece dimensions are made large to accommodate x-ray detectors for EDA. The STEM bright-field image of a $Ti_2Nb_{10}O_{29}$ crystal shown in Figure 12(a) has noticeably poorer resolution (~4.5 Å) than CTEM images taken several years ago (Figure 11(a)).

Figure 12(b) is a dark-field STEM image obtained with an annular detector; here the rows of metal atoms 3.8 Å apart are clearly resolved. Apparently for thin crystals, as for thin, weakly scattering specimens, the dark-field mode can provide a significant advantage in terms of resolution.

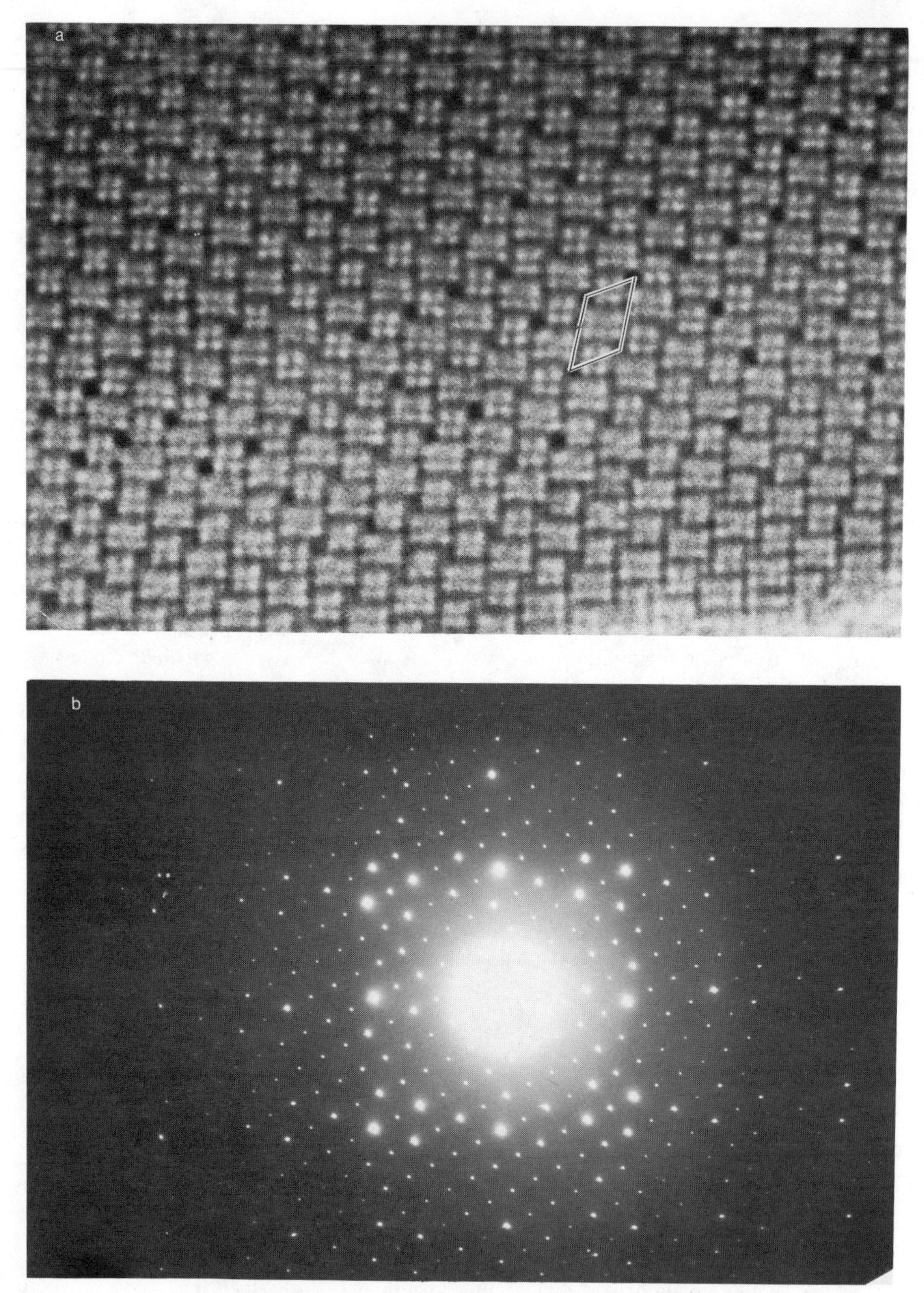

Figure 11 (a) Bright-field image and (b) selected-area diffraction pattern of a crystal of $Nb_{22}O_{54}$. The contrast at some position in the unit cell varies from cell to cell as a result of atomic disorder (courtesy of Sumio Iijima). The unit cell dimensions (indicated) are 21.2 Å and 15.6 Å.

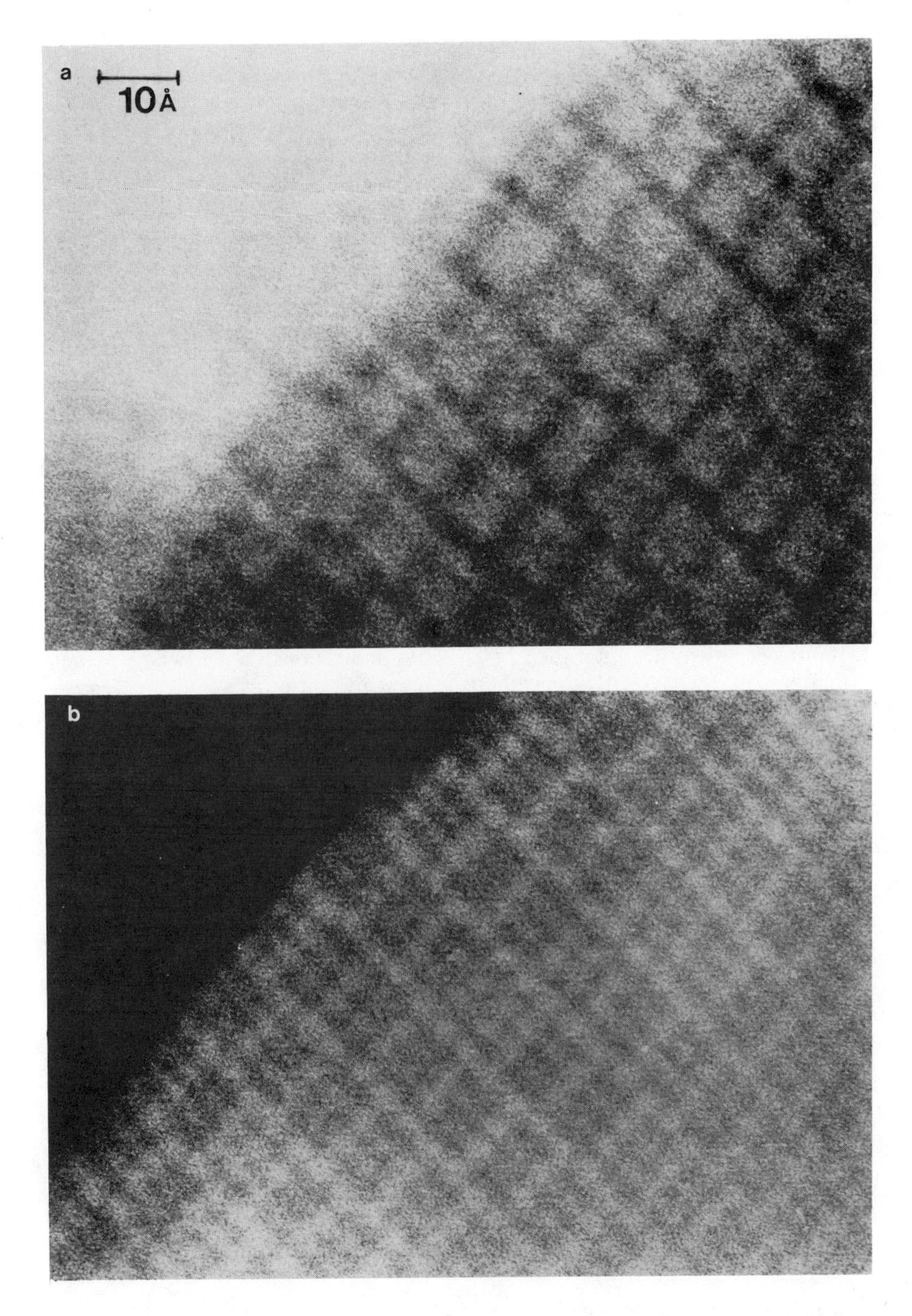

Figure 12 Images of a thin crystal of $Ti_2Nb_{10}O_{29}$ obtained with a STEM instrument. The resolution is seen to be appreciably better in (b) the dark-field image, obtained with an annular detector, than in (a) the bright-field image.

A. Special Imaging Conditions

It is well known that images of crystals can show details on a much finer scale than the point-to-point or Scherzer resolution limits. Fringes with spacings well below 1 Å have been observed. Hashimoto *et al.* (1977) have shown pictures of gold crystals having details on a scale approaching 0.5 Å within the intensity maxima at the positions of the rows of gold atoms. Pictures of silicon by Izui *et al.* (1977) show clearly separated spots near the projected positions of silicon atoms that are 1.36 Å apart.

These pictures are taken without the objective aperture limitation considered previously. They correspond to situations in which the diffraction pattern consists of only a few sharply defined diffraction spots. The requirement for clear imaging of the crystal periodicities is then that these fine diffraction spots should be recombined with maximum amplitude and well-defined relative phases. This is quite different from the requirement for a flat transfer function needed for imaging general nonperiodic objects or periodic objects with large unit cells. To maintain large amplitudes for the relatively high-angle diffraction spots, the effects of beam convergence and chromatic aberration must be minimized, not only by careful control of the instrumental parameters but also by choosing the values of the defocus that make the transfer function less sensitive to these effects.

It is possible to achieve these special imaging conditions only for crystals of simple structure in particular orientations. As Hashimoto *et al.* (1978) have pointed out, the image intensities may then be sensitive indications of the details of potential distributions in the crystals even though they are in no way to be regarded as providing direct pictures of the structure. The special imaging conditions, however, become rapidly more difficult to achieve as the size of the unit cell increases and are not relevant for the imaging of defects of the crystal structure.

On the other hand, an improved appreciation of the special conditions for imaging of periodic structures has led to the realization that some compromises are possible between the extreme situations. For crystals having defects that disrupt the periodicity to a limited extent (relatively small changes of lattice constant), one can image the structure with its defects to see detail that is finer than for a nonperiodic object, although not as fine as for a strictly periodic object. This is demonstrated, for example, by Spence and Kolar (1979) pictures of defects in silicon.

B. Mathematical Formulation

If the crystal is divided into very thin slices perpendicular to the incident beam, the transmission function of the nth slice is

$$q_n(xy) \simeq \exp[-\, i\sigma\phi_n(xy)] \tag{43}$$

where

$$\phi_n(xy) = \int_{z_n}^{z_{n+1}} \phi(R)dz.$$

Propagation through distance $\Delta = z_{n+1} - z_n$ to the $n+1^{th}$ slice is given by convolution with the propagation function (Eq. (12)). Then we have the recursion relationship

$$\psi_{n+1}(r) = [\psi_n(r) * p_\Delta(xy)]q_{n+1}(r) \tag{44}$$

or, in terms of Fourier transform,

$$\Psi_{n+1}(u) = [\Phi_n(u) \times P_{\Delta}(u)] * Q_{n+1}(u)$$

which, for a periodic object, can be written

$$\Psi_{n+1}(h,k) = \Sigma_{h_1,k_1} \Psi_n(h_1 k_1) P_{\Delta}(h_1 k_1) Q_{n+1}(h - h_1, k - k_1). \tag{45}$$

This is an operation readily programmed for a computer, with

$$P_{\Delta}(h,k) \simeq \exp\left[i\pi\Delta\lambda\left|\frac{h^2}{a^2} + \frac{k^2}{b^2}\right|\right] \tag{46}$$

for a unit cell with dimension a,b.

The image intensity is then calculated from the exit wave function or its Fourier transform $\Psi_N(u)$ as

$$I(r) = |\mathcal{F}\{\Psi_N(u) \times A(u) \exp[i\chi(u)]\}|^2. \tag{47}$$

The condition that the image wave amplitude should be identical with the wave at the exit face of a crystal is that $\exp[i\chi(u)] = 1$ for all diffracted beams $u = h/a$, $v = k/b$; i.e., that

$$\pi\Delta f\lambda\left|\frac{h^2}{a^2} + \frac{k^2}{b^2}\right| + \frac{1}{2}\pi C_s\lambda^3\left|\frac{h^2}{a^2} + \frac{k^2}{b^2}\right|^2 = N\pi. \tag{48}$$

The case for N odd is included because it represents the case of an identical wave function shifted by half the periodicity.

It is not possible to satisfy this condition unless a^2 and b^2 are in the ratio of integers. For the special case that $a = b$, Eq. (48) is satisfied for $\Delta f = na^2/\lambda$, and $C_s = 2ma^4/\lambda^3$ for integral n,m (Kuwabara, 1978). The extent to which these conditions may be relaxed is a measure of the limitations on the degree of crystal perfection or on the finite crystal dimensions for which the image may represent the ideal in-focus image of the crystal (Cowley, 1978c).

IX. THICKER CRYSTALS

When the crystal thickness exceeds 50 to 100 Å for 100-keV electrons or 200 to 500 Å for 1-MeV electrons, there is in general no relationship visible between the image intensities and the projected atom configurations (although for some particular cases the same thin-crystal image is repeated for greater thicknesses). It is suggested that the image intensities will be increasingly sensitive to details of crystal structure, such as the bond lengths, ionization or bonding of atoms, and thermal vibration parameters. However, since the images are also more sensitive to experimental parameters (such as the crystal alignment and the lens aberrations), the refinement of crystal structures by use of high-resolution, thick crystal images remains an interesting but unexploited possibility.

Most of the important work done on thicker crystals has involved study of the extended defects of crystals having relatively simple structures with medium resolution (5 Å or more) and no resolution of crystal structure periodicities. The extensive studies done on the form and behavior of dislocations and stacking faults in metals, semiconductors, and an increasingly wide range of inorganic materials—together with the theoretical basis for image interpretation in these cases—well described in such books as Hirsch *et al.* (1965) and Thomas and Goringe (1979).

Because dynamical diffraction effects are of overwhelming importance for these studies, it is essential to simplify the diffraction conditions as much as possible to make it relatively easy to calculate and to appreciate the nature of the diffraction contrast. For thin crystals with large unit cells, viewed in principal orientations, it may be necessary to take hundreds or even thousands of interacting diffracted beams into account (Figure 10(a)). For thicker crystals of relatively simple structure, it is often possible to choose orientations for which the two-beam approximation is reasonably good; namely, when the incident beam gives rise to only one diffracted beam of appreciable amplitude and these two beams interact coherently. The image can vary with crystal thickness because interference between the two beams gives both of them a sinusoidal variation with thickness. Crystal defects show up with strong contrast because changes in the relative phase of the two beams result in different interference effects and thus different intensities. The phase changes can be sudden, as when the crystal suffers a shear displacement at a stacking fault; or the phase change can be more gradual, as when the lattice is strained around a dislocation line or other defect and the deviation from the Bragg angle varies as the incident and diffracted beams travel through the strain field.

The standard dynamical diffraction theory, as originated by Bethe and developed by many other authors (Hirsch *et al.*, 1965), is the theory of the interaction of electron waves with a perfectly periodic potential distribution bounded by plane faces. To adapt this to the study of crystal defects, it is usual to make a simplifying "column approximation." For very thin crystals, we have made the assumption that the electron wave at a point on the exit face of the crystal is influenced only by the potential along a line through the crystal to that point in the beam direction. For thicker crystals, we may assume that the electron wave at a point on the exit face of the crystal is affected by the diffraction, not along a line, but within a thin column extending through the crystal in the beam direction.

The width of this column may be estimated in various ways, all of which agree that it may be surprisingly narrow. For 100-keV electrons, for example, and crystals several hundred angstroms thick, the column width may be taken as small as 5 to 10 Å, with errors that are not serious for most purposes. For a microscope resolution limit of 10 Å or more, the column approximation serves very well.

To calculate the image of a defect, it is necessary to calculate the amplitudes of the incident and diffracted beams for each column of crystal passing near the defect. The amplitudes for a column containing a particular sequence of lattice strains and disruptions can be calculated on the assumption that all surrounding columns are identical, i.e., that the crystal is perfectly periodic in directions perpendicular to the beam direction. The calculations are usually made by the difference equation method of Howie and Whelan (Hirsch *et al.*, 1965), in which the progressive changes of the wave amplitudes are followed as the waves progress through the crystal. Convenient computer techniques developed by Head *et al.* (1973) have provided systematic methods for identifying defects of many types. In many cases, especially for thick crystals, it is necessary to introduce the effects of inelastic scattering (mostly thermal diffuse scattering) on the elastically scattered waves. This is done usually by the simple expedient of adding a small imaginary part to make the structure amplitudes of the crystal complex. It leads to a variety of easily observable effects.

Refinements of these dynamical diffraction studies of crystal defects include the "weak beam method," which gives much finer details of defect structure at the expense of very low image intensities (Cockayne, Ray, and Whelan, 1969). This method relies on the fact that the weak beam intensities vary much more rapidly with crystal thickness or with change of incident-beam orientation than do the strong beams in a situation where both strong and weak diffracted beams, or only weak diffracted beams, are present. Thus, the details of dislocation structure in an image formed by allowing only a weak beam through the objective aperture may be on the scale of 10 to 15 Å, whereas the intensity variations for dark-field images formed with strong reflections may be stronger but show no detail finer than ~50 Å.

For the interpretation of detail on a very fine scale, the column approximation may not be sufficient. Anstis and Cockayne (1979) have reviewed the more exact treatments that avoid this approximation.

Because the image intensity modulations for crystal defects are strongly dependent on the angle of incidence of the electron beam in relation to the Bragg angle for the operative reflections, the visibility of defects may be reduced and the characteristic features of their images may be lost if the range of angles of incidence is too large in CTEM or if the collector aperture size is too large in STEM. For CTEM this usually does not constitute a serious restriction, but for STEM, especially if a high-brightness source is not used, it may produce an undesirable low intensity of the useful images (for example, Maher and Joy, 1976). With STEM some compromise is commonly necessary between contrast of the defect image and noisiness of the image.

As a rough estimate, it may be assumed that the same dynamical diffraction effects will be observed in STEM as in parallel-beam CTEM if the collection angle β for STEM is such that

$$\Delta w = \beta |g| \xi_g \lesssim 0.3$$

where w is a dimensionless parameter measuring the deviation from the Bragg condition for a reflection in terms of the reciprocal lattice vector g and the extinction length for the reflection. For example, for the (200) reflection from a Cu crystal for 100-keV electrons, the collector angle must be $<3 \times 10^{-4}$ rad.

The effect of the STEM collection angle varies for different diffraction effects. The positions of extinction contours can vary rapidly with CTEM incident-beam angles. Hence, the contrast of extinction contours is rapidly reduced for increasing collection angle β.

Thickness fringes for wedge-shaped crystal regions and stacking fault fringes vary in spacing and contrast with deviation from the Bragg angle. Since the fringe positions are fixed for small crystal thickness, however, the main effect of increasing β is to reduce the contrast of the fringes for the larger thicknesses. Dislocation contrast can be considered as made up of two components: (a) the black line component, which is due to local large-crystal distortion and is not greatly dependent on β; and (b) the thickness-dependent detail, which varies with β more rapidly, in much the same way as stacking fault contrast.

To some extent these various dependencies on β can be useful. For example, dislocation lines can sometimes be seen more readily if extinction contours are removed by choosing the collector size appropriately (Reimer and Hagemann, 1976).

A. Lattice Fringes

Within the limits set by incident-beam divergence and the mechanical and electrical stabilities of the electron microscope, it is possible to produce interference fringes in the image with the periodicity of the diffracting lattice planes for any crystal thickness, provided that the objective aperture allows two or more diffracted beams to contribute to the image. Under usual operating conditions, with no careful control of the experimental parameters of the electron microscope or of the specimen material, the information content of such images is very limited. The position of the dark or light fringes relative to the atomic planes is indeterminate since this is strongly dependent on the crystal orientation and thickness, the objective lens defocus, and the centering of the incident or diffracted beams with respect to the objective lens axis. The spacing of the fringes is usually close to that of the relevant lattice planes but may vary appreciably if the crystal varies in thickness or is bent or, in particular, if other strong reflections are excited locally.

If care is taken to avoid complications from all of these factors, however, some very useful data may be obtained by observations of lattice fringes. Variations of lattice plane spacings corresponding to variations in the composition or degree of ordering in alloys have been observed (Wu, Sinclair, and Thomas, 1978). Also, the presence of defects may be detected even though the perturbations of the fringe spacings or contrast can usually give no direct evidence on the defect structure.

Lattice fringes may, of course, be observed in STEM, as in CTEM. The accessibility of the convergent-beam diffraction pattern in a STEM instrument allows a clearer picture to be obtained of the conditions under which lattice fringes are formed (Spence and Cowley, 1978). If the disc-shaped diffraction spots corresponding to the individual reflections do not overlap, no interference is possible between incident-beam directions with sufficient angular separation to produce interference effects with the lattice plane periodicity. The region where diffraction-spot discs overlap is the region where such interference effects can take place, so that if the detector aperture includes the region of overlap, the image can show the lattice plane spacing.

It is not difficult to specify the detector aperture size and shape that will give maximum lattice fringe visibility and image intensity for any particular conditions of beam incidence, objective aperture size, and lens aberration.

B. Mathematical Considerations

To calculate the wave function at the exit face of a crystal, using the column approximation, the multislice formulation of Cowley and Moodie, previously described, may be used when there is a distortion of the crystal that is a continuous function of distance in the beam direction. The solution of the wave equation in each section of the crystal that can be considered as periodic is rarely feasible by following the Bethe formulation. Most commonly, the difference equation form of Howie and Whelan is used. The changes in incident and diffracted wave complex amplitudes caused by diffraction from other waves in the crystal, absorption effects, and excitation errors, can be written

$$\frac{d}{dz}\Phi = 2i(A+\beta)\Phi \tag{49}$$

where Φ is the column vector whose elements Ψ_h are the amplitudes of the diffracted waves, β is a diagonal matrix whose elements are $\beta_h = d[h \cdot R(z)]/dz$, and R(z) is the vector giving the displacements of the lattice points. The matrix A has diagonal and off-diagonal elements

$$A_{hh} = \zeta_h + i\sigma\Phi'_o/4\pi$$

$$A_{hg} = \sigma(\Phi_{h-g} + i\Phi'_{h-g})/4\pi \quad (50)$$

where ζ_h is the excitation error for the h reflection, and Φ'_h is the imaginary part added to the structure amplitude to represent the effect of absorption. For the two-beam case, this simplifies to a simple pair of coupled equations that may be integrated through successive slices of crystal.

For other than very thin crystals, it is not possible to represent by multiplication, with a scalar transmission function, the effect of the crystal on the incident wave function, as was assumed previously. Instead, we may consider the action of a crystal to be represented by the action of a scattering matrix on an incident wave vector, $\Phi_o(h)$, representing the Fourier coefficients of the incident wave. Thus,

$$\Psi = S\Psi_o \quad (51)$$

and the matrix $S = \exp[izM(h)/2k]$

where M(h) is similar to the matrix A of Eq. (49). This equation can also be iterated through successive slices. For n similar slices, $\Psi_n = S^n\Psi_o(h)$. This formulation follows the concepts developed by Sturkey (1962) and others and forms the basis for a number of sophisticated and powerful treatments of diffraction and imaging problems.

X. VERY THICK SPECIMENS

As the thickness of a specimen increases, the distribution of intensity in the diffraction pattern is dominated increasingly by multiple scattering effects. Diffracted beams traversing further regions of the specimen may be diffracted again and again, both elastically and inelastically. For crystals, the spot patterns given by thin crystals are gradually submerged under the diffuse background scattering. Diffraction of the diffusely scattered electrons by the crystal lattice gives complicated Kikuchi line configurations, which in turn gradually lose contrast and are lost in an overall broad background of scattering. For noncrystalline specimens, the initial featureless scattering distribution is successively broadened by multiple elastic scattering. Also, because the electrons lose energy through successive inelastic scatterings, the distribution of electron energies becomes broader, and the number of electrons that have not lost any energy becomes very small.

For CTEM the image resolution becomes poorer as a result of the increasing angular spread of the electrons because the position at which scattering occurs becomes less and less well defined as the electron beam spreads in the specimen. Further, as the energy spread of the transmitted electrons increases, the resolution suffers as a result of the chromatic aberration of the lenses. To improve the contrast, the objective aperture size is usually made small. This has the effect, however, of drastically reducing the image intensity because, as the angular range of the scattering is increased by multiple elastic scattering, the fraction of the transmitted radiation remaining near the central spot is rapidly reduced (Figure 13).

For STEM there is the same loss of resolution resulting from the angular spreading of the beam in the specimen. However, loss of electron energy by multiple inelastic scattering in the specimen does not affect the resolution since there are no imaging lenses after the specimen. Hence, the effect of specimen thickness on the resolution will be less severe for STEM than for CTEM. For 100-keV electrons and specimen thicknesses on the order of a few micrometers, estimates suggest that STEM will have an advantage over CTEM by a factor of ~ 3 (Sellar and Cowley, 1973). For special instrumental configurations, however, this factor has been estimated to be as high as 10 (Groves, 1975). With increasing accelerating voltage, this factor will decrease, being ~ 2 for 1-MeV electrons.

For STEM it has been shown that the best contrast is obtained for very thick specimens by using a very large detector aperture ($\sim 10^{-1}$ rad or more) (Figure 13). This has the added advantage that the image signal intensity can be as high as half the incident-beam intensity, so that there is no problem of decreasing image intensity (Smith and Cowley, 1975). For STEM it is also possible to get dark-field image contrast by using an energy filter to separate electrons that have lost less energy than the average from those that have lost more. The desirable energy cutoff for maximum contrast obtained in this way may be as high as several hundred volts energy loss for thick specimens (Pearce-Percy and Cowley, 1976; Rose and Fertig, 1976). Again, STEM appears to have a potential advantage in that its inherently flexible detection system, with associated signal processing possibilities, allows the optimum imaging conditions to be achieved in a relatively straightforward manner.

These conclusions apply for thick, amorphous specimens, such as thick sections of biological materials. In regard to detecting defects in crystalline materials, considerations are complicated by the requirements for diffraction contrast (Frazer, Jones, and Loretto, 1977). Conclusions derived from experimental observations suggest that, while STEM may have some advantages for light-atom crystals, these advantages disappear for medium-weight and heavy-atom materials.

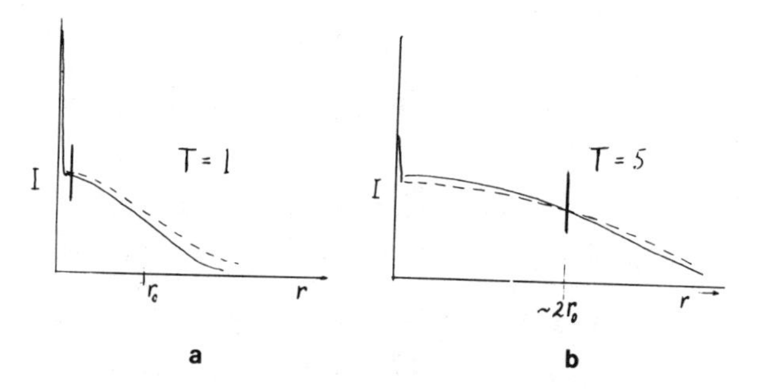

Figure 13 Diagrams suggesting the change in the angular distribution of scattered electrons and the reduction of the zero peak of unscattered electrons for amorphous specimens of thickness T equal to (a) the mean free path for elastic scattering and (b) five times this thickness. The optimum objective aperture size to give the greatest contrast in a STEM image for a small increase in thickness is indicated in each case, relative to r_o, the mean scattering angle for single scattering.

Mathematical Descriptions

In the incoherent scattering approximation, it is assumed that the intensity distribution from the first slice of an amorphous sample is spread further by scattering in a second slice, and so on. For a single slice of thickness Δz, the intensity distribution may be written

$$I_1(u) = \exp(-\mu\Delta z)\,[\delta(u) + \Delta z \times f^2(u)]$$

where $f^2(u)$ is the square of the scattering amplitude, and the absorption coefficient μ is given by $\int f^2(u)du$. Fourier transforming this in terms of arbitrary variable w,

$$G_1(w) = (-\mu\Delta z)\,[1 + \Delta z P(w)]\,.$$

The effect of transmission through n layers to give a total thickness $T = n\Delta z$ is to convolute $I_1(u)$ by itself n times or, correspondingly, to raise $G_1(w)$ to the nth power

$$G_1^n(w) = \exp(-\mu T)\,[1 + \Delta z P(w)]^n \rightarrow \exp[-\mu T + TP(w)] \qquad (52)$$

so that the angular distribution of scattered intensity becomes

$$I_T(u) = \mathcal{F}^{-1}\,(\exp\{T[P(w) - \mu]\})\,. \qquad (53)$$

The optimum detector aperture size for a STEM instrument with a very thick specimen is found by determing the value of $|u|$, for which the differential of the intensity $I_T(u)$ with respect to thickness T changes sign.

The distribution of the number of electrons with energy loss is found in the same way, in terms of the mean free path for inelastic scattering; the optimum energy cutoff for an energy filter is found by determining the energy-loss value for which the differential of this number with respect to T changes sign.

XI. CONCLUSIONS

One conclusion to be drawn from our discussions of the high-resolution imaging of thin specimens is that no simple definition of resolution is possible unless the concept of resolution is severely restricted. One can ask, for example: What is the smallest distance between two distinct maxima or minima of intensity in an image? This provides an operational definition of resolution of one kind, useful as a convenient criterion for instrument designers. It makes no reference, however, to the main function of an electron microscope, which is to provide information regarding the structure of the specimen. In practice, this type of definition requires qualification in that, as is well known, point-to-point imaging is different from lattice fringe imaging. In the latter, strong diffracted beams occur at large diffraction angles; therefore, some contrast may be seen, provided that the transfer function of the lens is not zero. For nonperiodic objects, the diffracted wave amplitudes fall off in a uniform manner; consequently, outer nonzero parts of the transfer function will multiply weak scattering amplitudes. The resultant contributions of the outer parts of the diffraction pattern to the image intensity will be so small that the corresponding fine detail of the image will be of too little contrast to be detected. Thus, to provide a reliable resolution criterion of this sort, it is necessary to specify the nature of the ordering and the degree of crystallinity in the specimen. Usually, however, no independent evidence on these questions, other than from electron microscopy, is available.

It is, of course, possible to consider only the extreme case of near-perfect periodicity and take the minimum observable lattice fringe spacing as a measure of the "resolution limit." This is, in fact, a good test of some instrumental parameters, such as the mechanical stability of the column, the specimen drift, interference from stray electrical or magnetic fields, and incident-beam convergence. It is not a sensitive measure of chromatic aberration and is insensitive to spherical aberration.

The use of a test object of completely random structure would provide a different basis for testing. But most "amorphous" materials, including amorphous carbon films, are known to contain small regions that are relatively well ordered to the extent that the diffraction patterns given by individual picture elements (of diameter comparable with the resolution limit) may contain strong maxima at high angles. Therefore, the assumption of a scattered amplitude falling off fairly uniformly with scattering angles is invalid.

From a different point of view, we may choose to measure resolution in terms of how well the electron microscope provides a recognizable image of the known structure of a test object. For very thin, weakly scattering objects, the usual Scherzer criterion applies. The resolution is assumed to be given by the reciprocal of the value of $u = 2\lambda^{-1} \sin(\Phi/2)$ at the outside limit of the flat part of transfer function $\sin[\chi(u)]$ for the optimum defocus. The objective aperture is chosen to eliminate all the radiation scattered at higher angles, for which the transfer function is oscillatory. As we have seen, this criterion for measuring resolution cannot be used for strongly scattering and thicker samples. Nor can it be used in general for dark-field imaging since, for detail near the resolution limit, the intensity distribution of dark-field images often has no direct relationship to the atomic arrangement in the specimen.

The situation is more favorable when the images of weakly scattering objects can be reasonably well interpreted by incoherent imaging theory. Such is the case of STEM with an annular dark-field detector, STEM bright-field imaging with a large detector aperture (Eq. (42)), or bright-field CTEM with a large angle of illumination (Nagata *et al.*, 1976). It is a well-known result of light optics (Born and Wolf, 1975) that the resolution can be better for incoherent illumination than for coherent illumination by a factor of ~ 1.5 (actually, $2^{1/2}$ for a Gaussian spread function, as is evident from Eq. (35)).

A. Measurement of Imaging Parameters

Probably the best way to characterize the performance of an electron microscope is to determine the transfer function for the objective lens for a thin phase object. This is consistent, although not identical, with the current practice in light optics. If the transfer function can be determined experimentally, the resolution and contrast of images produced by the instrument for any specimen can be evaluated according to any of the criteria that seem useful or by detailed calculation.

The most convenient method for achieving this is by using an optical diffractometer, which provides the Fourier transform (or, more accurately, the squared amplitude of the Fourier transform) of the image intensity distribution. Provided that the specimen is a thin, weakly scattering object, the optical diffraction pattern intensity will give, to a good approximation, the square of the transfer function $\sin[\chi(u)]$, multiplied by the intensity distribution in the diffraction pattern of the object. Usually the specimen used for this purpose is a thin, "amorphous" carbon film. Caution is necessary in practice to ensure that none of the restrictions we have mentioned on the use of this method are violated.

The difficulty with analysis by use of the optical diffractometer is, of course, that it is not possible to determine the astigmatism and defocus until well after this image has been recorded. The on-line analysis of image quality, using previously mentioned TV-rate recording and rapid, computerized Fourier transforms, overcomes this difficulty, allowing the image parameters to be determined before the recording is made. This, together with greatly improved methods for digital image simulation and image processing (Skarnulis *et al.*, 1981), represents a major advance toward improved quantification of specimen analysis by means of electron microscopy. The possibility of deriving accurate numerical data from images is probably a more desirable development at this time than the further improvement of resolution.

As instrumental resolution is improved, it becomes increasingly less likely that a straightforward, intuitive interpretation of the image contrast can be reliable. The derivation of reliable information from the image becomes increasingly dependent on computerized image simulation based on accurate values of the imaging parameters—those parameters that describe the objective lens performance as well as those that describe the specimen thickness and orientation and its physical and chemical environment. At present, image interpretation techniques seem to be lagging behind the improved resolution capabilities of recently produced electron microscopes.

B. CTEM and STEM

Finally, we summarize our conclusions as to the relative merits of the STEM and CTEM approaches as they have been established to date in theory and experiment.

1. Bright-Field Imaging

The CTEM provides better looking images of larger useful areas, and currently the resolution is better than for STEM, especially for crystalline samples manipulated in an adequate tilting stage. STEM images tend to be noisy, and the collection efficiency (depending on the ratio of the number of electrons detected to the number incident on the specimen) is poor. Possibilities exist for the use of multiple bright-field detectors to improve the resolution and contrast or to provide accurate digital intensity measurements, but such uses are still in the exploratory stages.

2. Dark-Field Imaging

Relative to the CTEM high-resolution modes, dark-field STEM with an annular detector has been shown to be more efficient in its use of the incident electrons, and the image intensity distributions are more readily interpreted. The attainable, interpretable resolution for very thin specimens is better than for bright-field images. The use of multiple detectors to detect separately the various parts of the diffraction pattern and the possibilities for on-line manipulation and combination of the various image signals provide many new possible imaging modes. Some of these are now being usefully exploited.

3. Thicker Specimens

For the images of defects in crystals, STEM offers a greater variety of imaging possibilities than does CTEM. In general, however, STEM is less useful in practice because of limitations of single strength. For very thick biological or other amorphous specimens, STEM has well-established advantages in image resolution and contrast. For very thick crystals, these advantages are less obvious.

4. Analytical Electron Microscopy

Because it is essential for microdiffraction and for microanalysis using ELS or EDS to illuminate the specimen with a focused beam of small diameter, STEM imaging is obviously more compatible with these AEM techniques. Practical difficulties in correlating microbeam data with CTEM images increase as the spatial resolution of the analytical methods is improved.

5. Quantitative Data Collection

Because of recent advances in electronics, CTEM and STEM capabilities for image intensity measurements and for image evaluation and processing are comparable, with the exception that STEM provides important advantages in allowing more direct slow-scan or fast-scan collection of data and the collection and processing of simultaneous multiple images.

TABLE OF CHAPTER VARIABLES

a, b, c	Periodicities of crystal in real space
A(uv)	Lens aperture function
c(xy), s(xy)	Real and imaginary parts of t(xy)
C_c	Chromatic aberration constant
C_s	Third-order spherical aberration contant
D(u)	Detector aperture function for STEM
E	Accelerating voltage for electron beam
F_h	Structure amplitude or coefficient of Fourier series representing $\phi(r)$
F(u)	Fourier transform operation; $F(u) = \mathcal{F}f(r)$, where $\mathcal{F}f(r) \equiv \int f(r) \exp\{2\pi i u \cdot r\} dr$
g(u)	Scattering amplitude as function of u
h, g	Reciprocal lattice vectors
hkl	Reciprocal lattice indices; Miller indices
I(r), I(x,y)	Intensity distribution in an image
$J_{DF}(R)$	Signal strength for a dark field (DF) image
k	Wave vector, magnitude $\lvert k \rvert = 2\pi/\lambda$
m, n, N	Integers
p(xy)	Fresnel propagation function in real space; $p(xy) \approx \exp\{-i\, k\, (x^2 + y^2)/2R\}$
P(uv)	Fourier transform of p(xy)
q(xy)	Phase object approximation transmission function; $q(xy) = \exp\{-i\, \sigma\, \phi\, (xy)\}$ where $\sigma = \pi/\lambda E$ and $\phi(xy) = \int \phi\, (xyz)\, dz$ = projected potential
Q(uv)	Fourier transform of q(xy)
r, R	Vectors in real space
r_i, r_j	Vectors to points i, j in real space
S(uv)	Source function, giving distribution of incident-beam amplitude for a finite source

S	Scattering matrix
t(xy)	Spread function for amplitudes in coherent image formation; $t(xy) = \mathcal{F}\, T(uv) = c(xy) + i\, s(xy)$
T(uv)	Lens transfer function; $T(uv) = A(uv) \exp\{i\, \chi(uv)\}$ where $\chi(uv) = \pi\, \Delta f\, \lambda(u^2 + v^2) + \frac{1}{2} \pi\, C_s\, \lambda^3\, (u^2 + v^2)^2$ and Δf = positive for overfocus (stronger lens)
u, U	Reciprocal space vector, magnitude $\lvert u \rvert = 2 \sin(\phi/2)/\lambda$
uvw	Coordinates, reciprocal space
w	Dimensionless parameter indicating deviation from Bragg condition
xyz	Coordinates, real space
α	Objective aperture angle
β	Collection angle for STEM
Δf	Defocus value for lens: positive for overfocus
ζ_g	Excitation error for the g reflection
θ_B	Bragg angle; $\theta_B = \phi/2$
λ	Wavelength
μ	Absorption coefficient
ξ_g	Extinction length for the g reflection
σ	Interaction constant for interaction of electrons with matter
ϕ	Scattering angle, components ϕ_x, ϕ_y
$\phi(r)$	Potential distribution in real space
$\phi(xy)$	Projection of $\phi(r)$ in z direction
$\Phi(u)$	Fourier transform of $\phi(r)$
Φ_h'	Imaginary part added to Φ_h to represent the effect of absorption
$\chi(uv)$	Phase factor in lens transfer function
$\psi(r)$	Wave function, real space
$\Psi(u)$	Fourier transform of $\psi(r)$

Symbols and Abbreviations

*	Convolution operation
1 Å	10^{-1} nm
AEM	Analytical electron microscopy
CBED	Convergent-beam electron diffraction
CTEM	Conventional TEM (fixed beam) as distinct from STEM
EDS	Energy-dispersive spectroscopy (of x-rays)
EELS	Electron energy-loss spectroscopy
keV	Kilo electron volts
STEM	Scanning transmission electron microscopy
TEM	Transmission electron microscopy

Note: Sign conventions are crystallographic conventions, consistent with plane wave

$$\psi(r, t) = \exp\, [i\{\omega t - k \cdot r\}]$$

as used in International Tables for Crystallography.

RECOMMENDED READING AND REFERENCE MATERIAL

Listed alphabetically are classical and general references either referred to in this chapter or recommended by the author, with some comments on each.

Anstis, G. R., and Cockayne, D. J. H. (1979), Acta Cryst. *A35*, 511.

An excellent analysis of theoretical approaches to the description of dynamical diffraction by crystals with defects.

Boersch, H. (1947), Z. Naturforsch. *2a*, 615.

Early work of this often-neglected scientist who contributed many important ideas to the subject of electron microscopy; includes discussion of the possibility of observing individual atoms.

Born, M., and Wolf, E. (1975), "Principles of Optics," 5th ed., Pergamon Press, London.

The standard reference work for many years on the contemporary approach to optics.

Bowen, D. K., and Hall, C. R. (1975), "Microscopy of Materials," John Wiley & Sons, New York.

A more modern treatment than Hirsch *et al.* (1965) but on a more introductory level, keeping as closely as possible to a nonmathematical, descriptive style; a little shaky on some points of fundamental physics.

Cowley, J. M. (1981), "Diffraction Physics," 2nd ed., North-Holland, Amsterdam.

A rather formidable book for the nontheorist. It attempts to correlate x-ray and electron diffraction with electron microscopy by use of a common theoretical basis. Needs to be read in conjunction with more detailed accounts, or prior knowledge, of the experimental situations.

Doyle, P. A., and Turner, P. S. (1968), Acta Cryst. *A24*, 390.

The earliest and, for many purposes, definitive discussion of the application of the reciprocity principle in electron diffraction.

Eyring, L. (ed.) (1982), Proc. ACS Symp. on High Resolution Electron Microscopy Applied to Chemical Problems. Ultramicroscopy.

An important compilation of papers giving an overview of the subject, with particular emphasis on solid-state chemistry.

Head, A. K., Humble, P., Clarebrough, L. M., Morton, A. J., and Forward, C. J. (1973), "Computer Electron Micrographs and Defect Identification," North-Holland, Amsterdam.

A clear, systematic account of the methods and typical results for calculating images of dislocations, stacking faults, etc., with computer programs.

Hirsch, P. B., Howie, A., Nicholson, R. B., Pashley, D. W., and Whelan, N. J. (1965), "Electron Microscopy of Thin Crystals," Butterworth's, London.

This is the "Yellow Bible" of materials-science electron microscopists. Produced as the result of a summer school, it contains an excellent account of the contrast effects for crystal defects in crystals. A second edition published in 1977 has a chapter added to summarize progress since 1965.

Kihlborg, L. (ed.) (1978-79), Chemica Scripta, vol. 14, Proc. of 1979 Nobel Symp. on Direct Imaging of Atoms in Crystals and Molecules.

A compilation of papers by leaders in the field; a valuable survey of the state of the art.

Lipson, S. G., and Lipson, H. (1981), "Optical Physics," 2nd ed., Cambridge Univ. Press, London.

A book biased by the fact that both authors are physicists and one is an outstanding crystallographer who contributed greatly to the use of optical diffraction analogues for x-ray diffraction processes.

Saxton, W. O. (1978), "Computer Techniques for Image Processing in Electron Microscopy," Acad. Press, New York.

A definitive but difficult book on the basic concepts and methods of image processing; includes a detailed description of the contributions of the author and his colleagues to this subject, with computer programs for the main operations.

Scherzer, O. (1949), J. Appl. Phys. *20*, 20.
The clear, original statement on how to produce the optimum phase contrast imaging for weakly scattering objects.

Spence, J. C. H. (1981), "Experimental High Resolution Electron Microscopy," Oxford Univ. Press, England.
An excellent introduction to the concepts, theoretical basis, and practical methods of modern high-resolution electron microscopy.

Thomas, G., and Goringe, M. J. (1979), "Transmission Electron Microscopy of Materials," John Wiley & Sons, New York.
A more recent introduction to electron microscopy for materials scientists, well presented and well illustrated.

Listed alphabetically are other references either referred to in this chapter or recommended by the author.

Agar, A. W., Alderson, R. H., and Chescoe, D. (1974), "Principles and Practice of Electron Microscope Operation," North-Holland, Amsterdam.

Anstis, G. R., and O'Keefe, M. A. (1976), Proc. 34th EMSA Mtg., Claitor's Pub. Div., Baton Rouge, p. 480.

Burge, R. E., and Van Toorn, P. (1980), in "Scanning Electron Microscopy/1980" (O. Johari, ed.), SEM Inc., Chicago, vol. I, p. 81.

Chan, I. Y. T., Cowley, J M., and Carpenter, R. W. (1981), in "Analytical Electron Microscopy 1981" (R. H. Geiss, ed.), San Francisco Press, San Francisco, p. 107.

Cockayne, D. J. H., Ray, I. L. F., and Whelan, J. M. (1969), Phil. Mag. *20*, 1265.

Cowley, J. M. (1969), Appl. Phys. Lett. *15*, 58.

——— (1976), Ultramicroscopy *2*, 3.

——— (1977), in "High Voltage Electron Microscopy 1977" (T. Imura and H. Hashimoto, ed.), Jap. Soc. Electron Microsc., Tokyo, p. 9.

——— (1978a), "Advances in Electronics and Electron Physics" (L. Matron, ed.), Acad. Press, New York, vol. 46, pp. 1-53.

——— (1978b), "Annual Reviews of Physical Chemistry" (B. S. Rabinovich, ed.), Annual Review, Inc., Palo Alto, vol. 29, pp. 251-83.

——— (1978c), in "Electron Microscopy 1978" (J. M. Sturgess, ed.), Microscopical Soc. of Canada, Toronto, vol. III, p. 207.

——— (1980), in "Scanning Electron Microscopy/1980" (O. Johari, ed.), SEM Inc., Chicago, p. 61.

Cowley, J. M., and Au, A. Y. (1978), in "Scanning Electron Microscopy/1978" (O. Johari, ed.), SEM Inc., AMF O'Hare, Illinois, vol. I, p. 53.

Cowley, J. M., and Bridges, R. E. (1979), Ultramicroscopy *4*, 419.

Cowley, J. M., and Iijima, S. (1977), Physics Today *30* (3):32.

Cowley, J. M., and Jap, B. K. (1976), in "Scanning Electron Microscopy/1976" (O. Johari, ed.), IITRI, Chicago, vol. I, p. 377.

Crewe, A. V., and Wall, J. (1970), J. Mol. Biol. *48*, 375.

Doyle, P. A., and Turner, P. S. (1968), Acta Cryst. *A24*, 390.

Frank, J. (1973), Optik *38*, 519.

Frazer, H. L., Jones, I. P., and Loretto, M. H. (1977), Phil. Mag. *35*, 159.

Groves, T. (1975), Ultramicroscopy *1*, 15.

Hashimoto, H., Endoh, H., Tanji, T., Ono, A., Watanabe, E. (1977), J. Phys. Soc. Japan *42*, 1073.

Hashimoto, H., Kumao, A., and Endoh, H. (1978), in "Electron Microscopy 1978" (J. M. Sturgess, ed.), Microscopical Soc. of Canada, Toronto, vol. III, p. 244.

Horiuchi, S., Matsui, Y., and Bando, Y. (1976), Jap. J. Appl. Phys. *15*, 2483.

Isaacson, M. S., Langmore, J., Parker, W. W., Kopf, D., and Utlaut, M. (1976), Ultramicroscopy *1*, 359.

Izui, K., Furono, S., Otsu, H. (1977), J. Electron Micros. *26*, 129.

Kuwabara, S. (1978), J. Electron Micros. *27*, 161.

Lamvik, M. K., and Langmore, J. P. (1977), in "Scanning Electron Microscopy/1977" (O. Johari, ed.), IITRI, Chicago, vol. 1., p. 401.

Langmore, J. P., Wall, J., and Isaacson, M. (1973), Optik *38*, 335.

Maher, D. M., and Joy, D. C. (1976), Ultramicroscopy *1*, 239.

Nagata, F., Matsuda, T., Komoda, T., and Hama, K. (1976), J. Electron Micros. *25*, 237.

Nussbaum, A., and Phillips, R. A. (1976), "Contemporary Optics for Scientists and Engineers," Prentice Hall, Englewood Cliffs, New Jersey.

Pearce-Percy, H. T., and Cowley, J. M. (1976), Optik *44*, 273.

Reimer, L., and Hagemann, P. (1976), in "Scanning Electron Microscopy/1976" (O. Johari, ed.), IITRI, Chicago, vol. II, p. 321.

Rose, H., and Fertig, J. (1976), Ultramicroscopy *2*, 77.

Sellar, J. R. (1977), in "High Voltage Electron Microscopy, 1977" (T. Imura and H. Hashimoto, ed.), Jap. Soc. Electron Microsc., Tokyo, p. 199.

Sellar, J. R., and Cowley, J. M. (1973), in "Scanning Electron Microscopy/1973" (O. Johari, ed.), IITRI, Chicago, p. 143.

Skarnulis, A. J., Wild, D. L., Anstis, G. R., Humphreys, C. J., and Spence, J. C. H. (1981), in "Electron Microscopy and Analysis 1981," Inst. of Physics, Bristol, p. 347.

Smith, D. J., and Cowley, J. M. (1975), Ultramicroscopy *1*, 127.

Spence, J. C. H., and Cowley, J. M. (1978), Optik *50*, 129.

Spence, J. C. H., and Kolar, H. (1979), Phil. Mag. *A39*, 59.

Strahm, M., and Butler, J. H. (1981), Rev. Sci. Instrum. *52*, 840.

Strojnik, A. (1973), in "Scanning Electron Microscopy/1973" (O. Johari and I. Corvin, ed.), IITRI, Chicago, pt. 1, p. 17.

Sturkey, L. (1962), Proc. Phys. Soc. *80*, 321.

Uyeda, N., Kobayashi, T., Ishizuka, K., and Fujiyoshi, Y. (1978-9), Chemica Scripta *14*, 47.

Wischnitzer, S. (1981), "Introduction to Electron Microscopy," 3rd ed., Pergamon Press, New York.

Wu, C. K., Sinclair, R., and Thomas, G. (1978), Metal Trans. *9A*, 381.

CHAPTER 4

PRINCIPLES OF X-RAY ENERGY-DISPERSIVE SPECTROMETRY IN THE ANALYTICAL ELECTRON MICROSCOPE

D. B. Williams and J. I. Goldstein

Department of Materials Science and Engineering
Lehigh University, Bethlehem, Pennsylvania

C. E. Fiori

National Institutes of Health
Bethesda, Maryland

I. INSTRUMENTATION

A. The Energy-Dispersive Spectrometer

The solid state x-ray detector or energy-dispersive spectrometer (EDS) was developed in the late 1960s and rapidly found use on electron-beam instruments (Fitzgerald *et al.*, 1968) because of its speed in collecting and simultaneously displaying x-ray data from a wide energy range. Its small size, but relatively large collection angle (see below), also gives it significant advantages over the traditional crystal or so-called wavelength-dispersive spectrometer (WDS), enabling it to interface easily to an AEM despite the severe space constraints of modern TEM stages. The drawbacks to the conventional EDS are its relatively poor energy resolution (~150 eV) and its inability to detect x-rays from elements below Na (Z = 11) in the periodic table. These limitations, however, are not major practical handicaps in most analyses, and no modern AEM (since the EMMA series in about 1970) has had a crystal spectrometer interfaced to the column. Hence, we will consider the EDS only.

The physics of the x-ray detection process are shown in Figure 1. X-ray photons generated by inner-shell ionizations in the specimen enter the detector, where they create electron-hole pairs in the lithium-drifted silicon crystal. The charge pulse thus created is proportional to the energy of the incoming x-ray. The charge is amplified and discriminated according to its energy, and a signal corresponding to the energy is processed into the appropriate channel of a multi-channel analyzer (MCA). The detector is enclosed in a high-vacuum tube protected from the atmosphere and stage environment by a thin (nominally 7.5 μm) beryllium sheet or "window." The whole assembly is cooled to liquid nitrogen temperatures to minimize electronic noise. This accounts for the large dewar attached to every EDS system and shown schematically in Figure 2.

Typical EDS spectra obtained from an AEM at ~100 keV are shown in Figure 3. The spectrum is a histogram of x-ray intensity versus energy and consists of several approximately Gaussian-shaped peaks characteristic of the elements present in the analyzed volume. The characteristic peaks are superimposed on a background or continuum intensity that rises from ~0.8 keV, peaks at ~1.5 keV, and then decreases slowly, with increasing energy, reaching zero at the incident-beam voltage (~100 keV).

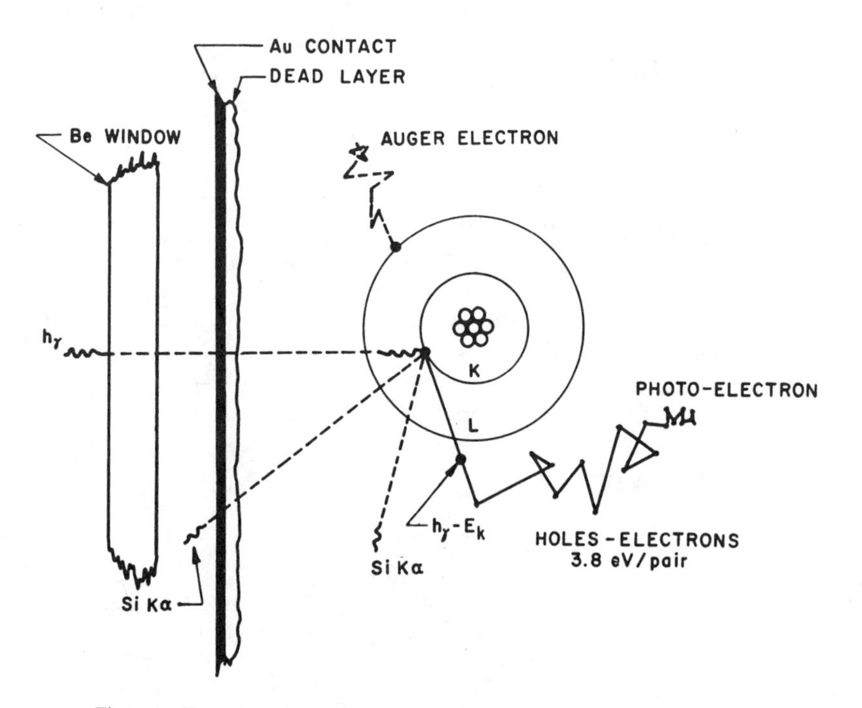

Figure 1 X-ray detection process in the Si(Li) detector (Goldstein *et al.*, 1981).

The characteristic peaks are due to x-rays emitted when ionized atoms return to the ground state; therefore, they contain the elemental information sought. The background is due to bremsstrahlung x-rays produced when electrons are slowed down by inelastic interactions with the nuclei in the specimen. Quantification of the spectrum involves removal of the background and determination of the true relative intensities of the characteristic peaks. This subject is addressed principally in the subsequent chapter. The rest of this chapter deals mainly with the precautions necessary to ensure that the resultant quantification is as accurate and precise as possible.

B. Interfacing the EDS to the AEM

The EDS is inserted into the stage of the AEM, usually in an orthogonal position with respect to the tilt axis, although this varies with manufacturer. While it is not essential to have the EDS in this configuration, it is important to know the position of the detector with respect to the region of the specimen from which microanalysis data are required. This position can be deduced from the stage geometry and a knowledge of the specimen traverse directions. The orthogonal position therefore has a geometrical advantage. The major variables of concern are those of the detector take-off angle, the solid angle of collection, and the resolution of the detector.

The x-ray take-off angle, α, is the angle between the plane of the specimen and the detector axis. If the detector sits in the plane of, and orthogonal to, the specimen holder (which then has to be tilted to permit detection of x-rays), the goniometer tilt angle is the take-off angle. However, the ideal configuration is a positive take-off angle

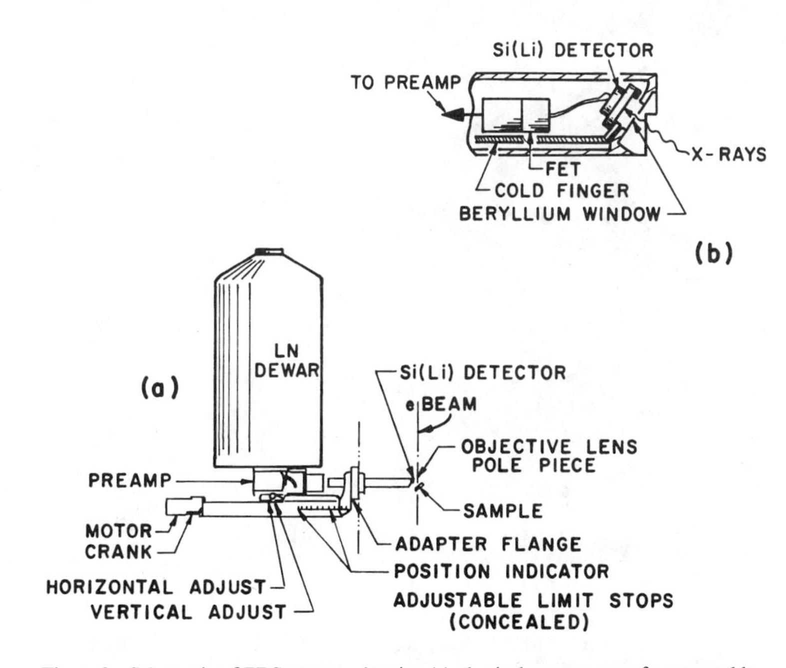

Figure 2 Schematic of EDS system, showing (a) physical appearance of a retractable detector and associated preamplifier electronics; (b) detail of Si(Li) mounting assembly (Goldstein *et al.*, 1981).

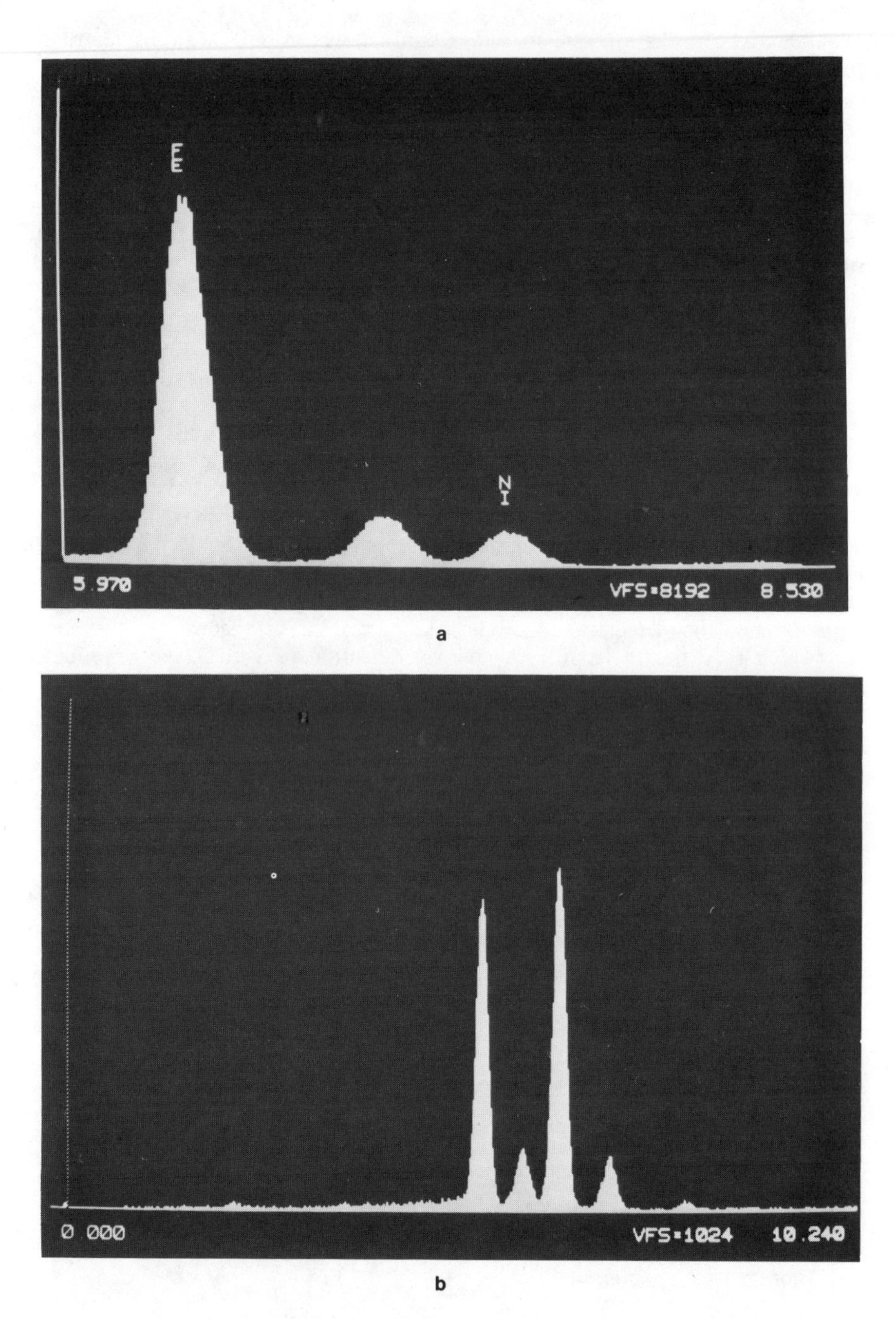

Figure 3 Typical EDS spectra from an AEM operated at ~100 keV: (a) Fe-10 wt% Ni alloy, spectrum displayed from 5.97 to 8.53 keV; (b) Fe-50 wt% Ni alloy, spectrum displayed from 0.0 to 10.24 keV.

such that the x-rays can be detected with the specimen at 0° tilt (i.e., normal to the beam). The specific value of α can clearly be changed by tilting the specimen, changing the specimen height, or retracting the detector. The exact value of α is critical for x-ray absorption corrections (Williams and Goldstein, 1981a), but rapid and accurate experimental determination is not straightforward, and reliance has to be placed on the manufacturer's estimate.

It has been shown by Zaluzec *et al.* (1981) that a take-off angle of 20° is a reasonable compromise position. Lower take-off angles introduce large path-length corrections for x-ray absorption (Chapter 5), and line-of-sight problems mean the specimen will probably have to be tilted away from 0° anyway. To go to higher take-off angles usually means that the detector has to avoid the upper objective polepiece. By suitable retraction of the detector and redesign of the polepiece, values of $\alpha \sim 70°$ are available. Because of the retraction, however, the detector collection efficiency drops markedly since the solid angle it subtends at the specimen decreases as the square of the distance away from the specimen. In fact, the collection angle of the detector Ω (in steradians) is given by the approximation

$$\Omega \approx \frac{A}{S^2} \tag{1}$$

where

A = active area of detector
S = distance from specimen to detector (not the collimator)

For example, with a detector of ~30 mm^2 active area, ~15 mm from the specimen, $\Omega \cong 0.13$ sr. (This approximate equation may be modified if the detector is in the plane of the specimen and not orthogonal to the specimen holder. See Zaluzec, 1981, for a full discussion of all the geometrical considerations.)

Any adjustment of the detector position will lower the value of Ω since ~15 mm is usually the closest distance of approach possible, without contact being made with the upper objective polepiece.

There are two detector variables that require calibration. First, the resolution and second, the Be window thickness. The resolution is customarily defined as the full width at half-maximum (FWHM) of the MnK_α peak (5.898 keV) and may be measured directly from the MCA output on a Mn specimen. The discrete nature of the MCA output leaves room for error since the chosen channels may not contain exactly half the counts in the maximum channel. Consequently, the MCA system often contains a program to calculate the effective resolution at the time of operation. Manufacturers' quoted resolutions vary from ~142 eV to ~155 eV, although in practice, a value of ~160 eV is reasonable to expect during routine EDS analysis.

As noted previously, the Be window thickness is usually assumed to be ~7.5 μm (0.3 mil), but recent difficulty in obtaining such thin Be sheet means that ~12 μm is being used more often. It is certain that the Be sheet is not perfectly parallel, nor is it perfectly flat, since SEM studies show it to contain pores (Freund *et al.*, 1972; Statham, 1981). Also, it may change with time because of interaction with high-keV electrons and/or strain induced by constant vacuum cycling or thermal cycling if the detector is warmed up. This latter step is particularly deleterious to the window integrity (because of thermal stresses at the seal with the detector); it should be carried out only when essential, and with the approval of the manufacturer.

Generally it is safe to assume that the window is not thinner than 7.5 μm, and a value of $>$10 μm is likely to be closer to the true figure. But the occasions when an accurate value is needed are few. In those circumstances (e.g., calculation of k factors and absorption corrections; Williams and Goldstein, 1981a), errors resulting from other factors usually outweigh the error in Be window thickness (Chapter 5).

C. Collimators

The collimator is the front portion of the detector assembly that limits the acceptance angle of radiation entering the detector. Little attention has been paid to the design and constituents of collimators, despite their important role. The fact that the collimators must invariably interact with high-energy electrons and x-rays means that care should be taken to ensure that this interaction does not affect the detected spectrum in any significant way. Nicholson *et al.* (1982) have described a carbon-coated Pb collimator containing Pb and Al baffles. The carbon-coated Pb design minimizes transmission of x-rays generated by radiation incident on the outside of the collimator. The Al baffles restrict the direct entry of backscattered electrons into the detector because the bending of the backscattered electron trajectories by the magnetic field of the objective lens can be used to ensure that a scattered electron following a path in the magnetic field will intercept the collimator wall or the baffles, which are lined with a low-atomic-number material. The exact design of such a collimator will depend on the particular instrument and detector configuration and may require a trial-and-error approach. Undoubtedly a heavy metal collimator is essential to limit the penetration of high-energy bremsstrahlung x-rays into the detector, and these are slowly becoming available commercially.

D. Windowless/Ultra-Thin Window (UTW) EDS

A major limitation of EDS is that the Be window is the primary absorber of low-energy x-rays in the system. This fact accounts in part for the drop-off in intensity of the spectral background below ~1.5 keV and the inability to detect characteristic peaks of energy below ~1 keV (NaK_α). As mentioned, the Be window is required to protect the cooled detector from the vagaries of the vacuum in the microscope stage. Removal of the window is possible only in an ultraclean environment; otherwise, contaminants will form on the cold detector and degrade the performance. In modern AEMs, the vacuum approaches acceptable levels of cleanliness, and windowless detectors have been placed in AEMs successfully, for short periods of time.

It is generally accepted, however, that so-called "ultra-thin window" (UTW) detectors are a better compromise. These use a gold-coated Formvar film, a few tens of nanometers thick, which protects the detector but absorbs fewer low-energy x-rays than the conventional Be window and thus permits detection of x-rays from boron ($Z = 5$) and higher-atomic-number materials. The drawback to these detectors is that the window cannot maintain the internal detector vacuum against atmospheric pressure; therefore, there has to be an airlock system to protect the window when the microscope goes to air. The airlock mechanism and the magnetic electron "traps" required to stop backscattered electrons from entering the detector prevent the placement of the EDS detector close to the specimen. Therefore, the collection efficiency is decreased. Also, the general detection efficiency of the spectrometer falls rapidly below ~1 keV. The fluorescence yield of the characteristic x-rays is low, absorption corrections are almost always required for the light-element x-rays, and overlap with the L lines of the transition metals often occurs. For example, the V L_α line is at 0.510 keV, and O K_α is 0.523 keV. These problems all mean that sophisticated data handling, background subtraction, and peak stripping routines are essential if light-element x-rays are to be detected and quantified by using windowless or UTW detectors. But such detectors do provide an alternative to electron energy-loss spectroscopy as a means of detecting the presence of low-Z elements (Thomas, 1980, 1984), and their increasing use is reflected in several papers on the subject presented at the 1984 AEM conference (Williams and Joy, 1984). An example of a spectrum obtained using a UTW detector is shown in Figure 4.

II. ANALYSIS PRECAUTIONS

Before analyzing the EDS spectrum, several precautions have to be taken to ensure that (a) the detected x-rays come only from the region of interest and (b) the region of interest in the thin specimen is compositionally unchanged from that of the bulk by the specimen-thinning procedure. We will discuss these two problems separately.

A. Instrumental Artifacts

This topic is of concern because instrumental effects may produce spurious x-rays that, although apparently from the specimen, may not have originated from the analysis point. Any such effect would reduce the accuracy of quantification and restrict

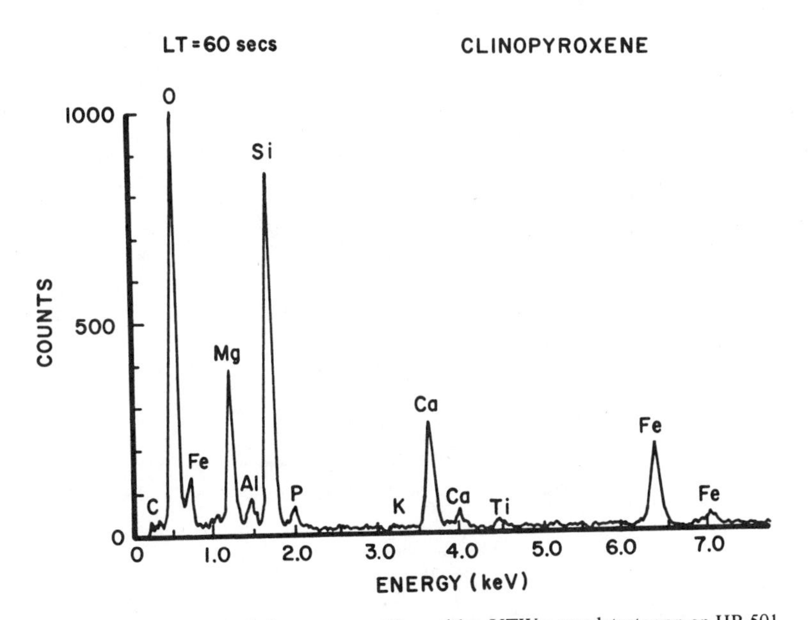

Figure 4 EDS spectrum of clinopyroxene taken with a UTW x-ray detector on an HB-501 STEM. The unprocessed spectrum shows an energy resolution of ~100 eV (FWHM) for oxygen K x-rays (spectrum courtesy of L. E. Thomas, Westinghouse Hanford, Richland, Washington).

the minimum mass fraction that is detectable. In particular, under these circumstances, unambiguous detection of small amounts of one element of interest in a matrix of another is impossible if the element of interest is also present elsewhere in the specimen or the AEM stage region in large amounts. AEM instrumental problems arise because of prespecimen illumination system effects and postspecimen effects in the AEM stage region. In addition, we must consider any artifacts resulting from the EDS system itself.

1. Illumination System Artifacts

Because the EDS detector is some distance from the specimen, the collimator has a wide angular view of the specimen and its surrounds. Therefore, generated x-rays should be confined to the region of interest if quantification is to be meaningful. However, this may not be the case in an AEM, either because uncollimated 100-keV primary electrons hit the specimen away from the point of interest, or more probably because bremsstrahlung radiation from the illumination system floods the specimen, fluorescing it, as shown in Figure 5. These two sources are instrument-dependent since illumination systems differ.

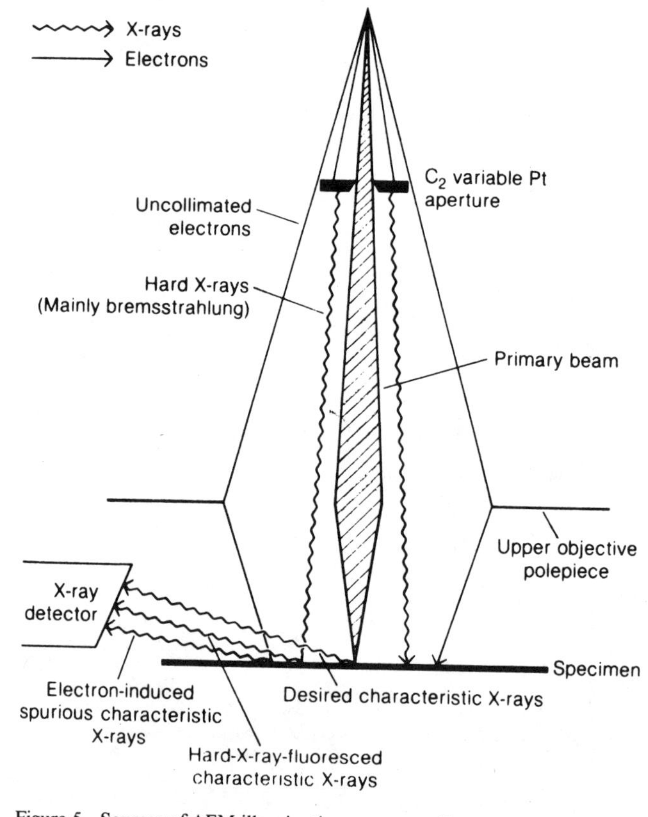

Figure 5 Sources of AEM illumination system artifacts: hard x-rays and uncollimated electrons. Hard x-rays that penetrate the C_2 aperture can fluoresce the specimen. Also, stray electrons that are poorly collimated or travel around the final C_2 aperture can generate spurious x-rays.

The existence of this type of artifact is easy to determine. By placing the primary beam down a hole in the specimen, a specimen-characteristic spectrum (often termed the "hole-count") will still be detected, as shown in Figure 6. Using a specimen with an EDS spectrum containing a low-energy L line (<5 keV) and a high-energy K line (>20 keV), the source of the spurious x-rays can be distinguished since a high L/K ratio is usually the result of electron excitation and a low L/K ratio, of x-ray excitation (Goldstein and Williams, 1978). In practice, the problem is usually due to high-energy bremsstrahlung generated in the illumination system (although stray electrons have been reported in early AEMs).

This problem has been recognized for a long time, and several reviews detail the solutions (Bentley *et al.*, 1979; Williams and Goldstein, 1981b; Allard and Blake, 1982). By now all manufacturers of modern AEMs minimize the effects by offering extra-thick, beam-defining apertures and/or nonbeam-defining "spray" apertures as options. These options are essential if the AEM is to be used for quantitative x-ray microanalysis; with their use, the illumination-system problem can be effectively solved (Figure 6). If the illumination-system bremsstrahlung remains a problem in any specific instrument, it is worth noting that thin-flake specimens (supported on Be grids), where the whole sample is electron-transparent or nearly so, can minimize the problem since the analyzed region is thinner than the average path length for fluorescence ($\sim$1 to 2 μm). Therefore, the amount of spurious x-rays generated is much smaller than for a disc specimen, where most of the specimen is relatively thick.

2. Postspecimen Effects

Assuming that the illumination-system problems are removed, it is still not certain that all the detected characteristic x-rays are generated solely from the point of interest. The primary reason for this is that the thin specimens allow both transmission of high-keV electrons and scatter of electrons in both the forward and backward directions.

These major "postspecimen interaction" sources may be summarized as follows (Figure 7):

(a) Incident high-energy electrons are backscattered into the microscope specimen chamber and generate x-rays from this region (e.g., cold trap, upper polepiece, EDS collimator). If the specimen is tilted away from 0°, interaction with the specimen is also possible since the electrons will be spiraling back up the column under the influence of the objective lens prefield. The possibility of direct backscatter into the EDS itself also cannot be discounted, although the strong prefield of the upper objective polepiece in the STEM mode helps to reduce this effect (Figure 7(a)).

(b) Incident high-energy electrons are transmitted, scattered, or diffracted into the specimen chamber (or specimen support grid if it is very close to the region of interest) (Figure 7(b)) and generate (i) characteristic and continuum x-rays and (ii) backscattered electrons from, e.g., the lower objective polepiece or objective aperture drive (which should *never* be inserted during microanalysis).

(c) Characteristic x-rays and high-energy bremsstrahlung from the point of interest on the specimen fluoresce the specimen environment. If the specimen is a disc and is tilted to some nonhorizontal orientation to permit detection of the characteristic x-rays, the possibility of self-fluorescence exists (i.e., fluorescence induced by the continuum radiation generated in the sample) (Figure 7(a)). This of course depends on the specimen shape, thickness, and microstructure,

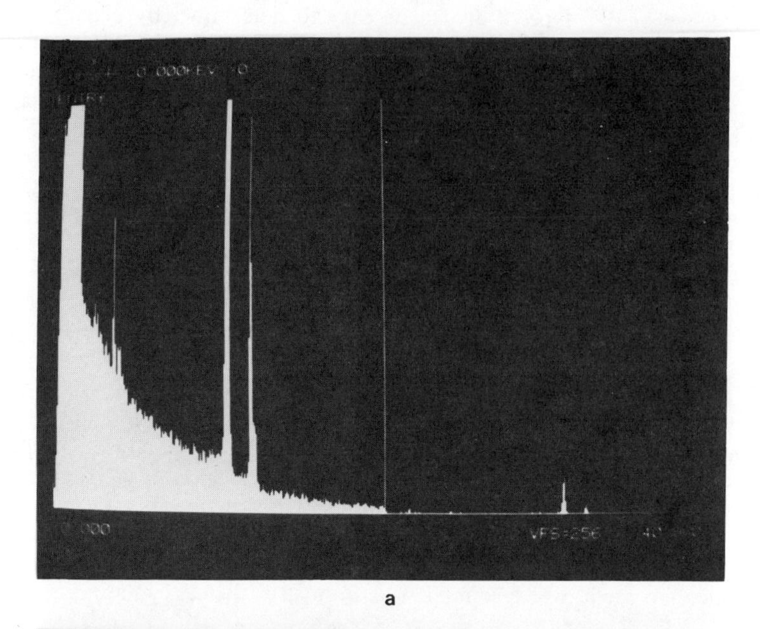

a

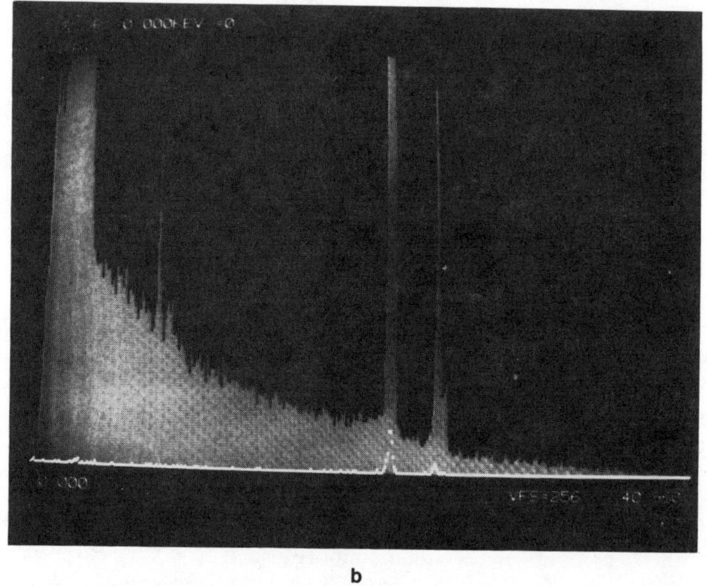

b

Figure 6 Hole-count on a Ag specimen: (a) Left-hand side: Ag spectrum down the hole in a Ag specimen using a conventional Pt C_2 aperture. Energy range 0 to 40.96 keV. Ag$K_{\alpha,\beta}$ and L_α x-rays are present. Right-hand side: Ag spectrum down the hole at the same intensity and energy scale (0 to 40.96 keV) after minimizing illumination system artifacts by use of an ultra-thick C_2 aperture. Small AgK_α and AgK_β peaks are still observed. (b) The two spectra of Figure 6(a) are superimposed. Scale: 0 to 40.96 keV.

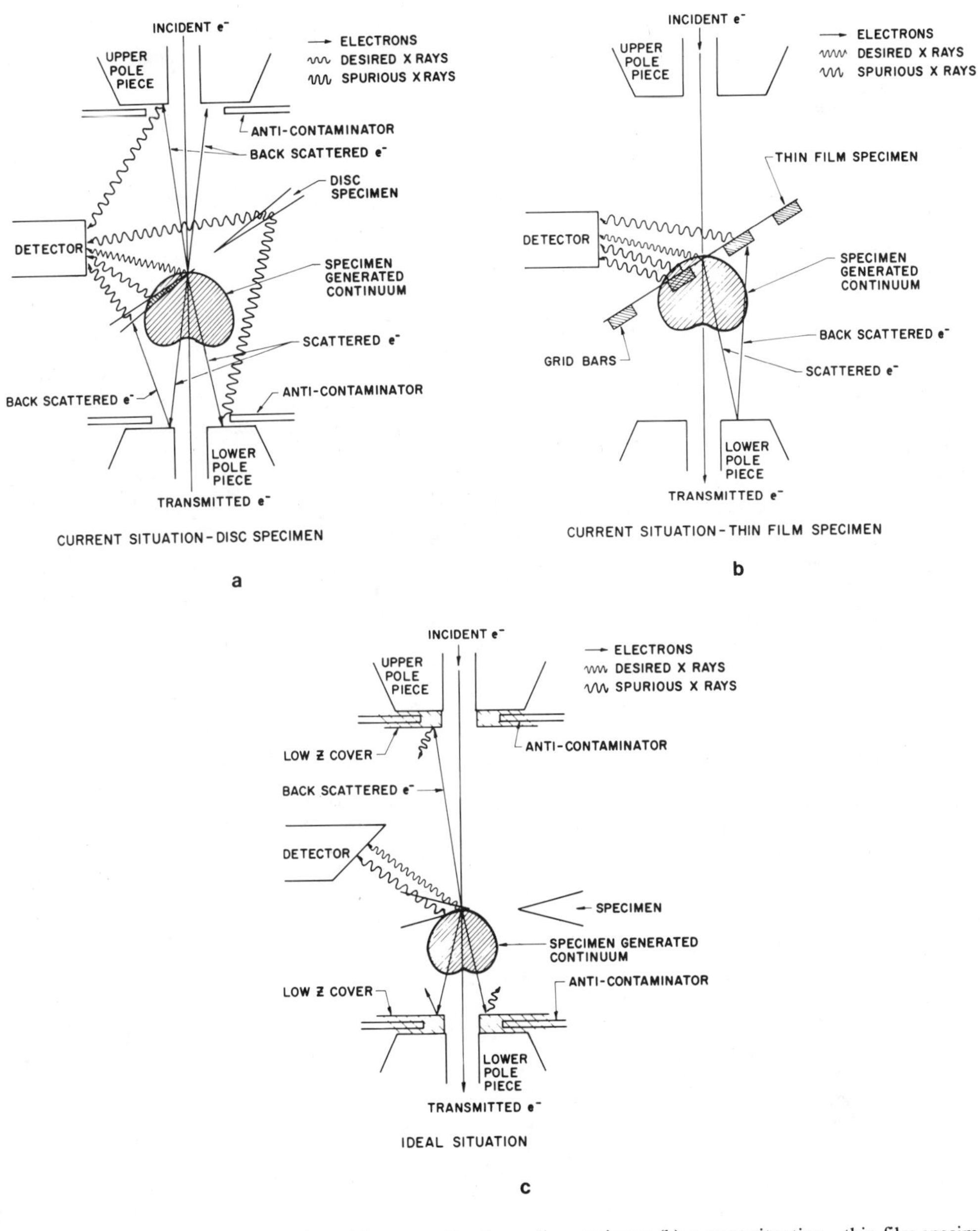

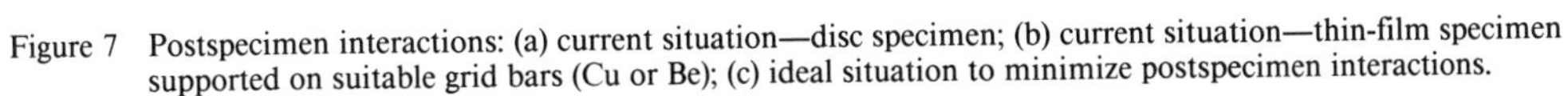

Figure 7 Postspecimen interactions: (a) current situation—disc specimen; (b) current situation—thin-film specimen supported on suitable grid bars (Cu or Be); (c) ideal situation to minimize postspecimen interactions.

and the actual energy and intensity distribution of the bremsstrahlung x-rays. If a specimen support grid is used, this radiation will fluoresce the grid (Figure 7(b)). A low-atomic-number cover for the lower polepiece and anticontaminator and the use of a low specimen tilt angle will minimize postspecimen interactions (Figure 7(c)).

Some combination of sources (b) and (c) probably accounts for the well-known observation of copper peaks in otherwise "clean" AEM systems when a copper support grid is used (Figure 8). Similarly, if a specimen on a Cu grid must be tilted because an image defect has to be aligned with respect to the beam, or if specific diffracting conditions have to be avoided, an increased Cu peak intensity will be noted, consistent with the increased interaction of the grid with postspecimen radiation.

Close examination of the background in the CuK_{α}/K_{β} region in Figure 8 reveals the presence of a copper absorption edge. The mass thickness of a typical thin film is not sufficient and does not produce enough absorption for an absorption edge to be observed. The copper absorption edge results from the fact that the grid has sufficient mass thickness to behave as a bulk target. Therefore, the copper is most probably excited by electrons that are elastically scattered from the film and rescattered from the microscope surfaces below the specimen, rather than by specimen-generated bremsstrahlung.

This scattering is essentially unavoidable, although it can be minimized by using low-atomic-number materials for the microscope surfaces (Nicholson *et al.*, 1982). Grids of low-atomic-number materials can eliminate the presence of anomalous characteristic peaks in the spectrum, but it must be recognized that a significant contribution to the bremsstrahlung can still be made by the low-Z grid and that specimen-characteristic x-rays will still be excited away from the point of interest. The spectrum thus measured should always be viewed as a possible composite of the spectrum of the specimen and of the surroundings. This is especially important in quantitative analysis since most methods depend on measurements of the peak and background for corrections. It should also be noted that some copper grids can contain several percent of elements such as Mn, Co, Fe, and Ni. The characteristic x-ray lines of these elements might then appear in the spectra.

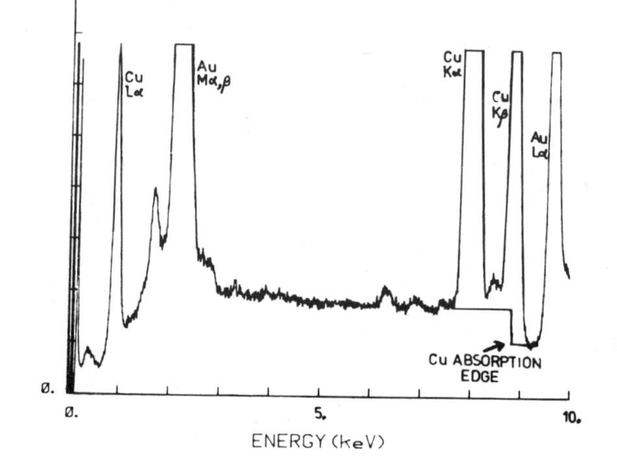

Figure 8 EDS spectrum obtained at 75 keV from a Au thin film supported on a Cu grid. The Cu absorption edge can be readily observed.

The use of "window"-polished foil fragments rather than self-supporting disc specimens will also reduce the possibility of specimen self-fluorescence. Furthermore, the use of high-take-off-angle detectors (standard on some AEMs and being developed on others) is an essential step since at zero tilt, specimen interaction with its own bremsstrahlung or its own backscattered electrons is minimized. In this configuration, however, increased backscatter directly into the detector is possible, further emphasizing the need for improved design of collimators.

B. Spectral Artifacts Caused by Operation at High Beam Energies

1. Processing High-Energy Pulses in the EDS

In addition to the well-known spectral artifacts such as the escape peak, sum peak (Goldstein *et al.*, 1981), and incomplete charge collection (Nicholson *et al.*, 1984), the very act of capturing a high-energy (e.g., $E > 20$ keV) photon in the EDS can introduce artifacts into the spectrum. The pulses that appear in the amplifier while processing low-energy photons are shown schematically in Figure 9. A pulse from a low-energy photon rises to a maximum and decreases smoothly to the baseline. However, when a high-energy photon is processed, the amplifier response is saturated. Saturation is the condition in which a further increase in the input to the amplifier does not result in an increase in the amplifier output. The return to the baseline from the saturated condition may not be smooth because of "ringing" or transient oscillations of the amplifier. These transients can disturb the electronic circuitry that first measures the voltage of the pulse and then classifies the photon in the appropriate channel of the MCA.

Processing saturated pulses can result in degradation of the detector resolution by as much as 5 eV or more (at MnK_α). Transient oscillations can disturb the dead time correction circuitry, resulting in incorrect spectral accumulation times. Another effect of transient oscillations is to disturb the return to the baseline. Moreover, the extreme width of the high-energy pulse relative to the pulses associated with low-energy photons increases the probability that a low-energy photon pulse will enter the detector before the amplifier has returned to the baseline. The low-energy photon pulse will thus be measured with apparently higher energy than normal, which will result in a significant deviation from the usual Gaussian shape on the high-energy side of the peak. The distortion of characteristic peak shapes varies as a function of peak energy, becoming more pronounced at low peak energies, since the influence of the tail of the high-energy pulse represents a larger fraction of a low-energy pulse.

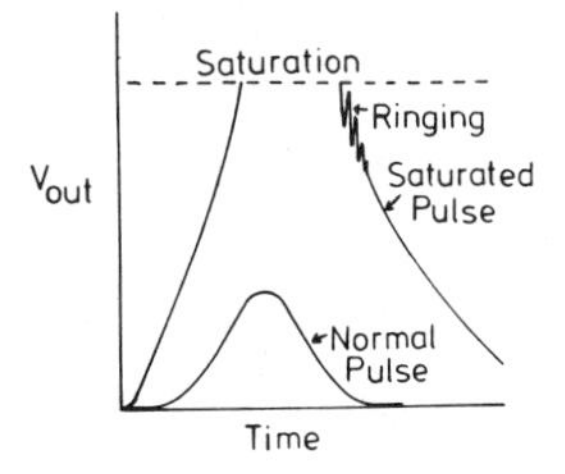

Figure 9 Schematic diagram of main amplifier response to normal and saturated pulses associated with low-energy and high-energy radiation.

2. Compton Scattering

The passage of high-energy x-rays through the detector introduces a new category of unavoidable artifacts not observed in the low-beam-energy EDS such as practiced in the SEM. The high-energy x-ray photons can scatter inelastically in the detector by means of the Compton effect, in which the photon interacts with a loosely bound electron, transferring energy to it. Energy is deposited in the silicon of the detector, in the form of the Compton recoil electron, while the x-ray photon, now of lower energy, escapes the detector. The detector will therefore register an apparent x-ray count because of the Compton electron.

The energy transferred to the recoil electron is ~10 keV or less, so that the apparent x-ray event will be recorded in the low-energy spectral range, which is usually the region of interest. These Compton events distort the shape of the spectrum from that which would be expected simply from window absorption of the generated spectrum. As shown in Figure 10, the cross sections for the photoelectric and

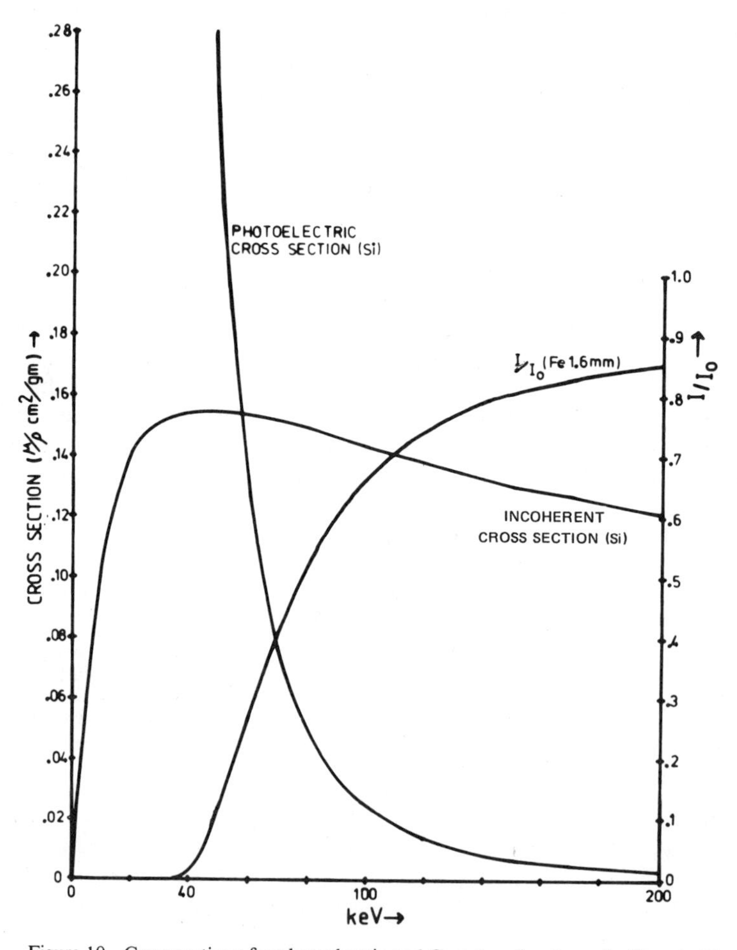

Figure 10 Cross sections for photoelectric and Compton (incoherent) effects in Si as a function of energy (keV).

Compton (incoherent) effects change rapidly at high energies, with Compton scattering becoming the dominant interaction process in the Si(Li) detector at photon energies above ~60 keV.

In considering sources of high-energy photons that can undergo Compton scattering in the detector, it is important to realize that the detector is vulnerable to other sources in addition to the x-rays generated in the specimen. The transmission of x-rays through the collimator of the detector becomes significant above 40 keV. Thus, electrons scattered in the microscope at the level of the detector or above can produce x-rays that can penetrate the wall of the collimator. In this region between the condenser and objective lenses, the electron flux may be quite high because of apertures that intercept the beam and cause scattering off all surfaces of the microscope. As we have seen, however, appropriate attention to collimator design may reduce any effects, and one way to do so is to install a heavy-element shroud around all parts of the detector housing. Note that this shroud should be used in addition to any heavy-element column liners and thick apertures that serve to protect the specimen from stray radiation in the illumination system. In fact, the very act of reducing the "in-hole" spectrum may actually increase the flux of hard radiation striking the detector through the walls as a result of placing efficient high-Z scattering materials above the specimen.

3. Electron Penetration

Penetration of high-energy-beam electrons through the beryllium window into the detector can occur above an electron energy of 20 to 25 keV. When an energetic electron strikes the silicon detector, it generates a voltage pulse through the production of electron-hole pairs in a manner similar to the process of x-ray detection. In penetrating the beryllium window, the energetic electron will lose ~20 keV of energy because of inelastic scattering. In the environment of the AEM, there will be a significant number of high-energy backscattered electrons with an energy range up to the accelerating voltage. Even with the loss of 20 keV in penetrating the beryllium window, these electrons can influence the measured spectrum over the full analytical range of interest, e.g., ~1 to 20 keV. Often, the influence of these backscattered electrons can go unnoticed unless the maximum energy range examined is increased to 40 or 80 keV or even greater.

An example of the distortion of the spectrum by electron penetration is shown in Figure 11, which was prepared from a bulk arsenic target bombarded with 40-keV electrons. In the situation shown in Figure 11, the energy loss of the electrons restricts their influence to the range 0 to 25 keV. Above 25 keV in this spectrum, the spectral intensity is due entirely to bremsstrahlung x-rays. By fitting this bremsstrahlung intensity distribution with a physical model, the bremsstrahlung contribution below 25 keV can be calculated, revealing directly the strong influence of the high-energy electrons entering the detector.

The influence of scattered electrons upon x-ray spectra measured in an AEM is difficult to predict because of the interaction of the electrons with the strong magnetic field of the upper objective polepiece. Therefore, the magnitude of the problem will vary with the instrument configuration and even with sample position in a particular instrument. Figure 12 shows two x-ray spectra of an aluminum-iron target obtained under constant-beam conditions but with the specimen at two different positions separated by ~3 mm in height. The intensity of the iron K_α peak is similar in both spectra.

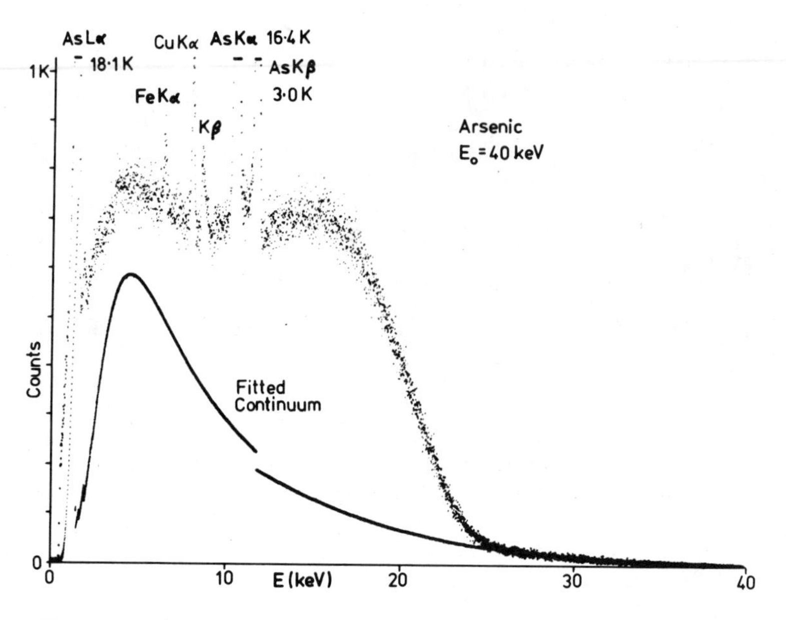

Figure 11 EDS spectrum of arsenic excited by a 40-keV electron beam, showing an anomalous background below 25 keV caused by direct entry of electrons scattered from the specimen into the detector. The x-ray continuum was fitted from the high-energy end of the spectrum (Goldstein *et al.*, 1981).

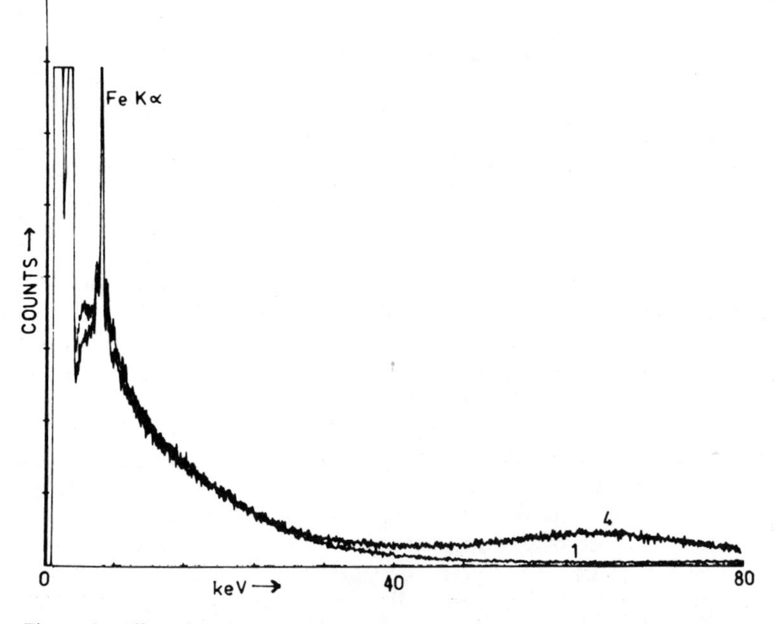

Figure 12 Effect of backscattered electrons on background shape. EDS spectra (0 to 80 keV) from an aluminum-iron specimen, with the specimen at two different heights in the instrument. The spectrum labeled "4" was obtained in the normal operating position and shows the effects of high-energy electron scatter.

The spectrum obtained in the lower sample position, labeled "1" in Figure 12, has a nearly ideal background shape, arising almost exclusively from bremsstrahlung x-rays. The spectrum labeled "4," obtained in the upper (normal operating) position, shows an anomalous background, with significant deviation above 30 keV caused by the entrance of high-keV electrons into the detector. A broad electron "hump" centered about 60 keV can be seen. Although the spectra in Figure 12 represent two different sample positions, similar scattered electron effects could be encountered at the same sample position if the spectra were recorded from different regions of the sample where the atomic number or mass thickness varied.

Upon initial examination of Figure 12, it appears that the spectrum in the analytical energy range of 1 to 20 keV is unperturbed by the electron effects above 30 keV. As we have noted, however, the high-energy electrons affect the analytical energy range in two deleterious ways. First, the system dead time correction may be adversely affected. Although the total live time may not be perturbed if the dead time correction circuit is functioning properly, the amount of collection time available for the 1- to 20-keV analytical energy range is necessarily reduced because of the number and length of the high-energy pulses from scattered electrons. If the spectra in Figure 12 represented low- and high-atomic-number regions of the same specimen, there would be an anomalous reduction of $\sim$25% in the live time available for x-ray counting in the analytical energy range of the spectrum as a result of the increased number of high-keV electrons entering the detector from the high-atomic-number region.

The second adverse effect of the scattered electrons is the perturbation of the characteristic x-ray peak shapes caused by the baseline restoration problem discussed previously. While high-energy x-ray pulses can also cause baseline restoration problems, the flux of electrons into the detector can be the dominant feature of the high-energy end of the spectrum, as shown in Figure 12.

Figure 13 shows an aluminum K_α x-ray peak obtained under the conditions shown in Figure 12 for the spectrum labeled "4," containing significant scattered electron effects. To quantify the data (Chapter 5), the peak is modeled with a procedure constrained to use a Gaussian function to describe the peak. Because the peak deviates significantly from a Gaussian shape, the "best fit" Gaussian would be substantially broader than the real peak. When the fitted peak is subtracted from the real peak, the resulting residuals, also shown in Figure 13, indicate a severe deviation from the

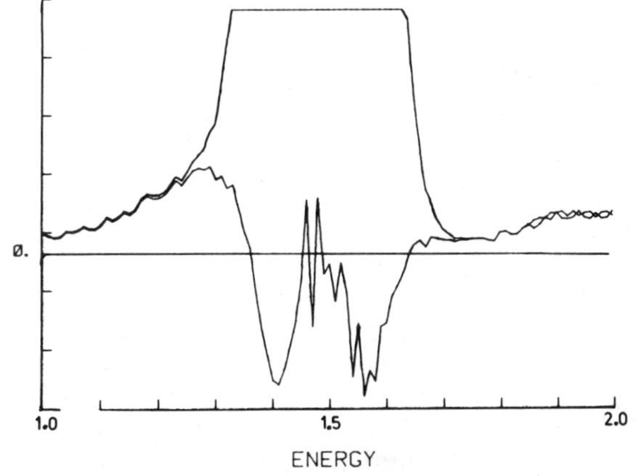

Figure 13 EDS spectrum of AlK_α peak from an aluminum-iron specimen obtained with a scattered electron component in the spectrum (position 4). The effect of the scattered electrons is to introduce strong deviations from the expected Gaussian profile (upper curve). These deviations are indicated by the residuals from the Gaussian peak-fitting operation.

expected Gaussian profile. The deviation from the Gaussian peak profile caused by the high-energy electrons leads to substantial errors in extracting intensities from the spectrum by any peak-fitting procedures used to solve spectral overlaps and may therefore influence quantification.

4. Detector Degradation

The operation of Si(Li) detectors in the AEM environment has resulted in a noticeable rate of detector failure. Detectors are often observed to undergo temporary or apparently permanent failure following exposure to high fluxes of electrons and x-rays when operating in the conventional TEM mode or during lens transitions from the STEM to conventional TEM modes. The detector malfunction may involve loss of resolution, generation of a "low-energy" noise peak, or even result in total failure, requiring return to the manufacturer. The mechanism of these problems is poorly understood at present.

When this phenomenon occurs, however, it is not necessarily indicative of permanent damage. For one detector system within the authors' experience, the following behavior was noted. After accidental exposure to a large radiation dose in the conventional TEM mode, the detector was found to be inoperative. After 72 hours in a quiescent state, the detector was found to be operating normally, with a slight degradation in resolution. Subsequent high-dose episodes produced failures with rapidly diminishing restoration periods. After about one month of exposure, the detector reached a condition in which it recovered within a few seconds after high-dose conditions associated with a STEM/TEM transient. The resolution degradation was $\sim$10 eV as a result of this sequence. The explanation for this behavior is currently unknown.

This experience suggests that when similar phenomena are observed, it would be wise to allow a recovery period including, perhaps, a brief excursion to room temperature before returning the detector to the manufacturer.

C. Unusual Sources of Interference

EDS detectors have been in extensive use in the field for more than 10 years. One would like to think that all of the artifacts arising from improper interfacing to the various instruments would have been identified by now, but sadly this is not the case. New artifacts are occasionally observed, and this will be especially true as EDS detectors are applied to high-voltage electron microscopes, high-energy proton microprobes, synchrotrons, etc. One example of a bizarre artifact encountered in a 200-keV AEM is illustrated in Figure 14. This artifact took the form of a peak with an apparent energy that varied in the range 0.5 to 2 keV, depending on the magnification of the STEM image!

In Figure 14(a), the three peaks were obtained by recording the spectrum for equal lengths of time at three different settings of the STEM magnification control, with the highest energy peak corresponding to the lowest magnification. This peak could also be caused to shift by altering the scan rate, with the peak disappearing at the slow photographic scan rate. No electrons struck the sample during the accumulation of this spectrum. The origin of this artifact is illustrated in Figure 14(b). Electromagnetic coupling occurs between the alternating current passing through the scan coils and the EDS detector assembly. This coupling produces a periodic signal in the EDS signal chain that is interpreted as a false x-ray pulse.

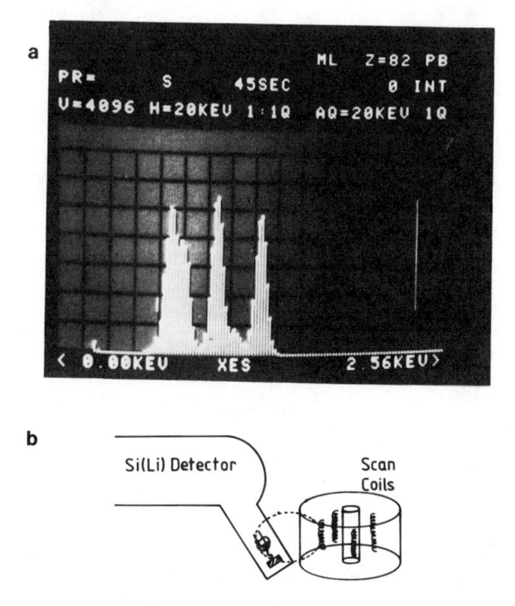

Figure 14 EDS artifact caused by electromagnetic coupling with the scan coils in a 200-keV AEM: (a) Apparent x-ray peaks measured: no electrons in the column; (b) Schematic of electromagnetic coupling between ac through the scan coils and the EDS detector assembly.

The problem is more likely to be observed in high-voltage scanning instruments for two reasons: (1) Larger currents are necessary to scan the beam as compared to an SEM operating at a beam energy of 20 keV. (2) A "top looking" detector in the AEM is normally located close to the scan coils and lens and is thus immersed in the electromagnetic field, whereas in the SEM the detector is located in the virtually field-free region of the sample chamber.

D. Specimen-Preparation Artifacts

The preparation of thin-foil TEM specimens has long been considered an art rather than an exact science. Because of the drastic measures (chemical attack, electropolishing, ion-beam thinning) used to generate electron-transparent specimens, much debate has occurred as to whether or not the thin-film microstructure is characteristic of the bulk. More recently, a similar debate has arisen concerning the chemistry of the thin film. The problem was first mentioned by Thompson *et al.* (1977), who reported composition variations with thickness in Al-Cu specimens that could not be explained by absorption, but were apparently due to redeposition of a thin layer of Cu on the foil surface during electropolishing. Morris *et al.* (1977) showed SEM image evidence for deposition of particles on the surface of Al-4% Cu and reported similar effects in Al-Zn and Al-Ag. The effect was present whether electropolishing or conventional ion-beam thinning was used and could only be removed by sputtering in an ultra-high-vacuum system. More recently, Pountney and Loretto (1980) observed the effect in electropolished Cu-Al by using Auger electron spectroscopy but reported that hand-polished specimens did not show preferential solute deposition on the surface. A review of the problem has been given by Fraser and McCarthy (1982).

This problem is clearly alloy-dependent and varies with specimen preparation technique. It can easily be detected by first observing nominally homogeneous thin-foil specimens and measuring x-ray intensity ratios as a function of thickness. Data such as those in Figure 15 indicate that surface-layer problems exist. Under these

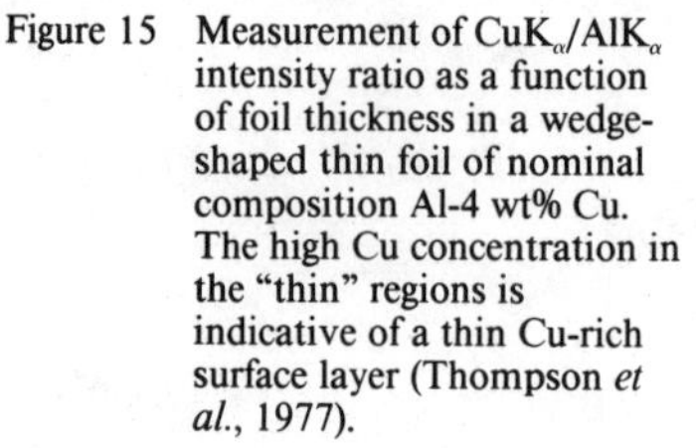

Figure 15 Measurement of $CuK_{\alpha}/AlK_{\alpha}$ intensity ratio as a function of foil thickness in a wedge-shaped thin foil of nominal composition Al-4 wt% Cu. The high Cu concentration in the "thin" regions is indicative of a thin Cu-rich surface layer (Thompson *et al.*, 1977).

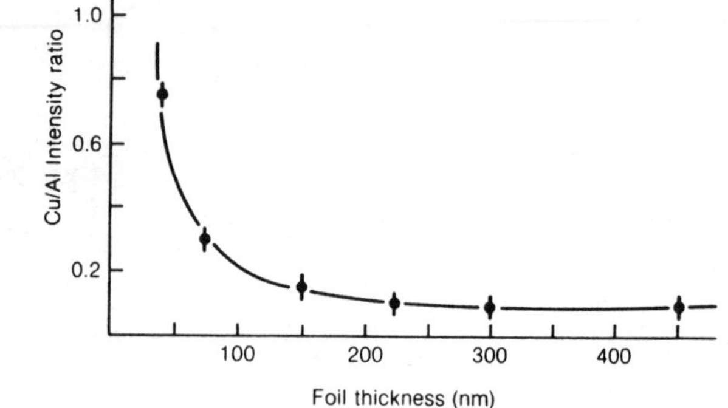

circumstances, alternative preparation techniques should be pursued, or a suitable surface-cleaning method should be found. Under the same circumstances, however, it is worth varying the preparation conditions to see the effect on surface-layer deposition. For example, in studies of Al-16% Ag (Merchant, 1982), severe surface-layer problems were found only if the electrolyte (nitric acid/methanol) was used on several occasions. Use of fresh electrolyte produced no detectable surface layers. Similarly, cleaning the thin foil with benzene removed any detected deposits very rapidly.

These data are contradictory to other reported results in Al-Ag (Morris *et al.*, 1977), but this serves only to emphasize variabilities inherent in the specimen-preparation process. Clearly, each specimen should be examined for such effects and analysis proceeded with or not, accordingly. There is always the possibility of proceeding to thicker areas of the foil where the surface-layer effect is too small to affect the bulk concentration. In Al-Cu, for example, this occurs at ~300 nm (Thompson *et al.*, 1977). The drawback to this, obviously, is that spatial resolution deteriorates probably to unacceptable levels. It has recently been proposed that ultramicrotomy be reexamined as a specimen-preparation technique for materials (Ball and Furneaux, 1982; Ball *et al.*, 1984). This technique is clearly unsuitable if the defect substructure is also to be studied. Otherwise, it might be a way around the chemical redistribution that apparently occurs in many specimens.

Extraction replication is another specimen preparation technique that is having somewhat of a resurgence due to the advent of AEM. This technique permits the removal of second-phase particles from the surrounding matrix, in principle allowing unambiguous phase identification. It is not yet known, however, whether the particles are unaffected by the vigorous chemical attack that is part of the process or whether, for example, a thin matrix film is left on the particle. These artifacts would limit use of this process in quantitative microanalysis.

III. SELECTION OF EXPERIMENTAL PARAMETERS

Prior to quantitative x-ray analysis in the AEM, a number of instrumental variables can be optimized to ensure that, for example, quantification is accurate or that the best conditions exist for detecting small amounts of a particular trace element or segregant. In these latter situations, x-ray counting statistics may be the limiting factor in quantification. Therefore, it is essential to maximize both the total counts and the count rate of the characteristic x-rays of interest. The time to acquire the x-ray counts must also be short enough, however, to ensure that specimen drift or contamination does not degrade the desired spatial resolution. As described in Chapter 5, spatial resolution is optimized by small probes and thin specimens, and these conditions are exactly the opposite of those required to generate high count rates. Therefore, under the situations where count rate is a limiting factor, a compromise experimental setup is often required whereby reasonably accurate quantification ($\sim\pm5\%$ relative accuracy) can be achieved at a reasonable spatial resolution ($\sim$10 to 30 nm). In situations where x-ray counting statistics are good, the accuracy of quantification or the spatial resolution can be improved accordingly.

As we shall see, the quantification process requires acquisition of the maximum characteristic x-ray intensity above the continuum background. Therefore, the peak-to-background ratio (P/B) should be maximized along with the absolute count rate. As a rule of thumb, the total x-ray count rate over the *whole* of the energy spectrum (up to E_0 the accelerating voltage) should not exceed $\sim$3000 counts per second, since the x-ray detector resolution may then be impaired. Even in typical thin-foil specimens using small (<10-nm) probes, this count rate can often be achieved in modern AEMs, so caution is required. However, even under these circumstances, the count rate in a small peak may be so low that long times (300 to 500 s) may be required to accumulate sufficient counts for quantification. Then spatial resolution may be lost, and it is more than ever essential to ensure that the AEM is optimized for x-ray analysis. The relevant instrumental variables over which the operator has control will now be discussed.

A. Choice of Accelerating Voltage

Several theoretical treatments (e.g., Joy and Maher, 1977) predict that P/B increases with keV, but until recently there was a lack of conclusive experimental evidence for this because of the increase in background x-rays from extraneous sources as the accelerating voltage increases. (Nicholson *et al.*, 1982, have reported increasing P/B in pure Co with increasing keV in a suitably "cleaned-up" AEM.) Given that this can be achieved in most AEMs, in combination with the fact that gun brightness increases and beam spreading decreases with increasing keV, there is a strong argument for invariably operating at maximum keV.

B. Choice of Probe Parameters

As well as emission current, which is a function of the gun, the operator has a choice of the probe size (C_1 lens strength) and the probe convergence angle, $2\alpha_s$ (C_2 aperture size). As shown in Figure 16, the current in the probe at the specimen can be varied over two orders of magnitude depending on the probe size (Williams, 1984). If spatial resolution is a secondary consideration, a large probe size will minimize any x-ray count problems. Similarly, increasing the value of $2\alpha_s$ increases the probe current (Figure 17) for a fixed probe size. In theory, probe size is independent of $2\alpha_s$, but in practice, spherical aberration at very high ($>10^{-2}$ rad) $2\alpha_s$ values and small (<4-nm) probe sizes may result in loss of resolution (Cliff and Kenway, 1982).

C. EDS Variables

The EDS detector itself has a fixed geometry with respect to active area ($\sim$30 mm^2) and take-off angle (0°, 20°, or $\sim$70°, depending on the specific instrument). The solid angle is usually maximized, again to maximize the number of detected x-ray

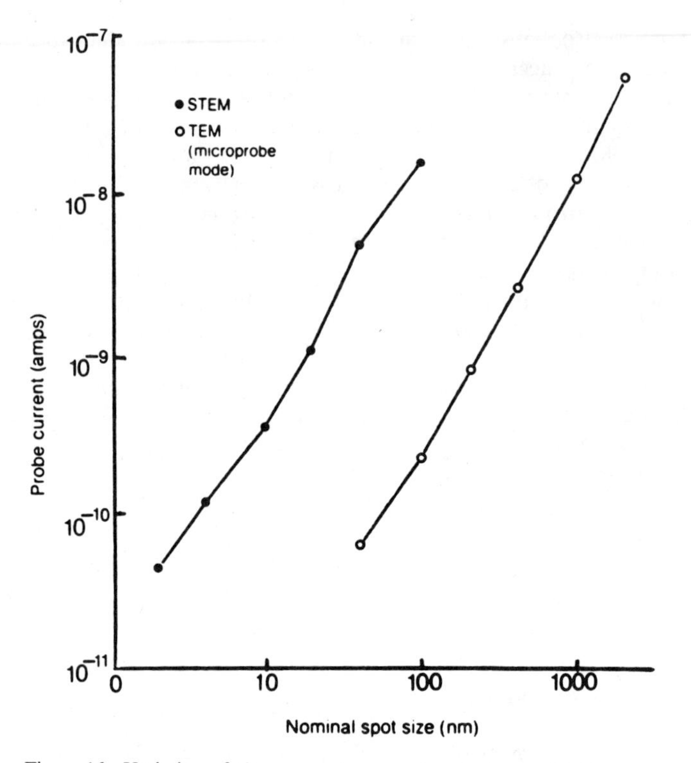

Figure 16 Variation of electron probe current in TEM or STEM modes with focused probe or spot size (courtesy Philips Electron Optics Publishing Group).

photons, by placing the detector as close (~15 mm) to the specimen as possible. The detector can be backed off mechanically if the count rate is too high or if it is considered that extraneous radiation (e.g., backscattered electrons or hard x-rays from the illumination system) is penetrating the collimator.

Attention should be given to the choice of spectral display variables, in particular, the energy range of the display on the MCA and the experimental counting time. The former should always be as large as possible (at least 40 keV) in the case of an unknown specimen where the existence of K and L lines at >10 keV may be essential in initial qualitative analysis. For example, as shown in Figure 18, it is possible to distinguish between MoL_α and SK_α overlap at 2.3 keV (Figure 18(b)) by observation of the MoK_α line at 17.5 keV (Figure 18(a)). Selection of the desired energy range for analysis should then maximize the resolution of the display to ~5 to 10 eV/channel if all the peaks of interest can still be displayed in the available channels. If the specimen is known, the appropriate energy range can, of course, be selected immediately.

Counting time is important insofar as it should be minimized to reduce the effects of contamination, specimen drift, and elemental volatilization (in the case of Na and other mobile species). This must be counterbalanced by the need to acquire enough counts for acceptable errors.

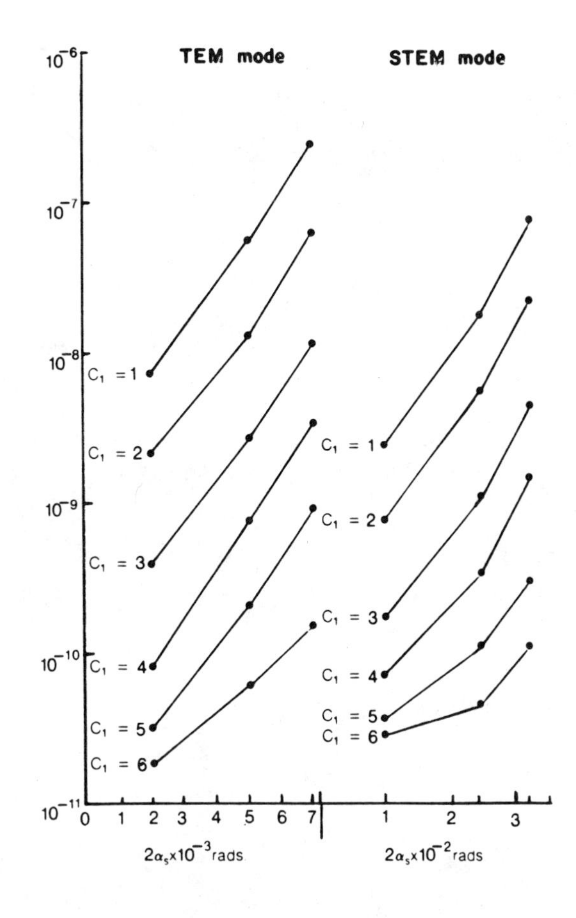

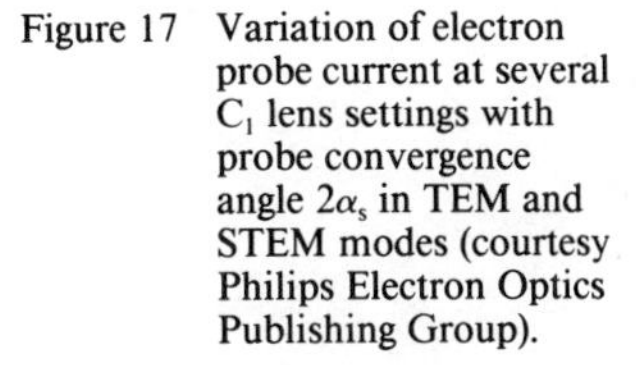
Figure 17 Variation of electron probe current at several C_1 lens settings with probe convergence angle $2\alpha_s$ in TEM and STEM modes (courtesy Philips Electron Optics Publishing Group).

D. Choice of Electron Gun

The choice for the average user is between a conventional W hairpin filament and a LaB_6 gun. The latter is brighter by a factor of ~10 times at 100 keV but substantially more expensive. However, if properly operated and maintained, a single LaB_6 source will operate for many months, which is significantly longer than the life of a W hairpin. The relatively poor brightness of the W gun can be offset partially by the fact that operation at emission currents of 100 μA or more is not unreasonable for short periods of time if a reduced filament life is acceptable. However, operating a LaB_6 gun much above 10-μA emission may result in premature breakdown. This difference in emission current does not totally offset the inherent brightness difference since the probe diameter formed by a LaB_6 source is at most half that from a W source. The field-emission gun as an alternative source is expensive, requires ultrahigh-vacuum conditions, and is available in very few laboratories. But, in situations where the highest spatial resolution in both images and microanalysis is demanded and minimum detectability problems are encountered, it is the best electron source to use.

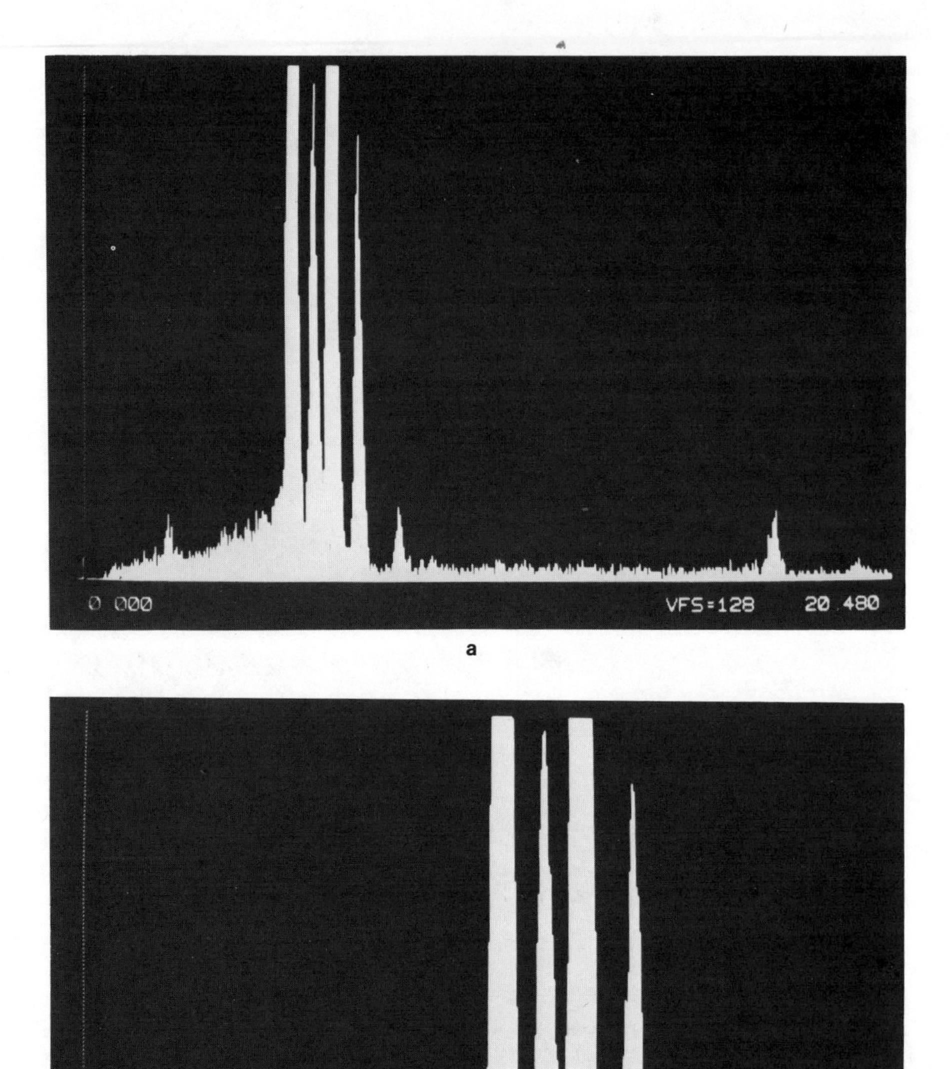

Figure 18 Identification of the presence of Mo in a NiCrFe weld metal: (a) EDS spectrum of NiCrFe weld metal, 0 to 20.48 keV. The major peaks are CrK_{α}/K_{β}, FeK_{α}/K_{β}, and NiK_{α}. A MoK_{α} peak is observed at 17.5 keV. (b) EDS spectrum of NiCrMo weld metal, 0 to 10.24 keV. The x-ray peak at 2.3 keV is MoL_{α} and not SK_{α} (see 0 to 20.48-keV spectrum in (a)).

IV. IMAGING AND DIFFRACTION CONDITIONS DURING ANALYSIS

The image and diffraction conditions for thin-foil crystalline specimens may affect the x-ray data because of the so-called "Borrmann effect," in which anomalously high electron penetration occurs when the specimen is close to the Bragg diffraction position ($\underline{s} \rightarrow 0$). Since this corresponds to the dynamical imaging conditions used in the TEM for maximum diffraction contrast ($\underline{s}$ small and positive), care should be taken not to perform x-ray microanalysis under the optimum imaging conditions.

This problem has been studied by several investigators, notably Cherns *et al.* (1973), using an EMMA 4, and more recently by Bourdillon (1982), using a modern AEM. Both studies concluded that the maximum x-ray generation anomalies occur near strong, low-index Bragg conditions, and a bend-center is the least desirable place from which to obtain x-ray data. However, by operating under kinematical conditions ($\underline{s} \gg 0$, i.e., well away from strong Bragg diffraction), the effect can be eliminated. Bourdillon also points out that using large probe-convergence angles ($2\alpha_s$) such as exist in STEM mode also reduces the effect. Similarly, if a ratio of intensities is used, as is almost always the case, the problem disappears.

However, the best approach experimentally is to observe the defect structure in TEM or STEM under appropriate dynamical conditions, but tilt slightly away from the Bragg condition before carrying out microanalysis. Although diffraction contrast will be reduced, it is usually possible to maintain enough contrast to localize the probe on the precipitate, defect, or other area of interest that is being analyzed. This problem does not exist in amorphous specimens.

It is worth noting, however, that under certain circumstances the variation of x-ray emission with orientation can be used to determine crystallographic site occupancy (Spence and Tafto, 1983). This is discussed in Chapter 5.

V. COHERENT BREMSSTRAHLUNG

Recently, a new phenomenon has been discovered in EDS spectra from thin crystalline specimens in the AEM that may give rise to errors in the interpretation of small intensity peaks from limited amounts (~1 to 3 wt%) of material. The phenomenon, termed "coherent bremsstrahlung" (CB), is not an artifact in the spectrum but the result of the regular crystal structure of a thin foil sample causing bremsstrahlung production in a regular, coherent manner rather than in random fashion. The net result is that, superimposed on the classical (random) bremsstrahlung background, are small Gaussian peaks at positions that depend on the beam energy and the specimen orientation.

The effect, discussed in detail by Spence *et al.* (1983) and Reese *et al.* (1984), can be predicted using the equation

$$E(\text{keV}) = 12.4\ \beta/L\ (1-\beta \cos (90 + \Omega) \tag{2}$$

where

E = energy of the CB peak
β = electron velocity/velocity of light
L = atomic plane spacing in the beam direction
Ω = take-off angle as defined in Eq. (1).

L can be determined from convergent-beam diffraction patterns.

The main reason for concern about the presence of CB peaks is that they may mask or be misinterpreted as genuine x-ray peaks from small amounts of a particular element in the specimen. For common accelerating voltages used in AEMs (~100 keV) and at atomic spacings (~0.3 nm) typical of many important materials, CB peaks occur at energies from ~1 to 8 keV. Often the CB peaks could be confused with the Si internal fluorescence peak, "impurity" peaks such as Cl and Ar from various specimen preparation techniques, or escape peaks/sum peaks. These facts may explain why it took so long to recognize the presence of CB peaks.

Typical examples of CB peaks from Cu over a range of accelerating voltages and Fe over a range of orientations are shown in Figures 19(a) and (b). The integers above the peaks correspond to the Laue zones in the crystal from which they originate. Vecchio (1985) has indicated methods by which CB peaks can be discriminated from elemental peaks for the specific purpose of detecting segregants in Cu and Fe. Basically, the effect of CB in overlapping elemental peaks can be minimized by careful choice of keV and orientation, using Eq. (2). If analysis is performed $>10^\circ$ away from a major zone axis, then CB peak intensities are lower (Figure 19(c)), although the general bremsstrahlung intensity appears to increase.

VI. MEASUREMENT OF X-RAY PEAK AND BACKGROUND INTENSITIES

When all the above precautions have been taken and the instrument optimized for the particular microanalysis problem at hand, acquiring the necessary x-ray spectra and identifying the characteristic peaks are straightforward matters since modern MCA systems are computerized and carry out most of the operations automatically.

The only experimental information that is required for quantification of the EDS spectrum is the characteristic peak intensities I_A, I_B, etc., of the elements of interest. As discussed in Chapter 5, these can then be converted directly into values of the wt% of each element C_A, C_B, or corrected for absorption, etc., where necessary. We will outline briefly the various options available to extract peak intensities. The subject is, in general, beyond the scope of this chapter, and we refer the reader to the literature for a more detailed explanation (Goldstein *et al.*, 1981; Heinrich *et al.*, 1981; and Fiori and Swyt, 1981).

The determination of the characteristic intensity requires first that the continuum background, or bremsstrahlung radiation contribution to the peak, be removed. Methods that separate characteristic x-ray peaks from the background can be classified into one of two categories: background modeling and background filtering.

Background modeling consists of calculating or measuring a continuum energy distribution and combining it with a mathematical description of the detector response function. The resulting function is then used to calculate an average background spectrum that can be subtracted from the observed spectral distribution. Background filtering ignores the physics of x-ray production, emission, and detection; the background is viewed as an undesirable signal to be removed by modification of the frequency distribution of the spectrum. Examples are digital filtering and Fourier analysis.

A method that does not require an explicit model of the background radiation is clearly advantageous for applications in the electron microscope operating in the energy range above 50 keV. One can, in principle, calculate background from a

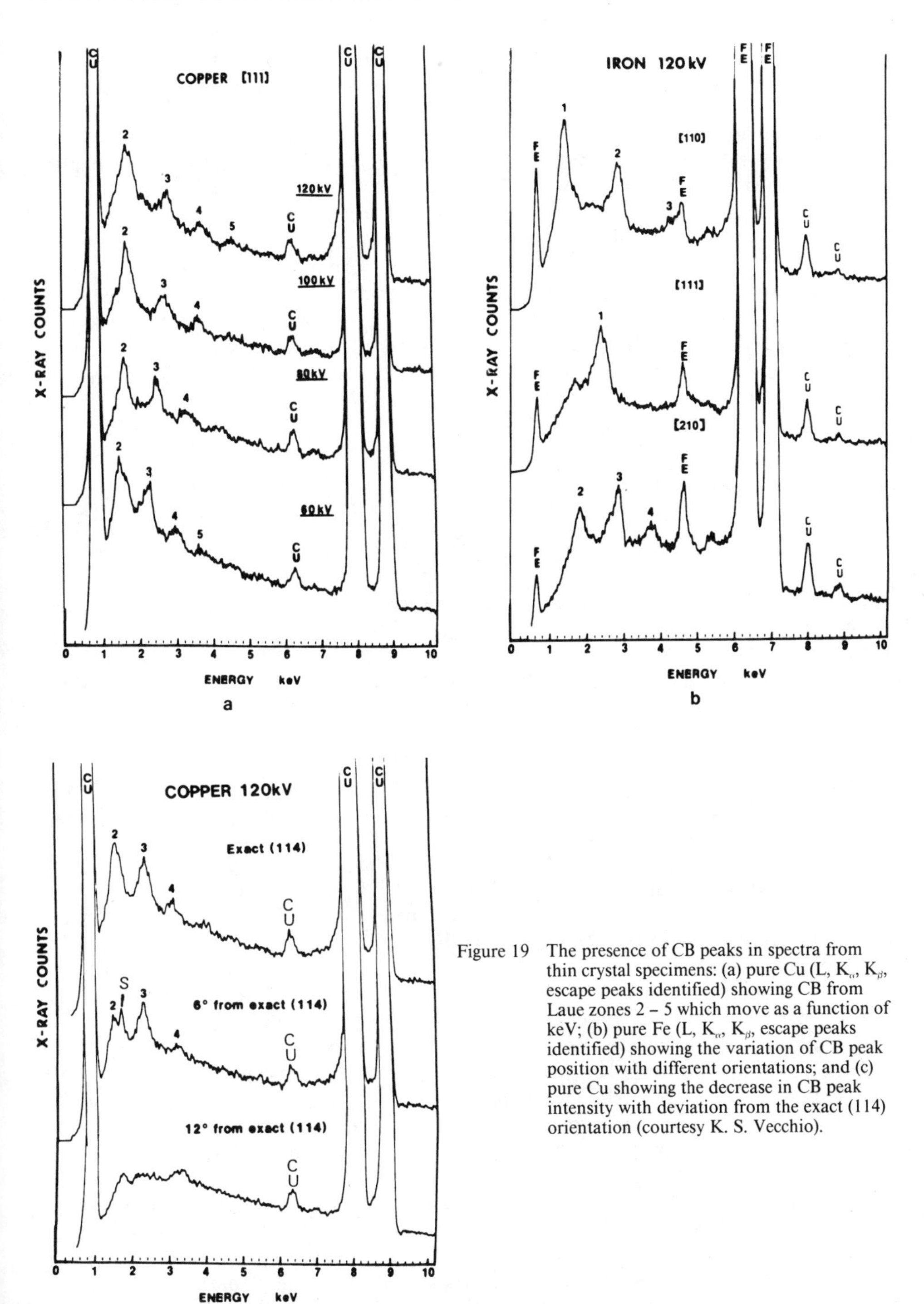

Figure 19 The presence of CB peaks in spectra from thin crystal specimens: (a) pure Cu (L, K_α, K_β, escape peaks identified) showing CB from Laue zones 2 – 5 which move as a function of keV; (b) pure Fe (L, K_α, K_β, escape peaks identified) showing the variation of CB peak position with different orientations; and (c) pure Cu showing the decrease in CB peak intensity with deviation from the exact (114) orientation (courtesy K. S. Vecchio).

continuum model. In practice, however, it is extremely difficult and often impossible to make such a prediction since a significant proportion of the background does not originate from the impact point of the primary electron beam but from grid bars or the specimen holder. Furthermore, Compton recoil electrons generated in the active volume of the silicon detector from high-energy x-rays (above 80 keV) further distort the expected shape of the background. Consequently, the following background filtering procedure is highly recommended for applications in the AEM.

The "top-hat" digital filter was first applied to energy-dispersive x-ray spectra by Schamber (1977). The algorithm is both simple and elegant. Briefly stated, counts in a group of adjacent channels of a spectrum are *averaged* and the *average* assigned to a channel equivalent to the center *channel* of the filter; the procedure is repeated as the filter is stepped through the spectrum and a new, filtered spectrum is created. The original spectrum is retained unaltered. One may describe the *averaging* by the following equation, using the notation of Statham (1977)

$$y_i' = \frac{1}{2M+1} \sum_{j=i-M}^{i+M} y_j - \frac{1}{2N} \left| \sum_{j=i-M-N}^{i-M-1} y_j + \sum_{j=i+M+1}^{i+M+N} y_j \right| \quad (3)$$

where y_i' is the content of the ith channel of the filtered spectrum and y_j is the content of the jth channel of the original spectrum.

The filter is divided into three sections: a positive, central section consisting of 2M + 1 channels and two side sections each containing N channels. The grand *average* of the side sections is subtracted from the grand *average* of the central section.

The effect of this particular averaging procedure is as follows: If the original spectrum is straight, across the width of the filter, then the *average* is zero. If the original spectrum is curved concave upward, across the width of the filter, the *average* is negative; if curved convex upward, the *average* is positive. The greater the curvature, the larger the *average*.

For the filter to respond with the greatest measure to the curvature found in spectral peaks, and with the least measure to the curvature found in the spectral background, the width of the filter must be carefully chosen. For a detailed treatment of the subject, see Schamber (1977) and Statham (1977). In general, the width of the filter for any given spectrometer system is chosen to be twice the FWHM of the MnK_α peak, with the number of channels in the central section equal to, or slightly more than, the combined number of channels in the side sections.

To measure the intensity of an x-ray peak in a spectrum, we must not only separate the peak from the average effect of the continuum but from the average effects of other peaks as well. If a spectral peak stands clear of other peaks in the spectrum, it is, of course, a trivial matter to sum a group of adjacent channels that straddle the peak and provide a measure of the peak intensity. The greatest peak-to-continuum ratio is obtained when the channels containing only the top half of the peak are summed. When peaks from other elements in the specimen or standard overlap the measured peak, we must consider ways to remove the average effects of the other peaks by some sort of mathematical procedure. Procedures that address themselves to the solution of this problem have a variety of names, such as curve fitting, deconvolution, peak stripping, peak unraveling, multiple linear least-squares curve fitting, etc. The essence of these procedures is as follows.

We start with a partial spectrum obtained from the multichannel analyzer. This partial spectrum is a group of adjacent channels fully encompassing the peak we wish to measure and all peaks that overlap this peak. We will call this partial spectrum our "data" spectrum. We next consider a second but corresponding spectrum that we will call a "calculated" spectrum. This spectrum can be created in several ways. One way is to use a mathematical model to describe each peak in the spectrum separately. At each channel in the calculated spectrum, the effects of all the peaks at that channel are summed together. This process is called convolution, and it is by this process that we can "generate" a spectrum. The mathematical model used to describe the shape of each peak has a minimum of three parameters—one to specify the amplitude of the peak, one to specify its position (energy), and one to specify its width. The Gaussian (normal) profile is the model most often used.

An alternative method of "calculating" the second spectrum is to add together simpler measured spectra (i.e., spectra obtained from standards that contain very few elements or are, preferably, pure elements). Each of these spectra is referred to as a "reference" spectrum.

Regardless of which method we use to create the "calculated" spectrum, we compare it to the data spectrum and adjust the parameters of the calculated spectrum until we obtain the "best" agreement between the two. This procedure provides us with a knowledge of the height of each individual peak, which, when added together will be the "data" spectrum.

The procedure by which the parameters in the "calculated" spectrum are adjusted is mainly responsible for the properties of a particular fitting procedure and, mainly therefore, its name. There are two major classifications: linear and nonlinear. In the linear procedures, only the amplitudes of the peaks in the "calculated" spectrum are adjusted, whereas in the nonlinear procedure, one can adjust also the width and position of each peak in the overlapped bundle. Each of the two methods has its advantages and disadvantages, but we must refer the interested reader to the references at the beginning of this section. A sufficiently detailed description is beyond the scope of this chapter.

VII. SUMMARY

In this chapter we have discussed the use of the energy-dispersive spectrometer to obtain characteristic x-ray intensities I_A, I_B, etc., of elements A, B, etc., present in a given specimen. These intensities are necessary for the measurement of chemical composition in thin-film specimens. Various factors that have been discussed include:

- Interfacing the EDS detector to the AEM
- Effect of collimators
- Special precautions necessary because of instrumental artifacts and of operation at high electron-beam energies
- Effect of specimen-preparation artifacts and imaging and diffraction conditions during analysis
- Selection of experimental parameters to maximize the quality of the x-ray data
- Coherent bremsstrahlung phenomena
- Method(s) to extract the desired characteristic x-ray intensities from the continuum background and from overlapping peaks

It appears that with sufficient attention to the EDS problems discussed in this chapter, pertinent characteristic intensity data can be obtained from specific specimen analysis areas when using the AEM.

TABLE OF CHAPTER VARIABLES

A	Active area of EDS detector
eV	Electron volts
E	Energy of coherent bremsstrahlung x-rays
E_0	Accelerating voltage
I_A	Characteristic x-ray intensity (above background) from element A
L	Atomic plane spacing in electron beam direction
mm	Millimeters
M, N	Number of channels in "top-hat" digital filter (M positive N negative)
nm	Nanometers
rad	Radians
s	Seconds
$\underline{s}$	Deviation parameter in electron diffraction pattern
S	Distance from specimen to detector
y_i	Content (counts) in i^{th} channel in filtered spectrum
y_j	Content (counts) in j^{th} channel of original spectrum
Z	Atomic number
α	X-ray take-off angle
α_s	Electron probe convergence semiangle
β	v/c = electron velocity/velocity of light
μm	Micrometers
Ω	X-ray detector collection angle

REFERENCES

Allard, L. F., and Blake, D. F. (1982), "Microbeam Analysis – 1982" (K. F. J. Heinrich, ed.), San Francisco Press, San Francisco, p. 8.

Ball, M. D., and Furneaux, R. C. (1982), "Electron Microscopy and Analysis – 1981" (M. J. Goringe, ed.), Inst. of Physics, Bristol and London, p. 179.

Ball, M. D.; Malis, T. F.; and Steele, D. (1984), "Analytical Electron Microscopy – 1984" (D. B. Williams and D. C. Joy, ed.), San Francisco Press, San Francisco, p. 189.

Bentley, J.; Zaluzec, N. J.; Kenik, E. A.; and Carpenter, R. W. (1979), SEM/1979/II (O. Johari, ed.), SEM Inc., AMF O'Hare, Chicago, p. 581.

Bourdillon, A. J. (1982), "Microbeam Analysis – 1982" (K. F. J. Heinrich, ed.), San Francisco Press, San Francisco, p. 84.

Cherns, D.; Howie, A.; and Jacobs, M. H. (1973), Z. Naturforsch. *28a*, 565.

Cliff, G., and Kenway, P. B. (1982), "Microbeam Analysis – 1982," (K. F. J. Heinrich, ed.), San Francisco Press, San Francisco, p. 107.

Fiori, C. E., and Swyt, C. R. (1981), "Microbeam Analysis – 1981" (R. H. Geiss, ed.), San Francisco Press, San Francisco, p. 320.

Fitzgerald, R.; Keil, K.; and Heinrich, K. F. J. (1968), Science *159*, 528.

Fraser, H. L., and McCarthy, J. P. (1982), "Microbeam Analysis – 1982" (K. F. J. Heinrich, ed.), San Francisco Press, San Francisco, p. 93.

Freund, H. U.; Hansen, J. S.; Kartunnen, E.; and Fink, R. W. (1972), Proc. 1969 Intl. Conf. Radioactivity and Nuclear Spectroscopy (J. H. Hamilton and J. C. Manthuruthil, ed.), Gordon and Breach, New York, p. 623.

Goldstein, J. I.; Newbury, D. E.; Echlin, P.; Joy, D. C.; Fiori, C. E.; and Lifshin, E. (1981), "Scanning Electron Microscopy and X-Ray Microanalysis," Plenum Press, New York.

Goldstein, J. I., and Williams, D. B. (1978), SEM/1978/I (O. Johari, ed.), SEM Inc., AMF O'Hare, Chicago, p. 427.

Heinrich, K. F. J.; Newbury, D. E.; Myklebust, R. L.; and Fiori, C. E. (ed.) (1981), "Energy Dispersive X-ray Spectrometry," NBS Special Pub. 604, US Dept. of Commerce, Washington, DC.

Joy, D. C., and Maher, D. M. (1977), SEM/1977/I (O. Johari, ed.), IITRI, Chicago, p. 325.

Merchant, S. M. (1982), M. S. Thesis, Lehigh University, Bethlehem, Pennsylvania.

Morris, P. L.; Davies, N. C.; and Treverton, J. A. (1977), "Developments in Electron Microscopy and Analysis 1977" (D. L. Misell, ed.), Inst. of Physics, Bristol and London, p. 377.

Nicholson, W. A. P.; Adam, P. F.; Craven, A. J.; and Steele, J. D. (1984), "Analytical Electron Microscopy – 1984" (D. B. Williams and D. C. Joy, ed.), San Francisco Press, San Francisco, p. 258.

Nicholson, W. A. P.; Gray, C. C.; Chapman, J. N.; and Robertson, B. W. (1982), J. Micros. *125*, 25.

Pountney, J. M., and Loretto, M. H. (1980), "Electron Microscopy 1980," Proc. 7th Euro. Cong. Elec. Mic. *3*, 180 (P. Brederoo and V. E. Cosslett, ed.), Elec. Mic. Fdn., Leyden.

Reese, G. M.; Spence, J. C. H.; and Yamamoto, N. (1984), Phil. Mag. A. 49, 697.

Schamber, R. H. (1977), "X-ray Fluorescence Analysis of Environmental Samples" (T. G. Dzubay, ed.), Ann Arbor Sci. Pub., p. 241.

Spence, J. C. H.; Reese, G. M.; Yamamoto, N.; and Kurizki, G. (1983), Phil. Mag. B. *48*, L39.

Spence, J. C. H., and Tafto, J. (1983), J. Microsc. 130, 147.

Statham, P. J. (1977), Anal. Chem. *49*, 2149.

——— (1981), "Energy Dispersive X-ray Spectrometry" (K. F. J. Heinrich, D. E. Newbury, R. L. Myklebust, and C. E. Fiori, ed.), NBS Special Pub. 604, US Dept. of Commerce, Washington, DC, p. 127.

Thomas, L. E. (1980), Proc. 38th EMSA Mtg. (G. W. Bailey, ed.), Claitor's Pub. Div., Baton Rouge, p. 90.

——— (1984), "Analytical Electron Microscopy – 1984" (D. B. Williams and D. C. Joy, ed.), San Francisco Press, San Francisco, p. 358.

Thompson, M. N.; Doig, P.; Edington, J. W.; and Flewitt, P. E. J. (1977), Phil. Mag. *35*, 1537.

Vecchio, K. S. (1985), Proc. 43rd EMSA Mtg. (G. W. Bailey, ed.), San Francisco Press, San Francisco, p. 248.

Williams, D. B. (1984), "Practical Analytical Electron Microscopy in Materials Science," Philips Electron Optics, Mahwah, New Jersey.

Williams, D. B., and Goldstein, J. I. (1981a), "Analytical Electron Microscopy 1981" (R. H. Geiss, ed.), San Francisco Press, San Francisco, p. 39.

——— (1981b), "Energy Dispersive X-ray Spectrometry" (K. F. J. Heinrich, D. E. Newbury, R. L. Myklebust, and C. E. Fiori, ed.), NBS Special Pub. 604, US Dept. of Commerce, Washington, DC, p. 341.

Williams, D. B., and Joy, D. C. (ed.) (1984), "Analytical Electron Microscopy – 1984," San Francisco Press, San Francisco.

Zaluzec, N. J.; Maher, D. M.; and Mochel, P. E. (1981), "Analytical Electron Microscopy – 1981" (R. H. Geiss, ed.), San Francisco Press, San Francisco, p. 25.

CHAPTER 5

QUANTITATIVE X-RAY ANALYSIS

J. I. Goldstein and D. B. Williams

Department of Materials Science and Engineering
Lehigh University, Bethlehem, Pennsylvania

G. Cliff

Department of Metallurgy and Materials Science
Joint University/UMIST, Manchester, England

I. QUANTIFICATION SCHEMES

A. Ratio Method

1. Formulation

In the preceding chapter, methods were described to measure the intensity of characteristic x-rays of the elements A, B, C, etc., present in a specimen of interest. Having obtained x-ray spectra from the desired regions of the specimen, the values of I_A and I_B can be converted to values of C_A and C_B by using either a ratio method or thin-film standards. The former method is more widespread and will be discussed in detail, along with the necessary corrections, when absorption and fluorescence are significant. We shall also discuss the principles and limits of quantification. Detailed theory and derivations of formulae will not be given. These can be found in Philibert and Tixier (1975), Goldstein *et al.* (1977), and Zaluzec (1979).

In electron-transparent thin films, electrons lose only a small fraction of their energy in the film (~5 eV/nm). In addition, few electrons are backscattered (note Chapter 2), and the trajectory of the electrons can be assumed to be the same as the thickness of the specimen film, t. Under these circumstances, the generated characteristic x-ray intensity, I_A^* for element A, can be given by a simplified formula

$$I_A^* = \text{const. } C_A \, \omega_A \, Q_A \, a_A \, t/A_A \tag{1}$$

where

C_A = weight fraction of element A
ω_A = fluorescence yield (the fraction of ionizations that result in x-ray emission) for the K, L, or M characteristic x-ray line of interest
a_A = fraction of the total K, L, or M line intensity that is measured
A_A = atomic weight of A
Q_A = ionization cross section, related to the probability of an electron of a given energy causing ionization of a particular K-, L-, or M-shell of atom A in the specimen.

As shown in Chapter 2, the ionization cross section is given by the general form (Powell, 1976)

$$\sigma = Q = \frac{6.51 \times 10^{-20}}{E_c^2 U} n_s \, b_s \ln c_s U \tag{2}$$

where the dimensions are ionizations/e$^-$/(atom/cm^2). In this equation,

n_s = number of electrons in a shell or subshell (e.g., n_s = 2 for a K-shell, n_s = 8 for an L-shell, and n_s = 18 for an M-shell)
b_s,c_s = constants for a particular shell
E_c = ionization energy for the K-, L-, or M-shell (keV) of given element A
U = overvoltage equal to E_o/E_c where E_o is the operating voltage of the AEM.

In Eq. (1) only Q varies with E_o.

If we assume that the analyzed film is "infinitely" thin, the effects of x-ray absorption and fluorescence can be neglected, and the generated x-ray intensity and the x-ray intensity leaving the film are identical. This assumption is known as the thin-film criterion, the validity of which we will discuss in later sections. However, measured intensity I_A from the EDS spectrum may be different from generated intensity I_A^* because, as already mentioned, the generated x-rays may be absorbed as they enter the EDS detector in the Be window, Au surface layer, and Si dead layer. This absorption process is described briefly in Chapter 4. In addition, if the incoming x-rays are very energetic, they may not be totally absorbed in the active area of the detector. Therefore, the measured intensity is related to the generated intensity by

$$I_A = I_A^* \epsilon_A \tag{3}$$

where

$$\epsilon_A = \{\exp[-(\mu/\rho)_{Be}^{A} \rho_{Be} X_{Be} - (\mu/\rho)_{Au}^{A} \rho_{Au} X_{Au} - (\mu/\rho)_{Si}^{A} \rho_{Si} X_{Si}]\} \times \{1 - \exp[-(\mu/\rho)_{Si}^{A} \rho_{Si} Y_{Si}]\} \tag{4}$$

and

μ/ρ = appropriate mass absorption coefficients of element A in Be, Au, and Si
ρ = densities of Be, Au, and Si
X = thickness of Be window, Au surface layer, and Si dead layer
Y_{Si} = EDS active layer thickness.

From Eq. (1) through (4), it appears that the composition of element A in an analyzed region can be obtained simply by measuring the x-ray intensity I_A and by calculating the constant and other terms. In practice however, this cannot be done easily because many of the geometric factors and constants cannot be obtained exactly. In addition, the specimen thickness varies from one point to another, making it inconvenient to measure t at every analysis point.

If the x-ray intensities (I_A, I_B) of two elements A and B can be measured simultaneously, the procedure for obtaining the concentrations of elements A and B can be greatly simplified. Combining Eq. (1 - 4) to calculate the intensity ratio I_A/I_B, the following relationship for the concentration ratio C_A/C_B is obtained

$$\frac{C_A}{C_B} = \left[\frac{(Q\,\omega\,a/A)_B \times \epsilon_B}{(Q\,\omega\,a/A)_A \times \epsilon_A}\right] \times \frac{I_A}{I_B} \,. \tag{5}$$

In this case the constants and film thickness t drop out of the equation, and the mass concentration ratio is directly related to the intensity ratio. The term in the brackets of Eq. (5) is a constant at a given operating voltage and is referred to as the k_{AB} factor or Cliff-Lorimer factor. Equation (5) is usually given in a simplified form as

$$C_A/C_B = k_{AB} \times I_A/I_B \,. \tag{6}$$

This relationship was applied initially by Cliff and Lorimer (1975) using an EDS detector. The Cliff-Lorimer ratio method has gained great popularity because of its simplicity. It must be borne in mind, however, that the technique is based on the assumption of the thin-film criterion.

In a binary system, using Eq. (6), we can determine C_A and C_B independently since

$$C_A + C_B = 1.0 \,. \tag{7}$$

In ternary and higher order systems, the intensities of all the elements whose mass concentrations are unknown must be measured, or the weight fraction of a particular element in the sample must be known in advance of the analysis. For example, in a three-element analysis

$$C_A/C_B = k_{AB} \times I_A/I_B \tag{8a}$$

$$C_B/C_C = k_{BC} \times I_B/I_C \tag{8b}$$

$$C_A + C_B + C_C = 1 \ . \tag{8c}$$

The ratio method is often referred to as a standardless technique. However, this description is strictly true only when k_{AB} is determined by calculation of the Q, ω, etc., terms given in Eq. (5). More often, the k_{AB} factor is determined using standards where the concentrations C_A, C_B, etc., are known. In this case, the characteristic x-ray intensities are measured, and the Cliff-Lorimer k_{AB} factors are determined directly by using Eq. (6). The standards approach is often more accurate, particularly because ϵ_A, ϵ_B vary from one instrument to another and, as we have observed, may vary even in a single instrument over a period of time.

By convention, k_{AB} factors are compared to Si; that is, tabulation usually occurs as k_{ASi}, k_{BSi}, etc., and the values are called k factors. The relationship between k factors and k_{AB} factors is given by

$$k_{AB} = \frac{k_{ASi}}{k_{BSi}} = \frac{k_A}{k_B} \ . \tag{9}$$

2. Standards or Experimental k Factor Approach

Until recently the only comprehensive ranges of measured k factors for K_α lines of a number of elements were reported by Cliff and Lorimer (1975) and Lorimer *et al.* (1977). These results were obtained by using an EMMA-4 analytical instrument, which has a relatively large probe size ($\sim 10^2$ nm). Limited work has been carried out on modern analytical instruments, but most reported results refer to a single k_{AB} determination for a system of interest to the investigator.

Two new sets of measured k factors were obtained by Wood *et al.* (1981, 1984) and Schreiber and Wims (1981b). Both studies give k factors for K and L lines. The Wood *et al.* data were measured on a Philips EM400T operated at 120 keV, whereas the Schreiber and Wims data were taken on a JEOL-JEM 200C operated at 100 and 200 keV. Figure 1 shows measured k factors plotted as a function of the K_α characteristic x-ray energy of various elements (Wood *et al.*, 1984). The values of k_{ASi} = k are all close to 1.0.

The k_{ASi} or k factor plot of Figure 1 is most useful to geologists where silicon is often the major component element after oxygen. For metallurgists, it is often more useful to display k factors as k_{AFe} factors where element B is iron. Figure 2 shows such experimental k_{AFe} factors obtained by Wood *et al.* (1984) at 120-keV operating potential. The experimental alloys were relatively easy to obtain because a large number of common metals can be alloyed in a convenient, homogeneous manner with the transition metals. Direct determination of k_{AFe} is then possible, and few interpolations are required (Figure 2).

When no convenient alloy is available, for example, Cu-Fe or Zn-Fe, an indirect determination of k_{AFe} is needed. For example, to obtain k_{AFe}, k_{AB} was measured from an A-B alloy, and k_{AFe} was calculated using the measured k_{FeB} factor, so that

$$k_{AFe} = \frac{k_{AB}}{k_{FeB}} \ . \tag{10}$$

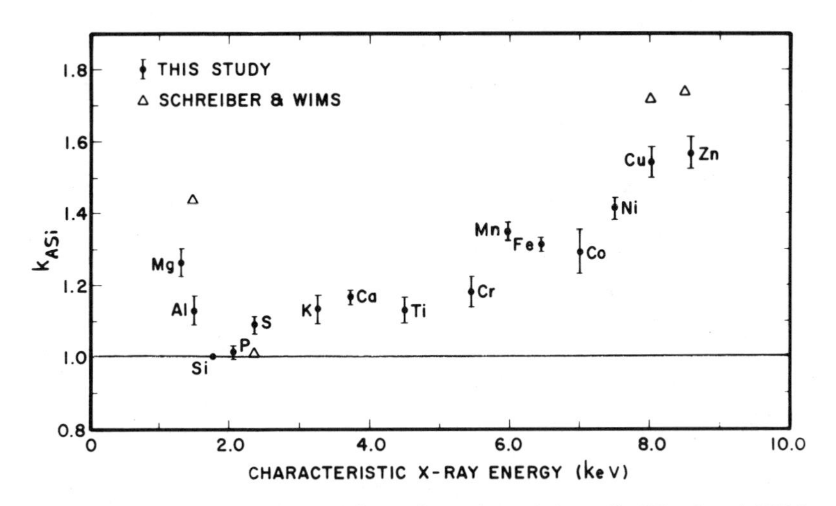

Figure 1 Comparison of measured k factors for K_α lines relative to Si of Wood *et al.* (1984) and Schreiber and Wims (1981a).

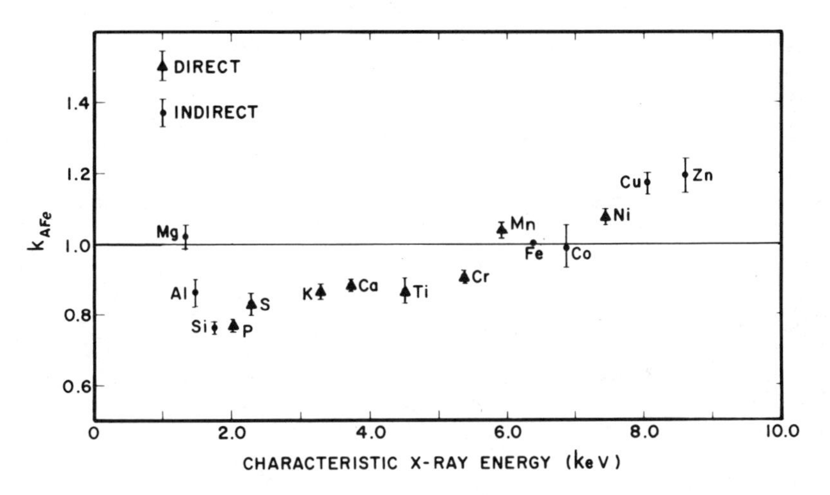

Figure 2 Experimental k_{AFe} factors for K_α lines of Wood *et al.* (1984) for a 120-keV operating potential.

The interpolation error is the sum of the errors in the measurements of k_{AB} and k_{FeB}. The measurement errors for k_{AFe} (Figure 2) approach $\pm 1\%$ relative in a few selected cases. More typically, measurement errors are in the range $\pm 1\%$ to $\pm 4\%$. Since the measurement of concentration C_A is directly related to the k_{AB} factor (Eq. (6)), the accuracy of the ratio method is limited to the systematic error in the measurement of k_{AB}. Figure 3 shows selected experimental k_{AFe} values at 100 and 120 keV as a function of the L_α characteristic energy of various elements (Wood *et al.*, 1984; Goldstein *et al.*, 1977). None of these k_{AFe} values actually used Fe as element B and are therefore interpolated. Table I summarizes measured k_{AFe} factors for K and L lines.

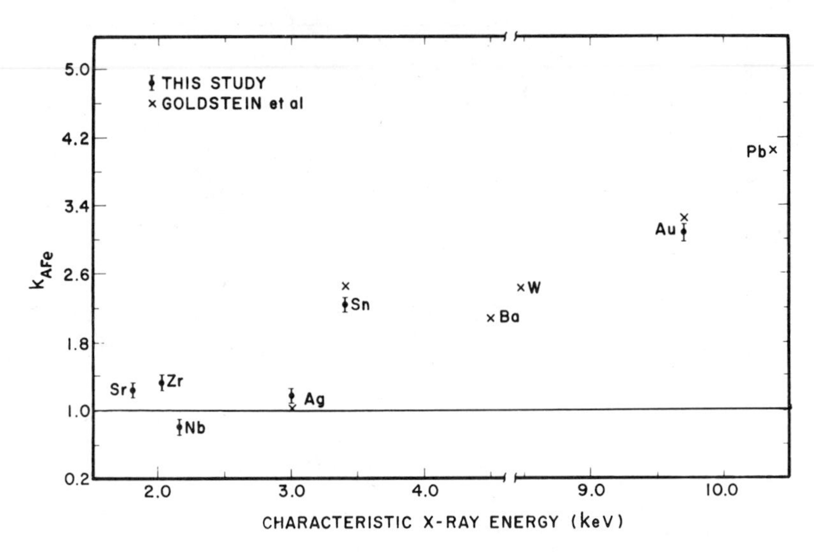

Figure 3 Experimental k_{AFe} factors for L lines of Wood *et al.* (1984) (120 keV) and Goldstein *et al.* (1977) (100 keV). Below a characteristic x-ray energy of 3.2 keV, the k_{AFe} factor is calculated by using the intensity ratio of the L_α and L_β lines of element A.

For elements with characteristic K_α or L_α line energies <4.5 keV (TiK_α), it may not be possible to use measured k factors, since the EDS detector efficiencies may vary significantly from one instrument to another. In addition, many k factors (particularly for L and M lines or for operating voltages other than 100 or 120 keV) have not been measured.

Furthermore, unless the characteristic intensities I_A, I_B, etc., have been measured and the background intensities subtracted in an identical manner, the experimental k factors from different instruments will vary. Given the range of MCA systems used on AEMs, it is difficult to assume that identical procedures have been used. Therefore, it is often advisable for the analyst to measure the necessary k_{AB} factors directly. In many cases, well-characterized alloys or stoichiometric oxides containing the elements of interest are available. When direct k_{AB} factor measurement is not possible, the analyst must resort to using the standardless ratio method and calculate k_{AB} factors directly.

3. Standardless k Factor Approach

To calculate k_{AB} directly, the various terms in the brackets of Eq. (5), ω, a, ϵ, and Q, must be obtained. To calculate fluorescence yield ω for K, L, and M lines, the Burhop (1955) equation can be employed

$$[\omega/(1 - \omega)]^{1/4} = A + BZ + CZ^3 \tag{11}$$

where A, B, and C are constants for a given line, and Z is the atomic number. The ω values obtained from this equation are more reliable at higher atomic numbers, $Z > 24$. For K lines, the fitted ω_K values of Bambynek *et al.* (1972) can be used.

TABLE I

Experimental k_{AFe} Factors for K Lines.

Element	Lorimer *et al.* (1977)* (100 keV)	McGill and Hubbard (1981) (100 keV)	Wood *et al.* (1984) (120 keV)
Na	2.46		
Mg	1.23 ± 0.08	1.16	0.96 ± 0.03
Al	0.92 ± 0.08	0.8	0.86 ± 0.04
Si	0.76 ± 0.08	0.71	0.76 ± 0.004
P			0.77 ± 0.005
S			0.83 ± 0.03
K	0.79	0.77	0.86 ± 0.014
Ca	0.81 ± 0.05	0.75	0.88 ± 0.005
Ti	0.86 ± 0.05		0.86 ± 0.03
Cr	0.91 ± 0.05		0.90 ± 0.006
Mn	0.95 ± 0.05		1.04 ± 0.025
Fe	1.0		1.0
Co	1.05		0.98 ± 0.06
Ni	1.14 ± 0.05		1.07 ± 0.06
Cu	1.23 ± 0.05		1.17 ± 0.03
Zn	1.24		1.19 ± 0.04
Nb			2.14 ± 0.06
Mo	3.38†		3.80 ± 0.09
Ag	6.65†		9.52 ± 0.03

Experimental k_{AFe} Factors for L Lines.

Element	Wood *et al.* (1984) (120 keV)	Goldstein *et al.* (1977) (100 keV)
Sr‡	1.21 ± 0.06	
Zr‡	1.35 ± 0.01	
Nb‡	0.90 ± 0.06	
Ag‡	1.18 ± 0.06	1.04
Sn	2.21 ± 0.07	2.39
Ba		2.18
W		2.43
Au	3.10 ± 0.09	3.27
Pb		4.14

* Error bars from 1975 study.
† Data from 1975 study.
‡ k factors determined for the combined L_α and L_β intensities of these elements.

For L- and M-shell x-rays, the empirical fit for A, B, and C of Colby (1968) to the experimental data of Fink *et al.* (1966) can be used. Table II lists the appropriate values of constants A, B, and C.

The relative intensity factor a is the fraction of the total x-ray emission from a given atomic shell. For K_α, a is expressed as $I_{K\alpha}/(I_{K\alpha} + I_{K\beta})$. For L_α lines, a is given as $I_{L\alpha}/(I_{L\alpha} + I_{L\beta} + ...)$. The a values for K lines can be obtained from Slivinsky and Ebert (1972) for atomic numbers 22 and above and from Heinrich *et al.* (1979) for atomic numbers 15 to 20. Recent measurements of the a factor for K, L, and M lines have been made by Schreiber and Wims (1981a, 1982). The relative intensity values of Schreiber and Wims (1981a) and Slivinsky and Ebert (1972), for K lines, are very similar, differing at most by only 0.2%. The appropriate constants for the calculation of a are given in Table III.

TABLE II

Constants for Calculating the Fluorescence Yield.

	K line (Bambynek *et al.*, 1972)	L line (Colby, 1968)	M line (Colby, 1968)
A	0.015	−0.11107	−0.00036
B	0.0327	0.01368	0.00386
C	-0.64×10^{-6}	-0.21772×10^{-6}	0.20101×10^{-6}

TABLE III

Equations for Calculating a_K, a_L, and a_M Values. (Schreiber and Wims, 1981a, 1982)

X-Ray Line	Atomic Number	Equations
K	11 to 19	$a_K = 1.052 - 4.39 \times 10^{-4} (Z^2)$
	20 to 29	$a_K = 0.896 - 6.575 \times 10^{-4} (Z)$
	20 to 60	$a_K = 1.0366 - 6.82 \times 10^{-3} (Z) + 4.815 \times 10^{-5} (Z^2)$
L	27 to 50	$a_L = 1.617 - 0.0398 (Z) + 3.766 \times 10^{-4} (Z^2)$
	51 to 92	$a_L = 0.609 - 1.619 \times 10^{-3} (Z) - 0.03248 \sin [0.161 (Z - 51)]$
M	60 to 92	$a_M \backsimeq 0.65$

The effect of x-ray absorption in the EDS detector, ϵ, can be calculated by using Eq. (4). The appropriate mass absorption coefficients (μ/ρ) for elements A and B in Be, Si, and Au can be obtained from the tables of Heinrich (1966). Unless specific values are available, the EDS detector parameters as suggested by Zaluzec (1979) can be used, namely, $X_{Be} = 7.6\ \mu m$ (0.3 mil), $X_{Au} = 0.02\ \mu m$, and $X_{Si} = 0.1\ \mu m$. The Si active layer thickness (Y_{Si}) is a function of bias voltage and is nominally 0.3 cm. Maher *et al.* (1981) have considered the relative accuracy of the calculation of ϵ. They found that: (a) low-atomic-number elements are most strongly influenced by the assumptions made for the Be window and Au layer thickness; (b) the calculations are essentially unaffected by the assumed thickness of the Si dead layer; and (c) the analysis of high-energy x-ray lines will be in error if the thickness of the Si active layer is significantly smaller than the assumed value.

The general form of ionization cross section Q is given by Eq. (2). Table IV collects the various values of constants b_s and c_s for K- and L-shells. Zaluzec (1979) suggests the use of a relativistic cross section originally derived by Williams (1933), where the term [ln c_s U] in Eq. (2) is replaced by

$$[\ln (c_s U) - \ln (1 - \beta_\ell^2) - \beta_\ell^2] \ . \tag{12}$$

For K and L lines,

$$c_s = 0.8\ U/[1 - \exp(-\gamma)] \times [1 - \exp(-\delta_c)] \tag{13}$$

where

γ = $1250/(EU^2)$
E = energy of the x-ray line of interest in keV*
δ_c = $E_c/2$ in keV.

The term β_ℓ (V/C) is the velocity of the electrons at E_o (V) relative to that of the speed of light (C). Another form of relativistic correction originally suggested by Dupouy (1968) has been used (Goldstein *et al.*, 1977). This correction can, however, be used only in electron optical calculations (Trebbia, 1984). Therefore, the original Williams (1933) equation is recommended.

Schreiber and Wims (1981b) have used measured k_{AB} values to back-calculate best-fit ionization cross sections and have found that Eq. (2) can be more accurately expressed as

$$Q = \frac{6.51 \times 10^{-20}}{E_c^2\, U^{d_s}}\, n_s\, b_s \ln c_s\, U \ . \tag{14}$$

The d_s term indicates the overvoltage effect that was observed. For K lines, $d_K = 1.0667 - 0.00476(Z)$. For L and M lines, $d_L = d_M = 1.0$. The other terms, b_s and c_s, are given in Table IV.

Table V lists values of ω, a, and ϵ for K and L lines of selected elements. Table VI lists calculated K and L ionization cross sections at 120 keV for selected elements. All the various expressions for Q listed in Table IV were evaluated. The values of Tables V and VI can be combined to calculate k_{AFe} factors (Table VII).

*E_c, the critical ionization energy, is more appropriate here.

TABLE IV

Summary of Constants Used in Various Expressions for the Ionization Cross Section Q_K and Q_L.

Reference	Line	b_s	c_s	Comments
Mott & Massey (1949)	K	0.35	2.42	—
	L	0.25	2.42	—
Green & Cosslett (1961)	K	0.61	1.0	—
Powell (1976)	K	0.9	0.65	$4 \leq U \leq 25$
	L	0.75	0.60	See discussion, Goldstein *et al.* (1977)
Brown (1974); Powell (1976)	K	0.52 + 0.0029 (Z)	1.0	—
	L	0.44 + 0.0020 (Z)	1.0	—
Zaluzec (1979)	K	0.35	Relativistic correction, see text	
	L	0.25	Relativistic correction, see text	
Schreiber & Wims (1981b)	K	$8.874 - 8.158 \ln(Z) + 2.9055 (\ln Z)^2 - 0.35778 (\ln Z)^3$, $Z \leq 30$ 0.661, $Z > 30$	1.0	Investigators found overvoltage effect, see text
	L	$0.2704 + 0.007256 (\ln Z)^3$	1.0	—

TABLE V

Fluorescence Yields (ω), Relative Intensity Factors (a), and Detector Efficiencies (ϵ) for Selected Elements (from Wood *et al.*, 1984).

K Lines

Element	Z	ω_K	a_K	ϵ
Na	11	0.0192	0.999	0.437
Mg	12	0.0265	0.989	0.604
Al	13	0.0357	0.978	0.727
Si	14	0.0469	0.966	0.811
P	15	0.0603	0.953	0.816
S	16	0.0760	0.940	0.792
K	17	0.138	0.894	0.898
Ca	20	0.163	0.883	0.915
Ti	22	0.219	0.882	0.949
Cr	24	0.281	0.880	0.968
Mn	25	0.314	0.880	0.975
Fe	26	0.347	0.879	0.980
Co	27	0.381	0.878	0.983
Ni	28	0.414	0.878	0.983
Cu	29	0.446	0.877	0.987
Zn	30	0.479	0.875	0.991
Mo	42	0.764	0.835	0.993
Ag	47	0.830	0.822	0.995

L Lines

Element	Z	ω_K	a_K	ϵ
Sr	38	0.0242	0.648	0.80
Zr	40	0.0308	0.628	0.821
Nb	41	0.0345	0.618	0.845
Ag	47	0.0630	0.578	0.898
Sn	50	0.0815	0.569	0.898
Ba	56	0.126	0.495	0.948
W	74	0.304	0.506	0.99
Au	79	0.356	0.513	0.993
Pb	82	0.386	0.507	0.995

TABLE VI

Calculated Ionization Cross Sections (Q) at 120 keV for Selected Elements (From Wood *et al.*, 1984).

K Lines

Element	Q_{MM}*	Q_{GC}*	Q_P*	Q_{BP}*	Q_{S+W}*	Q_Z*
Na	1.811×10^{-21}	2.668×10^{-21}	3.585×10^{-21}	2.414×10^{-21}	4.840×10^{-21}	4.450×10^{-21}
Mg	1.439×10^{-21}	2.106×10^{-21}	2.819×10^{-21}	1.916×10^{-21}	3.798×10^{-21}	3.450×10^{-21}
Al	1.163×10^{-21}	1.691×10^{-21}	2.254×10^{-21}	1.550×10^{-21}	3.037×10^{-21}	2.720×10^{-21}
Si	9.554×10^{-22}	1.381×10^{-21}	1.832×10^{-21}	1.269×10^{-21}	2.473×10^{-21}	2.183×10^{-21}
P	7.957×10^{-22}	1.142×10^{-21}	1.510×10^{-21}	1.550×10^{-21}	2.042×10^{-21}	1.780×10^{-21}
S	6.712×10^{-22}	9.578×10^{-22}	1.261×10^{-21}	8.893×10^{-22}	1.708×10^{-21}	1.470×10^{-21}
K	4.238×10^{-22}	5.935×10^{-22}	7.710×10^{-22}	5.600×10^{-22}	1.042×10^{-21}	8.718×10^{-22}
Ca	3.692×10^{-22}	5.138×10^{-22}	6.648×10^{-22}	4.870×10^{-22}	8.950×10^{-22}	7.450×10^{-22}
Ti	2.861×10^{-22}	3.932×10^{-22}	5.043×10^{-22}	3.760×10^{-22}	6.690×10^{-22}	5.570×10^{-22}
Cr	2.265×10^{-22}	3.074×10^{-22}	3.907×10^{-22}	2.970×10^{-22}	5.065×10^{-22}	4.260×10^{-22}
Mn	2.030×10^{-22}	2.737×10^{-22}	3.462×10^{-22}	2.660×10^{-22}	4.421×10^{-22}	3.760×10^{-22}
Fe	1.826×10^{-22}	2.446×10^{-22}	3.080×10^{-22}	2.390×10^{-22}	3.865×10^{-22}	3.330×10^{-22}
Co	1.648×10^{-22}	2.194×10^{-22}	2.749×10^{-22}	2.150×10^{-22}	3.381×10^{-22}	2.960×10^{-22}
Ni	1.494×10^{-22}	1.975×10^{-22}	2.462×10^{-22}	1.950×10^{-22}	2.961×10^{-22}	2.371×10^{-22}
Cu	1.358×10^{-22}	1.783×10^{-22}	2.212×10^{-22}	1.770×10^{-22}	2.592×10^{-22}	2.360×10^{-22}
Zn	1.236×10^{-22}	1.613×10^{-22}	1.990×10^{-22}	1.605×10^{-22}	2.266×10^{-22}	2.120×10^{-22}
Mo	0.473×10^{-22}	0.564×10^{-22}	0.643×10^{-22}	0.593×10^{-22}	0.816×10^{-22}	0.680×10^{-22}
Ag	0.399×10^{-22}	0.385×10^{-22}	0.421×10^{-22}	0.414×10^{-22}	0.555×10^{-22}	0.451×10^{-22}

L Lines

Element	Q_{MM}*	Q_P*	Q_{BP}*	Q_{S+W}*	Q_Z*
Sr	2.560×10^{-21}	5.590×10^{-21}	4.37×10^{-21}	5.714×10^{-21}	5.800×10^{-21}
Zr	2.177×10^{-21}	4.703×10^{-21}	3.72×10^{-21}	4.942×10^{-21}	4.830×10^{-21}
Nb	2.014×10^{-21}	4.330×10^{-21}	3.45×10^{-21}	4.611×10^{-21}	4.423×10^{-21}
Ag	1.325×10^{-21}	2.763×10^{-21}	2.28×10^{-21}	3.172×10^{-21}	2.750×10^{-21}
Sn	1.090×10^{-21}	2.239×10^{-21}	1.88×10^{-21}	2.662×10^{-21}	2.200×10^{-21}
Ba	7.630×10^{-22}	1.516×10^{-21}	1.32×10^{-21}	1.925×10^{-21}	1.460×10^{-21}
W	3.290×10^{-22}	5.898×10^{-22}	5.77×10^{-22}	8.904×10^{-22}	5.540×10^{-22}
Au	2.690×10^{-22}	4.666×10^{-22}	4.72×10^{-22}	7.363×10^{-22}	4.370×10^{-22}
Pb	2.393×10^{-22}	4.065×10^{-22}	4.19×10^{-22}	6.584×10^{-22}	3.810×10^{-22}

*MM - Mott-Massey
GC - Green-Cosslett
P - Powell
BP - Brown-Powell
Z - Zaluzec
S+W - Schreiber and Wims

TABLE VII

Calculated k_{AFe} Factors Evaluated at 120 keV for Selected Elements (from Wood *et al.*, 1984).

Theoretical k_{AFe} Factors for K Lines

Element	k_{MM}*	k_{GC}*	k_P*	k_{BP}*	k_{S+W}*	k_Z*
Na	1.42	1.34	1.26	1.45	1.17	1.09
Mg	1.043	0.954	0.898	1.03	0.836	0.793
Al	0.893	0.822	0.777	0.877	0.723	0.696
Si	0.781	0.723	0.687	0.769	0.638	0.623
P	0.813	0.759	0.723	0.803	0.671	0.663
S	0.827	0.776	0.743	0.817	0.688	0.689
K	0.814	0.779	0.755	0.807	0.701	0.722
Ca	0.804	0.774	0.753	0.788	0.702	0.727
Ti	0.892	0.869	0.853	0.888	0.807	0.835
Cr	0.938	0.925	0.917	0.936	0.887	0.909
Mn	0.98	0.974	0.970	0.979	0.953	0.965
Fe	1.0	1.0	1.0	1.0	1.0	1.0
Co	1.063	1.069	1.074	1.066	1.096	1.079
Ni	1.071	1.085	1.096	1.074	1.143	1.23
Cu	1.185	1.209	1.227	1.19	1.31	1.24
Zn	1.245	1.278	1.305	1.255	1.44	1.32
Mo	3.13	3.52	3.88	3.27	3.84	3.97
Ag	4.58	5.41	6.23	4.91	5.93	6.28

k_{AFe} Factors for L Lines

Element	k_{MM}	k_P	k_{BP}	k_{S+W}	k_Z
Sr†	1.73	1.33	1.32	1.64	1.39
Zr†	1.62	1.26	1.24	1.51	1.33
Nb†	1.54	1.21	1.18	1.43	1.28
Ag†	1.43	1.16	1.09	1.26	1.26
Sn	2.55	2.09	1.93	2.21	2.3
Ba	2.97	2.52	2.25	2.49	2.83
W	3.59	3.37	2.68	2.8	3.88
Au	3.94	3.84	2.94	3.05	4.43
Pb	4.34	4.31	3.25	3.34	4.97

*MM – Mott-Massey
GC – Green-Cosslett
P – Powell
BP – Brown-Powell
S+W – Schreiber and Wims
Z – Zaluzec
†k_{AFe} evaluated for combined L_α and L_β intensities

Figures 4 and 5 show a comparison between measured and calculated k_{AFe} factors for K lines at a 120-keV operating potential (Wood *et al.*, 1984). Figure 4 shows the range of the calculated k_{AFe} factors for K_α lines using the six expressions available to calculate ionization cross sections as listed in Table VII. The data of Wood *et al.* (1984) are also included. Theoretical k factors using the Brown (1974)-Powell (1976), Green and Cosslett (1961), and Mott and Massey (1949) expressions for the ionization cross section show the closest agreement with the most recent experimental study (Figure 5). The average difference between these calculated values and the experimental data is $\sim\pm4\%$ (relative). The Brown (1974)-Powell (1976) cross section yields k factors that agree very well with the measured data and is the best expression available to date. By contrast, k factors calculated by using the ionization cross section expression of Schreiber and Wims (1981b) and Zaluzec (1979) show an average difference with experimental data of $\sim\pm14\%$ (relative). The general merits of calculating k factors, particularly in terms of the choice of ionization cross section, have been discussed in detail by Williams *et al.* (1984).

Earlier studies have also pointed out the range of calculated k_{AB} factors. Maher *et al.* (1981) show that the uncertainties in using the Mott and Massey vs the Zaluzec modified Q equations lead to an error in the calculated k_{AB} of $\lesssim\pm15\%$. Goldstein *et al.* (1977) show a similar uncertainty when comparing calculated k_{AB} values using the cross sections of Mott and Massey (1949), Green and Cosslett (1961), and Powell (1976). However, by using the Brown (1974) and Powell (1976) ionization cross section expression, it may be possible to calculate k_{AB} factors for K_α lines with the same precision as the measurement of selected k_{AB} factors.

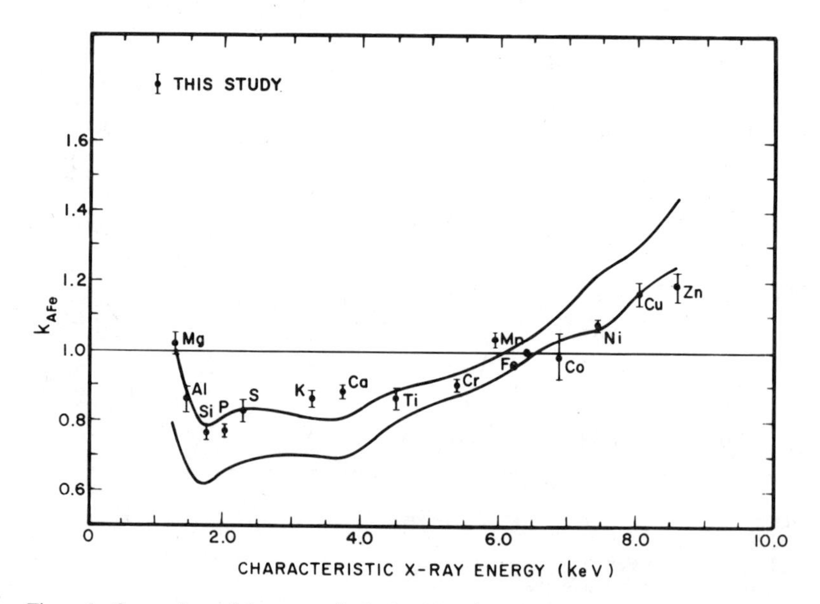

Figure 4 Comparison of the range of calculated k_{AFe} factors with measured k_{AFe} factors for K_α lines at a 120-keV operating potential (Wood *et al.*, 1984).

The discrepancies between measured and calculated k_{AB} factors for L lines are much larger than those for K lines. Figure 6 shows the range of the calculated k_{AFe} factors for L_α lines using the five expressions available to calculate ionization cross sections as listed in Table VII. The data of Wood *et al.* (1984) and Goldstein *et al.*

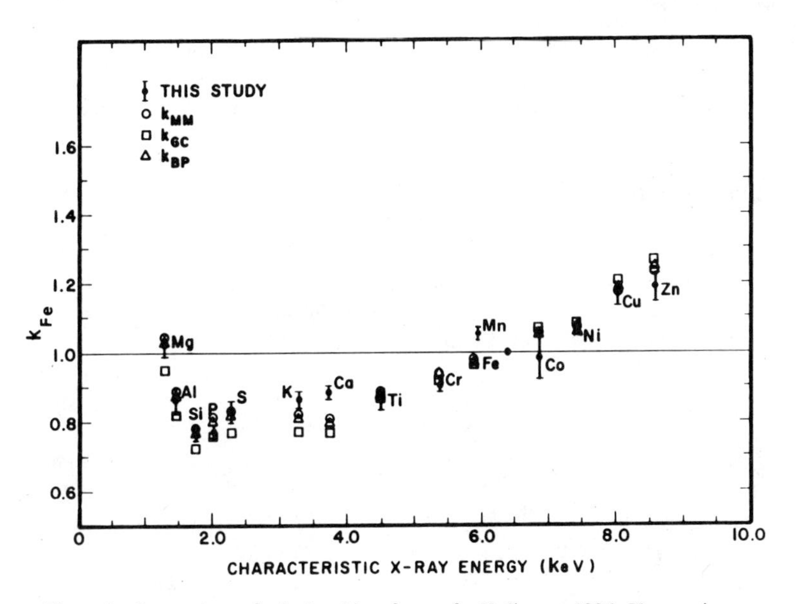

Figure 5 Comparison of calculated k_{AFe} factors for K_α lines at 120-keV operating potential of Brown (1974)-Powell (1976), Green-Cosslett (1961), and Mott-Massey (1949) with the measured k_{AFe} factors of Wood *et al.* (1984).

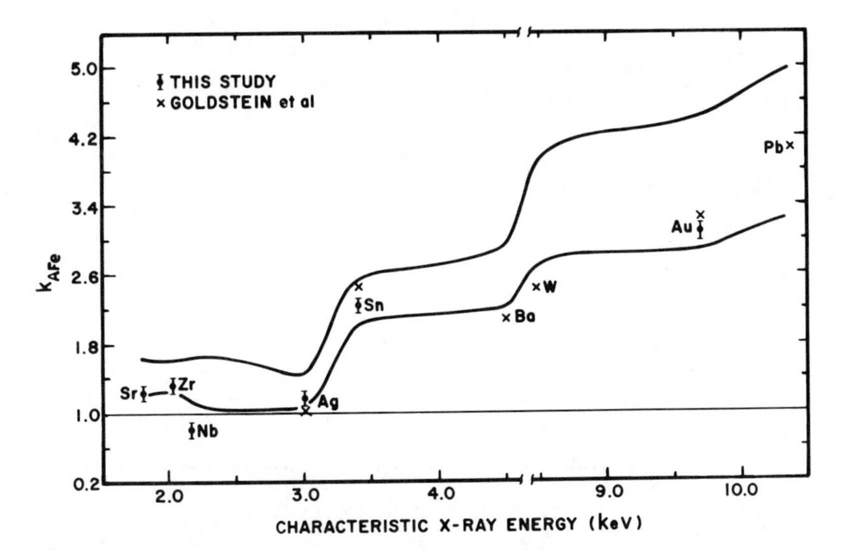

Figure 6 Comparison of the range of calculated k_{AFe} factors with measured k_{AFe} factors for L lines at 120-keV operating potential (Wood *et al.*, 1984) and at a 100-keV operating potential (Goldstein *et al.*, 1977).

(1977) at 100 keV are also included. Theoretical k factors using the Brown (1974) and Powell (1976), and Powell (1976) expressions for the ionization cross section show the closest agreement with the most recent experimental study (Figure 7). The average difference between these calculated values and the experimental data is $\sim\pm 12\%$ (relative). By contrast, k factors calculated by using the ionization cross section expression of Mott and Massey (1949) show an average difference with the experimental data of $\sim\pm 36\%$ (relative). At this time it appears that calculated k_{AB} factors for L_α lines are too inaccurate for quantitative compositional analysis. It is necessary to produce k_{AB} standards if analyses with errors of $<\pm 10\%$ (relative) are desired.

B. Thin-Film Standards

An alternative procedure to that of the ratio method uses thin-film standards. As discussed by Zaluzec (1981a), the absolute intensity of element A measured in an unknown thin-film specimen I_A^U relative to the absolute intensity of the same x-ray line in a thin-film standard I_A^S is given by an equation of the form

$$\frac{I_A^U}{I_A^S} = \frac{(\eta\,\rho t\,C_A)^U}{(\eta\,\rho t\,C_A)^S} \tag{15}$$

where

η = absolute electron flux
ρ = specimen density
t = thickness
C_A = concentration

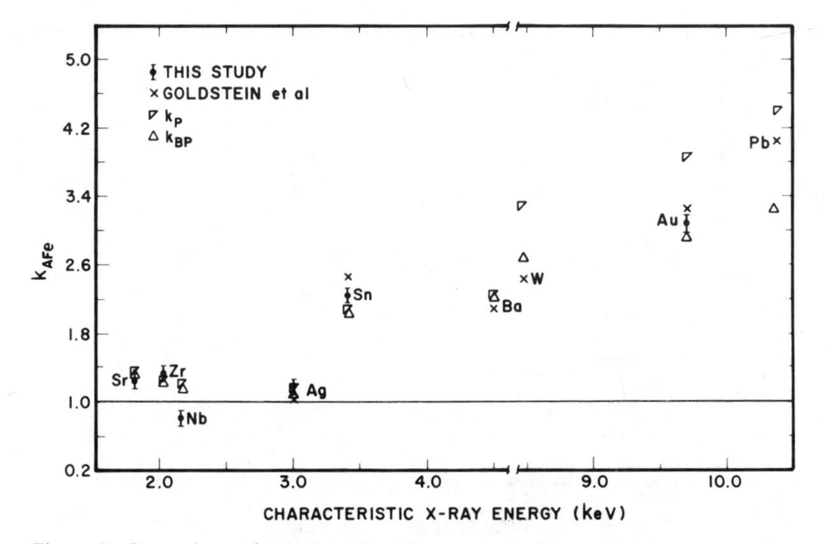

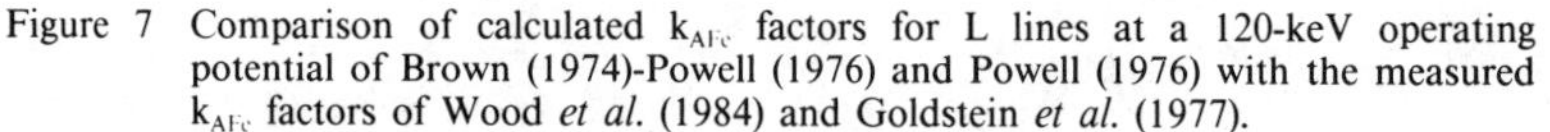

Figure 7 Comparison of calculated k_{AFe} factors for L lines at a 120-keV operating potential of Brown (1974)-Powell (1976) and Powell (1976) with the measured k_{AFe} factors of Wood *et al.* (1984) and Goldstein *et al.* (1977).

in the standard S or the unknown U. In this case, neither k_{AB} or ϵ_A values need be calculated. Unfortunately, one must know η, ρ, and t for both standard and unknown and have a standard for each element measured in the unknown. As discussed in the next section on absorption, accurate measurements of t are difficult to make. In this case, t must be measured for both the unknown and for all the standards. These disadvantages are so scrious that this technique has seen little practical use. Brown *et al.* (1981) have used pure element Al, Mo, and Ni standards to determine the composition of β-NiAl and δ-NiMo alloys. In addition to difficulties in preparing some pure metal standards, changes in beam current lead to instabilities in the absolute electron flux. Unless preparation of standards becomes easier and thickness measurements become more accurate, the thin-film standards technique will not be widely used. The ratio method will continue to be employed by the vast majority of AEM users.

II. ABSORPTION CORRECTION

A. Formulation

The thin-film criterion states that the effects of x-ray absorption and fluorescence can be neglected. However, this is not always possible. Based on the original work of Tixier and Philibert (1969) and König (1976), Goldstein *et al.* (1977) have derived an equation to correct the k_{AB} factor for the preferential absorption of x-rays from elements A and B

$$k_{AB} = k_{AB})_{TF} \left[\frac{\int_o^t \phi_B (\rho t) \exp\{-\mu/\rho]^B_{SPEC} \operatorname{cosec} \alpha(\rho t)\} \, dt}{\int_o^t \phi_A (\rho t) \exp\{-\mu/\rho]^A_{SPEC} \operatorname{cosec} \alpha(\rho t)\} \, dt} \right] \tag{16}$$

where

$k_{AB})_{TF}$ = absolute value of k_{AB} when there is no absorption or fluorescence (thickness t = 0)

$\phi_{A,B}(\rho t)$ = depth distribution of x-ray production from element A or B as a function of mass thickness (ρt)

$\mu/\rho]^{A,B}_{SPEC}$ = mass absorption coefficient for x-rays from element A or B in the specimen

α = x-ray take-off angle.

The term t $\times$ cosec α is the path length for x-ray absorption in a parallel-sided thin foil of thickness t when the thin foil is normal to the incident electron beam (Figure 8). The absorption correction factor is calculated for each analysis point. The factor approaches 1.0 when no absorption occurs.

Assuming $\phi_A(\rho t) = \phi_B(\rho t) \backsimeq 1$ (i.e., x-ray production is uniform throughout the specimen), the integrals in Eq. (16) can be evaluated to give

$$k_{AB} = k_{AB})_{TF} \left[\frac{\mu/\rho]^A_{SPEC}}{\mu/\rho]^B_{SPEC}} \right] \times \left[\frac{1 - \exp - \{\mu/\rho]^B_{SPEC} \operatorname{cosec} \alpha(\rho t)\}}{1 - \exp - \{\mu/\rho]^A_{SPEC} \operatorname{cosec} \alpha(\rho t)\}} \right]. \tag{17}$$

ABSORPTION CORRECTION FACTOR

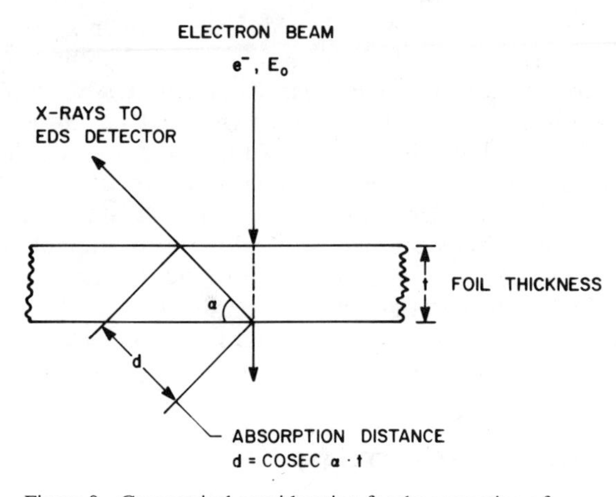

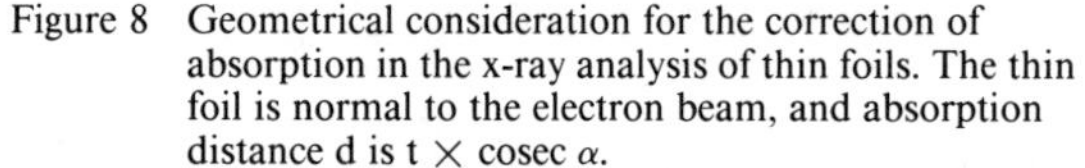
Figure 8 Geometrical consideration for the correction of absorption in the x-ray analysis of thin foils. The thin foil is normal to the electron beam, and absorption distance d is t $\times$ cosec α.

The assumption that absorption is significant only if I_A/I_B or the absorption correction factor is changed by $> \pm 10\%$ was justified originally by Goldstein *et al.* (1977) because this was the level of accuracy with which k_{AB} factors and the values of I_A, I_B could be experimentally determined. Tixier and Philibert (1969) developed a similar criterion for absorption of either I_A or I_B. With recent improvements in STEM electron optics and the development of clean specimen environments, measurements of k_{AB} factors and I_A, I_B can be made more accurately. As we have already discussed, k_{AB} factors can be routinely determined with an error of between $\pm \sim 1\%$ and 4%. Given this precision, the arbitrary definition of significant absorption should be lowered accordingly so that the correction factor in Eq. (17) is as low as 3%. In many cases (Table VIII), such a new definition of significant absorption will reduce the thickness of specimens required to avoid absorption corrections to small values (e.g., <10 nm in NiAl), which are difficult to obtain in practice. Also, the absorption correction must now be applied in many analyses where it was previously thought to be unnecessary.

To apply the correction factor for absorption to k_{AB}, the values of (μ/ρ), ρ, and the absorption path length (which involves take-off angle α and film thickness t) must be accurately known. The following sections discuss the limitations in applying Eq. (17) and make some suggestions for minimizing the correction.

B. Mass Absorption Coefficient

Values of the mass absorption coefficient are accurately known from earlier studies of x-ray absorption and may be obtained from one of several sets of tables available in the literature (Heinrich, 1966; Henke and Ebisu, 1974). It is worth noting here that when $\mu/\rho]^A_{SPEC} \left(= \sum_i C_i \, \mu/\rho]^A_i \right)$ is computed for any given system, *every* component i in that system has to be included, even if the x-rays from certain elements are not detected by conventional EDS. Significant absorption of x-rays from elements Na through Si can occur if O, N, and/or C are present in substantial amounts. This problem has been studied by Bender *et al.* (1980) in NiO-MgO, where correcting the Mg intensity for absorption by Ni does not fully remove the absorption effect (Figure 9). Only when absorption of Mg K_α x-rays by oxygen is considered is the variation in x-ray intensities approximately constant with thickness. Similar effects are to be expected in many ceramic systems. Clearly, when thin-window or windowless EDS systems are available on AEM instruments (Thomas, 1980), consideration of absorption by, and absorption of, x-rays from elements Z = 5 and above will be more important.

TABLE VIII

Thickness Limitations for Application of the Thin-Film Criterion. Limitation is defined in terms of (a) 10% and (b) 3% change in the k_{AB} factor (from Williams, 1984).

	Thickness (nm)		Absorbed
Material	a	b	X-Ray Line
Al-7% Zn	336	94	Al K_α
$CuAl_2$	40	12	Al K_α
NiAl	32	9	Al K_α
Ni_3Al	20	7	Al K_α
Al_6Fe	155	43	Al K_α
Ag_2Al	33	10	Al K_α, Ag L_α
Ag_3Al	31	9	Al K_α, Ag L_α
FeS	180	50	S K_α
Fe_2S	104	36	S K_α
FeS_2	286	79	S K_α
FeP	119	34	P K_α
Fe_3P	77	22	P K_α
Fe-5% Ni	322	89	Ni K_α
Cu_3Au	36	10	Cu K_α, Au M_α
CuAu	38	11	Cu K_α, Au M_α
MgO	304	25	Mg K_α, O K_α
Al_2O_3	113	14	Al K_α, O K_α
SiO_2	167	14	Si K_α, O K_α
SiC	13	3	Si K_α, C K_α
Si_3N_4	413	6	Si K_α, N K_α

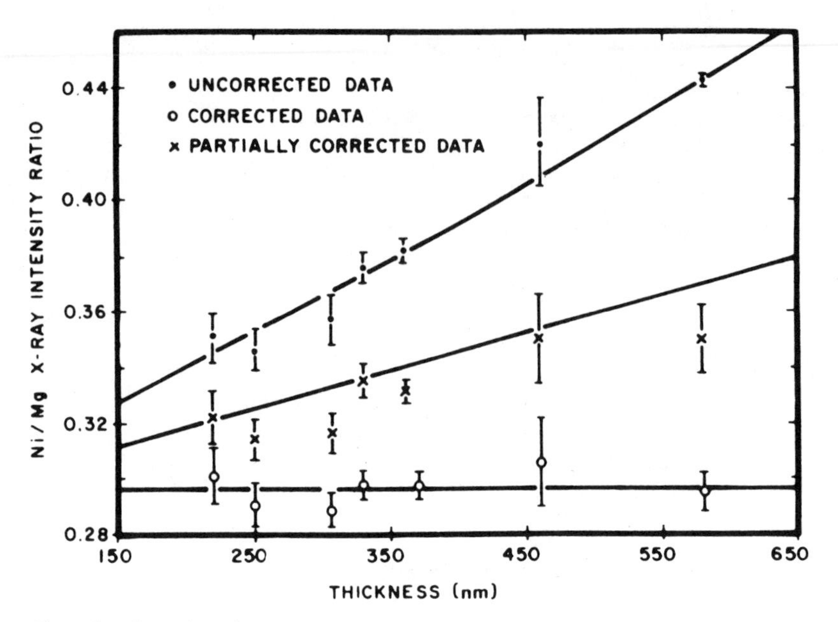

Figure 9 Absorption of Mg K_α x-rays in Mg-10% NiO as a function of thickness; the upper line is the uncorrected Ni K_α/Mg K_α intensity ratio, the middle line shows results of correcting for absorption by Ni, and the lower line shows the results of correcting for absorption by Ni and O. The lower line slightly overcorrects for absorption as thickness increases (Bender *et al.*, 1980).

C. Depth Distribution of X-Ray Production

The depth distribution of x-ray production [$\phi(\rho t)$] is usually assumed to be constant throughout the thin film, but this is by no means a valid assumption in all materials. For example, Kyser (1979) has calculated that $\phi(\rho t)$ should increase by 5% and 20% in 400 nm of Cu and Au, respectively, thus indicating that Eq. (16), and therefore (17), are not valid descriptions of any absorption correction involving these elements in foils of this thickness. Stenton *et al.* (1981) have experimentally determined $\phi(\rho t)$ for up to 380 nm of Ni and found that at 120 keV the assumption of a constant value of $\phi(\rho t)$ is valid only up to ~80 nm. Above this value, $\phi(\rho t)$ increases up to ~30% at 380 nm. These experimental data were greater than the 10% increase predicted up to 400 nm by using Monte Carlo techniques at 120 keV (Newbury, 1981). Both these data are shown in Figure 10. Until more $\phi(\rho t)$ data are available, however, Eq. (17) should be used, but with the understanding that (particularly for Z $\gtrsim$ 25) the assumption of a constant $\phi(\rho t)$ may be incorrect at most reasonable specimen thicknesses, and this may affect significantly the value of any absorption-corrected k_{AB} factor. If an increase in $\phi(\rho t)$ is not taken into account, Eq. (17) will overcompensate for absorption.

D. Specimen Density (ρ)

Specimen density is clearly a function of composition, and any absorption correction procedure is an iterative process in which an approximate density and μ/ρ value are assumed before determining C_A and C_B. When these are determined, the corrected values should be used on a second iteration, etc., until a self-consistent result is obtained. In this process, it should be noted that the density of a particular phase is not always a linear function of composition, and in such cases the variation must be known or calculated.

E. X-Ray Absorption Path Length

1. Thin Foil Normal to the Primary Electron Beam

Assuming a parallel-sided thin foil of thickness t, with the primary beam at normal incidence, the x-ray path length through the specimen is t $\times$ cosec α, where α is the x-ray detector take-off angle as shown in Figure 8. Clearly, the higher the take-off angle α, the shorter will be the path length and the less will be the absorption correction necessary for a given thickness (Eq. (17)). This fact, together with the improved peak-to-background ratio expected for a high take-off angle detector (Zaluzec, 1979), is a strong argument in favor of orienting the detector to look down on the specimen.

Microscope stage constraints usually result in a loss of collection angle with increasing α, and some instruments compromise with a relatively low value of $\sim$20°. Nevertheless, as shown in Figure 11, the variation in the absorption correction factor as a function of α begins to become much more significant when α is 15° or less.

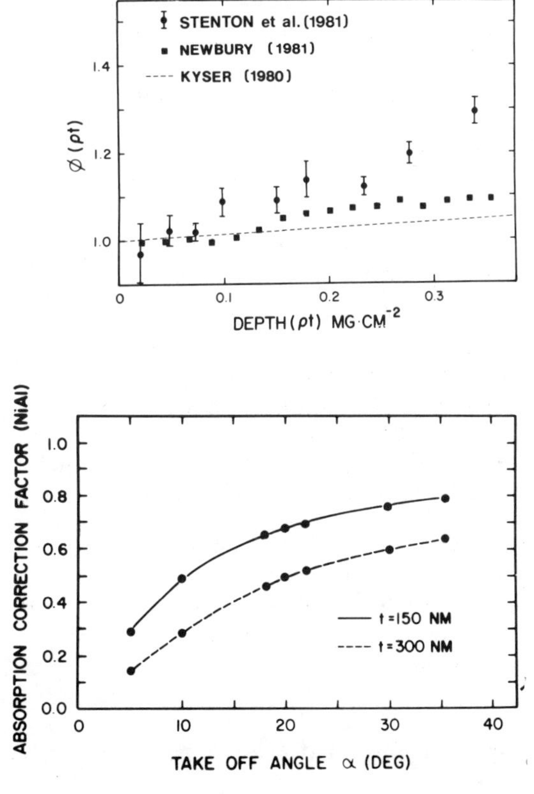

Figure 10 Increase in Ni K_α x-ray production as a function of mass thickness at 120 keV determined experimentally by Stenton *et al.* (1981) and calculated by Monte Carlo techniques by Newbury (1981). Also shown are Monte Carlo calculations by Kyser (1979) for Cu at 100 keV (Stenton *et al.*, 1981).

Figure 11 Variation in the absorption correction factor (Eq. (17)) for two thicknesses of NiAl as a function of take-off angle α.

At low α, relatively small shifts in the specimen with respect to the detector (and more importantly, vice-versa) will result in significant changes in α and therefore in the absorption correction factor (Williams and Goldstein, 1981). For significant absorption effects, it is necessary that the value of α should be known accurately.

2. Thin Foil Tilted With Respect to the Primary Electron Beam

In some AEMs, the x-ray detector axis lies in the same plane as the specimen, and therefore one must tilt the specimen holder by as much as 20° to 30° before characteristic x-rays from the specimen may be measured. If the thin foil is tilted with respect to the primary beam, the absorption path length is not equal to $t \times \text{cosec}\ \alpha$. Zaluzec (1979) and Zaluzec *et al.* (1981) have developed equations for calculating the actual path length caused by nonnormal incidence of the electron beam on parallel-sided thin foils of thickness t. Figure 12 shows the configuration that was developed.

A plane section through the specimen containing the incident-beam direction and the EDS detector is shown. Take-off angle α is the same as shown in Figure 8 and defined as the angle between the perpendicular to the electron beam, at the eucentric height of the specimen stage, and the detector axis. Angle β is measured between the incident probe and the sample surface in the plane of the figure. Distance t^* is the projected specimen thickness along the incident-beam direction to the eucentric height halfway through the foil and is equal to $t/2 \sin \beta$. As β decreases, t^* will increase. Absorption distance d from the eucentric point is given by

$$\frac{t^* \sin \beta}{\cos (\beta - \alpha)} = \frac{t/2}{\cos (\beta - \alpha)} \,. \tag{18}$$

The total absorption distance from the bottom of the foil to the top of the foil is twice the distance from the eucentric point, 2d. Therefore, the total absorption distance equals

$$\frac{2t^* \sin \beta}{\cos (\beta - \alpha)} = \frac{t}{\cos (\beta - \alpha)} \,. \tag{19}$$

At normal incidence, $\beta = 90°$, and therefore, the total absorption distance $= 2t^* \times \text{cosec}\ \alpha = t \times \text{cosec}\ \alpha$ since at p $= 90°$, $2t^* = t$.

Upon examination of Eq. (16) and (17), the integrals from the absorption correction are evaluated from $t = 0$ to $t = t$, where t is the foil thickness. For nonnormal electron-beam incidence, the absorption correction is evaluated along the specimen thickness parallel to the beam direction, from $t^* = 0$ to $t^* = 2t^*$. The absorption correction factor (Eq. (17)) is modified so that the path length $t \times \text{cosec}\ \alpha$ is replaced by the term $2t^* \sin \beta/\cos (\beta - \alpha)$, which is equal to $t/\cos (\beta - \alpha)$ (Eq. (19)). For 45° beam incidence and a 45° take-off angle, the absorption path length is decreased by ~40% over that for 90° beam incidence and a 45° take-off angle. The value of t^* is increased, however, by ~40%. As long as $\phi_A \simeq \phi_B \simeq 1.0$, the increased excitation distance does not affect the absorption correction.

If the EDS detector axis is perpendicular to the tilt axis, β is equal to 90° minus the tilt angle. However, in a number of AEM instruments, the EDS detector axis is not perpendicular to the tilt axis. Therefore, angle β in the electron-beam detector axis plane must be calculated. The solid geometry for the correction has been developed by Zaluzec (1981b) and Zaluzec *et al.* (1981) and is given in the correct form in the latter paper.

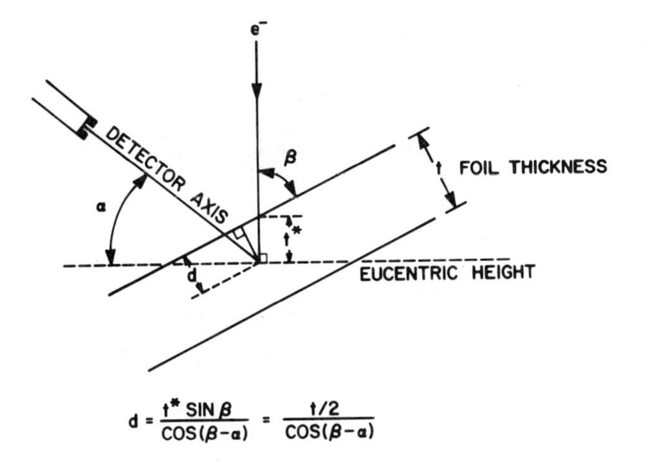

Figure 12 Geometrical consideration for the correction of absorption when the thin foil is not perpendicular to the electron beam (Zaluzec *et al.*, 1981).

The effects of the tilt angle and the take-off angle on the absorption correction for thin films have been reviewed by Zaluzec *et al.* (1981). Overall, it appears that intermediate take-off angles of 15° to 20° with tilted specimens are the most versatile and optimum configuration.

3. Thin-Foil Thickness t

The measurement of the thickness of t at each analysis point is a major problem in correcting for absorption. There are several ways to measure t as reported in the literature, none of which is universally applicable and accurate. The contamination-spot-separation method first used by Lorimer *et al.* (1976) is universally applicable if contamination can be generated. During analysis, a contamination spot is produced on the top and the bottom of the foil (Figure 13). The foil is then tilted through an appropriate angle, and a photograph is taken of the two contamination spots now observed. The distance between the two spots is measured, and by simple geometry and a knowledge of the sample magnification, the thickness t can be calculated. Unfortunately, the accuracy of the technique has been called into question on several occasions (Love *et al.*, 1977; Stenton *et al.*, 1981; Rae *et al.*, 1981).

Stenton *et al.* (1981) carried out contamination-spot-separation measurements on evaporated thin films and compared the data with interferometry measurements. Errors from ~50% to ~200% were reported, with the contamination spots invariably overestimating the thickness, thus causing an overcorrection in any absorption calculation. Rae *et al.* (1981) have reported the presence of relatively broad contamination deposits beneath the sharp cones sometimes apparent on the specimen and describe the overall configuration as that of a "witch's hat," as shown in Figure 14(a). By comparison with Figure 14(b), which is the generally assumed morphology from which measurements are made, it is easy to see the reason for overestimation. This effect is clearly demonstrated in Figure 14(c), which shows contamination spots deposited above and below a grain boundary. The boundary fringe contrast, which is faintly visible, does not intercept the top and bottom contamination deposits. Errors associated with this measurement technique may also contribute to any overcorrection observed in data that have been adjusted for absorption.

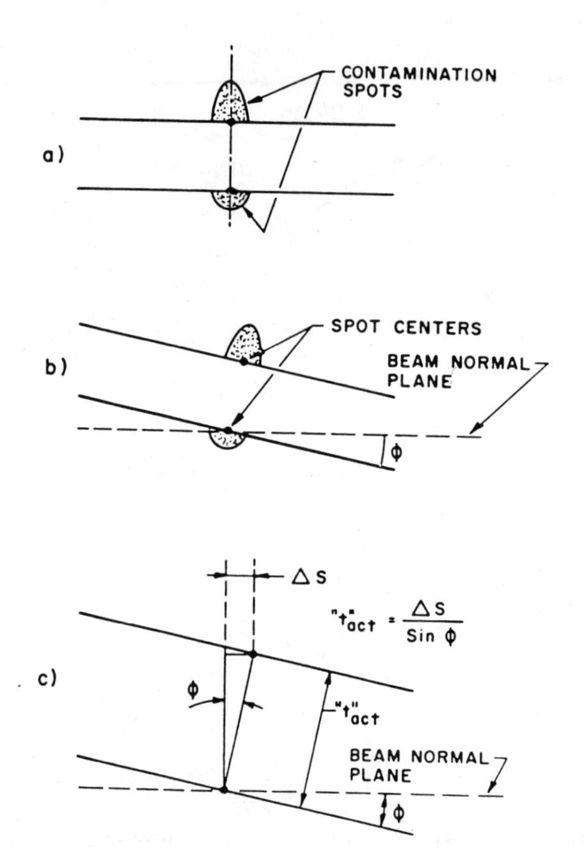

Figure 13 Illustration of the steps used for thickness determination.

The second most popular and the most accurate method of determining thin-foil thickness for crystalline materials is that of measuring the spacing of the Kossel-Möllenstedt fringes in convergent-beam diffraction patterns (Kelly *et al.*, 1975). This method often requires a double-tilt holder to set up the correct diffraction conditions, with a symmetrical fringe pattern inside the hkl disc. Also, a minimum of three fringes on each side of the central bright fringe must be present since a straight-line interpolation is used to measure t, unless the extinction distance ξ_g is known. This results in a lower limit for thickness determination, and the range of applicability of the technique has been examined in detail by Allen (1981). In practice, the required conditions are not always easy to obtain at precisely the region of interest. To make the technique practical, a low-background, double-tilt holder and contamination-free instrument are required so that data acquisition and thickness determination can be carried out concurrently.

Alternative methods of thickness determination are invariably less universal. For example, if extinction distance ξ_g is known accurately (Edington, 1976), thickness can be measured by counting the fringes under two-beam diffracting conditions (which are *not* the conditions under which microanalysis should be performed because of possible anomalous x-ray generation (Cherns *et al.*, 1973). The relative transmission of

contamination deposits

actual thickness

thin foil

measured thickness

(a)

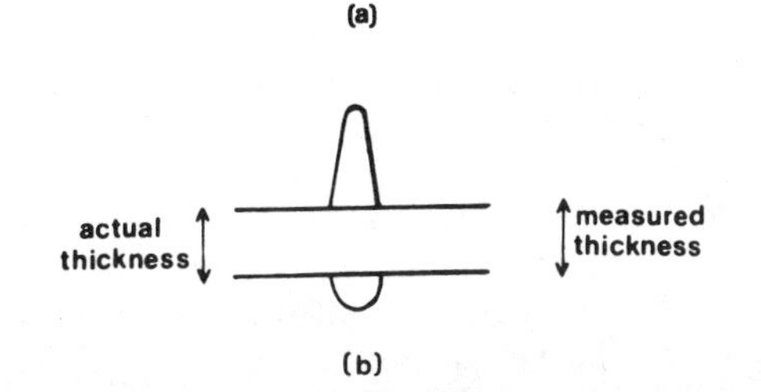

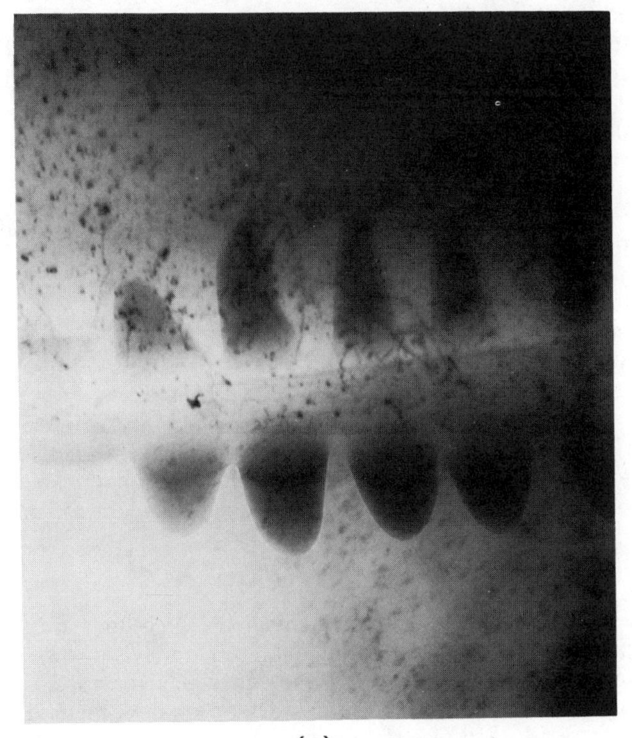

Figure 14 (a) Schematic diagram of the proposed morphology of contamination deposits (Rae *et al.*, 1981, that would account for the overestimation of thickness often observed using the contamination spot technique. (b) Morphology of contamination deposits apparently observed in the AEM and used to measure thickness. (c) TEM image of contamination deposits placed along a grain boundary in Cu showing a discrepancy between the boundary fringe position and the apparent base of the contamination deposits (courtesy J. R. Michael).

electrons is useful for amorphous specimen studies, and Joy and Maher (1975) have reported its use for crystalline materials. This requires careful calibration of transmitted currents with a Faraday cup and is clearly a function of diffracting conditions. The reported accuracy of $\pm 10\%$ for $t > 0.2\ \mu m$ is reasonable. In many practical situations, however, thicknesses less than this give rise to absorption (Table VIII), and it is not always possible to reproduce (or know in the first place) the exact diffraction conditions, particularly if Kikuchi lines are not visible because of specimen dimensions or local deformation. Other techniques, such as the latex ball method (von Heimendahl and Willig, 1980) or projected width of planar defects (Hall and Vander Sande, 1975), are limited in their general applicability. Also, measurement of the relative heights of the zero-loss and first-plasmon-loss peaks in an EEL spectrum (Joy, 1979) requires that the plasmon mean free path be known, and this information is not available for many materials.

The difficulties in measuring ρ and t separately can be overcome by obtaining the mass thickness of the specimen (ρt) directly by using the x-ray data. Two intensity-ratio methods have been described by Morris *et al.* (1979). The first uses the observed ratio of K to L intensity for a given element to deduce the absorption path length and is suitable provided both peaks are visible in the spectrum (usually for $Z > 27$). In practice, the method is difficult because an iterative procedure (composition vs ρt) must be used, and the L line intensity is often quite low and difficult to separate from the continuum background. The second method is more general and requires several spectra to be recorded from the same area of the specimen at different tilt angles. The composition of the analysis point remains the same, and all that is changed is the relative x-ray absorption path lengths. An iterative procedure is used, adjusting ρt and composition values until the corrected analyses at each tilt approach the same value. This method is somewhat inconvenient as several spectra (>3) must be obtained from the same area at several tilt angles. Successful measurements using this method were given by Statham and Ball (1980). The major source of error for this method is the uncertainty of the sample geometry.

Nockolds *et al.* (1979) have proposed an expression for estimating specimen thickness directly from the characteristic x-ray intensity data obtained during the analysis. The method involves the measurement of characteristic x-ray intensity from the specimen relative to that from a *bulk* standard. Cliff and Lorimer (1980) have modified the expression to provide a measure of ρt directly. This method involves atomic-number, absorption, and fluorescence (ZAF) corrections of bulk specimen data taken at high voltages. Such corrections have their own inaccuracies, leading to uncertainties in this method. A better approach as proposed by Porter and Westengen (1981) is to use a thin foil of known mass thickness and composition (preferably a pure element) to calibrate intensity in terms of thickness. The beam current must be constant as measurements of sample and standard are made.

F. Specimen Geometry

The assumption, implied in all the preceding discussion, that the specimen is a parallel-sided thin film (Figure 8) is rarely, if ever, the case in practical x-ray microanalysis because of the way in which materials specimens are characteristically produced. Electropolishing (twin-jet or window method) and ion-beam thinning both give rise to a wedge-shaped specimen (Figure 15(a)), which complicates the situation considerably. As first pointed out by Glitz *et al.* (1981), it is essential to know (a) the configuration of the EDS detector with respect to the specimen and (b) the exact specimen dimensions in the plane parallel to the axis of the detector. If the specimen is of a simple geometry (constant wedge angle), then the correction is relatively simple (Zaluzec, 1981b; Glitz *et al.*, 1981), but this is rarely the case, and as shown in Figure 15(b), the situation can be quite complex, particularly in multiphase samples that are thinned at different rates.

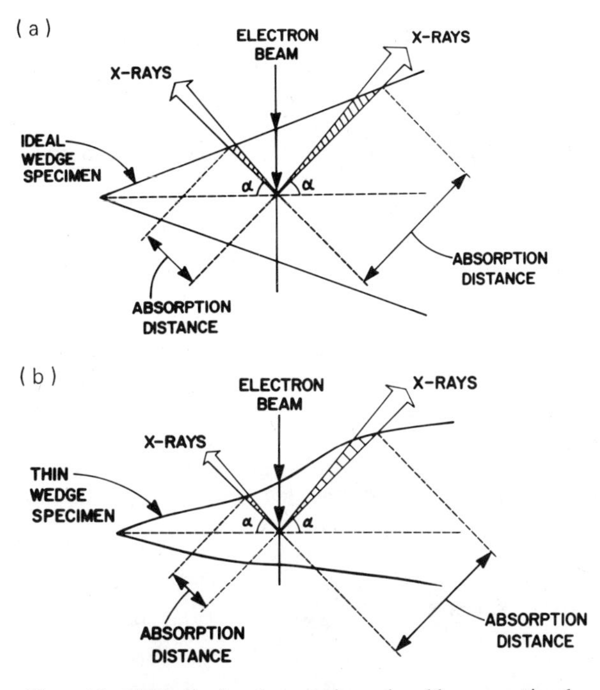

Figure 15 (a) Idealized wedge sample produced by conventional thin-foil preparation techniques. (b) More realistic sample cross section showing how the absorption path length may vary as a function of the specimen geometry and detector position with respect to the specimen (from Williams, 1984).

The effect of the variation of absorption path length with detector orientation was studied by Glitz *et al.* (1981). A highly absorbing ion-beam thinned foil of NiAl was used. Initially, an analysis was taken with the NiAl foil oriented such that the detected x-rays passed through the thinner portion of the wedge-shaped cross section. The foil was then physically rotated 180° and the analysis repeated. Foil thicknesses were measured by the contamination-spot method.

The geometry of the NiAl thin-foil cross section used in this demonstration is illustrated in Figure 16. Analysis 1 was taken at a foil thickness of ~270 nm, and x-rays from point 1 passed through the thinner portion of the wedge en route to the detector. Analysis 2 was taken at a foil thickness of ~260 nm, and the x-rays from point 2 passed through the thicker portion of the wedge. The ratio of Ni K_α to Al K_α peak intensities, foil thicknesses, and x-ray exit path lengths are listed in Table IX. One path length is calculated as $t \times \text{cosec}\ \alpha$, and the other is the actual path length determined from the foil geometry.

In spite of the fact that Analysis 1 was taken in a slightly thicker portion of the foil, the measured Ni K_α/Al K_α intensity ratio from point 1 is 0.50 that from point 2, indicating that the Al K_α x-rays from point 1 were absorbed less than those from point 2. Absorption calculations using Eq. (17) and based on the calculated path length show that the intensity ratio at point 1 should be 1.04 times that at point 2. However, calculations using the actual path lengths that were based on foil geometry/detector orientation considerations show that the intensity ratio at point 1 should be 0.465 times that at point 2. From this measurement, it is obvious that absorption calculations based on foil thickness, without consideration of foil geometry, can lead to erroneous answers in highly absorbing systems.

Clearly under these circumstances, the specimen must be homogeneous, not only through-thickness (as is invariably assumed in all analytical electron microscopy), but also along the path length for absorption. In the case of a multiphase specimen, any interfaces should be aligned parallel to the EDS detector axis, as shown in Figure 17, and this would be facilitated by the use of a tilt-rotation holder, preferably a low-background model.

The specimen geometry problem may be minimized by using thin-flake specimens prepared by window polishing and sandwiched between Be grids. The specimen wedge angle is usually smaller than for a self-supporting disc. These specimens have an added advantage in that they minimize spurious effects caused by stray radiation. Also, high ($\gtrsim 20°$) take-off angle detectors are essential as has long been recognized in the microprobe field, where absorption is invariably present. However, higher take-off angle detectors should not be used if the detector must be placed so far from the specimen that detector sensitivity is greatly decreased.

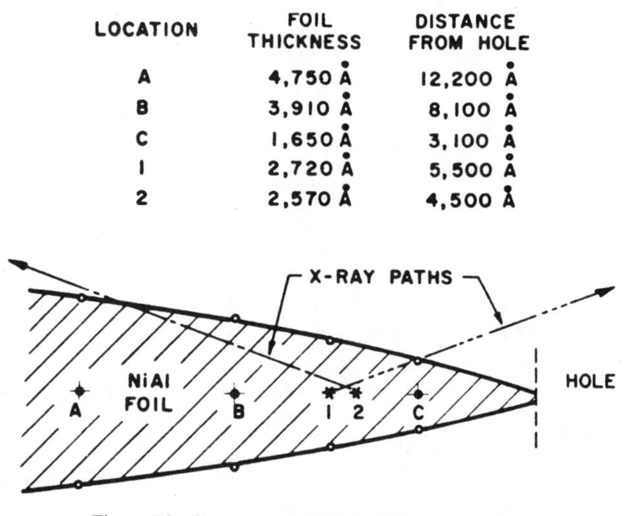

LOCATION
FOIL THICKNESS
DISTANCE FROM HOLE
A
4,750 Å
12,200 Å
B
3,910 Å
8,100 Å
C
1,650 Å
3,100 Å
1
2,720 Å
5,500 Å
2
2,570 Å
4,500 Å
X-RAY PATHS
NiAl FOIL
A
B
1 2
C
HOLE

Figure 16 Geometry of NiAl thin-foil cross section.

TABLE IX

X-Ray Microanalysis Measurements From a Wedge-Shaped NiAl Foil (Glitz *et al.*, 1981).

Analysis	$I_{NiK\alpha}/I_{AlK\alpha}$	Foil Thickness (nm)	Path Length (nm)	
			Calculated	Actual
No. 1	2.64	270	398	270
No. 2	5.18	260	376	670

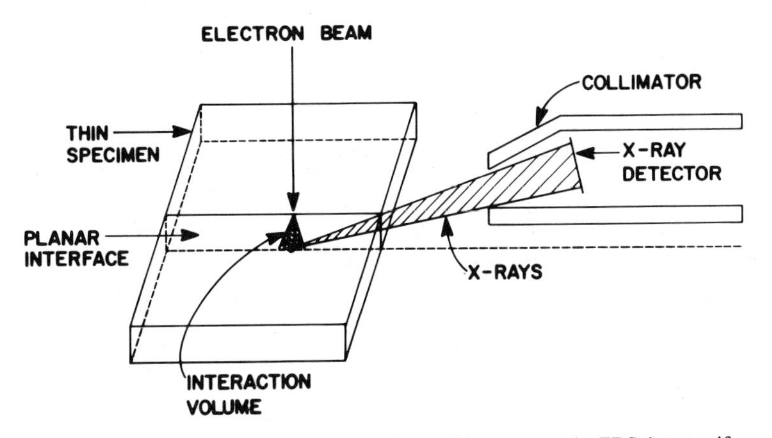

Figure 17 Correct alignment of the specimen with respect to the EDS detector if absorption conditions vary across the planar interface.

G. Summary

Most of the factors currently used in the standard absorption-correction equation are not accurately known and can contribute to significant errors in quantification. Specifically, thickness determination and specimen geometry are critical factors that probably cause the largest error at present. As these factors are determined more accurately, the variation in $\phi(\rho t)$ will become significant and have to be accounted for. Finally, in light of the improved accuracy of k_{AB} factor determination, the currently accepted definition of "significant" absorption should be revised downward to $\leq 3\%$ in specific circumstances.

III. FLUORESCENCE CORRECTION

X-rays produced by electron ionization within the specimen may themselves be sufficiently energetic to excite and ionize other atoms. This effect, known as x-ray fluorescence, should be less significant in thin specimens than in bulk specimens simply because there is less material available for fluorescence. Fluorescence is very sensitive to the particular elements and their concentration in the material being analyzed. For example, in an Fe-Mn steel, Fe K_α characteristic x-rays will not fluoresce the Mn K_α, but if Cr is substituted for Mn, the Fe will strongly fluoresce the Cr. This results from the excitation energy E_c of Mn (6.537 keV) being greater than the energy of Fe K_α x-rays (6.403 keV). Therefore, Fe K_α cannot excite Mn, but the excitation energy E_c of Cr (5.988 keV) is just less than the energy of Fe K_α and consequently is very strongly fluoresced. Note, however, that it is possible for Fe K_β at 7.057 to excite Mn, but because of its low intensity (only ~14% of Fe K_α), its influence is small.

Although fluorescence effects in thin specimens can be expected to be less than those in bulk samples, they may be significant if strongly fluorescing alloy systems are analyzed. The typical system exhibiting strong bulk fluorescence effects mentioned above (that of the Fe-Cr alloys) has been analyzed by Lorimer *et al.* (1977). Thin-foil specimens of these alloys were used to assess experimentally the extent of the problem. Figure 18 shows the measured variation in percentage increase of Cr as a function of Cr concentration for bulk and thin-foil specimens at 100 keV. For Cr concentrations <15 wt%, the experimentally observed percentage increase in Cr is >5% and is, therefore, significant if analysis errors of <5% are required.

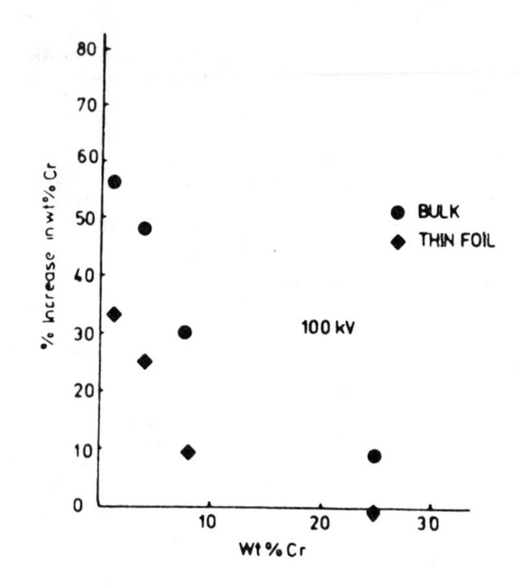

Figure 18 Variation in % increase in Cr concentration as a function of Cr concentration for bulk and thin specimens at 100 keV (Nockolds *et al.*, 1979).

Philibert and Tixier (1975) have analyzed the problem of thin-specimen fluorescence theoretically and used the model shown in Figure 19. They considered a thin foil of mass thickness t composed of elements A and B, with element B fluorescing A. It was assumed that all the x-rays originated from point P in the middle of foil on the beam axis. For the theoretical details of the model in Figure 19, the reader is referred to the original source. It is sufficient to show that from this model, the following expression for fluorescence enhancement can be developed

$$\frac{I^A}{I_A} = 2\omega_B C_B \left(\frac{r_A - 1}{r_A}\right) \frac{A_A}{A_B} (\mu/\rho]\,^B_A)\,(\mu/\rho]\,^B_{SPEC}) \left(\frac{E_{c_A}}{E_{c_B}}\right) (\rho t)^2 \tag{20}$$

where

I^A/I_A = ratio of fluorescence intensity to primary intensity
ω_B = fluorescence yield of element B
r_A = absorption edge jump ratio of element A
$\mu/\rho]\,^B_A, \mu/\rho]\,^B_{SPEC}$ = mass absorption coefficients of x-rays from element B in element A and the specimen, respectively
A_A, A_B = atomic weights of elements A and B
E_{c_A}, E_{c_B} = critical excitation energies for the characteristic radiation of A and B.

Although Eq. (20) predicts very small fluorescence enhancements in typical thin foils, it does not predict the observed effects in Figure 18.

An alternative formulation for fluorescence in thin foils was proposed by Nockolds *et al.* (1979), using the model shown in Figure 20, where x-ray generation is assumed to be uniform through the specimen along the line of the incident beam. This model requires an additional integral to allow for the fact that point P is now no longer fixed at the midpoint of the specimen but moves along the beam path through the specimen. Also, the beam path length depends on tilt, requiring an additional term (sec α) if the sample is tilted toward the detector through an angle (α). This effect is not apparent if Figure 19 is used. Again the reader is referred to the original source for details. The expression of Nockolds *et al.* (1979) is

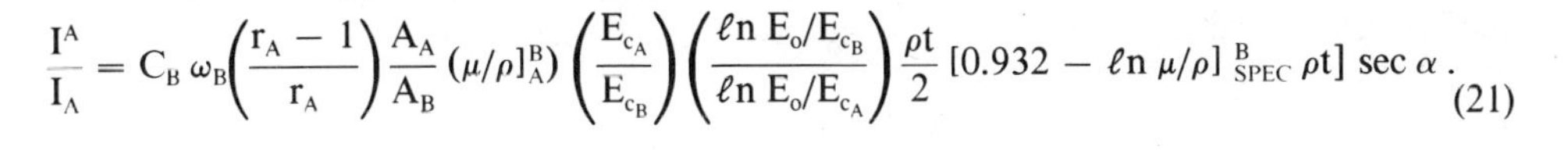

$$\frac{I^A}{I_A} = C_B\, \omega_B \left(\frac{r_A - 1}{r_A}\right) \frac{A_A}{A_B} (\mu/\rho]^B_A) \left(\frac{E_{c_A}}{E_{c_B}}\right) \left(\frac{\ell n\, E_o/E_{c_B}}{\ell n\, E_o/E_{c_A}}\right) \frac{\rho t}{2} [0.932 - \ell n\, \mu/\rho]^{B}_{SPEC}\, \rho t] \sec \alpha \,. \tag{21}$$

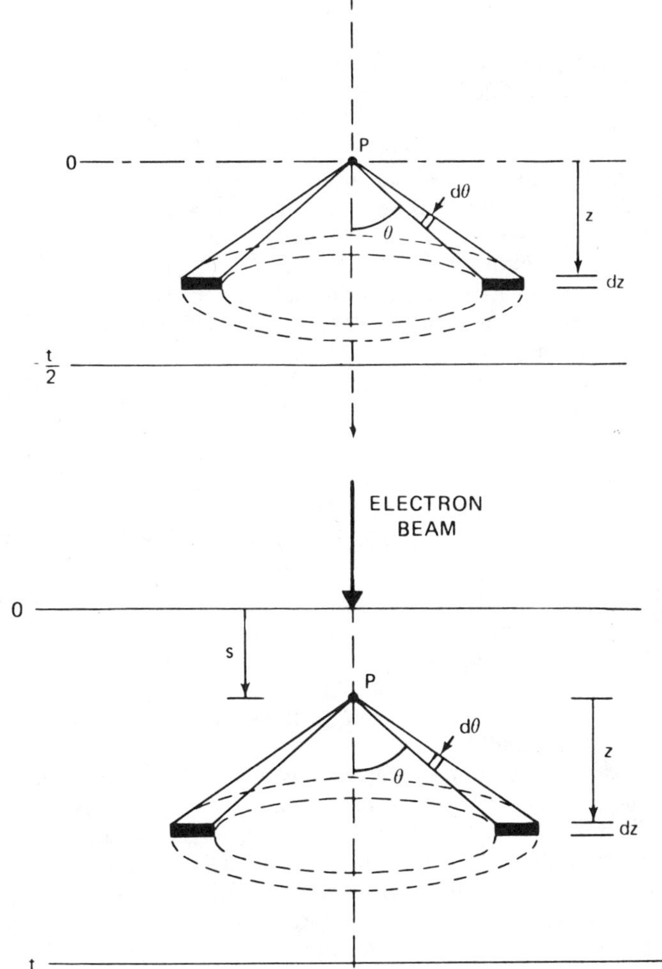

Figure 19 Philibert-Tixier model for fluorescence. P is the point source of B x-rays, t is the foil thickness, Z is the distance of the annulus from the foil center, dZ is the thickness of a small sample of the annulus in which a small fraction of the fluorescence occurs, θ is the semiangle subtended by the annulus, and $d\theta$ is the incremental angle of the annulus.

Figure 20 Nockolds *et al.* (1979) model for fluorescence. Same terms as used in Figure 19 except Z is the distance of the annulus from point source P and S is the distance of point P from the specimen surface.

Nockolds *et al.* (1979) showed that Eq. (21) could be used to correct analysis data at 100 keV from a thin specimen sample of Fe-10.5 wt% Cr up to specimen thicknesses in excess of 600 nm. In Figure 21, the uncorrected results calculated from the x-ray intensity data using an Fe/Cr k factor of 1.09 are compared with the fluorescence-corrected data as a function of specimen thickness. Note that an absorption correction to the Cr/Fe intensity ratio data is not necessary in this particular system. The mass absorption coefficients for Cr and Fe x-rays in the specimen are so similar that the x-ray intensity ratio would change by $<1\%$ for a sample 16 μm thick. The fluorescence-corrected results show good agreement with the bulk microprobe analysis (obtained from the sample before thin specimen preparation).

A comparison of the corrections predicted by Eq. (20) and (21) for thin-specimen fluorescence enhancement in 10 wt% Cr-90 wt% Fe is shown in Figure 22. The expression of Philibert and Tixier does not predict the nearly linear fluorescence effects in this thin-foil specimen that the data of Figure 21 clearly show. Note, however, that in this specimen, the typical enhancement is $<10\%$ at reasonable specimen thicknesses ($\sim$200 nm).

Work by Tixier *et al.* (1981) on a 17 wt% Cr stainless steel, using an EMMA 4 with crystal spectrometers, has shown that, within the limits of precision of their measurements, the effects of fluorescence in their alloy were negligible. It is clear, therefore, that the effect is small, but the available experimental evidence does not, so far, allow any definitive judgment to be made on the general validity of the two formulas for correction of thin-specimen fluorescence.

An alternative theoretical analysis has been made by Twigg and Fraser (1982). These authors re-evaluated the model of Philibert and Tixier (1975) (Figure 19) and showed that the mathematical execution of the correction was inadequate, whereas that of Nockolds *et al.* (1979) appeared to be correct. When corrected, the version of the Philibert and Tixier model yielded the expression

$$\frac{I^{A}}{I_{A}} = C_{B}\,\omega_{B}\left(\frac{r_{A}-1}{r_{A}}\right)\frac{A_{A}}{A_{B}}\left(\frac{E_{c_A}}{E_{c_B}}\right)\mu/\rho]^{B}_{A}\,\frac{\rho t}{2}\left(1.12+\mu/\rho]^{B}_{SPEC}\,\frac{\rho t}{4}-\ell n(\mu/\rho]^{B}_{SPEC}\,\rho t)\right). \tag{22}$$

Twigg and Fraser (1982) showed that results from this equation do not differ significantly from results obtained by using Eq. (21). It would appear that the expression of Nockolds *et al.* is theoretically more suitable than the earlier expression of Philibert and Tixier.

Any discussion of fluorescence effects, either thin or bulk, would be incomplete without some mention of fluorescence caused by the inevitable continuum, or "bremsstrahlung," radiation. Philibert and Tixier (1975) attempted to describe theoretically the influence of continuum fluorescence in thin specimens, but Twigg and Fraser (1982) concluded that the Philibert and Tixier equation for continuum fluorescence is incorrect. It must be observed, however, that even in many popular schemes for correcting x-ray intensity data in bulk specimens, where the effect would be strongest (e.g., FRAME), no allowance is made for continuum fluorescence. Until experimental evidence is available to prove otherwise, it may be safely concluded that continuum fluorescence generated in a thin specimen by the primary beam can be neglected.

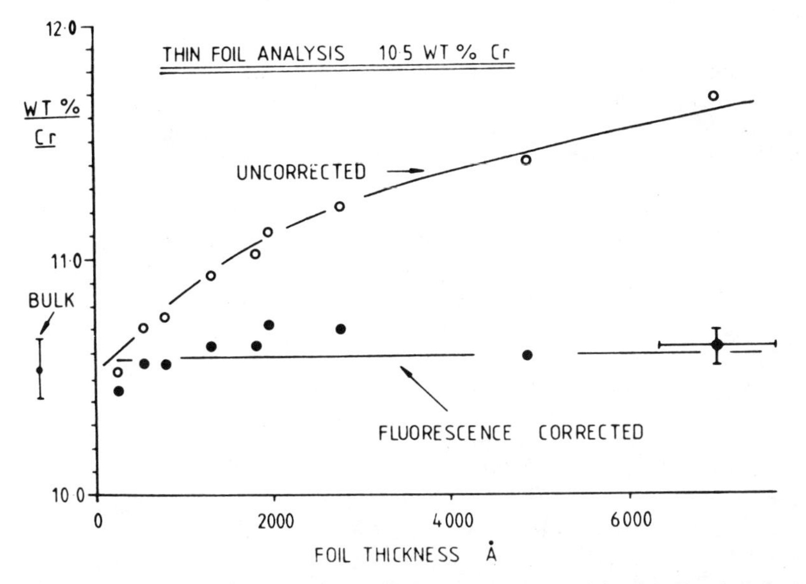

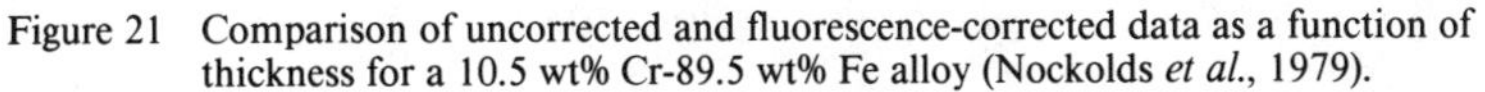
Figure 21 Comparison of uncorrected and fluorescence-corrected data as a function of thickness for a 10.5 wt% Cr-89.5 wt% Fe alloy (Nockolds *et al.*, 1979).

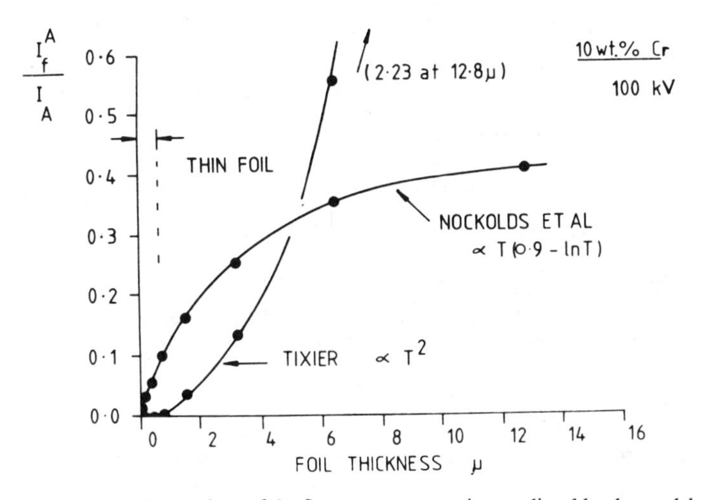

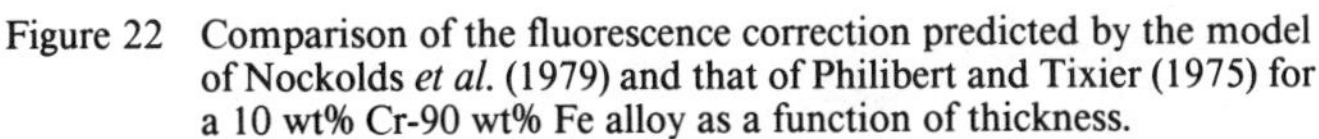
Figure 22 Comparison of the fluorescence correction predicted by the model of Nockolds *et al.* (1979) and that of Philibert and Tixier (1975) for a 10 wt% Cr-90 wt% Fe alloy as a function of thickness.

IV. CORRECTION PROCEDURES FOR THIN-FOIL ANALYSIS

For an "infinitely" thin foil, the effects of x-ray absorption and fluorescence can be neglected. For specimens in which the thin-film criterion is fulfilled, the Cliff-Lorimer ratio method can be applied (Eq. (6) through (8)), namely

$$C_A/C_B = k_{AB} \times I_A/I_B \ .$$

In considering the application of the thin-film criterion, one should note that k_{AB} factors can be routinely determined with an error of between $\pm 1\%$ to 4% relative. Therefore, if x-ray absorption is $> \sim 3\%$ relative, the thin-film criterion is no longer fulfilled, and the Cliff-Lorimer method, Eq. (6), cannot be applied.

To determine whether an absorption correction is needed, one can calculate the absorption correction factor (ACF) as given in Eq. (17). An approximate absorption path length is usually assumed for the calculation. If the absorption correction factor is <0.97 or >1.03, the thin-film criterion is no longer fulfilled, and a full absorption correction must be made. This correction necessitates the measurement of thin-foil thickness t at every analysis point.

A fluorescence correction may be necessary if the primary radiation of element A excites that of element B or vice versa when the ratio of C_A/C_B is considered. If fluorescence occurs, the expression of Nockolds (Eq. (21)) can be applied to calculate the ratio of fluorescence intensity to primary intensity I^A/I_A for element A, I^B/I_B for element B. If this ratio is $>5\%$, the thin-film criterion is no longer fulfilled, and a full fluorescence correction must be made. As in the case of absorption, this correction necessitates the measurement of the thin-foil thickness t at every analysis point.

If the thin-film criterion is not fulfilled (absorption $>3\%$ rel. and/or fluorescence $>5\%$ rel.), a full correction of the following form must be made, namely

$$C_A/C_B = k_{AB} \frac{I_A}{I_B} \times [\mathrm{ACF}] \times \frac{1}{\left(1 + \frac{I^A}{I_A}\right)} \ . \tag{23}$$

In this equation, element A is assumed to be fluoresced by element B in the specimen. The absorption correction factor is given by Eq. (17), namely

$$\mathrm{ACF} = \left[\frac{\mu/\rho)^A_{SPEC}}{\mu/\rho)^B_{SPEC}}\right]\left[\frac{1 - \exp - (\mu/\rho)^B_{SPEC} \operatorname{cosec} \alpha(\rho t)}{1 - \exp - (\mu/\rho)^A_{SPEC} \operatorname{cosec} \alpha(\rho t)}\right] .$$

The fluorescence correction factor is given by Eq. (21).

A full calculation of C_A/C_B requires the measurement of t at each analysis point. In addition, the values of C_A, C_B, and other elements in the specimen must be available to calculate the mass absorption coefficients. An iterative technique is used in which values of C_A and C_B are first assumed. The values of C_A and C_B are adjusted until the calculated and assumed values are in agreement. Examples of the use of these correction procedures for thin-foil analysis are given in Section VII of this chapter.

V. X-RAY STATISTICS

A. Errors in X-Ray Analysis in the AEM

The most significant error in AEM x-ray analysis is due to the poor x-ray counting statistics obtained from the x-ray energy-dispersive data. Experimental observations show that AEM x-ray counting statistics obey Gaussian behavior. Hence, $\sigma = \sqrt{N}$, where σ is the standard deviation and N is the number of accumulated counts at each analysis point. At the 3σ confidence level, the error in the number of accumulated counts would be $3\sqrt{N}$. For a single measurement, there is a 99.73% chance that the value of N will be within 3σ of the true mean $\overline{N}$. The relative error in the number of counts is $\frac{3\sqrt{N}}{N} \times 100$, so as N decreases, the relative error increases. The Cliff-Lorimer relation uses the x-ray ratio, I_A/I_B. The relative error for this ratio is the sum of the relative errors for I_A and I_B. Therefore, the total relative error in C_A/C_B for any one measurement would be the sum of the errors for I_A, I_B, and k_{AB}. Error bars for each data point can be determined by using this procedure. To minimize the statistical error in quantification, it is best to accumulate at least 10,000 counts above background in each peak that is to be used. Under these circumstances, the 3σ error will be $\pm 3\%$ relative.

The error in the composition of a phase is substantially smaller than that of an individual measurement since these compositions are determined from n individual determinations. To establish the error for a phase composition, the following procedure can be used. The total absolute error in I_A/I_B at the 99% confidence level is $\pm \frac{t_{99}^{n-1}S}{\sqrt{n}}$, where t_{99}^{n-1} is the Student t value and S is the standard deviation, calculated for n values of the ratio I_A/I_B. Standard deviation S is given by

$$S = \left[\sum_{i=1}^{n} \frac{(N_i - \overline{N})^2}{n-1} \right]^{1/2} . \qquad (24)$$

The relative error in I_A/I_B is $\frac{t_{99}^{n-1}S}{\sqrt{n}} / \overline{(I_A/I_B)} \times 100$, where $\overline{(I_A/I_B)}$ is the average intensity ratio calculated from n values of I_A/I_B. Although the relative error in I_A/I_B can be reduced by increasing the number of measurements, the error in k_{AB} is fixed and cannot be reduced. With typical errors in k_{AB} of $\pm 1\%$ to $\pm 4\%$ rel., the errors in x-ray analysis in the AEM are often $> \pm 5\%$ rel.

B. Minimum Detectability Limit or Minimum Mass Fraction

Ziebold (1967) has shown that trace-element sensitivity or the minimum mass fraction (MMF) limit can be expressed as

$$\text{MMF} \sim \frac{1}{(P/B \times P \times \tau)^{1/2}} \qquad (25)$$

where

P = pure element counting rate
P/B = peak/background ratio of the pure element (the ratio of the counting rate of the pure element to the background counting rate of the pure element)
τ = counting time.

To improve the MMF, all three terms in Eq. (25) should be maximized. As discussed previously, τ is limited by contamination and electronic-mechanical stability of the instrument. Peak intensity P can be increased by increasing the electron current density and by optimizing the x-ray detector configuration (see Chapter 4).

Several authors have developed expressions for P/B ratios in thin-film analysis (Joy and Maher, 1977; Zaluzec, 1979). Because these expressions are quite complex, the interested reader is directed to the original papers for specific details. These calculations show a direct increase in P/B for pure elements with E_o for all elements. Clearly then, P/B ratios can be increased and MMF improved at higher operating voltages. Calculations by Zaluzec (1979) also show that P/B increases with x-ray detector observation angle and levels out after ~125° (relative to the forward beam direction) at E_o = 100 keV. Therefore, the P/B ratio will be increased and the MMF improved if the x-ray detector is placed thus. In this configuration, the detector looks down at the specimen, and the thin film can be placed at 0° tilt. Several AEM manufacturers have adapted their instruments to incorporate these high-take-off-angle x-ray detectors.

The analyst can determine the MMF directly for a sample of interest by making several relatively simple measurements on the sample thin film. The following discussion develops the necessary equations. As elemental composition C approaches low values, the number of x-ray counts of element A (I_A) from the sample is no longer much larger than the number of x-ray counts from the background continuum radiation (I_b^A) for element A. The analysis requirement is to detect significant differences between the sample and continuum background generated from the sample. To say that I_A is "significantly" larger than I_b^A, the value of I_A must exceed I_b^A by $3(2I_b^A)^{1/2}$. To convert the number of x-ray counts to the MMF in wt%, we use the ratio technique, where

$$\frac{I_A - I_b^A}{I_B - I_b^B} \geq \frac{3(2I_b^A)^{1/2}}{I_B - I_b^B} = k_{AB}\frac{\text{MMF}}{C_B}\,. \tag{26}$$

The term I_b^B is the background continuum radiation for element B. The x-ray intensity for element B in the above equation is expressed in terms of total accumulated x-ray counts above background, $I_B - I_b^B$ (not counts/s).

To use Eq. (26), it is assumed that the concentration of element B (C_B) is accurately known. Since I_A and I_B vary with foil thickness, there are some difficulties in establishing the MMF experimentally. However, if I_A, I_B, and I_b are measured simultaneously and the ratio technique, Eq. (26), is used, a value of the MMF can be established at each film thickness.

Another approach to determining the MMF is to use a thin film where element A is present and its concentration in the sample (C_A), is known. For this type of analysis

$$\frac{3(2I_b^A)^{1/2}}{(I_A - I_b^A)} \times C_A = \text{MMF}\,. \tag{27}$$

Experience in numerous AEM laboratories using thermal-emission guns shows that a MMF of ~0.5 wt% can be expected under practical analysis conditions. Oppolzer and Knauer (1979) have used a dedicated STEM with a field-emission gun to measure the minimum mass fraction of a foil of an amorphous metallic glass. The instrument was operated at 100 keV, with a 100-Å probe spot and a 100-s counting time. The foil was 500 Å thick. Using Eq. (27), the MMF was ~0.15 wt% for the major elements (P, Cr, Fe, Ni) that comprised the glass. The higher count rates obtained at small beam diameters with the field-emission gun allowed for the improved MMF.

Although the MMF for thin-film x-ray microanalysis (with a $\leq$20-nm spot size) is generally $<$1 wt%, several experimental problems may occur. Assuming that the spurious x-ray signal is minimized, if the EDS spectrum is relatively clean and has few characteristic peaks, background subtraction and the determination of I_b^A is relatively straightforward. In some cases, peak overlaps, continuum background subtraction, and EDS spectrometer artifacts must be considered. If peak overlaps occur or backgrounds cannot be easily determined so that they can be subtracted from peak intensities, the actual MMF will be much larger than the values given by Eq. (26). Methods for treating continuum background subtraction, peak overlaps, and artifacts in EDS detectors have been reviewed by various authors (see Chapter 4) and will not be discussed here. A statistical treatment of peak overlap and background measurement has been given by Statham (1982). Other statistical treatments of quantitative x-ray analysis are given by Tixier *et al.* (1981) and Blake *et al.* (1983); the interested reader is referred to these works for a more complete treatment of the pertinent statistics. In practice, if the x-ray EDS spectrum contains a large number of peaks or the x-ray peak of interest is at low energies where the continuum is a maximum, a more sophisticated treatment of the data is necessary.

VI. BEAM BROADENING AND X-RAY SPATIAL RESOLUTION

A. Formulation

The concept of the spatial resolution of x-ray microanalysis would at first sight appear trivial, but a generally acceptable definition has yet to emerge from the numerous definitions that exist in the literature. Although all the definitions agree that the x-ray spatial resolution is a function of the interaction volume of the incident beam with the specimen, the dependence of that volume on such parameters as accelerating voltage, atomic number, probe size, specimen thickness, etc., remains an area of lively debate (Jones and Loretto, 1981). Even the defined percentage of x-ray generation within the interaction volume varies from author to author, as shown in Table X.

TABLE X

X-Ray Spatial Resolution
(% of x-ray generation).

	% Definition
Reed (1966)	99
Goldstein *et al.* (1977)	90
Doig *et al.* (1980)	50
Faulkner & Norrgard (1978)	95
Kyser (1979)	90
Venables & Janssen (1980)	80

The probe diameters of early analytical electron microscopes, such as the EMMA 4, were comparable with the dimensions of specimen thickness, usually 100 nm to 500 nm, and therefore the effects of beam broadening in the specimen were small in comparison with the probe size. The simple observation that contamination spots—formed on both surfaces of the specimen during microanalysis—had similar diameters suggested that broadening within a thin specimen was insignificant in Al and Au at 100 keV (Lorimer *et al.*, 1976), and the probe size therefore determined the resolution. However, the development of dedicated STEM systems, capable of providing high-resolution scanning images using ultrafine small probes (apparently $\ll$2 nm in diameter), has created the possibility of producing high x-ray spatial resolution on a scale apparently limited only by beam broadening in the specimen (Bovey *et al.*, 1981). When the probe size is $\sim$100 nm, it should be noted that the extent of beam spreading in the foil remains small when compared with the incident probe, as can be seen in Figure 23, taken from the work of Stenton *et al.* (1981). The diameters of contamination spots at the specimen surface are very similar, showing that little broadening has occurred in the specimen.

Goldstein *et al.* (1977) derived an analytical expression for the beam broadening of small probes by assuming that each electron suffered a single large-angle Rutherford elastic scattering at the center of the foil (Figure 24(a)). The expression for a broadening parameter, b, containing 90% of the scattered electrons (and therefore for a thin specimen, 90% of the x-ray generation) was given as

$$b = 6.25 \times 10^5 \times \frac{Z}{E_0} \times \left(\frac{\rho}{A}\right)^{1/2} \times t^{3/2}\ \text{cm} \tag{28}$$

where

Z = atomic number
A = atomic weight
E_0 = accelerating voltage (in eV)
ρ = density (g/cm^{-3})
t = specimen thickness (cm).

This expression has gained the title of the "single-scattering equation" and as such has been assumed to apply only to foils having thicknesses equal to the mean free path for Rutherford scattering (Newbury and Myklebust, 1979). Surprisingly, it can be shown that the expression derived by Goldstein *et al.* is identical with that expected to apply beyond both single scattering and even plural scattering, at the onset of multiple scattering in specimens too thick to be suitable for electron microscopy (Cliff and Lorimer, 1981).

The plural-scattering model used by Cliff and Lorimer (1981), which is shown in Figure 24(b), enclosed the extreme limits of the 90% cone of plurally scattered electrons. It was argued that the cone width at half the foil thickness was equivalent to a plural-scattering beam-broadening parameter. Unfortunately, because the modeling described the extreme limit of broadening, it would not satisfactorily estimate the mean beam width at the specimen exit surface. This problem can be overcome by assuming that the mean depth of plural scattering occurs at the foil center; then the model of Figure 24(a) can be used to estimate a mean plural-scattering exit width at the bottom of the specimen. The beam-broadening parameter in both the single-scattering model of Goldstein *et al.* (1977) and the plural-scattering model of Cliff and Lorimer (1981) therefore becomes identical in all respects. The surprising feature of this conclusion is that any expression based on single scattering should fail beyond specimen thicknesses equivalent to one mean free path for Rutherford scattering, and yet Eq. (28), based on single scattering, adequately describes the broadening parameter, b, beyond the single-scattering limit.

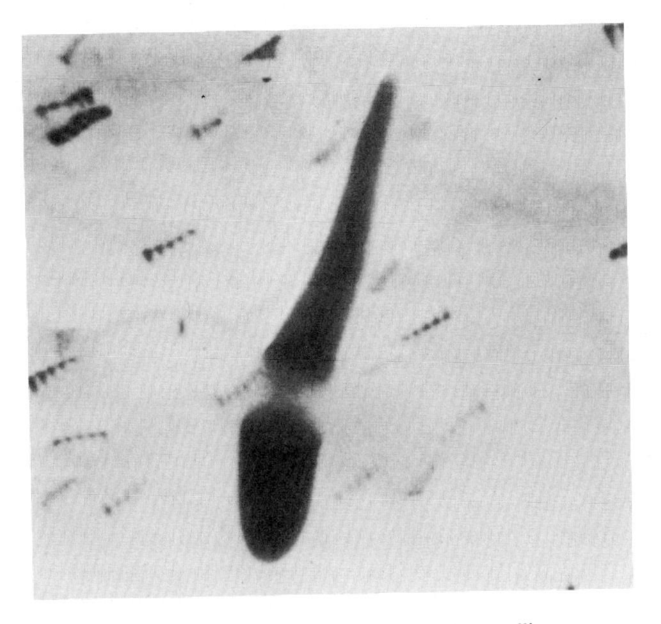

Figure 23 Contamination spot deposited on polycrystalline, vapor-deposited nickel at normal incidence. Foil tilted 45° (X100,000), Stenton *et al.* (1981).

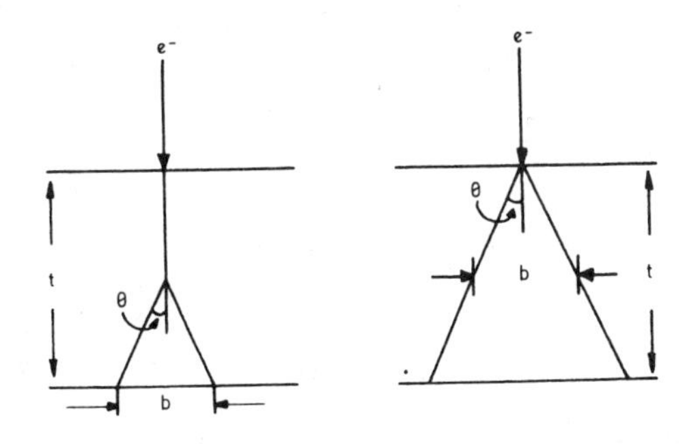

Figure 24 (a) The original single-scattering model of Goldstein *et al.* (1977). (b) The plural-scattering model of Cliff and Lorimer (1981). The specimen thickess is t and the beam-broadening parameter is b in both models.

It was demonstrated by Cliff and Lorimer (1981) that the angular distribution of electrons undergoing single scattering alone should be independent of foil thickness. However, the broadening parameter b depends on both the angle of scatter and the foil thickness. If the scattering event occurred singly, it can be shown (using Figure 24(a)) that the expression

$$b = \frac{\theta_0 t}{(P_\theta)^{1/2}} \tag{29}$$

should correctly define a "true" single-scattering parameter, b, where θ_0 is the screening parameter for the nucleus and P_θ is the probability that the scattering has occurred beyond the angle θ defined in Figure 24(a). Equation (29) has the expected linear dependence on t (not the $t^{3/2}$ dependence of Eq. (28)) and more importantly has a dependence on θ_0, the screening parameter, which is the median angle of single scattering. An estimate of θ_0 can be obtained from $\theta_0 = 3.69\ Z^{1/3}\ E_0^{-1/2}$ rad, if E_0 is in eV. From this approach, it is expected that the expression of Goldstein *et al.* (1977) would overestimate the extent of beam broadening, even in specimens that were sufficiently thick to invalidate the single-scattering assumption.

The variation of b as a function of thickness in Al and Au at 100 keV is plotted in Figures 25 and 26 for values of b calculated by (a) the expression of Goldstein *et al.* (Eq. (28)); (b) a modified version derived by Jones and Loretto (1981) that multiplies the constant by $2/\sqrt{3}$; (c) the Monte Carlo work of Newbury and Myklebust (see Chapter 1 by Newbury for a discussion of Monte Carlo techniques); (d) the Monte Carlo work of Geiss and Kyser (1979); and (e) and (f), the single-scattering and Monte Carlo work, respectively, of Cliff and Lorimer (1981). The expression of Goldstein *et al.* is seen to apply from single ($P_e = 1$) through plural ($P_e < 20$) to multiple scattering ($P_e > 20$), where P_e is the mean number of scattering events. It lies above most of the Monte Carlo results, supporting the conclusion that the expression should overestimate broadening. However, the true single-scattering theory (Eq. (29); Cliff and Lorimer, 1981), does not agree with Monte Carlo work beyond a foil thickness equivalent to one mean free path.

Key: + Single-scattering model of Goldstein *et al.* (1977)
O Modified single-scattering equation of Jones and Loretto (1981)
X Monte Carlo theory of Geiss and Kyser (1979)
∇ Monte Carlo theory of Newbury and Myklebust (1979)
□ Monte Carlo theory of Cliff and Lorimer (1981)
● Single-scattering equation of Cliff and Lorimer (1981)

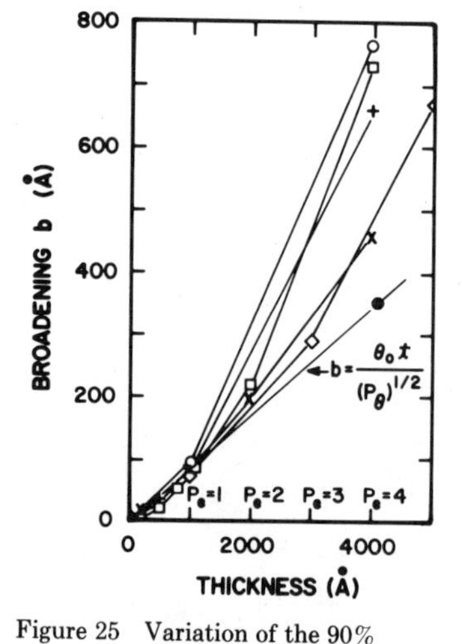

Figure 25 Variation of the 90% beam-broadening parameter b in Al with foil thickness at 100 keV.

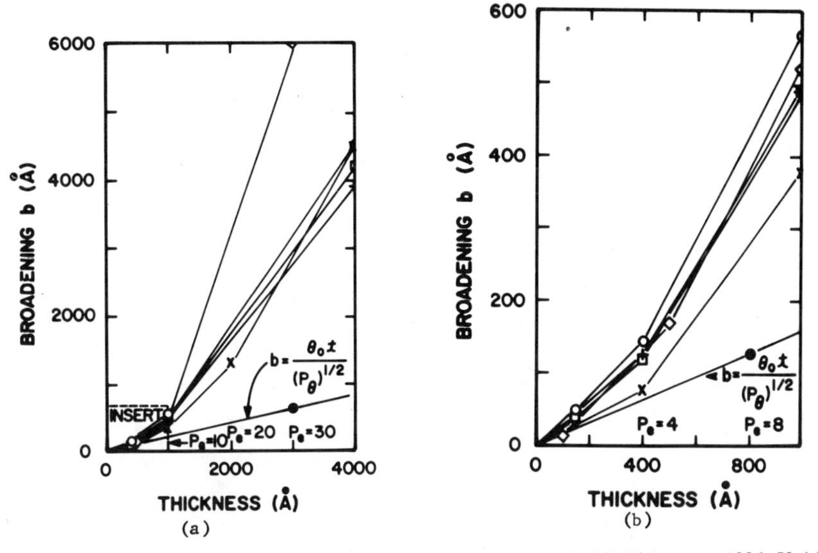

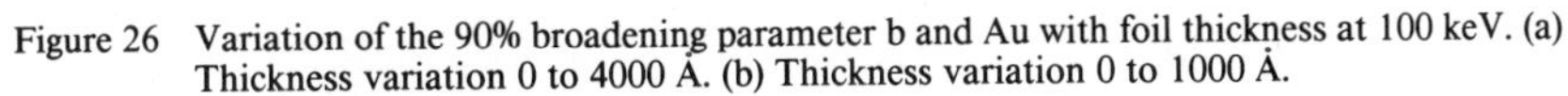

Figure 26 Variation of the 90% broadening parameter b and Au with foil thickness at 100 keV. (a) Thickness variation 0 to 4000 Å. (b) Thickness variation 0 to 1000 Å.

Key: + Single-scattering model of Goldstein *et al.* (1977)
O Modified single-scattering equation of Jones and Loretto (1981)
X Monte Carlo theory of Geiss and Kyser (1979)
▽ Monte Carlo theory of Newbury and Myklebust (1979)
□ Monte Carlo theory of Cliff and Lorimer (1981)
● Single-scattering equation of Cliff and Lorimer (1981)

The only experimental data so far reported that are specifically designed to evaluate the width of beam broadening at the exit surface of the specimen, at 40 keV and 100 keV, are those of Hutchings *et al.* (1979) and Stephenson *et al.* (1981) for the element Si. The data are compared in Figures 27 and 28 with the theoretical treatments of Goldstein *et al.* (1977), the single-scattering equation and screened Rutherford Monte Carlo work of Cliff and Lorimer (1981), and two analytical multiple-scattering theories developed by Doig *et al.* (1980 and 1981). Poor agreement is shown with the first of the two theories of Doig *et al.*, but their second version is a significant improvement and now agrees very well with the simple model of Goldstein *et al.* and the more sophisticated Monte Carlo results. The most important point illustrated by Figures 27 and 28 is that, given the reported error bars, the experimental data of Hutchings *et al.* (1979) and Stephenson *et al.* (1981) cannot distinguish between the better theoretical treatments. The original, simple analytical expression of Goldstein *et al.* (1977) can be used as a convenient and reasonably quantitative alternative to Monte Carlo calculations of beam broadening.

We can arbitrarily add the effect of beam broadening b to probe diameter d_p (the electron-beam crossover FWHM at the specimen plane) to estimate a conservative value of x-ray resolution as $d_p + b$. Given that b contains 90% of the scattered electrons, but d is the FWHM, we should multiply d by 1.82 so that both terms refer to 90% of the electrons. Then a more reasonable estimate of x-ray resolution (Reed, 1982) would be obtained by adding the contributions from b and d in quadrature to yield $(1.82^2 d_p^2 + b^2)^{1/2}$. In both cases, to optimize x-ray spatial resolution, we need to use the smallest practical probe size d (consistent with sufficient beam current to generate statistically significant counts), the thinnest specimens possible, and the highest available accelerating voltage, all of which will reduce the influence of beam broadening b.

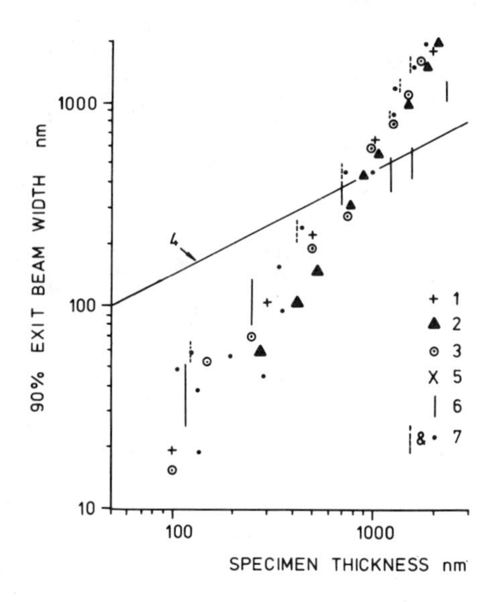

Figure 27 Experimental and theorectical 90% exit-beam widths in Si at 40 keV according to:

(1) Single-scattering theory of Goldstein *et al.* (1977)

(2) Monte Carlo theory of Jones and Loretto (1981)

(3) Monte Carlo theory of Cliff and Lorimer (1981)

(4) Original multiple-scattering theory of Doig *et al.* (1979)

(5) Modified multiple-scattering theory of Doig *et al.* (1981)

(6) Experimental data of Hutchings *et al.* (1979)

(7) Experimental data of Stephenson *et al.* (1981)

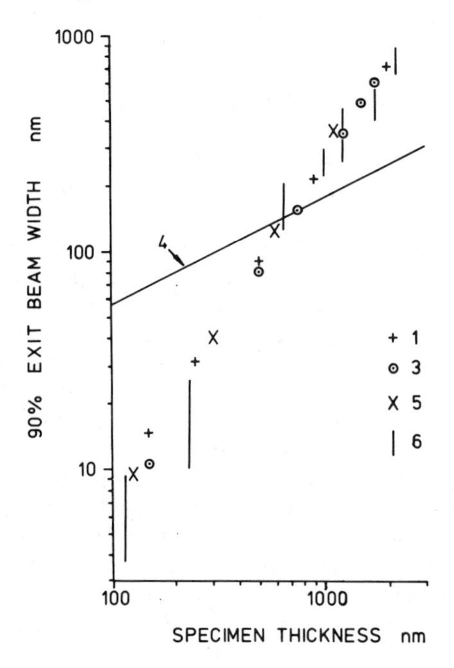

Figure 28 Experimental and theoretical exit-beam widths in Si at 100 keV. The theoretical models and experimental data are those listed in the caption of Figure 27.

B. Measurement of X-Ray Spatial Resolution

The most successful technique for evaluating x-ray spatial resolution involves measuring the width between predefined concentration limits in composition profiles. The profiles should have abrupt concentration discontinuities such as occur at boundaries in two-phase specimens. This technique was originally developed for bulk specimens (Goldstein and Yakowitz, 1975) but is readily adaptable to the measurement of x-ray resolution in thin specimens. Its merits lie in the simple nature of the measurement and the relative simplicity of its interpretation. Its disadvantages lie primarily in obtaining a suitable specimen known to have the required abrupt concentration change at a two-phase boundary.

The model used to evaluate resolution is shown in Figure 29, which presents a cross section through a two-phase specimen at an interface. The interface is assumed to be parallel to the incident beam. As the electron beam is moved, in a direction normal to the boundary, the measured solute concentration is plotted versus perpendicular distance from that boundary. The expected result for a Gaussian electron probe is shown in Figure 30. Ideally, Phase 1 would be a single-element phase different from Phase 2, so that the measured element concentrations from the two phases could be distinguished, but they should not differ too greatly in atomic number, atomic weight, and density—it is important that Phase 1 and Phase 2 be physically similar but distinguishable.

From a profile such as Figure 30, we can quantitatively assess an x-ray resolution if we have a suitable definition. In Table X a number of arbitrary choices for a percentage definition of x-ray generation within the beam/specimen-interaction volume were given, with values of $\sim$90% being popular. As the concentration profile in Figure 30 is evaluated directly from the x-ray intensity data, a 90% definition for x-ray resolution can be obtained by measuring the distance between the 5% and 95% concentration levels of Figure 30. These levels are seen to be close to "knees" in the concentration profiles, and as a result the resolution is simple to measure. Inspection of Figure 30 shows that percentage definitions $\gg$90% will cause resolution measurements to be very sensitive to the accuracy of the measurement technique in the tails of the concentration profile, whereas definitions $\ll$90% will cause relatively small changes in the measured resolution to result from large changes in the percentage definition. The 90% definition, though arbitrary, is clearly a convenient compromise.

Several experiments have been reported using this technique, and the data are summarized in Table XI. The work on the Fe/Ni system has generally produced concentration profiles similar to Figure 31, taken from the results of Romig (1979) and reported by Goldstein (1979). This figure clearly shows a diffusion profile up to the abrupt concentration discontinuity at the boundary between the α and γ phases, which are not shown schematically in the ideal profile of Figure 30. The concentration change at the boundary does, however, allow estimates of a 90% width to be made from Figure 31. The resolution in this experiment is such that an accurate measure of the diffusion profile concentrations could be made within 100 nm of the boundary.

Two alternative methods of assessing x-ray spatial resolution in thin specimens have also been proposed, but these both suffer from practical difficulties associated with their interpretation. The first technique measures the x-ray intensity ratio from particles of a known size in specimens of a known thickness and has been used by Faulkner *et al.* (1977) to estimate resolution in Fe alloys containing TiC particles.

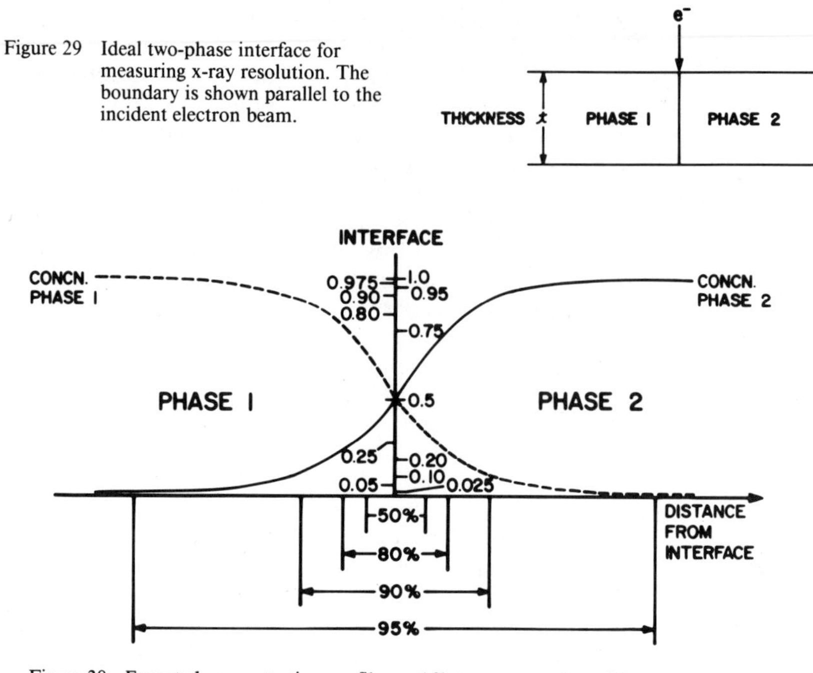

Figure 29 Ideal two-phase interface for measuring x-ray resolution. The boundary is shown parallel to the incident electron beam.

Figure 30 Expected concentration profiles and % x-ray generation widths obtained as a Gaussian electron probe is moved across a two-phase interface (see Figure 29).

TABLE XI

Experimentally Determined Minimum 90% X-Ray Resolution at Two-Phase Interfaces.

Reference	Two-Phase System	keV	Thickness (nm)	90% Resolution (nm)
Goldstein *et al.* (1976)	α/γ in Fe/Ni Meteorite	100	<500	<200*
Easterling (1977)	α/γ Al-Ag	200	Thin	<50
Rao & Lifshin (1977)	Carbide/Matrix Interface in Cr, Ni Steel	100	<200	<50
Romig (1979)	α/γ in Fe/Ni Meteorite	100	<150	<100
Champness *et al.* (1980)	Orthopyroxene Lamella	100	200	30
			400	60
Romig *et al.* (1982)	α/γ in U-6% Nb	100	60	40
			100	90
	α/γ in U-2% Mo	100	55	35
			115	80

* Using an AEI EMMA 4. The other results were obtained using TEM/STEM systems and probe size <20 nm.

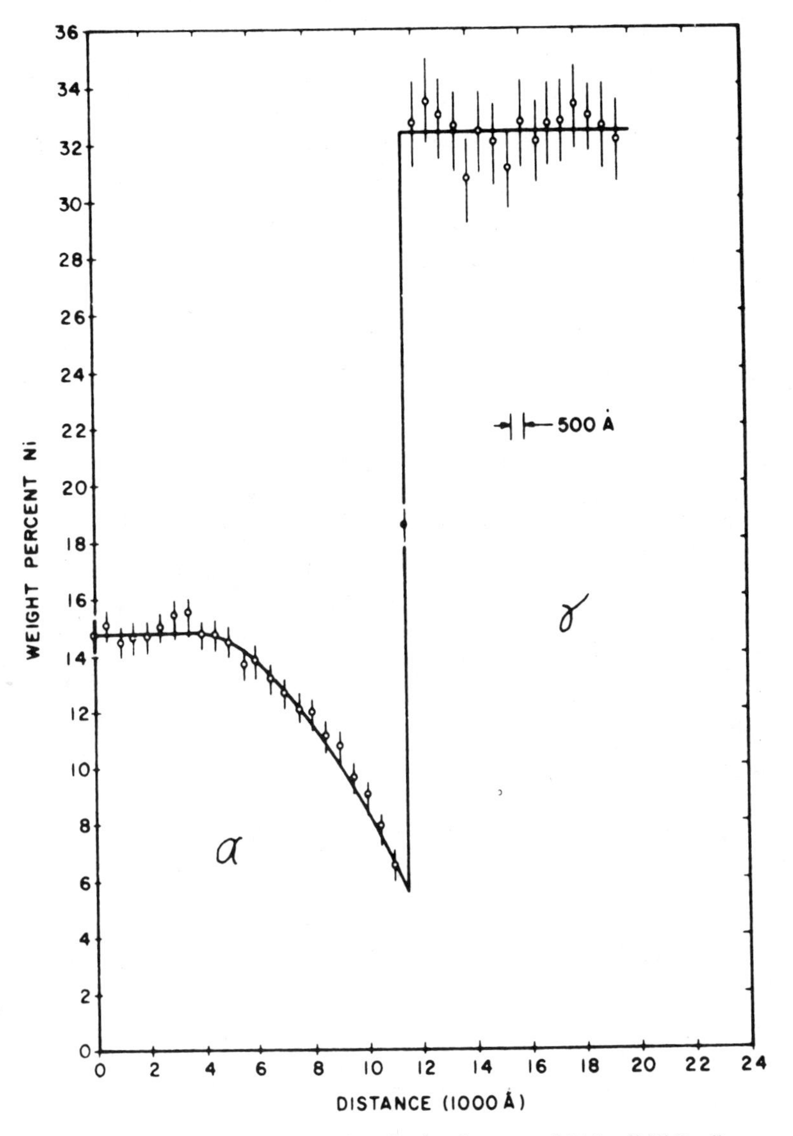

Figure 31 Ni profile across an α/γ interface in a heat-treated 14.7 wt% Ni-Fe alloy (Romig, 1979).

The model of Figure 32 illustrates the concept of the technique. The x-ray resolution defined as $2r_2$ is given by

$$2r_2 = 2r_1 \left[\frac{4r_1^3}{3t} \left(1 + \frac{V_2}{V_1} \right) \right]^{1/2} \tag{30}$$

where V_2/V_1 is the ratio of x-ray emission volumes of the matrix and particle. This ratio is measured from the intensity ratios of one element uniquely present in the matrix and one element unique to the particle. These ratios are proportional to the concentration ratios and hence the volume of the phase.

The model assumes that the beam/specimen interaction volume is cylindrical, that this volume does not broaden with thickness and furthermore that no absorption or fluorescence corrections are necessary in evaluating V_2/V_1. Faulkner *et al.* used this technique to measure resolution in austenitic alloys ~150 nm thick containing TiC particles ~60 nm in diameter at 200 keV and 100 keV, obtaining values for $2r_2$ of 50 nm and 10 nm, respectively. The variation in $2r_2$ as a function of keV is as predicted by the equation of Goldstein *et al.* (1977), but the broadening parameter b calculated by the "single-scattering equation" is half that of $2r_2$. This disparity is a result of the two definitions for resolution being intrinsically different. The quantitative comparison of b and $2r_2$ is therefore perhaps unnecessary but does serve to illustrate that the concept of "resolution" is equally a function of measurement and definition.

The second technique for estimating an x-ray spatial resolution measures concentration profiles at a grain boundary, at which segregation has occurred. Element segregation to grain boundaries is generally believed to be very localized at the grain boundary with the solute elements extending to distances <10 nm from the grain boundary. In principle, the width of the measured profiles can be used to determine resolution, but this technique has been used extensively only by Doig and Flewitt (1977) and by Hall *et al.* (1981). In both cases, the solute concentration at the boundary and the true form of the concentration profile were unknowns, and each group was attempting to obtain the "true" profiles. This technique is satisfactory only for broad profiles that are larger than the beam/specimen interaction volume. The situation then has features similar to the diffusion profiles seen in Figure 31, and the experiment becomes similar to the technique previously described at an abrupt interface. The interpretation, however, is complicated by the absence of a known, sharp change in concentration. For narrow profiles, when the profile width is smaller than the probe size, a measure of resolution can be defined only if both probe size and concentration profile are known.

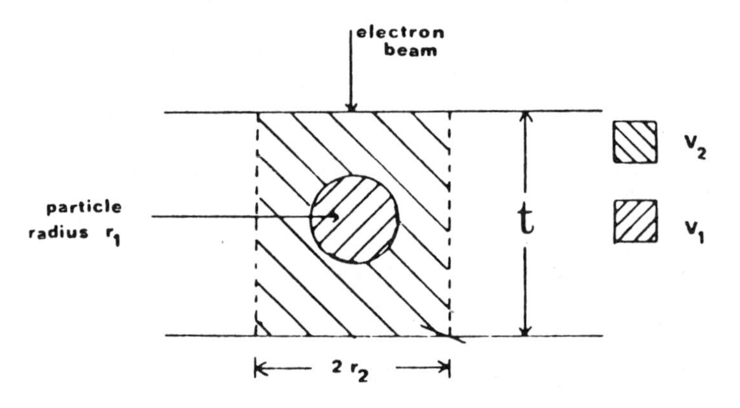

Figure 32 A method for determining the x-ray spatial resolution (Faulkner *et al.*, 1977).

Doig and Flewitt (1977) used the model shown in Figure 33 to assess resolution indirectly by analyzing the grain-boundary concentration of Sn and P in embrittled Fe-3 wt% Ni alloys. The Sn-rich grain-boundary layer had an assumed width of δ, and the beam/specimen interaction volume, assumed to be a cylinder, had a diameter D. The specimen thickness was t. If C_M is the measured concentration on the grain boundary of solute element P (or Sn), C_{GB} is the concentration of the same element in the boundary phase of width δ, and C_{MTX} is the concentration for that element in the matrix, then mass balance in the interaction volume requires that

$$C_M = \left[C_{GB}(D\,t\,\delta) + C_{MTX}\left(\frac{\pi D^2 t}{4} - Dt\delta\right)\right] / \frac{\pi D^2 t}{4}\,. \tag{31}$$

Equation (31), however, has two unknown quantities: profile width δ and diameter D of the interaction volume. By assuming a probe size $d \approx 15$ nm and the value for b using the single-scattering equation [$b = 65$ nm if $t = 200$ nm (Fe-3 wt% Ni)], D could be estimated as 80 nm. From the C_M value for P of 1.4 wt%, δ could be shown to be ~4 nm. This value is consistent with the expectation that the profile at the boundary extended over <10 nm. The problem with this result, however, is that an interaction volume (or "resolution" as previously defined for an interface) of 80 nm has apparently "resolved" a 4-nm feature.

In contrast, Vander Sande *et al.* (1980) have defined resolution as the measured width of solute profiles at grain boundaries, analyzing the Fe concentration in Fe-doped MgO by AEM techniques. They have reported the measured profiles shown in Figure 34. They concluded from their data that the resolution was unaffected by specimen thickness and beam broadening because the profile width did not change as a function of thickness. Like Doig and Flewitt, however, they did not know the solute concentration profile at the boundary and could not unambiguously interpret the data. Calculation of the beam-broadening parameter for the thickest samples gave $b = 50$ nm when the measured profile widths were suggesting a spatial resolution of 5 nm. Probe size d was claimed to be 2.5 nm (VG HB5 FEG/STEM); therefore, the diameter of the interaction volume $(d + b)$ was ≈ 50 nm, and this volume had apparently resolved detail on a scale of 5 nm.

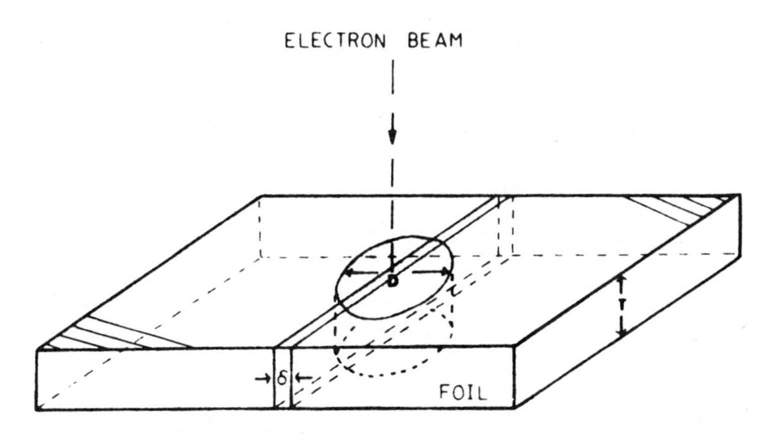

Figure 33 Schematic diagram showing the microanalysis of a grain-boundary region (Doig and Flewitt, 1977).

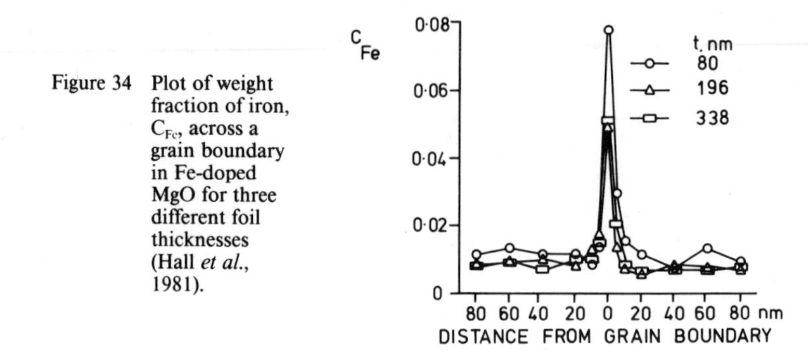

Figure 34 Plot of weight fraction of iron, C_{Fe}, across a grain boundary in Fe-doped MgO for three different foil thicknesses (Hall *et al.*, 1981).

Again, it should be recognized that for very narrow solute layers on grain boundaries, the boundary width is much smaller than the interaction volume in thin specimens, and the measured concentration profile in this case will be different from the profile measured at a planar interphase between two phases. A profile obtained from a *narrow* grain-boundary layer (δ smaller than the probe size) does not measure beam broadening or resolution directly. If the boundary width becomes much smaller than the beam size, the unknown, true concentration profile can be mathematically approximated to a delta function. When convolved with any other function (in this case, the function describing the shape of the electron probe), a delta function has the interesting effect of yielding a third function that is simply the shape of the other function, i.e., the probe.

The effect is well illustrated in Figure 35, taken from the early work of Rapperport (1969), who attempted to resolve (or deconvolve) "true" concentration profiles from measured data obtained by using electron beams with diameters much larger than the dimensions of the "true" profile. Examination of Figure 35 shows that the shape of the measured data h(x) bears a clear resemblance to the shape of the probe profile f(x). In the limit, as the true concentration profile g(x) tends to very small widths with respect to the probe width, the shape of h(x) will more closely approximate that of f(x). It seems possible that the data of Vander Sande *et al.* (1980), instead of accounting for either the concentration profile or the x-ray spatial resolution, yields nothing more than the shape of the probe.

Another approach to obtaining a measure of x-ray spatial resolution, in fact the beam-broadening parameter b at the exit surface, was originally developed by Hutchings *et al.* (1979) and improved by Stephenson *et al.* (1981). The principle of the experiment is shown in Figure 36. Thin samples of Si were prepared with either Au or Cr strips deposited on one side of the specimen. The x-ray intensity from a strip was monitored as the probe was stepped across the strip while the strip was on either the electron entrance (Figure 36(b)) or exit (Figure 36(a)) side of the specimen. The difference between the 90% widths of the two measurements was used as a measure of beam spreading at various specimen thicknesses and beam voltages. The beam-spreading measurements are shown in Figures 27 and 28. Good agreement exists between theory and experiment. However, the measurement errors, even in Si, remain large.

The last two techniques for measuring resolution at boundaries imply that it may be possible to resolve measured detail on a scale smaller than the beam/specimen interaction volume. By means of appropriate deconvolution techniques and careful modeling of the beam/specimen interaction volume, it should be possible to resolve

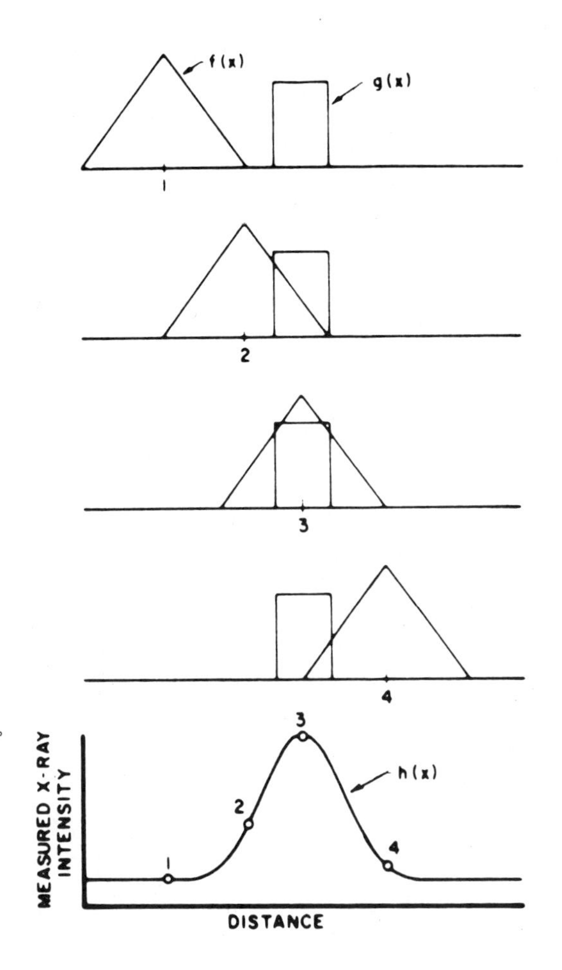

Figure 35 The convolution technique. A concentration step g(x) measured by a triangular-shaped x-ray distribution f(x) will result in the observed profile h(x) as a beam passes over the concentration step (from Rapperport, 1969).

specimen analysis details on a scale very much less than the dimensions of the interaction volume. However, the resolution of x-ray microanalysis is then defined by the scale of the phenomenon being studied (Twigg *et al.*, 1981). This would seem to be a less acceptable definition of resolution, particularly from the point of view of making comparisons between experiments in different systems.

In a practical analysis, it is of interest to decrease or minimize the amount of beam broadening. One of the most straightforward methods to accomplish this is to produce very thin foils. Figure 37 shows a Cr concentration profile taken with a HB5 instrument over two ferrite/cementite boundaries in an ion-milled sample of steel (Garratt-Reed, 1982). The analysis region of the foil of the α iron (ferrite) and carbide (cementite) was only 20 nm thick. Beam-broadening calculations (Goldstein *et al.*, 1977) yield a value of b that is only 1.72 nm. From Figure 37, the x-ray spatial resolution using a 90% definition is $\sim$2 nm. Since the probe diameter is nominally 0.25 nm, the x-ray spatial resolution is essentially b, the amount of beam broadening. It is hoped that more progress in developing foil-thinning procedures can be made in the future. Progress in this area will minimize the need for using deconvolution techniques to "improve" spatial resolution.

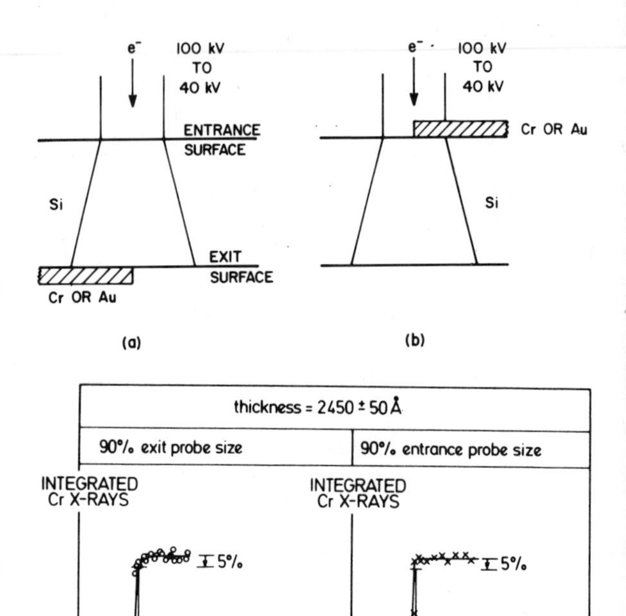

Figure 36 A method for the measurement of x-ray spatial resolution (Stephenson *et al.*, 1981).

(a) Technique for the measurement of beam broadening and the beam diameter-exit surface in Si.

(b) Technique for the measurement of the beam diameter-entrance surface in Si.

(c) Experimental data – Cr x-ray profiles obtained at 40 keV from a Si specimen 245 nm thick, together with measurements of the 90% x-ray source size obtained when the Cr was on the electron exit (o) or entrance (x) side of the Si.

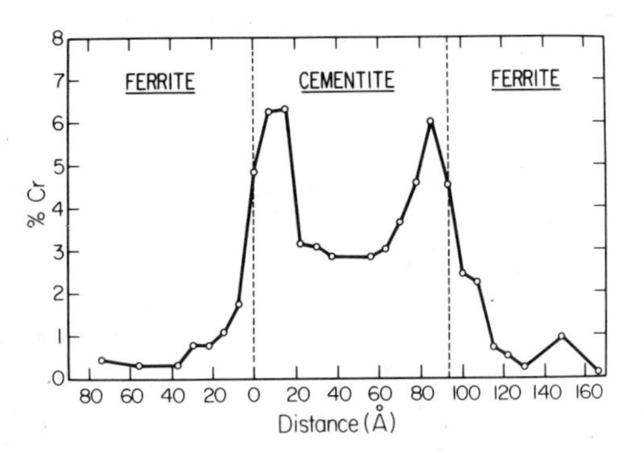

Figure 37 Cr concentration profile taken across two ferrite-cementite interfaces in an Fe-base alloy. The thin foil was 20 nm thick (Garratt-Reed, 1982).

VII. EXAMPLES OF QUANTITATIVE EDS ANALYSIS IN THE AEM

A. Determination of the k_{TiFe} Factor for the TiFe System (courtesy J. E. Wood)

Experimental determination of k_{AB} factors is the most accurate method available for quantification of EDS data and is the first step in obtaining quantitative data for any particular system, AB. It is first necessary to obtain a suitable thin-foil standard, several discussions of which are in the literature. A standard specimen should be chemically homogeneous at the level of spatial resolution used (<100 nm), should not degrade under the beam, and ideally not require any absorption or fluorescence correction, i.e., it should obey the thin-film criterion. The bulk specimen composition should be a suitable mineral or stoichiometric compound or should be determined previously by other well-calibrated methods, such as EPMA, wet chemistry, or mass spectrometry techniques.

After a suitable specimen is obtained, the x-ray spectrum should be accumulated under the same conditions as for subsequent microanalyses. Specifically, the operating keV should be identical since k_{AB} factors vary with keV. To minimize statistical error in quantification, it is best to accumulate at least 10,000 counts above background in each peak that is to be used. Under these circumstances, the 3σ error bars will be $\pm 3\%$ relative. More counts will decrease the error accordingly. If contamination or etching or some other problem prevents counting on one point for a long time, many data points from different regions of the same specimen can minimize such effects and still give reasonably small errors in k_{AB}. For example, using a homogeneous specimen of ilmenite (36.36% Fe, 27.41% Ti, 3.87% Mn, balance O), 30 spectra were obtained from different thin regions at 120 keV. A typical spectrum after 100 seconds had 32,570 Fe K_α counts and 27,830 Ti K_α counts, giving $I_{Ti}/I_{Fe} = 0.88$. Knowing $C_{Ti}/C_{Fe} = 0.75$, $k_{TiFe} = 0.857$. Combining this value with another 29 similar spectra yields $k_{TiFe} = 0.85$, with a relative error of $\pm 0.9\%$. Relative errors are determined in the following manner

$$\% \text{ error} = \left[\frac{t_{99}^{n-1}}{\sqrt{n}} \frac{S}{\bar{k}_{AB}}\right] \times 100 \tag{32}$$

where

t_{99}^{n-1} = Student t value of n readings (n = 30 in this case) at a 99% confidence level
S = standard deviation from 30 readings
$\bar{k}_{AB}$ = average value of k_{TiFe} from each of the 30 spectra.

Thus, high accuracy may be achieved if specimens are suitably homogeneous and sufficient x-ray counts are accumulated.

B. Determination of Ag Composition Gradients in a Heat-Treated Al-16 wt% Ag Alloy (courtesy S. M. Merchant)

After a suitable k_{AB} factor for a particular system has been obtained, analysis of inhomogeneous regions with high resolution can be achieved by simply obtaining suitable spectra to give I_A and I_B, and calculating C_A and C_B by using the Cliff-Lorimer equation. For example, in studying precipitate-free zones (PFZ) in Al-Ag alloys, a value of $k_{AlAg} = 0.73 \pm 0.03$ (4.3% rel. error) was determined by using a homogenized Al-9.4 wt% Ag alloy in a manner similar to the previous description for k_{TiFe}. This k_{AlAg} value was used to determine C_{Al} and C_{Ag} across a PFZ in Al-16 wt% Ag.

Before microanalysis, the absorption correction factor (ACF; Eq. (17)) was calculated for the range of foil thicknesses (150 to 300 nm) used in the study and for Ag contents of $\leq$ 16 wt% in Al. The absorption correction was <3% relative, and in this example the thin-film criterion can be applied. Therefore

$$\frac{C_{Al}}{C_{Ag}} = k_{AlAg}\,\frac{I_{Al}}{I_{Ag}}\,.$$

A typical Ag gradient near the grain boundary in a heat-treated Al-16 wt% Ag sample is shown in Figure 38. The error bars on each point were determined in the following manner. If N counts were accumulated in either the Al or Ag peaks, the 3σ ($\approx$99%) confidence limit is given by $3\sigma = 3\sqrt{N}/N \times 100\%$. The total error in C_{Al}/C_{Ag} is then the sum of the 3σ error in I_{Al} and I_{Ag} plus the error in the k_{AlAg} value $\pm 4.3\%$. The error bars in Figure 38 are a measure of this value.

C. Determination of the Minimum Detectability Limit for Mn in a Cu-Mn Alloy (courtesy J. R. Michael)

Using an approach originally developed by Ziebold (1967), Romig and Goldstein (1978) have shown that a Gaussian-shaped x-ray characteristic peak is a statistically valid peak if

$$I_A - I_b^A \geq 3\sqrt{2I_b^A}$$

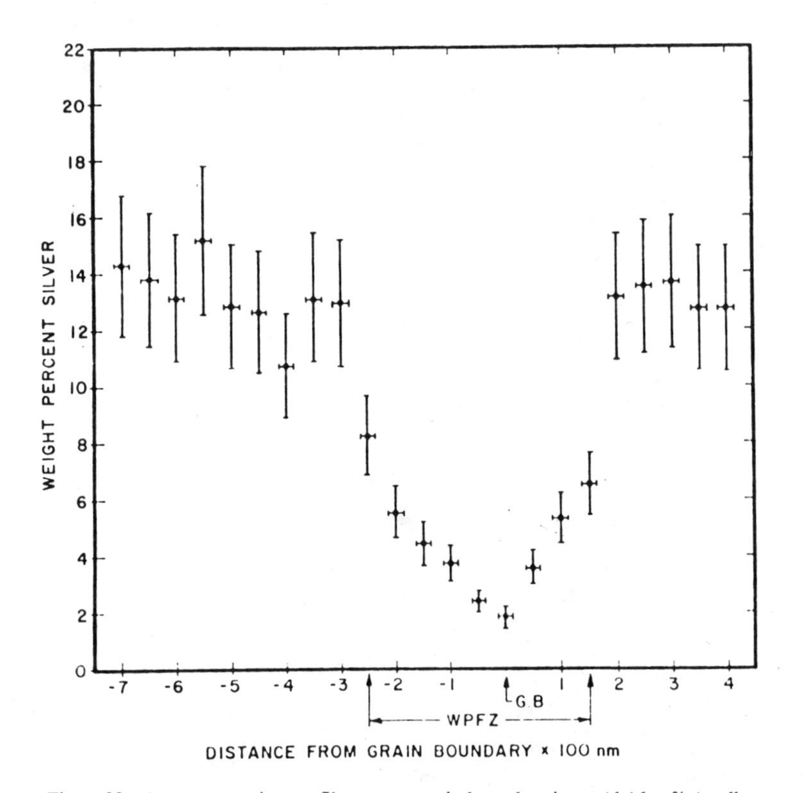

Figure 38 Ag concentration profile across a grain boundary in an Al-16 wt% Ag alloy aged at 523 K for 0.5 h. WPFZ is the white precipitate-free zone (courtesy S. M. Merchant).

where I_A is the intensity in the peak of interest from element A, and I_b^A is the background intensity under the same peak. Thus, from any detected x-ray peak, the minimum detectable amount of any element can be deduced. This equation gives a confidence limit of 95% that a peak is indeed present. A sample of Cu-3.36 wt% Mn was used to determine the detectability limit of Mn in Cu (C_{DL})

$$C_{DL} = C_{Mn} \times \frac{3\sqrt{2I_b^{Mn}}}{I_{Mn} - I_b^{Mn}} \text{ wt\%} .$$

At 120 keV, with a 20-nm spot size, a typical spectrum accumulated 11,964 counts in the Mn K_α peak ($=I_{Mn}$) and 1995 counts in the background ($=I_b^{Mn}$) over 200 seconds of live time. The detectability limit from Eq. (2) $=$ 0.064%, which is 640 ppm of Mn. Under favorable conditions of counting time, lack of contamination, a well-characterized sample, and no peak overlaps, absorption, etc., to worry about, the theoretical sensitivity of x-ray analysis in the AEM can approach the performance of EPMA of bulk specimens. Since this value is obtained from a thin-foil specimen, the absolute mass that is detected is significantly smaller, of the order 10^{-18} to 10^{-19} g.

D. Determination of the Mg and Al Content of Oxide Ceramics (courtesy J. E. Wood)

During a study of precipitation in oxide ceramics, the k factor for Mg-Al had to be determined, and a suitable stoichiometric standard was the spinel $MgAl_2O_4$ with density $\rho = 3.6 \text{ g/cm}^{-3}$ and a composition of the following weight fractions

$$C_{Mg} = 0.171$$
$$C_{Al} = 0.379$$
$$C_O = 0.45$$

In a region of the specimen where thickness t $=$ 180 nm (determined by convergent-beam diffraction), the following integrated characteristic intensities were obtained above background at 120 keV, with take-off angle $\alpha = 20°$

$$I_{Mg} = 9{,}200 \text{ (Mg } K_\alpha); \; I_{Al} = 25{,}240 \text{ (Al } K_\alpha) .$$

If these data were used alone, the k factor would be incorrectly determined because of absorption of Al K_α x-rays by Mg and O and the absorption of Mg K_α x-rays by O. The uncorrected k_{MgAl} value is given from the Cliff-Lorimer equation by

$$k_{MgAl} = \frac{C_{Mg}}{C_{Al}} \frac{I_{Al}}{I_{Mg}} = \frac{17.1}{37.9} \times \frac{25{,}240}{9{,}200} = 1.24 .$$

The ACF for absorption is given by Eq. (17), namely

$$ACF = \left[\frac{\mu/\rho]^A_{SPEC}}{\mu/\rho]^B_{SPEC}}\right]\left[\frac{1 - \exp - \{\mu/\rho]^B_{SPEC} \times \operatorname{cosec} \alpha(\rho t)\}}{1 - \exp - \{\mu/\rho]^A_{SPEC} \times \operatorname{cosec} \alpha(\rho t)\}}\right]$$

where

Mg $=$ A
Al $=$ B
$\mu/\rho]^{A/B}_{SPEC}$ $=$ mass absorption coefficient of A or B in the specimen
α, ρ, and t have been defined previously.

To determine $\mu/\rho]^{Mg}_{SPEC}$ and $\mu/\rho]^{Al}_{SPEC}$, respectively, we need to know from Heinrich's tables of mass absorption coefficients (Heinrich, 1966) that

$$\mu/\rho]^{Mg}_{Mg} = 463.6 \text{ cm}^2 \text{ g}^{-1} \qquad \mu/\rho]^{Al}_{Mg} = 4376.5 \text{ cm}^2 \text{ g}^{-1}$$

$$\mu/\rho]^{Mg}_{Al} = 614.7 \text{ cm}^2 \text{ g}^{-1} \qquad \mu/\rho]^{Al}_{Mg} = 385.7 \text{ cm}^2 \text{ g}^{-1}$$

$$\mu/\rho]^{Mg}_{O} = 2432.8 \text{ cm}^2 \text{ g}^{-1} \qquad \mu/\rho]^{Al}_{O} = 1503.3 \text{ cm}^2 \text{ g}^{-1}$$

and from these data we find

$$\begin{aligned} \mu/\rho]^{Mg}_{SPEC} &= \mu/\rho]^{Mg}_{Mg} C_{Mg} + \mu/\rho]^{Mg}_{Al} C_{Al} + \mu/\rho]^{Mg}_{O} C_{O} \\ &= (463.6 \times 0.171) + (614.7 \times 0.397) + (2432.8 \times 0.45) \\ &= 1407.00 \text{ cm}^2 \text{ g}^{-1} . \end{aligned}$$

A similar calculation for $\mu/\rho]^{Al}_{SPEC}$ gives 887.45 cm^2 g^{-1}. Inserting these values in the ACF equation, we find

$$\text{ACF} = \left[\frac{1407}{887.45}\right]\left[\frac{1 - \exp - (887.45 \times 2.9238 \times 3.6 \times 180 \times 10^{-7})}{1 - \exp - (1407 \times 2.9238 \times 3.6 \times 180 \times 10^{-7})}\right]$$

$$\text{ACF} = 1.585\left[\frac{1 - \exp(-0.168)}{1 - \exp(-0.267)}\right]$$

$$\text{ACF} = 1.045 .$$

The full correction for absorption and fluorescence is given in Eq. (23). Since fluorescence is insignificant in the Mg-Al-O system, the following equation can be used to obtain k_{MgAl}, namely

$$k_{MgAl} = \left(\frac{C_{Mg}}{C_{Al}}\right)\left(\frac{I_{Al}}{I_{Mg}}\right) / \text{ACF} .$$

Therefore, the actual k_{MgAl} value is 1.24/1.045 = 1.18, which can be used in appropriate calculations.

E. Determination of the Cr and Fe Content of CrFe Alloys

In this example, the specimen was a homogeneous, single-phase solid solution of nominal composition Fe-10 wt% Cr (C_{Fe} = 0.9, C_{Cr} = 0.1), in which Cr K_α x-rays (energy = 5.414 keV) are fluoresced by Fe K_α x-rays (energy = 6.403 keV), resulting in an apparently higher Cr content. The specimen was examined in an AEM at 100 keV, with take-off angle α = 45°, using a previously determined k_{FeCr} factor of 1.09. The thin-film specimen was unusually thick (~900 nm). The integrated intensities above background were I_{Fe} = 77,169 and I_{Cr} = 10,756. Assuming that the thin-film criterion is satisfied, these data are used in the Cliff-Lorimer equation so that

$$\frac{C_{Fe}}{C_{Cr}} = k_{FeCr}\,\frac{I_{Fe}}{I_{Cr}}$$

$$\frac{C_{Fe}}{C_{Cr}} = 1.09\left(\frac{77{,}169}{10{,}756}\right)$$

$$\frac{C_{Fe}}{C_{Cr}} = 7.82.$$

Since $C_{Fe} + C_{Cr} = 1$,

$$C_{Cr} = 0.113 = 11.3\ \text{wt\%};\ C_{Fe} = 88.7\ \text{wt\%}\ .$$

The mass absorption coefficients of Cr and Fe are very similar in a nominal Fe-10 wt% Cr alloy. The amount of the absorption correction even for a 900-nm-thick foil is <1% relative, and the absorption correction factor is 1.006. Therefore, the thin-film criterion is fulfilled for absorption, and the Cliff-Lorimer equation, now modified to handle fluorescence (Eq. (23)), can be applied, namely

$$\frac{C_{Fe}}{C_{Cr}} = k_{FeCr}\,\frac{I_{Fe}}{I_{Cr}}\left(1 + \frac{I^{Cr}}{I_{Cr}}\right)\ .$$

The fluorescence correction is given by

$$\frac{I^{Cr}}{I_{Cr}} = C_{Fe}\,\omega_{Fe}\left(\frac{r_{Cr}-1}{r_{Cr}}\right)\frac{A_{Cr}}{A_{Fe}} \times \mu/\rho]^{Fe}_{Cr} \times \frac{E_{c_{Cr}}}{E_{c_{Fe}}} \times \frac{\ell n\,(E_o/E_{c_{Fe}})}{\ell n\,(E_o/E_{c_{Cr}})} \times \frac{\rho t}{2}$$

$$\times\ [0.923 - \ell n\,(\mu/\rho]^{Fe}_{SPEC}\,\rho t) \times \sec\alpha\ .$$

The various terms have already been defined in the text (Section III and Eq. (21)). The values of the terms used in the calculation are

$C_{Cr} = 0.1$	$r_{Cr}-1 = 0.88$	$\mu/\rho]^{Fe}_{Cr} = 474.2\ cm^2\ g^{-1}$
$C_{Fe} = 0.9$	$A_{Cr} = 52.01$	$E_{c_{Cr}} = 5.988$ keV
$\omega_{Fe} = 0.3136$	$A_{Fe} = 55.85$	$E_{c_{Fe}} = 7.111$ keV
$E_o = 100$ keV	$\rho t = 7.074 \times 10^{-4}\ g\ cm^{-2}$	$\mu/\rho]^{Fe}_{SPEC} = 111.7\ cm^2\ g^{-1}$

where C_{Cr} and C_{Fe} are estimated.

Inserting these values in the fluorescence correction

$$\frac{I^{Cr}}{I_{Cr}} = 0.1062\ .$$

This result shows that ~10% of the Cr K_α intensity is due to fluorescence. Using this ratio of I^{Cr}/I_{Cr}

$$\frac{C_{Fe}}{C_{Cr}} = 1.09\left(\frac{77{,}169}{10{,}756}\right)(1 + 0.1062)$$

$$= 8.65\ .$$

Therefore, the true concentration corrected for fluorescence is

$$C_{Cr} = 10.4 \text{ wt\%}; \qquad C_{Fe} = 89.6 \text{ wt\%} .$$

A slightly more accurate answer can be obtained if the fluorescence correction is recalculated using these values of C_{Cr} and C_{Fe}.

F. ALCHEMI (Atom Location Using Channeling-Enhanced Microanalysis) (courtesy H. M. Chan)

In Chapter 4 it was recommended that x-ray analysis be performed away from strong diffracting conditions because of the possibility of anomalous x-ray generation. This phenomenon can, however, be used to locate the position of certain atoms in atomic structures. This technique, known as ALCHEMI (Spence and Tafto, 1983), measures the change in x-ray emission from different atoms as a function of specimen orientation. In this particular example, Ca impurity atom site positions were determined in $BaTiO_3$ (Chan *et al.*, 1984).

ALCHEMI is well suited to investigating the site occupancy of impurity atoms in $BaTiO_3$ because, in this material, the barium and titanium atoms lie on alternate, parallel planes (<h00>). Thus, depending on the incident-beam orientation, the electron intensity will be concentrated at either the A or B planes; hence, x-ray emission from one element will be enhanced relative to the other. By comparing the ratio of the characteristic x-ray intensity of an impurity element to that of either Ba or Ti for the two different orientations, it is possible to determine the fraction of impurity atoms on each type of site.

The specimen was tilted so that a systematic row of <h00> reflections was visible in the diffraction pattern. It was found that the most pronounced channeling effect occurred for slightly positive and slightly negative excitations of the [100] reflection. These two orientations were used for collecting the x-ray spectra. It is worth noting, however, that in the ALCHEMI technique, the precise orientations within the systematic row are unimportant, provided that sufficient channeling is obtained. The following two compositions were studied: $(Ba_{1-x}Ca_xO)_yTiO_2$, where $y = 1.02$ and 0.98; $x = 0.15$.

The results of the ALCHEMI determination are shown in Table XII. C_x represents the fraction of Ca atoms occupying Ti sites and is calculated from the following expression:

$$C_x = (R - 1) / (R - 1 + \gamma - \beta R)$$

where

$$R = [N_{Ba}^{(1)} / N_{Ca}^{(1)}] / [N_{Ba}^{(2)} / N_{Ca}^{(2)}]$$

$$\beta = N_{Ti}^{(1)} / kN_{Ba}^{(1)}$$

$$\gamma = N_{Ti}^{(2)} / kN_{Ba}^{(2)}$$

$$k = N_{Ti}/N_{Ba}$$ for an orientation in which there is no channeling.

The superscripts $^{(1)}$ and $^{(2)}$ represent the two orientations, and N is the number of counts in the x-ray peak.

TABLE XII

Results of ALCHEMI Determination.

Specimen	C_x(%)	Number of Determinations
$(Ba_{0.85}Ca_{0.15}O)_{0.98}TiO_2$	9.0 ± 9.7	10
$(Ba_{0.85}Ca_{0.15}O)_{1.02}TiO_2$	26.3 ± 13.2	18

It can be seen that the spread in C_x values is quite high. However, by applying a Student's t-test to the two sets of data, it can be shown that the difference in C_x values is indeed "significant" (for a 95% confidence limit) and cannot be accounted for by the spread in the data alone. Thus, it can be concluded that for the A site excess specimen, a definite fraction of calcium atoms is occupying Ti sites.

The strength of the channeling effect that can be obtained is also an important factor, and this is dependent on the crystal structure and chemical composition. The channeling that can be obtained in $BaTiO_3$ is relatively weak (~30% increase in the N_{Ti}/N_{Ba} ratio for the two orientations)—hence, the sensitivity of the C_x value to the count rate. Clearly, for materials where the channeling is more pronounced, the accuracy will be improved.

VIII. SUMMARY

The various approaches to quantification of x-ray data in the AEM have been developed. Criteria for the application of absorption and fluorescence corrections have also been given. In many cases the ratio method, developed by Cliff and Lorimer, can be applied. Although the direct measurement of k_{AB} factors is preferred, methods have been detailed for the calculation of these factors where data cannot be obtained.

In addition to quantification techniques, statistical methods for determining analytical errors and minimum detectability limits have been developed for the user. Approaches to defining and measuring x-ray spatial resolution have also been discussed. Finally, several examples have been given illustrating the use of the quantification techniques and the x-ray statistical methods that were discussed in the chapter.

TABLE OF CHAPTER VARIABLES

a_A	Fraction of the total K, L, or M line intensity measured
A_A,A_B	Atomic weight of elements A, B, C, etc.
A,B,C	Constants in Burhop (1955) equation for ω
b	Beam-broadening parameter
b_s, c_s	Constants for a particular shell in the ionization cross section equation
C_A,C_B	Composition (wt. fraction) of elements A, B, C, etc., in a specimen
C_{GB}	Concentration of a given element in the grain boundary

C_M	Measured concentration of a given element on the grain boundary
C_{MTX}	Concentration of a given element in the matrix
d	Absorption distance
d_p	Electron probe diameter
d_s	Term used in Schreiber and Wims (1981b) cross section formulation
D	Diameter of cylinder beam/specimen interaction volume (Doig and Flewitt, 1977)
E	Energy of an x-ray line of interest
E_c	Ionization energy for the K, L, or M shell of a given element A, B, C, etc.
E_{c_A}, E_{c_B}	E_c for the characteristic radiation of A and B
E_o	Operating voltage of the instrument
E_o^*	Relativistic energy
f(x)	Shape of the probe profile
g(x)	Shape of the true concentration profile
h(x)	Shape of measured concentration data
I_A,I_B	Intensity of characteristic x-rays of the elements A, B, C, etc.
I^A/I_A	Ratio of fluorescence intensity to primary intensity
$1+I^A/I_A$	Fluorescence correction for A
I_A^S	Absolute intensity of element A measured in a thin-film standard
I_A^U	Absolute intensity of element A measured in an unknown thin-film specimen
I_A^*	Generated characteristic x-ray intensity for element A
I_b^A	Number of x-ray counts of element A from the background continuum for element A
I_b^B	Number of x-ray counts of element B from the background continuum for element B
k	k_{ASi}
k_{AB}	Cliff-Lorimer factor
$k_{AB})_{TF}$	Absolute value of k_{AB} when $t = o$
n	Number of intensity measurements
n_s	Number of electrons in a shell or subshell
N	Number of accumulated counts at each analysis point
P	Pure element counting rate
P_θ	Probability that scattering has occurred beyond the angle θ
Q_A,Q_B	Ionization cross section, related to the probability of an electron of a given energy causing ionization of a particular K, L, or M shell of atom A, B, C, etc., in the specimen
r_A	Absorption edge jump ratio of element A
r_1	Particle radius
$2r_2$	X-ray resolution
S	Standard deviation
t	Specimen thickness
t*	Projected specimen thickness
$t_{1-\alpha}^{n-1}$	Student's t value, where n is the number of replicates for $1-\alpha$ degree of confidence
U	Overvoltage equal to E_o/E_c
V_1	X-ray emission volume of particle
V_2	X-ray emission volume of matrix
x	Thickness of Be window, or Au surface layer, or Si dead layer

Y_{Si}	EDS active layer thickness
Z	Atomic number of element A, B, C, etc.
α	X-ray take-off angle
β	Angle measured between the incident probe and the sample surface
β_ℓ	Velocity of electron/speed of light (V/C)
γ	$1250/(EU^2)$
δ	Width of grain boundary
δ_c	$E_c/2$
ϵ_A	Efficiency of the EDS x-ray detector due to x-ray absorption
η	Absolute electron flux
θ_o	Screening parameter for the nucleus
μ/ρ	Mass absorption coefficient
$\mu/\rho]^{A,B}_{SPEC}$	Mass absorption coefficient for x-rays from element A, B, C, etc., in the specimen
ξ_g	Extinction distance
ρ	Specimen density
ρt	Mass thickness of specimen
σ	Standard deviation
τ	Counting time
$\phi_{A,B}(\rho t)$	Depth distribution of x-ray production from element A, B, C, etc., as a function of mass thickness (ρt)
ω_A	Fluorescence yield for the K, L, or M characteristic x-ray line of interest for element A

Abbreviations

ACF	Absorption correction factor
MMF	Minimum mass fraction
P/B	Peak/background ratio of the pure element

REFERENCES

Allen, S. M. (1981), Phil. Mag. *A43*, 325.

Bambynek, W.; Crasemann, B.; Fink, R. W.; Freund, H.-U.; Mark, H.; Swift, C. D.; Price, R. E.; and Venugopala Rao, P. (1972), Rev. Mod. Phys. *44*, 716.

Bender, B. A.; Williams, D. B.; and Notis, M. R. (1980), J. Am. Ceram. Soc. *63*, 149.

Blake, D. F.; Isaacs, A. M.; and Kushler, R. H. (1983), J. Micros. *131*, 249.

Bovey, P. E.; Long, N. J.; and Wardell, I. R. M. (1981), in "Quantitative Microanalysis with High Spatial Resolution" (G. W. Lorimer, M. H. Jacobs, and P. Doig, ed.), The Metals Soc., London, Book 277, p. 182.

Brown, D. B. (1974), in "Handbook of Spectroscopy" (J. W. Robinson, ed.), CRC Press, Cleveland, vol. 1, p. 248.

Brown, J. M.; Loretto, M. H.; and Fraser, H. L. (1981), in "Analytical Electron Microscopy – 1981" (R. H. Geiss, ed.), San Francisco Press, San Francisco, p. 61.

Burhop, E. H. S. (1955), J. Phys. Radium *16*, 625.

Champness, P.; Cliff, G.; and Lorimer, G. W. (1980), Bull. de Min. *104*, 236.

Chan, H. M.; Harmer, M. P.; Lal, M.; and Smyth, D. M. (1984), Mat. Res. Soc. Symp. Proc. *31*, Elsevier Science Pub. Co., New York, p. 345.

Cherns, D.; Howie, A.; and Jacobs, M. H. (1973), Z. Naturforsch. *28a, 565*.

Cliff, G., and Lorimer, G. W. (1975), J. Micros. *103,* 203.

——— (1980), in "Electron Microscopy 1980," 7th Euro. Cong. Elec. Mic. (P. Brederoo and V. E. Cosslett, ed.), Leyden, *3*, 182.

——— (1981), in "Quantitative Microanalysis with High Spatial Resolution" (G. W. Lorimer, M. H. Jacobs, and P. Doig, ed.), The Metals Soc., London, Book 277, p. 47.

Colby, J. W. (1968), in "Advances in X-ray Analysis" (J. Newkirk, G. Mallett, and H. Pfeiffer, ed.), Plenum Press, New York, vol. 11, p. 287.

Doig, P., and Flewitt, P. E. J. (1977), J. Micros. *110*, 107.

Doig, P.; Lonsdale, D.; and Flewitt, P. E. J. (1980), Phil. Mag. *A41*, 761.

——— (1981) in "Quantitative Microanalysis with High Spatial Resolution" (G. W. Lorimer, M. H. Jacobs, and P. Doig, ed.), The Metals Soc., London, Book 277, p. 41.

Dupouy, G. (1968), Adv. in Optical and Electron Microscopy *2*, 168.

Easterling, K. E. (1977), J. Mat. Sci. *12*, 857.

Edington, J. W. (1976), "Practical Electron Microscopy in Materials Science," Van Nostrand Reinhold, New York, p. 329.

Faulkner, R. G.; Hopkins, T. C.; and Norrgard, K. (1977), X-ray Spectrometry *6*, 73.

Faulkner, R. G., and Norrgard, K. (1978), X-ray Spectrometry *7*, 184.

Fink, R. W.; Jopson, R. C.; Mark, H.; and Swift, C. D. (1966), Rev. Mod. Phys. *38*, 513.

Garratt-Reed, A. J. (1982), private communication.

Geiss, R. H., and Kyser, D. F. (1979), Ultramicroscopy *3*, 397.

Glitz, R. W.; Notis, M. R.; Williams, D. B.; and Goldstein, J. I. (1981), in "Microbeam Analysis - 1981" (R. H. Geiss, ed.), San Francisco Press, San Francisco, p. 309.

Goldstein, J. I. (1979), in "Introduction to Analytical Electron Microscopy" (J. J. Hren, J. I. Goldstein, and D. C. Joy, ed.), Plenum Press, New York, p. 83.

Goldstein, J. I.; Costley, J. L.; Lorimer, G. W.; and Reed, S. J. B. (1977), SEM/77 (O. Johari, ed.), IITRI, Chicago, *1*, 315.

Goldstein, J. I.; Lorimer, G. W.; and Cliff, G. (1976), Proc. 6th Euro. Cong. Elec. Mic. (D. G. Brandon, ed.), Tal International, Jerusalem, *1*, 56.

Goldstein, J. I., and Yakowitz, H. (1975), "Practical Scanning Electron Microscopy," Plenum Press, New York, p. 457.

Green, M., and Cosslett, V. E. (1961), Proc. Phys. Soc. *78*, 1206.

Hall, E. L.; Imeson, D.; and Vander Sande, J. B. (1981), Phil. Mag. *A43*, 6, 1569.

Hall, E. L., and Vander Sande, J. B. (1975), Phil. Mag. *32*, 1289.

Heinrich, K. F. J. (1966), in "The Electron Microprobe" (T. D. McKinley, K. F. J. Heinrich, and D. B. Wittry, ed.), J. Wiley, New York, p. 351.

Heinrich, K. F. J.; Fiori, C. E.; and Myklebust, R. L. (1979), J. Appl. Phys. *50*, 5589.

Henke, B. L., and Ebisu, E. S. (1974), in "Advances in X-ray Analysis," Plenum Press, New York, vol. 17, p. 150.

Hutchings, R.; Loretto, M. H.; Jones, I. P.; and Smallman, R. E. (1979), Ultramicroscopy *3*, 401.

Jones, I. P., and Loretto, M. H. (1981), J. Micros. *124*, 3.

Joy, D. C. (1979), "Introduction to Analytical Electron Microscopy" (J. J. Hren, J. I. Goldstein, and D. C. Joy, ed.), Plenum Press, New York, p. 223.

Joy, D. C., and Maher, D. M. (1975), Proc. 33rd Annl. EMSA Mtg. (G. W. Bailey, ed.), Claitor's Pub. Div., Baton Rouge, p. 242.

——— (1977), SEM/1977, (O Johari, ed.), IITRI, Chicago, *1*, 325.

Kelly, P. M.; Jostsons, A.; Blake, R. G.; and Napier, J. G. (1975), Phys. Stat. Sol. *31*, 771.

König, R. (1976), in "Electron Microscopy in Mineralogy" (H. R. Wenk, ed.), Springer-Verlag, Berlin, p. 526.

Kyser, D. F. (1979), in "Introduction to Analytical Electron Microscopy" (J. J. Hren, J. I. Goldstein, and D. C. Joy, ed.), Plenum Press, New York, p. 199.

Lorimer, G. W.; Al-Salman, S. A.; and Cliff, G. (1977), in "Developments in Electron Microscopy and Analysis" (D. L. Misell, ed.), Inst. Phys. Conf. Ser. No. 36, Inst. of Physics, Bristol and London, p. 369.

Lorimer, G. W.; Cliff, G.; and Clark, J. N. (1976), in "Developments in Electron Microscopy and Analysis" (J. A. Venables, ed.), Acad. Press, London, p. 153.

Love, G.; Cox, M. G. C.; and Scott, V. D. (1977), in "Developments in Electron Microscopy and Analysis" (D. L. Misell, ed.), Inst. Phys. Conf. Ser. No. 36, Inst. of Physics, Bristol and London, p. 347.

Maher, D. M.; Joy, D. C.; Ellington, M. B.; Zaluzec, N. J.; and Mochel, P. E. (1981), in "Analytical Electron Microscopy - 1981" (R. H. Geiss, ed.), San Francisco Press, San Francisco, p. 33.

McGill, R. J., and Hubbard, F. H. (1981), "Quantitative Microanalysis with High Spatial Resolution" (G. W. Lorimer, M. H. Jacobs, and P. Doig, ed.), The Metals Soc., London, Book 277, p. 30.

Morris, P. L.; Ball, M. D.; and Statham, P. J. (1979), in "Electron Microscopy and Analysis" (T. Mulvey, ed.), Inst. Phys. Conf. Ser. No. 52, Inst. of Physics, Bristol and London, p. 413.

Mott, N. F., and Massey, H. S. W. (1949), "The Theory of Atomic Collisions" (2nd ed.), Oxford Univ. Press, London, p. 243.

Newbury, D. E. (1981), private communication.

Newbury, D. E., and Myklebust, R. L. (1979), in "Microbeam Analysis - 1979" (D. B. Wittry, ed.), San Francisco Press, San Francisco, p. 173.

Nockolds, C.; Nasir, M. J.; Cliff, G.; and Lorimer, G. W. (1979), in "Electron Microscopy and Analysis" (T. Mulvey, ed.), Inst. Phys. Conf. Ser. No. 52, Inst. of Physics, Bristol and London, p. 417.

Oppolzer, H., and Knauer, U. (1979), SEM/1979 (O. Johari, ed.), SEM Inc., Chicago, *1*, 111.

Philibert, J., and Tixier, R. (1975), "Physical Aspects of Electron Microscopy and Microbeam Analysis" (B. M. Siegel and D. R. Beaman, ed.), J. Wiley, New York, p. 333.

Porter, D. A., and Westengen, H. (1981), "Quantitative Microanalysis with High Spatial Resolution" (G. W. Lorimer, M. H. Jacobs, and P. Doig, ed.), The Metals Soc., London, Book 277, p. 94.

Powell, C. J. (1976), NBS Special Pub. 460 (K. F. J. Heinrich, D. E. Newbury, and H. Yakowitz, ed.), p. 97.

Rae, D. A.; Scott, V. D.; and Love, G. (1981), in "Quantitative Microanalysis with High Spatial Resolution" (G. W. Lorimer, M. H. Jacobs, and P. Doig, ed.), The Metals Soc., London, p. 57.

Rao, P., and Lifshin, E. (1977), Proc. 12th Annl. Conf. MAS, p. 118.

Rapperport, E. J. (1969), in "Electron Probe Microanalysis" (A. J. Tousimis and L. Marton, ed.), Advances in Electronics and Electron Physics, Supplement G, Acad. Press, New York, p. 117.

Reed, S. J. B. (1966), Proc. 5th Intl. Cong. X-ray Optics and Microanalysis (R. Castaing, R. Deschamps, and J. Philibert, ed.), Herman, Paris, p. 339.

——— (1982), Ultramicroscopy *7*, 405.

Romig, A. D. (1979), Ph.D. Thesis, Lehigh Univ., Bethlehem, PA.

Romig, A. D., and Goldstein, J. I. (1978), Met. Trans. *9A*, 1599.

Romig, A. D.; Newbury, D. E.; and Myklebust, R. L. (1982), in "Microbeam Analysis - 1982" (K. F. J. Heinrich, ed.), San Francisco Press, San Francisco, p. 88.

Schreiber, T. P., and Wims, A. M. (1981a), in "Microbeam Analysis - 1981" (R. H. Geiss, ed.), San Francisco Press, San Francisco, p. 317.

——— (1981b), Ultramicroscopy *6*, 323.

——— (1982), X-ray Spectrometry *11*, 41.

Slivinsky, V. W., and Ebert, P. J. (1972), Phys. Rev. *A5*, 1581.

Spence, J. C. H., and Tafto, J. (1983), J. Micros. *130*, 147.

Statham, P. J. (1982), in "Microbeam Analysis - 1982" (K. F. J. Heinrich, ed.), San Francisco Press, San Francisco, p. 1.

Statham, P. J., and Ball, M. D. (1980), in "Microbeam Analysis - 1980" (D. B. Wittry, ed.), San Francisco Press, San Francisco, p. 165.

Stenton, N.; Notis, M. R.; Goldstein, J. I.; and Williams, D. B. (1981), in "Quantitative Microanalysis with High Spatial Resolution" (G. W. Lorimer, M. H. Jacobs, and P. Doig, ed.), The Metals Soc., London, Book 277, p. 35.

Stephenson, T. A.; Loretto, M. H.; and Jones, I. P. (1981), in "Quantitative Microanalysis with High Spatial Resolution" (G. W. Lorimer, M. H. Jacobs, and P. Doig, ed.), The Metals Soc., London, Book 277, p. 53.

Thomas, L. E. (1980), Proc. 38th EMSA Mtg. (G. W. Bailey, ed.), Claitor's Pub. Div., Baton Rouge, p. 90.

Tixier, R., and Philibert, J. (1969), Proc. 5th Intl. Cong. X-ray Optics and Microanalysis (G. Möllenstedt and K. H. Gaukler, ed.), Springer-Verlag, Berlin, p. 180.

Tixier, R.; Thomas, B.; and Bourgeot, J. (1981), in "Quantitative Microanalysis with High Spatial Resolution" (G. W. Lorimer, M. H. Jacobs, and P. Doig, ed.), The Metals Soc., London, Book 277, p. 15.

Trebbia, P. (1984), J. Micros. Spectrosc. Electron *9*, 101.

Twigg, M. E., and Fraser, H. L. (1982), "Microbeam Analysis - 1982" (K. F. J. Heinrich, ed.), San Francisco Press, p. 37.

Twigg, M. E.; Loretto, M. H.; and Fraser, H. L. (1981), Phil. Mag. *43*, 1587.

Vander Sande, J. B.; Kelly, T. F.; and Imeson, D. (1980), Proc. 38th Annl. EMSA Mtg. (G. W. Bailey, ed.), Claitor's Pub. Div., Baton Rouge, p. 356.

Venables, J. A., and Janssen, A. P. (1980), Ultramicroscopy *5*, 297.

von Heimendahl, M., and Willig, V. (1980), Norelco Reporter *27*, 22.

Williams, D. B. (1984), "Practical Analytical Electron Microscopy in Materials Science," Philips Electron Optics Publishing Group, Mahwah, New Jersey.

Williams, D. B., and Goldstein, J. I. (1981), in "Analytical Electron Microscopy - 1981" (R. H. Geiss, ed.), San Francisco Press, San Francisco, p. 39.

Williams, D. B.; Newbury, D. E.; Goldstein, J. I.; and Fiori, C. E. (1984), J. Micros. *136*, 209.

Williams, E. J. (1933), Proc. Roy. Soc. *A139*, 163.

Wood, J. E.; Williams, D. B.; and Goldstein, J. I. (1981), in "Quantitative Microanalysis with High Spatial Resolution" (G. W. Lorimer, M. H. Jacobs, and P. Doig, ed.), The Metals Soc., London, Book 277, p. 24.

——— (1984), J. Micros. *133*, 255.

Zaluzec, N. J. (1979), in "Introduction to Analytical Electron Microscopy" (J. J. Hren, J. I. Goldstein, and D. C. Joy, ed.), Plenum Press, New York, p. 121.

——— (1981a), in "Analytical Electron Microscopy - 1981" (R. H. Geiss, ed.), San Francisco Press, San Francisco, p. 47.

——— (1981b), in "Microbeam Analysis - 1981" (R. H. Geiss, ed.), San Francisco Press, San Francisco, p. 325.

Zaluzec, N. J.; Maher, D. M.; and Mochel, P. E. (1981), in "Analytical Electron Microscopy - 1981" (R. H. Geiss, ed.), San Francisco Press, San Francisco, p. 25.

Ziebold, T. O. (1967), Anal. Chem. *39*, 858.

CHAPTER 6

EDS QUANTITATION AND APPLICATION TO BIOLOGY

T. A. Hall and B. L. Gupta

Department of Zoology
University of Cambridge, England

I. INTRODUCTION

The quantitative x-ray microanalysis of biological tissue sections differs strikingly from that of inorganic specimens in several ways:

(a) Measurements of local concentrations in biological tissue sections may be misleading because the elemental distributions may have been altered inadvertently during the preparation of the specimens—elements may have been largely lost, or displaced, during preparation.
(b) Even if a section as prepared is accepted as a proper object for measurement, the procedures available for measurements in thin organic specimens are subject to systematic errors arising from assumptions about section uniformity or from beam damage.

(c) A given procedure may produce a measurement of concentration that is technically correct but biologically irrelevant or ambiguous. For example, in a section stained with a heavy metal, the measurement of amount of element per kilogram of stained specimen is not in itself useful when much of the mass may consist of the metal stain. A less blatant example: For electrolyte elements not bound to organic matrix but almost entirely "free" in aqueous solution in living tissue, the relevant concentration for the physiologist is millimoles (mmol) of element per liter of water. If millimoles of element per total mass of tissue are simply measured, quite different values may be obtained for regions where there are actually identical values of mmol of element per liter of water but different aqueous fractions (i.e., differing values for mass of water/total mass).

(d) Biological tissues are quite variable. One must realize that there may be inhomogeneities within a tissue—gradients, for example, within the cytoplasm of a single cell; concentrations within an organism may vary with time or physiological condition; differences will exist among individuals of the same species.

Because of these complicating factors, the role of quantitative microanalysis in biology is restricted, and the analyst must be clear about the significance of intended measurements.

II. MEASUREMENTS ON THIN OR ULTRATHIN SECTIONS MOUNTED ON THIN FILMS

A. Elemental Ratios

The ratio of the concentrations of two elements A and B can be determined from the equation

$$\frac{C_A}{C_B} = \frac{I_A}{I_B}\left[\frac{I_B}{I_A} \times \frac{C_A}{C_B}\right]_r \tag{1}$$

where

C_A = the mass fraction of element A (which can be expressed either as mass of element A per unit total mass of specimen, or as millimoles (mmol) of element A per kilogram of total specimen mass)

I_A = the x-ray peak integral obtained for a selected characteristic emission from element A

$\left[\frac{I_B}{I_A} \times \frac{C_A}{C_B}\right]_r$ refers to a thin standard of known composition

r refers to a standard used for ratio measurements.*

(The quantities "I" refer to the genuine characteristic signal after correction of the peak integrals for background. The means of background correction, and spectral deconvolution in general, are outside the scope of this chapter.)

---

* A list of symbols with full definitions is given at the end of this chapter.

An increase in section thickness can affect the validity of Eq. (1) in two ways: A correction may be needed for the absorption of soft radiation within the specimen and, because of electron energy-loss, the relative excitation efficiencies for elements A and B may not be the same in specimen and standard. For the analysis of inorganic inclusions within sections of soft tissue, the situation is just like the case of inorganic thin films, while the effects of thickness are actually much less restrictive for the case of elements distributed within the soft tissue matrix. A correction for absorption is outlined below. With respect to excitation efficiency, it is sufficient to ensure that the probe electrons lose on average only a small part of their energy in passing through the section (or standard). This condition is easily satisfied. For example, 50-keV electrons lose on average <1 keV while passing through an organic film with density $1\ g \cdot cm^{-3}$ and thickness 1 μm.

As thin standards for use with Eq. (1), one may use preparations developed for the analysis of inorganic specimens, for example, finely ground pieces of well-characterized mineral (Cliff and Lorimer, 1972) or even slurry-mixes (Rowse *et al.*, 1974). But the effects of absorption and energy degradation are less if one uses an organic-matrix standard developed for absolute measurement of concentration (Section II.H, "Standards"), or a preparation of salt solution in the form of dried droplets (Morgan *et al.*, 1975) or dried streaks (Krefting *et al.*, 1981). It is also possible to use Eq. (1) without standards by deriving the bracketed quantity from theory (Russ, 1974). Finally, it is worth noting that the relative sensitivities for different elements are, in general, fairly similar for EDS systems even under varied operating conditions, except for soft radiations where detector window absorption is important. Therefore, crude estimates can be made on the basis of published sensitivity curves like those of Cliff and Lorimer (1975).

While ratio measurements in biological tissue sections are basically like those in inorganic films, and are indeed simpler in practice, generally one cannot deduce absolute concentrations from a set of ratios. In metallic films, one can often measure the relative amount of every constituent element, and it is then easy to deduce the absolute concentrations, given that they must add up to 100%. A similar calculation is possible in mineralogical specimens when the only missing element is oxygen and its stoichiometry is known. In soft tissues, the main constituents are water plus organic material consisting predominantly of the elements C, N, O, and H; if absolute concentrations are required, the ratio method cannot suffice.

B. Millimoles of Element per Unit Volume

Absolute measurements of elemental mass per unit volume are based on the presence of a "peripheral standard." Before sectioning, the tissue must be in contact with a medium of known composition. This medium serves as the standard; it must be sectioned with the tissue and must be present in the section alongside the tissue itself.

Since a characteristic x-ray signal from a thin specimen is proportional to the local amount of the emitting element per unit area, one may write

$$M_A = I_A(M_A/I_A)_p \tag{2}$$

where M_A is the local mass of element A per unit area in the analyzed region of the specimen and subscript p refers to the peripheral standard. It is assumed that the analyzed region of the tissue and the utilized region of the standard are equally thick.

If both sides of Eq. (2) are divided by the thickness, we get (since area $\times$ thickness = volume)

$$C'_A = I_A(C'_A/I_A)_p \tag{3}$$

where C'_A is the mass of element A per unit volume, which is commonly expressed by physiologists in units of mmol $\cdot$ L^{-1} (millimoles of element per liter). But it must be noted that the quantity C' is mmol per liter of specimen, *not* mmol per liter of water.

Equation (3) has been used extensively for the analysis of frozen-dried sections of quench-frozen epithelial tissues by Dörge and co-workers (Dörge *et al.*, 1975; Rick *et al.*, 1978; Rick *et al.*, 1982), and it has been used for measurements on freeze-substituted, epoxy-embedded botanical tissues by Pallaghy (1973).

Even if the standard contains only one element in known amount per unit volume, other elements as well can be assayed absolutely in the specimen by combination with the ratio method. A combination of Eq. (1) and (3) gives

$$C'_A = I_A \left[\frac{C'_B}{I_B}\right]_p \left[\frac{I_B}{I_A}\frac{C'_A}{C'_B}\right]_r \tag{4}$$

(the ratios C'_A/C'_B and C_A/C_B being, of course, the same). This equation gives the concentration of element A in the specimen, C'_A, in terms of the concentration of element B in the peripheral standard and the relative sensitivity for A and B as determined from a ratio standard or from theory. Pallaghy (1973) has used this approach with chlorine in the epoxy medium as the reference element.

With increasing section thickness, electron path length in the section increases disproportionately (because of electron scatter); excitation cross sections change because of degradation of the electron energy; the proportionality between I_A and M_A breaks down, and Eq. (3) becomes invalid. The upper limit on thickness set by these factors is not very restrictive for organic material. Dörge *et al.*, (1978) have shown with frozen-dried organic standards that even at a probe voltage of 15 kV, the proportionality holds up to a "mass thickness" of $\sim$40 μg $\cdot$ cm^{-2}, while at 27 kV the signals are proportional to thickness (except for an absorption effect for soft radiation) up to at least 60 μg $\cdot$ cm^{-2}. On theoretical grounds, the upper limit on thickness should increase approximately in proportion to probe voltage (except for absorption effects).

The method embodied in Eq. (3) is not invalidated in principle by the use of heavy-metal stains. Of course, the use of liquid staining media during specimen preparation is likely to produce intolerable elemental translocations, but the mere presence of a heavy-metal stain does not in itself upset the validity of the equation. Thus, quantitative analysis by means of Eq. (3) may still be possible after staining with osmium vapor, which may cause only negligible translocations while substantially improving on the poor image contrast that bedevils studies of unstained sections (A. V. Somlyo *et al.*, 1977).

The main limitations of the method are:

(a) In the case of cryosections of quench-frozen material, the peripheral standard and the analyzed region of tissue must be equally thick just after the moment of sectioning ("true" sectioning without chipping, fracturing, or edge-compression). This condition may be quite difficult to satisfy, depending on the tissue and on sectioning technique, and it becomes progressively more difficult as lower sectioning temperatures come into use in order to guard against elemental diffusion (Saubermann *et al.*, 1977).

(b) If the cryosections are freeze-dried before analysis, there must be no shrinkage during dehydration (or else the peripheral standard and the analyzed tissue must shrink to the same extent). Shrinkage changes mass per unit area, and nonuniform shrinkage would therefore invalidate Eq. (3).

(c) Uniformity of section thickness and lack of shrinkage are more readily achieved in epoxy-embedded sections, but the dangers of elemental shifts during embedding (and during freeze-substitution) are greater than in the case of cryosections of quench-frozen tissues.

(d) The method is generally used with frozen-dried or freeze-substituted material. With both of these preparative modes, there must be a considerable displacement of elements that had been in the aqueous phase in extracellular spaces lacking an organic matrix, so that the elemental concentrations in such spaces cannot be determined.

(e) In principle, the method may be applied to frozen-hydrated sections for the analysis of extracellular spaces. However, even in sections otherwise uniform in thickness, the local section thickness in free-fluid extracellular spaces may be different, presumably because a discontinuity in mechanical properties may promote chipping or fracturing during sectioning. This effect has been documented in individual sections (p. 36 and Table 1 in Gupta and Hall, 1982; Hall and Gupta, 1982a).

For analyses of *cytoplasm*, the thickness differences between peripheral standard and analyzed field in individual sections may well average out to give reliable concentration values from Eq. (3) when the data from many sections are pooled (as practiced by Rick *et al.*, 1982). But such averaging may be of no use for measurements in free-fluid spaces, where fragmentation or dropout can lead to a *systematic* difference in thickness among the sections that are suitable for analysis. This is illustrated by Table I, abstracted from a study by the authors of frozen-hydrated sections of cockroach salivary gland (Gupta and Hall, 1983). The tabulated concentrations, C', are based on Eq. (3); concentrations C are obtained from a different formulation that takes account of local variation in mass thickness (Eq. (7)). The continuum x-ray count shows that average mass thickness was nearly equal in peripheral medium and in the microvillar space (P-cell cavities), and the results from the two formulations are in good agreement there. But in the secretory ducts, because the average section mass thickness is much larger, as indicated by the continuum count, Eq. (3) leads to concentration values that are too high to be physiologically sensible. In the study of another tissue, locust midgut, we obtained more uniform average thicknesses and consistent results in all analyzed compartments (Dow *et al.*, 1981). But any general reliance on Eq. (3) for the analysis of frozen-hydrated sections will have to be based on a quality of sectioning higher than has been achieved at least so far in the authors' laboratory.

C. Millimoles of Element per Kilogram of Dried Tissue (Continuum Method)

The requirement for equal thickness of specimen and standard is removed when the analytical method includes the determination of local variations in "mass thickness" (i.e., mass per unit area). The proportionality between signal I_A and local mass of element A per unit area can be written in the form

$$\frac{I_A}{M\,C_A} = \left[\frac{I_A}{M\,C_A}\right]_o \tag{5}$$

where M is the local total specimen mass per unit area and subscript o denotes a thin standard. (Recall that C_A is the mass fraction of element A expressed either as mass of A/total mass or as mmol of A per kg of total mass.)

The x-ray continuum can provide the measure of local mass. According to the analysis of Kramers (1923), the x-ray continuum intensity generated in a thin film depends on M and on atomic composition in accord with the equation

$$I_w = k\,M \sum_i (f_i\,Z_i^2/A_i) \equiv k\,M\,G\,. \tag{6}$$

Here

k = a constant
Z_i = atomic number of element i
A_i = atomic weight of element i
f_i = mass fraction of element i expressed as elemental mass/total mass.

TABLE I

Effect of Unequally Thick Compartments on Concentration Measurements[a].

Fields		Contin[b]	Element				
			Na	P	S	Cl	K
Periph. Medium[c] (n = 17)	Avg. Counts[d]	10516	160	60	79	1572	156
	C[e]		83±7	5±2	6±2	109±2	10±1
P-Cell Cavities (n = 8)	Avg. Counts	10144	51	1324	546	1317	1997
	C′[e]		32±3	115±6	42±4	94±8	139±9
	C		27±6	118±5	42±2	94±6	137±8
Secretory Ducts (n = 12)	Avg. Counts	17565	208	184	154	2380	348
	C′		130±8	16±1	12±1	170±9	24±1
	C		78±3	10±1	7±4	100±1	14±1

(a) Measurements on frozen-hydrated sections of cockroach salivary glands (from Gupta and Hall, 1983).
(b) Continuum x-rays, counted in the range 4.5 to 18 keV (I_w in Eq. (8)).
(c) Ringer-Dextran solution. Prepared to be 16.8% Dextran by weight with concentrations (mmol · kg^{-1}) Na 100, P 2, S 4, Cl 112.5, K 8.3.
(d) Average counts per 100-s run (probe current 0.4 nA).
(e) Concentrations C′ obtained from Eq. (3) for Na and Cl, and from Eq. (4) for P, S, and K, with Cl as the reference element. Concentrations C obtained by continuum method as described below (Eq. (7)), with pre-filed standards data. Tabulated concentration values C and C′ obtained directly from the average counts. Standard errors of the mean obtained from the concentrations calculated for each run separately.

The sum is taken over all constituent elements, and I_w is an x-ray continuum count taken in a fixed quantum-energy band (most conveniently, a band free of characteristic peaks, although such peaks, if present, can be stripped out by deconvolution to give the continuum signal proper). The weighted average of Z^2/A is represented in the second line of the equation, and in the remainder of this text, by the symbol, G.

Kramers' analysis of continuum generation was very approximate, and the proportionality of continuum intensity per unit mass per unit area to Z^2/A cannot be taken as precise. However, the adequacy of the expression for the analysis of soft tissues has been confirmed in several studies, including Shuman *et al.* (1976), Hall and Werba (1971), and Krefting *et al.* (1981). Krefting *et al.* judge that a factor, $Z^{2.1}$, might be more accurate than Z^2, but the difference is negligible for the range of compositions commonly encountered in soft tissues and similar standards. To judge the errors that may arise from the Z^2/A approximation, readers may compare it with a comprehensive paper on continuum generation based on modern theory and recent experiments (Chapman *et al.*, 1983; Fiori *et al.*, 1982; Fiori *et al.*, Chapter 13, this volume).

Substitution of Eq. (6) into Eq. (5) and rearrangement of terms gives

$$C_A = \frac{I_A}{I_w} G \left[\frac{C_A}{(I_A/I_w)\, G} \right]_o . \qquad (7)$$

This equation expresses mass fraction C_A in terms of a ratio (I_A/I_w) that is independent of thickness in thin specimens. The upper limit to thickness is much less stringent than for the method described earlier in this section under part B, "Millimoles of Element per Unit Volume," since I_A and I_w are affected to almost the same extent by the factors that upset the proportionality between I_A and M. Hall (1979a) has shown that with a probe voltage of 30 kV or more, Eq. (7) may be used for organic sections with mass thicknesses up to at least 200 $\mu g/cm^2$ (given suitable correction, again, for absorption when soft radiations are observed).

The use of Eq. (7) has been discussed in detail in several articles (Hall, 1971, 1979a; Hall *et al.*, 1973; Shuman *et al.*, 1976, 1977). With respect to technical difficulties, we must note the following points here:

(a) I_w is the continuum signal generated in the specimen itself. To determine this quantity, one should correct the observed continuum count for extraneous contributions from the supporting film and from surrounding bulk material. Continuum from the film can be estimated by putting the beam directly on the bare film. Continuum from bulk surroundings (grid bars or specimen rod) can be greatly reduced if these surroundings are entirely of low atomic number, but this background can probably be accounted for more accurately if there is a heavier element (subscript e below) present in the bulk material and not in the specimen. One can then use this element as a monitor of bulk-generated continuum and estimate the bulk correction from data obtained with the beam put directly onto the bulk material. Thus, in the correction procedure described here, three runs are involved: the run on the specimen itself, a run with the beam directed onto the surrounding bulk material, and a run with the beam on the supporting film.

The run with the beam on the bulk material establishes a ratio, $(I_w/I_e)_b$; here subscript b designates the bulk-run, and I_w and I_e within parentheses are, respectively, the count in the continuum band and the peak-integral count for the monitor element, e. We assume that the ratio of *bulk-generated* continuum to the characteristic count, I_e, is the same whether the beam is directed onto the specimen or onto the bulk material. (The geometry for self-absorption effects may differ in the two cases, and it is therefore best to use a high-energy continuum band and a high-energy monitor line to minimize the effect of absorption on this ratio.) It follows that for the run on the specimen, the bulk-generated continuum background is given by $(I_w/I_e)_b\, I_e$, where co-factor I_e is the peak-integral count for element e.

When the beam is directed onto the film, the continuum is generated within the film and also within the bulk surroundings by electrons scattered from the film. Hence, the continuum generated within the film itself is given by $[I_w - (I_w/I_e)_b\, I_e]_f$, where subscript f denotes the run with the beam directed onto the film, and the second term represents the bulk contribution during this run.

The corrected signal, I_w, representing continuum generated within the specimen itself, is hence given by

$$I_w = I_{wt} - (I_w/I_e)_b\, I_e - [I_w - (I_w/I_e)_b\, I_e]_f \tag{8}$$

where I_{wt} is the uncorrected continuum count for the specimen run, and the film run is done with live-time and probe current matching the specimen run.

The significant bulk background may come from either a grid or a specimen rod. For cases where *both* contribute significantly, a similar though longer equation is readily formulated.

No distinction has been made here between the bulk backgrounds generated by electrons scattering off the specimen and by *stray* radiations that do not strike the specimen at all (the "hole-count" background). Roomans has provided a procedure and formalism incorporating this distinction (Roomans, 1980a; Roomans and Kuypers, 1980). However, the procedure described above and Eq. (8) already take account of bulk excitation by stray electrons just as well as bulk excitation by electrons scattered from the specimen; the main source of remaining error is likely to be from interaction of the bulk surround with x-rays generated in lens apertures, an effect that can be made negligible by the use of "top-hat" apertures. A test of the relatively simple Eq. (8), in the geometry of a scanning electron microscope, has been reported by Hall (1979a), while an alternative empirical correction procedure has been described by Grote and Fromme (1981). The entire problem of extraneous background has been discussed thoroughly by Nicholson *et al.* (1982).

(b) Equation (7) requires knowledge of the quantities Z^2/A (or G). This is not a problem in the case of the standard, where the composition is known, but an estimate must be made for the specimen. The specimen must be regarded as a composite of organic matrix, whose elemental composition cannot be fully known, plus heavier elements whose characteristic radiations are detected. The problem is simply to guess the value, G, for the organic matrix. Once this value is postulated, Eq. (7) can be solved iteratively with successive corrections to the overall value, G, to account for the effect of the heavier elements, or one may use the closed form given by Hall (1971, p. 238). The concentrations reached in this way are a consistent set, but they are still erroneous if the assumed value of G for the organic matrix is wrong. Fortunately, G does not vary much among organic compounds (Hall, 1979a). A typical value for protein (excluding sulfur, which is assayed by its x-ray emission) is $G = 3.2$. Uncertainty over the real G value of the organic matrix in an analyzed region of dehydrated soft tissue introduces a probable error of approximately a few percent.

(c) Beam damage seriously complicates the use of Eq. (7). Discussion of this problem is deferred to Section IV, "Effects of Beam Damage."

Apart from technical problems, we must note two basic limitations in the use of the continuum method for measurement of mmol of element per kg in dehydrated specimens:

(a) Measurement of mmol of element per kg of "dry weight" has an obvious biological significance for elements that are mainly bound to the organic matrix. For elements that were predominantly in aqueous solution before dehydration, the interpretation is less direct, and the significance is less definite (as noted previously in Section I).

(b) Whatever the method of analysis, one must expect dislocation of the elements in matrix-free extracellular spaces when dehydration is part of the preparative procedure. Measurements in such spaces in frozen-dried sections are therefore dubious.

The continuum method has been extensively used for the microanalysis of frozen-dried sections by A. P. Somlyo and co-workers (A. V. Somlyo *et al.*, 1977, 1981; A. P. Somlyo *et al.*, 1979; Stewart *et al.*, 1980; James-Kracke *et al.*, 1980); by Roomans and co-workers (Roomans and Sevéus, 1976; Roomans, 1978, 1979, 1980b); and by Cameron and co-workers (Cameron *et al.*, 1977; Smith and Cameron, 1981).

D. Millimoles of Element per Kilogram Wet Weight (Continuum Method, Frozen-Hydrated Sections)

The continuum method and Eq. (7) may be applied to frozen-hydrated sections for the direct determination of mmol of element per kg "wet weight." The method requires the use of a cold-stage to maintain the hydration even during analysis. The major advantages of this difficult technique are the ability to analyze small, matrix-free extracellular spaces and to measure local "dry-weight fraction."

The continuum method does not formally require the presence of a peripheral standard. But the method can be applied accurately to hydrated material only if it is known that the analyzed fields have remained fully hydrated, and full hydration of a section can generally be confirmed by incorporating and successfully analyzing a peripheral standard (Gupta and Hall, 1981a). In the event of inadvertent partial dehydration before analysis, the peripheral standard can give some measure of degree of dehydration so that an approximate correction can be made. An alternative means of confirmation of hydration, when no peripheral standard is available, has been described by Hall and Gupta (1982b).

A complication arises in the estimation of the value G for the matrix, since the tissue matrix in frozen-hydrated material is a composite of frozen water (with $G = 3.67$) and organic material ($G \sim 3.2$). This is not a serious problem in practice, since the organic ("dry-weight") fraction in most soft tissue is only 10% to 30%, and a fairly accurate match in matrix G is provided by a standard consisting of a frozen aqueous solution containing organic material at a level of $\sim$20%. If the local dry-weight fraction in an analyzed region is actually determined, as described in the next section, then a value for the composite matrix can actually be worked out from the observations.

Discussion of the major problem of beam damage is again deferred to Section IV.

The continuum method has been used extensively for the microanalysis of frozen-hydrated sections by Gupta and co-workers (Gupta *et al.*, 1977; Gupta *et al.*, 1978a; Gupta and Hall, 1979, 1981a; Lubbock *et al.*, 1981).

Saubermann and co-workers have used a different procedure for the analysis of cryosections (Bulger *et al.*, 1981). In this procedure, a section is brought onto the cold-stage of the microanalyzer in the hydrated state and continuum measurements made. The section is then dehydrated by warming the stage, and the stage is re-cooled; characteristic and continuum x-ray measurements are then made. The concentrations are obtained first in terms of mmol · kg^{-1} *dry* weight, using the data from the dehydrated state alone, by the continuum method just as described above. These concentrations are then converted to mmol · kg^{-1} *wet* weight by multiplying by the dry-weight fraction, obtained as the ratio of continuum intensities before and after dehydration (Eq. (9)).

Although this procedure has been described as the analysis of frozen-hydrated sections, it cannot achieve one of the major goals in the study of frozen-hydrated material, i.e., the analysis of matrix-free extracellular spaces, since the contents of these spaces are dislocated by dehydration before any characteristic x-rays are recorded. The procedure of Bulger *et al.* (1981) seems liable to several other weaknesses as well. Determination of the dry-weight fraction, F_d, from Eq. (9) is subject to error if shrinkage occurs on dehydration. Also, for their continuum measurements in the hydrated state, Saubermann and co-workers have generally circumvented the difficulties of compartment identification by making only one measurement over a large field, or several sampling measurements without compartmental correlation (Saubermann, personal communication); this can lead to erroneous values of F_d in small compartments where the thickness in the hydrated section actually differed from that in the sampled areas. Finally, the procedure has been used with no peripheral standard and no apparent means of confirming that any particular section was actually *fully* hydrated before deliberate dehydration.

E. Dry-Weight and Aqueous Fractions

A local dry-weight fraction can be measured in a frozen-hydrated section by observing the continuum intensity, dehydrating the section within the microscope column by removal from the cold-stage, and observing the continuum intensity from the same region again after dehydration (Gupta *et al.*, 1980). If there were no shrinkage during dehydration, and the continuum signal were proportional to mass per unit area, dry-weight fraction F_d would be given by

$$F_d = \frac{I_{wd}}{I_{wh}} \tag{9}$$

where I_{wh} is the continuum signal in the hydrated state and I_{wd} is the continuum signal after dehydration. In the authors' experience, Eq. (9) is unreliable because dehydration is often accompanied by substantial shrinkage that tends to increase mass per unit area and hence increase the continuum signal. Since shrinkage must affect elemental and continuum signals in the same way, this problem can be managed by normalizing against the characteristic signal from any element that is prominent in the specimen. The problem of the altered proportionality between continuum intensity and mass per unit area is precisely accounted for by inclusion of the G factors. Hence, Eq. (9) should be replaced by

$$F_d = \frac{I_{wd}/I_{Ad}}{I_{wh}/I_{Ah}} \frac{G_h}{G_d} \tag{10}$$

where I_{Ah} and I_{Ad} refer to the characteristic intensities for element A before and after dehydration (subscripts h and d signify "hydrated" and "dehydrated").

Another formula for dry-weight fraction follows directly from the definition of F_d:

$$F_d = C_{Ah}/C_{Ad} \,. \tag{11}$$

This formula can also be derived from the combination of Eq. (7) and (10) by applying Eq. (7) to both hydrated and dehydrated states, bearing in mind that the bracketed quantity is invariant. Thus, the local dry-weight fraction is simply the ratio of elemental concentrations measured before and after dehydration by means of Eq. (7).

Gupta (1979) has noted that it is better to make the measurements before and after dehydration on adjacent fields rather than on the identical field. The initial field is prone to severe beam damage on re-irradiation after dehydration, due perhaps to the expression of latent beam damage when the specimen temperature is raised for dehydration.

The importance of the normalization to a characteristic signal has been shown with data from biological measurements by Gupta (1979) and with data from standards by Saubermann *et al.* (1981). Biologically significant applications of the method for measuring local dry-weight fractions have appeared in several studies (Gupta *et al.*, 1978a; Gupta *et al.*, 1978b; Gupta and Hall, 1981b; Civan *et al.*, 1980).

In the case of cryosections that have been frozen-dried before acquisition of any x-ray data, a different formula may be used to estimate local dry-weight fractions, provided that a peripheral standard has been incorporated in the section:

$$F_d = I_{wd}(F_d/I_{wd})_p \tag{12}$$

(the symbol, p, again referring to the peripheral standard). This equation is a natural extension of the formalism outlined above for measurement of mmol per unit volume, and the same assumptions are inherent: uniformity of section thickness and absence of differential shrinkage during drying. Again, the equation can be refined to take account of any difference in G-value between standard and local analyzed field. The more accurate form is

$$F_d = \frac{I_{wd}}{G}\left[\frac{F_d G}{I_{wd}}\right]_p \,. \tag{13}$$

Whichever method is used to obtain the dry-weight fraction, F_d, the aqueous fraction, F_A, is of course simply

$$F_A = 1 - F_d \,. \tag{14}$$

F. Conversion to Millimoles of Element per Liter of Water

In the case of physiological studies involving electrolyte elements, one generally wants to know their concentrations in the aqueous phase, i.e., mmol of element in a given volume of water. In cases where it may be shown or assumed that an electrolyte element is virtually entirely in the aqueous phase (i.e., that it is negligibly bound), one can deduce local mmol per liter of water from the measurement of local mmol per kg of tissue in the frozen-hydrated state (C) and the measurement of the local aqueous fraction, F_A. Given negligible binding, the concentration of element B in the aqueous phase, C''_B, is simply

$$C''_B = \frac{C_B}{F_A}. \tag{15}$$

Equation (15) has been used to compare microprobe results with aqueous-phase measurements by ion-selective microelectrodes (Gupta *et al.*, 1978b; Gupta and Hall, 1981b).

G. Absorption Corrections

Because the mean atomic number and the x-ray absorption cross sections in soft tissues are low, absorption in thin sections is generally negligible. When a correction is necessary for a low-energy radiation, it may be calculated from a simple expression based on the assumption that x-ray generation is uniform throughout the depth of the section. The formula derived by Hall (1971, p. 240) is

$$\frac{I'_A}{I_A} = \frac{P\,M\,\operatorname{cosec}\alpha}{1 - \exp(-P\,M\,\operatorname{cosec}\alpha)} \tag{16}$$

where

I_A = observed peak integral for the characteristic radiation of element A
I'_A = integral that would be obtained if there were no absorption in the specimen
P = effective x-ray absorption coefficient in units $cm^2 \cdot g^{-1}$
M = local section mass per unit area in units $g \cdot cm^{-2}$
α = take-off angle.

The effective coefficient, P, is obtained by summing over the constituent elements, i:

$$P = \sum_i (f_i P_i). \tag{17}$$

To use Eq. (16), one must have a value for local mass thickness M. This can be measured by comparison with the continuum signal from a mass standard, a thin organic film of known thickness and composition, by means of the equation

$$M = M_o \frac{I_w}{I_{wo}} \frac{G_o}{G} \tag{18}$$

the subscript, o, referring here to the mass standard. For this standard the authors use polycarbonate film ("Makrofol," supplied by Siemens), thickness 2 μm, mass thickness 240 $\mu g \cdot cm^{-2}$, composition $C_{16}O_3H_{14}$, and G-value 3.08. To avoid beam damage, the continuum signal from the mass standard should be recorded with the beam spread enough to keep the integrated beam loading below 10^{-10} $C/\mu m^2$.

Equation (16) also requires assumptions about specimen composition in order to estimate P and G. If one calculates absorption for sodium x-rays in a section of ice with $M = 100$ $\mu g \cdot cm^{-2}$ (thickness 1 μm and density 1 $g \cdot cm^{-3}$), with a take-off angle of 35°, the result is a correction factor, I'_A/I_A, equal to 1.3. It is clear from this that absorption is generally quite negligible for ultrathin sections of organic material, that it is generally unimportant as well for more penetrating radiations in 1-μm sections, and that the requisite assumptions need not be very accurate when the correction has to be made. When peripheral standards are used, the absorptions in standard and tissue, even for sodium, may be similar enough to make an explicit correction unnecessary. Absorption corrections based on an equation similar to (16), but in the context of inorganic thin films, are discussed in greater detail by Goldstein in a previous chapter of this book.

H. Standards

To minimize the importance of corrections, the most suitable standards for the analysis of soft-tissue sections are standards of similar thickness and composition. In the case of epoxy-embedded tissue sections, there are good ways to prepare sections of epoxy standards "doped" with known concentrations of many cations and anions (Spurr, 1974, 1975; Roomans and Van Gaal, 1977). In the case of cryosections of quench-frozen tissue, the standards are generally cryosections of quench-frozen aqueous solutions containing known concentrations of salts and some 10% to 20% of organic macromolecular material. Typical macromolecular additives are albumin (Dörge *et al.*, 1975), gelatin (Roomans and Sevéus, 1977), high-molecular-weight Dextran (Gupta *et al.*, 1977), and polyvinylpyrrolidone (PVP) (Saubermann and Echlin, 1975). Roomans and Sevéus (1977) have compared several of these materials as to their suitability for quantitative standards.

With cryosections, the aqueous standards are best used in the peripheral form, i.e., tissue and standard are in contact and are quench-frozen and sectioned together. The standard then remains hydrated during the analysis of frozen-hydrated tissue sections and will dehydrate with the tissue when the section is to be analyzed frozen-dried. For physiological studies, the peripheral standard is actually the extracellular medium within which the functioning and composition of the tissue are to be determined. It is then essential to establish that the macromolecular component does not interfere with physiological performance. Albumin, Dextran, and PVP have all been used in peripheral standards, but tolerance for them varies with the type of tissue (Echlin *et al.*, 1977; Gupta and Hall, 1982). Unfavorable effects of PVP have been noted in certain insect tissues (Forer *et al.*, 1979; Gupta *et al.*, 1977) and certain plant tissues (Wilson and Robards, 1982).

Very small pieces of tissue plus peripheral medium should not be left long in an unsaturated atmosphere before quench-freezing because evaporation can rapidly change the elemental concentrations in the thin layer of medium. For example, concentrations increase by ~20% in 20 seconds in the case of a hemispherical droplet of an aqueous solution of 100-mM KCl and 20% Dextran, diameter 0.5 mm, at room temperature in an atmosphere with relative humidity 77% (Hall and Gupta, 1982b).

Correct preparation of aqueous solutions for peripheral standards involves a minor point that generally escapes attention. When the method of measurement is to provide results in terms of mmol of element per unit volume (see above), the standard solution should be prepared with measured amounts of element per unit volume of solution; when the method will measure mmol of element per kg wet weight (see above), the solution should be prepared with measured amounts of element per unit total mass of solution. (Either procedure is all right for measurements of mmol of element per kg dry weight since the essential datum in that case is mmol of element per kg of nonaqueous material.)

Another minor difficulty stems from the change of volume on freezing, which has a confusing effect on the measurement of amount of element per unit volume. This effect will not be considered here.

I. Comparison of Quantitation Schemes

The choice of analytical scheme (continuum method or use of characteristic x-rays alone; analysis of dehydrated or hydrated sections) depends largely on the requirements of the particular study (Hall and Gupta, 1982b). But these schemes are not contradictory or mutually exclusive. Civan *et al.* (1980) have compared analyses of toad urinary bladder done in two laboratories, one analyzing both hydrated and dehydrated sections by the continuum method and the other measuring frozen-dried sections on the basis of the characteristic signals alone. Dow *et al.* (1981; reviewed in Hall and Gupta, 1982b), studying locust midgut, have applied both x-ray formalisms to the same data. The degree of agreement has been generally satisfactory.

III. EFFECTS OF CONTAMINATION WITHIN THE MICROSCOPE COLUMN

Contamination deposited on the section within the microscope may include elements of analytical interest or may simply add to local specimen mass. The extent of the problem depends on the nature of the microscope vacuum and also on the thickness of the analyzed sections.

Contamination seems to be insignificant in microscopes operated at ultrahigh vacuum.

For microscopes run at ordinary vacuum, in the authors' experience sulfur and chlorine may occasionally deposit onto specimens mounted on a cold-stage, but this contamination can be reduced by effective trapping above the diffusion pump and regular outgassing of the foreline trap in the roughing-pump line. The effect can then be prevented by use of a protective cold-cap fitted closely above the specimen. However, analysts should be aware that in the absence of these precautions, chlorine contamination can occur insidiously: chlorine may be picked up from the environment and retained in particular tissue compartments even when it does not accumulate on the supporting film or in the standards or other compartments (Dow *et al.*, 1981; Hall and Gupta, 1982b).

"Mass" contamination—carbon or organic debris—affects the methods of quantitation that rely on the measurement of local mass, specifically the continuum method. With an ordinary vacuum system, the rate of contamination is usually not high enough to add significantly to the mass of a 1-μm section during analysis, but the effect may be large for an ultrathin specimen exposed to a finely focused beam. Deposition rates are notoriously nonreproducible, and the dependence of rate on beam current is not monotonic, but under poor conditions local mass in ultrathin specimens may be increased on the order of 50% during a 100-second analytical run. Hence, the application of the continuum method to the quantitative analysis of ultrathin specimens requires either ultrahigh vacuum or a very effective anticontamination shield.

IV. EFFECTS OF BEAM DAMAGE

Beam-induced loss of "mass" (material of low atomic number) does not affect the analytical scheme based on characteristic radiation alone, but it does pose a serious problem in the use of the continuum method. The beam breaks chemical bonds, leading to the escape of volatile fragments and volatile reaction products. In uncooled organic films, typically 20% to 30% of the mass is lost under a beam loading of $\sim 10^{-10}$ C · μm^{-2}, and the specimen then becomes fairly stable against further loss (Stenn and Bahr, 1970). A dose of 10^{-10} C · μm^{-2} is almost invariably exceeded manyfold during microanalysis.

One approach to the problem of mass loss is to assume that specimen and standards are affected to the same degree. I_w and I_{wo} would then be changed by the same factor, and the basic Eq. (7) would remain valid. In practice, this assumption may be all right most of the time, but it is risky and must be false at times, in view of the fact that the end-point for mass loss actually varies from ~10% to ~90% in a range of organic films (Bahr *et al.*, 1965).

A more promising approach to the control of mass loss is the use of low temperature. It is difficult to determine accurately the relationship between temperature and beam-induced loss of mass, probably because of the problem of establishing the temperature of the specimen itself, but it is known that stage-cooling by liquid nitrogen substantially reduces and/or slows the loss (Hall and Gupta, 1974; Shuman *et al.*, 1976; Egerton, 1982), and loss is slight or nil near the temperature of liquid helium (Ramamurti, 1977; Ramamurti *et al.*, 1975; Dubochet, 1975).

In the authors' laboratory, where 1-μm frozen-hydrated or frozen-dried sections are analyzed on a stage held at −170°C, mass loss during analysis occurs inconsistently. We attribute the inconsistency to variability in the temperature of the specimen, caused by uneven contact between the sections and the conducting support film. However, one can tell *after* an analytical run whether substantial loss has occurred. Mass loss makes the analyzed region appear relatively "bleached" against a darker surround in the scanning transmission image. Losses of only a few percent produce visible bleaching. Another sign of loss is a fall in the continuum count rate during the analysis. On the basis of these criteria, we judge that analyses can be done with only slight mass loss, albeit inconsistently, under our conditions of operation (50-keV beam, beam currents near 1 nA, analyzed areas 0.1 to 1 μm^2, real time of run 100 to 200 seconds, sections frozen-hydrated or frozen-dried and 1 μm thick, mounted on thin nylon film coated with 15 nm of evaporated aluminum).

There is also a danger that the beam may remove certain elements, generally by volatilization. The literature is too fragmentary and dispersed to be reviewed here, but losses of Na, S, Cl, and K have been observed. The process of elemental loss is usually slow enough to be followed as a progressive drop in characteristic-line intensity during analysis. Since mass loss and elemental loss are both manifestations of the mobility of small volatile molecules, conditions that permit loss of mass are likely to lead to elemental losses as well.

V. SPECIMEN PREPARATION

In the preparation of tissue sections for electron-microscopical morphology, liquid media are commonly used for fixation, dehydration, staining, precipitation of soluble components of interest, and/or embedding. In quantitative analytical microscopy, such media are generally avoided because of the risk of altering the distributions of the elements of interest, and cryopreparative techniques are preferable.

The cryopreparation of tissue sections begins with the quench-freezing of a small piece of tissue (for a list of quench-freezing methods, see Hall, 1979b). The pathway then follows the alternative routes of freeze-substitution or cryosectioning.

Freeze-substitution entails replacement of the ice by a cold solvent, replacement of the solvent by embedding medium, and finally sectioning. A great deal of experimentation has still to be done to establish just how successful this technique can be in preserving elemental distributions. The effect of infiltrating with the liquid embedding medium is particularly suspect. But several workers have developed and used freeze-substitution for analytical microscopy (Lauchli *et al.*, 1970; Forrest and Marshall, 1976; Van Zyl *et al.*, 1976; Harvey *et al.*, 1976; Burovina *et al.*, 1978). Lechene and Warner (1977) and Chandler and Battersby (1979) have assessed the potential of freeze-substitution unoptimistically. Hirokawa and Heuser (1981) have shown that localization of a particular element (calcium) can be sharpened by adding a precipitating agent (oxalic acid) to the cold solvent, but this technique cannot open the way to a simultaneous quantitative analysis of several elements.

The cryosectioning route entails sectioning of the frozen block and then freeze-drying or transfer of the hydrated section onto a cold-stage in the analytical microscope for analysis in the hydrated state. The combination of quench-freezing and cryosectioning is in principle the simplest and "cleanest" preparative procedure for analytical microscopy, but again a great deal remains to be learned about the requisite conditions, especially with regard to temperatures. We do not know how much diffusion may occur in a block of frozen tissue at the usual sectioning temperatures (which range from −30°C in some laboratories to −110°C in others), and displacements also occur above −70°C in some freeze-drying routines (Barnard and Sevéus, 1978).

Another area of relative ignorance is that of ice-crystal formation. Quench-freezing should be as rapid as possible to minimize ice-crystal formation, which may shift diffusible elements and limit the analytical spatial resolution, but the poor thermal conductivity of tissues restricts rapid freezing to the neighborhood of the surface when a piece of tissue is quench-frozen (Sjöström, 1975). Cryoprotectants like glycerol can inhibit crystal formation but are not used because they penetrate cells and lead to elemental shifts. However, the addition of certain nonpenetrating macromolecules (e.g., Dextran, PVP) to the extracellular medium can inhibit ice-crystal formation, even within cells, to a distance of 50 to 100 μm from the surface (Franks *et al.*, 1977). The explanation of this inhibition is not known.

There is an enormous literature on the tricky subject of cryopreparation. Detailed descriptions of the techniques used in some laboratories have been given by Dörge *et al.* (1978), A. V. Somlyo *et al.* (1977), Gupta and Hall (1981a, 1982), and Saubermann *et al.* (1981). Cryopreparation was the main subject of a conference held in 1977 (Echlin, 1979) and another in 1981 (see J. Microscopy *125*(2) and *126*(1) and (3)).

VI. SPECIMENS OTHER THAN SECTIONS MOUNTED ON THIN FILMS

Quantitative x-ray microanalysis is applied to many types of biological specimen other than tissue sections mounted on thin films. Such analyses are mostly outside the scope of "analytical microscopy," but the interested reader can refer to the following papers:

Tissue sections on bulk supports. The method of Colby (1968) has been adapted for biological use by Warner and Coleman (1974, 1975).

Tissue in bulk form. Frozen-hydrated: Marshall (1977, 1980); Fuchs and Fuchs (1980); Lechene *et al.* (1979); Echlin *et al.* (1982). Frozen-dried and frozen-hydrated: Zs.-Nagy *et al.* (1982).

Tissue homogenates. Morgan *et al.* (1975) describe the analysis of sections of tissue homogenates by means of a cobalt internal standard. A known concentration of cobalt is added to the homogenate, and the other elements in the sections are then assayed by means of the ratio method, with cobalt as the reference element.

Fluids obtained by micropuncture. Specimens on bulk supports, analysis by diffracting spectrometer (WDS): Ingram and Hogben (1967); Lechene (1974); Lechene and Warner (1979a); Morel (1975); Roinel (1977); Garland *et al.* (1978). Thin-film support and EDS: Quinton (1978, 1979). Thin-film support, WDS, and EDS compared: Van Eekelen *et al.* (1980). WDS, thin-film, and bulk supports compared: Quinton *et al.* (1979).

VII. SAMPLE CALCULATIONS

What follows is an example of one set of microprobe data worked out in different ways. We compare concentrations deduced by the continuum method (Eq. (7)) and from characteristic x-ray intensities alone (Eq. (3)). In the case of the continuum method, we compare calculations including or ignoring the effects of matrix composition and heavier elements on the mean atomic-number parameter, G. These effects are taken into account automatically in some current computer programs, but the calculations are done here "by hand" for the sake of exposition. Also to simplify exposition, the "data" are hypothetical and idealized; the numbers are related as one would expect if there were no statistical fluctuations or instrumental error. For detailed examinations of actual experimental data from the authors' laboratory, the reader may refer to Gupta (1979) and Gupta and Hall (1979, 1982).

Table II presents the hypothetical experimental data plus a final column giving the continuum counts corrected for extraneous background by means of Eq. (8). For calculations of mmol of element per liter, it is assumed that there was a peripheral standard ("periph" in the table), made by sectioning a frozen solution containing 150 g Dextran, 140 mmol Na, 150 mmol Cl, and 10 mmol K per liter of solution; for calculations of mmol per kg, it is assumed that the standard solution was made up to contain these amounts in an aliquot with a total mass of 1 kg. The hypothetical ratio standard is a frozen section of saline solution containing 15% Dextran and 100 mM KCl. The mass standard is a polycarbonate film, $C_{16}H_{14}O_3$, with mass per unit area of $2.4\ 10^{-4}$ g · cm^{-2}. Live-time per run and probe current are assumed constant (except for reduced current to avoid overload when the beam is directly on the holder, "bulk" in the table). The holders are Dural, containing Al plus a few percent Cu. The tissue section with its peripheral standard is on one holder, and the mass standard is on another. The section support film is nylon, cast in the laboratory and coated with evaporated aluminum. It is assumed, in the first calculations below, that the runs are done first with the specimen frozen-hydrated and that the section is then dehydrated in the microscope column preceding the final set of runs.

A. Calculations

1. Millimoles of Element per Kilogram Wet Weight

The calculation is based mainly on Eq. (7). With the continuum method, one does not necessarily use the peripheral medium as the standard; the necessary standardization may come from a file of averaged standards runs, and analysis of the known peripheral medium as a specimen-field can then provide a check on the hydration of the section. In this example, however, the peripheral medium does serve as the standard.

(a) Approximation assuming G(specimen) = G(standard), and neglecting absorption:

$$C_A \cong (I_A/I_w)\, C_{Ao}\, (I_w/I_A)_o$$

$$C_{Na} = (51/2700)\ 140\ (3000/560) = 14.2 \text{ mmol per kg}$$

$$C_{Cl} = (1800/2700)\ 150\ (3000/5400) = 55.6 \text{ mmol per kg}$$

$$C_K = (2600/2700)\ 10\ (3000/378) = 76.4 \text{ mmol per kg} \qquad (19)$$

(b) Calculation of C_K by ratio method:

The potassium count of 378 from the 10 mmol of K in the hydrated standard would be subject in practice to a large probable error. While the numbers in the present example have been contrived as if statistical fluctuations did not exist, in practice it would be better to calculate C_K through use of the ratio method, with Cl as the reference element. The analogue of Eq. (4) is valid:

$$C_K = C_{Cl}\,(I_K/I_{Cl})\,[(I_{Cl}/I_K)\,(C_K/C_{Cl})]_r\ . \qquad (20)$$

Substitution of the data and the calculated value of C_{Cl} gives

$$C_K = 55.6\ (2600/1800)\ (2000/2100)\ (100/100) = 76.4 \text{ mmol per kg.}$$

TABLE II

Initial Data Plus Corrected Continuum Count.

Field	Peak Integrals					
	Na	Cl	K	Cu	I_{wt}	I_w
Ratio Standard		2000	2100			
Mass Standard				300	6500	6000
Mass-Bulk				6000	10000	
Hydrated Section						
Periph	560	5400	378	75	3650	3000
Film				50	600	500
Cytoplasm	51	1800	2600	60	3320	2700
Bulk				5000	10000	
Dehydrated Section						
Periph	724	6000	420	30	1022	462
Cytoplasm	62	1800	2600	40	1327	747

(c) More accurate calculation using matrix G-values:

If we regard the cytoplasm as consisting of water (with G = 3.67) plus proteinaceous matter (matrix G taken ~3.28, based on assumed elemental mass fractions H 0.07, C 0.50, N 0.16, O 0.25, and [P + S] 0.02) with dry-weight fraction F_d, we may write

$$\begin{aligned} G(\text{cytoplasm}) &= (1 - F_d)\, G(\text{water}) + F_d\, G(\text{proteinaceous}) \\ &= (1 - F_d)\, 3.67 + F_d\, 3.28 \\ &= 3.67 - 0.39\, F_d\,. \end{aligned}$$

F_d can be estimated adequately here from Eq. (10) with the G-factors omitted. Using potassium as normalizing element, we estimate

$$F_d \cong \frac{747/2600}{2700/2600} = 0.28$$

implying G(cytoplasm) = 3.56. A similar calculation for the medium of known composition gives G(periph) = 3.34 dehydrated and 3.62 hydrated. Hence, the concentrations above should be corrected by a factor 3.56/3.62 = 0.98, so the preferred values become

$$C_{Na} = 0.98\,(14.2) = 13.9 \text{ mmol per kg}$$

$$C_{Cl} = 0.98\,(55.6) = 54.5 \text{ mmol per kg}$$

$$C_K = 0.98\,(76.4) = 74.9 \text{ mmol per kg}\,.$$

It is evident that in this case the matrix G-correction is slight in hydrated tissue when $F_d \simeq 0.3$, while a lower F_d would involve an even smaller correction since the match to the G-value of the medium would be even closer. The effect of the heavier elements on G is not evaluated here since it too is usually slight in hydrated material, but this effect is estimated below in the case of dehydrated tissue.

(d) Absorption correction for sodium:

The absorption correction is calculated from Eq. (16). We have to work out the G-values, total mass per unit area, M, and effective absorption coefficients, P, for the analyzed field and the peripheral standard.

As worked out above, in this case G(cytoplasm) = 3.56, and G(periph) = 3.62. For the mass standard, the composition $C_{16}H_{14}O_3$ gives a G-value 3.08. Then Eq. (18) gives

$$M(\text{cytoplasm}) = 2.4\ 10^{4}\ (2700/6000)\ (3.08/3.56) = 0.937\ 10^{-4}\ \text{g} \cdot \text{cm}^{-2}$$

$$M(\text{periph}) = 2.4\ 10^{-4}\ (3000/6000)\ (3.08/3.62) = 1.02\ 10^{-4}\ \text{g} \cdot \text{cm}^{-2}.$$

The absorption coefficients are worked out from Eq. (17), using published coefficients for the constituent elements. Table III gives the effective coefficients, P, for water, dehydrated tissue, and Dextran.

The effective coefficients for the tissue and for the peripheral standard are calculated from these component coefficients, using $F_d = 0.3$ for the tissue and $F_d = 0.15$ for the peripheral standard

$$P(\text{cytoplasm}) = 0.3\ (2228) + 0.7\ (3652) = 3225$$

$$P(\text{periph}) = 0.15(2688) + 0.85(3652) = 3507$$

The example is worked out here for a take-off angle of 35°. Table IV summarizes the other values inserted into Eq. (16) and gives the values worked out in the same way for the dehydrated section.

TABLE III

Absorption Cross Sections of Water, Tissue, and Dextran for Sodium Radiation (see Eq. (17)).

		Material					
		Water		Dry Tissue		Dextran	
Absorbing Element	P_i	f_i	f_iP_i	f_i	f_iP_i	f_i	f_iP_i
H		0.11	—	0.07	—	0.07	—
C	1534	0	0	0.50	767	0.44	675
N	2450	0	0	0.16	392	0	0
O	4109	0.89	3652	0.25	1027	0.49	2013
S	2103	0	0	0.02	42	0	0
Total ($cm^2 \cdot g^{-1}$)			3652		2228		2688

TABLE IV

Values of the Main Quantities in the Absorption Correction for Sodium.

Field	10^4M	P	Correction Factor (I_A/I'_A)
Hydrated Tissue	0.937	3225	0.77
Hydrated Periph	1.02	3507	0.75
Dehydrated Tissue	0.280	2228	0.94
Dehydrated Periph	0.153	2688	0.97

The application of the absorption corrections to the sodium counts from the hydrated tissue and standard changes the calculated sodium concentration from 13.9 to 13.9(0.75/0.77) = 13.5 mmol per kg.

With respect to this correction, the following points should be noted:

(a) With a suitable standard, it is seen that the correction almost cancels out. If the local value of F_d were not measured, it would be adequate for most purposes to estimate the correction by using a guessed value for F_d.
(b) In the hydrated section, tissue and standard both consist predominantly of water; therefore, similar absorption can be expected. Tissue and standard may be less similar after dehydration, but of course there is much less absorption after dehydration.
(c) The calculations presented here are for section thicknesses of ~1 μm. It is clear that for ultrathin sections (100 nm thick), the absorption correction is usually negligible.
(d) Again, the heavier elements have been left out of the example for the sake of simplicity, but they can be included readily in an iterative routine.

2. Millimoles per Kilogram Dry Weight

Only a few laboratories are analyzing frozen sections in the hydrated state. More often, the sections are dehydrated by freeze-drying before they are put into the analytical microscope. Measurement by the continuum method then gives concentrations in terms of mmol per kg of dry weight. As an example, we take Table II again, simply neglecting the runs on the section before dehydration. The data for the dehydrated section can be processed directly by Eq. (7). We now have a dehydrated peripheral standard containing 140 mmol Na, 150 mmol Cl, and 10 mmol K per 150 g Dextran; the concentrations work out to be Na 881, Cl 944, and K 62.9 mmol per total kg. G(cytoplasm) is now taken to be 3.28 and G(periph) to be 3.34. To show the computation more neatly, Eq. (7) is written in the form

$$C_A = C_{Ao}\,(I_A/I_w)\,(I_{wo}/I_{Ao})\,(G/G_o).$$

Substitution of the values above and of the observed counts gives

$$C_{Na} = 881\,(62/747)\,(462/724)\,(3.28/3.34) = 45.8 \text{ mmol per kg dry}$$

$$C_{Cl} = 944\,(1800/747)\,(462/6000)\,(3.28/3.34) = 172 \text{ mmol per kg dry}$$

$$C_K = 62.9\,(2600/747)\,(462/420)\,(3.28/3.34) = 236 \text{ mmol per kg dry.}$$

With respect to this calculation, a few fine points should be noted:

(a) Again, one may correct for sodium absorption by the method given in Calculation Section 1(d) above. The calculated correction factors are included in Table IV.
(b) We have already noted that in measurements by the continuum method, an uncertainty is associated with the limited knowledge of the local value of G in the specimen. The quantity G/G_o is quite near unity in frozen-hydrated material where water is the main constituent of both specimen and standard; the incomplete knowledge of the composition of the organic matrix leads to a greater uncertainty of the value of G/G_o in frozen-dried material.

(c) Although one cannot know the G-value for the organic matrix precisely, one *can* correct for the effect on G of the measured heavier elements, and this should be done at least for dehydrated specimens. For this correction, the concentrations are expressed as mass fractions f, and the values of $G = \Sigma(f_i Z_i^2/A_i)$ are recalculated. In the present case, the mass fractions are as shown in Table V.

The corrected G-values are then:

$$\begin{aligned} G(\text{periph}) &= 0.9438(3.34) + 0.0203\ (121/23) + 0.0335(289/35.5) \\ &\quad + 0.0024(361/39) = 3.55 \\ G(\text{cyto}) &= 0.9836(3.28) + 0.0011\ (121/23) + 0.0061(289/35.5) \\ &\quad + 0.0092(361/39) = 3.37. \end{aligned}$$

The corrected ratio G/G_o is thus 3.37/3.55 = 0.949 in place of 3.28/3.34 = 0.982 in Eq. (7). The resulting corrected concentrations are $C_{Na} = 44.3$, $C_{Cl} = 166$, and $C_K = 228$ mmol per kg dry weight.

3. Dry-Weight Fractions

The cytoplasmic dry-weight fraction has already been estimated roughly above for the purpose of making matrix and absorption corrections. Eq. (10) is used now for more accurate estimations.

(a) Peripheral standard:

Eq. (10) is used with chlorine as the normalizing element. G-values, as estimated above, are $G_h = 3.62$ and $G_d = 3.34$ (neither of these values has been corrected for the heavier elements). The data of Table II then give for the standard

$$F_d = \frac{462/6000}{3000/5400} \frac{3.62}{3.34} = 0.15\,.$$

TABLE V

Mass Fractions f in the Dehydrated Material
$f_A(g/g) = 10^{-6}\ A\ C_A\ (\text{mmol} \cdot \text{kg}^{-1})$.

	Periph		Cytoplasm	
	C_A	f_A	C_A	f_A
Na	881	0.0203	45.8	0.0011
Cl	944	0.0335	172	0.0061
K	62.9	0.0024	236	0.0092
Dextran		0.9438		
"Protein"				0.9836

It is important that this value agrees well with the known composition. A result much above the true value of 15% would suggest that the section might have been far from fully hydrated during the initial analysis; the values obtained by the continuum method for the hydrated tissue in the same section would then tend to be too high.

(b) Cytoplasm:

Eq. (10) may be used with potassium as the normalizing element. Using G-values 3.28 and 3.56 as estimated above, we get

$$F_d = \frac{747/2600}{2700/2600}\,\frac{3.56}{3.28} = 0.300\,.$$

A different estimate may be obtained from Eq. (11), using the potassium concentrations obtained above (74.9 mmol per kg wet, 236 mmol per kg dry, G-values not corrected for the heavier elements):

$$F_d = 74.9/236 = 0.317.$$

These two estimates for F_d are both preferable to the crude estimate in Calculation Section 1(c) above, $F_d = 0.28$, which was obtained with no correction for G-values. The discrepancy between 0.300 and 0.317 is due to inaccuracies in the calculations of the G-corrections; Eq. (10) and (11) must give the same result with a computer routine that provides sufficient iteration and corrects fully for the effects of both matrix and heavier elements on G. For further illustrative calculations below, we take F_d (cytoplasm) $= 0.300$.

4. Magnitudes of Typical Continuum-Method Corrections

In Sections 1 through 3 above, we have processed a hypothetical set of data using various approximations and corrections. To show clearly the quantitative effects of the corrections, the results of the calculations are recapitulated in Table VI.

5. Millimoles of Element per Liter of Water

If it is assumed that the elements Na, Cl, and K are entirely in the aqueous phase in the analyzed region of tissue, the values of mmol per kg wet weight may be converted to mmol per liter of water by dividing by the local aqueous fraction, $F_A = 1 - F_d$.

In the present example this gives

$$C''_{Na} = 13.5/0.7 = 19.3 \text{ mmol per liter of water}$$

$$C''_{Cl} = 54.5/0.7 = 77.9 \text{ mmol per liter of water}$$

$$C''_{K} = 74.9/0.7 = 107 \text{ mmol per liter of water.}$$

6. Millimoles of Element per Unit Volume of Specimen

Equation (3) can be used to measure mmol of element per unit volume in both frozen-hydrated and dehydrated sections.

TABLE VI

Typical Magnitudes of Corrections.

		Element ($mmol \cdot kg^{-1}$)			
Correction	Equation	Na	Cl	K	F_d
<u>Hydrated State</u>					
None	19	14.2	55.6	76.4	
Matrix G	7	13.9	54.5	74.9	
Na Absorption	16	13.5	—	—	
None	*				0.28
Matrix G	10				0.30
<u>Dehydrated State</u>					
Matrix G	7	45.8	172	236	
Na Absorption	16	44.4	—	—	
G: Matrix *and* Heavy Elements	7	42.9	166	228	

*Eq. (10) with G-factors omitted.

When the equation is applied to the data for the frozen-hydrated section in Table II, the results are

$$C'_{Na} = 51\ (140/560) = 12.7 \text{ mmol per liter}$$

$$C'_{Cl} = 1800\ (150/5400) = 50.0 \text{ mmol per liter}$$

$$C'_{K} = 2600\ (10/378) = 68.8 \text{ mmol per liter.}$$

When the equation is applied to the data obtained from the same section after dehydration, the results are

$$C'_{Na} = 62\ (140/724) = 12.0 \text{ mmol per liter}$$

$$C'_{Cl} = 1800\ (150/6000) = 45.0 \text{ mmol per liter}$$

$$C'_{K} = 2600\ (10/420) = 61.9 \text{ mmol per liter.}$$

These two sets of results should be compared with the results from the continuum method applied to the same hydrated specimen (Calculation Section 1(c)):

$$C_{Na} = 13.9 \text{ mmol per kg wet}$$

$$C_{Cl} = 54.5 \text{ mmol per kg wet}$$

$$C_{K} = 74.9 \text{ mmol per kg wet.}$$

The numerical differences have nothing to do with the difference in units (mmol per liter vs mmol per kg). The difference in results for the hydrated section arises from an assumption built into the hypothetical Table II, namely, that the section thickness is not quite uniform. This is manifest in the value $I_w = 2700$ compared to $I_{wo} = 3000$. If such a nonuniformity exists, Eq. (7) is still valid, but an error arises from the use of Eq. (3).

A further source of error has been introduced into the hypothetical data for the dehydrated sections: for K and Cl, the tissue counts in the dehydrated state are the same as in the hydrated state, but dehydration has led to changes in these counts in the peripheral standard, indicating nonuniform shrinkage during dehydration (the sodium count changes in the tissue as well, but only because of reduced absorption). If nonuniform shrinkage occurs on dehydration, an error results when Eq. (3) is applied to the analysis of frozen-dried sections. (The actual occurrence of nonuniform section thickness or shrinkage must depend on the techniques of sectioning, mounting, and dehydration, and on the nature of the tissue.)

B. Summary

These calculations show how various results may be obtained from different formulations, all applied to one body of data. (Here we neglect the possible difference in the standard prepared volumetrically or gravimetrically.) The results are summarized in Table VII.

TABLE VII

Summary of the Results of the Calculations.

Specimen	Equation Numbers	Concentrations Na	Cl	K	F_d	Units
Hydrated	3	12.7	50.0	68.8		mmol per liter of *tissue*
Dehydrated	3	12.0	45.0	61.9		mmol per liter of *tissue*
Hydrated	7	13.5	54.5	74.9		mmol per kg wet tissue
Dehydrated	7	44.3	166.	228.		mmol per kg dry tissue
Combined Data	10	—	—	—	0.3	dry mass/wet mass
Combined Data	7, 10	19.3	77.9	107.		mmol per liter of *water*

TABLE OF CHAPTER VARIABLES

A_i Atomic weight of element i
C_A Mass fraction of element A (units can be either mmol of element per kg of specimen or g of element per g of specimen, to suit convenience)
C'_A Amount of element A per unit volume of specimen
C''_A Amount of element A per liter of tissue water
f_A Mass fraction of element A, units g of element per g of specimen
F_d Local dry-weight fraction (nonaqueous mass/total mass in the hydrated tissue)
F_A Local aqueous fraction ($F_A = 1 - F_d$)

G The weighted average of Z^2/A, $\sum_i f_i Z_i^2/A_i$, for a specimen or for any multielement component such as water or organic matrix

I_A Spectral peak integral for characteristic radiation of element A (except with subscript w, which refers to continuum radiation)

I'_A Peak integral after correction for self-absorption in the specimen (see Eq. (16))

I_{Ah} Value obtained for I_A in a hydrated section

I_{Ad} Value obtained for I_A in the section after dehydration

I_w Quantum count generated within the specimen itself, taken within some fixed quantum-energy band of the x-ray continuum, excluding any contribution from characteristic radiations

I_{wh} Value obtained for I_w in a hydrated section

I_{wd} Value obtained for I_w in the section after dehydration

I_{wt} Total continuum count before correction for extraneous sources

k A constant

M Local total specimen mass per unit area (area normal to the electron beam). This quantity is the same as the quantity ρt defined by Goldstein in Chapter 5 of this book.

M_A Local mass of element A per unit area

M_o Mass per unit area in a mass standard (Eq. (18))

P_i X-ray absorption coefficient for (absorbing) element i, $cm^2 \cdot g^{-1}$. This quantity is the same as the quantity $(\mu/\rho)_i$ defined by Goldstein in Chapter 5 of this book.

P Effective x-ray absorption coefficient (Eq. (17))

Z_A Atomic number of element A

α Take-off angle

Subscripts

b Bulk material in the neighborhood of the specimen

e Monitor element in the bulk material

f Specimen supporting film

o Thin standard, general

p Thin standard, peripheral

r Thin standard used for elemental ratios

RECOMMENDED READING AND REFERENCE MATERIAL

Listed alphabetically are references either included in this chapter or recommended by the authors.

Bahr, G. F.; Johnson, F. B.; and Zeitler, E. (1965), Lab. Invest. *14*, 1115.

Barnard, T., and Sevéus, L. (1978), J. Micros. *112*, 281.

Bulger, R. E.; Beeuwkes III, R.; and Saubermann, A. J. (1981), J. Cell Biol. *88*, 274.

Burovina, I. V.; Gribakin, F. G.; Petrosyan, A. M.; Pivovarova, N. B.; Pogorelov, A. G.; and Polyanovsky, A. D. (1978), J. Comp. Physiol. *127*, 245.

Cameron, I. L.; Sparks, R. L.; Horn, K. L.; and Smith, N. R. (1977), J. Cell Biol. *73*, 193.

Chandler, J. A. (1977), "X-Ray Microanalysis in the Electron Microscope," Elsevier/North-Holland Biomedical Press, Amsterdam.

Chandler, J. A., and Battersby, S. (1979), "Microbeam Analysis in Biology" (C. Lechene and R. Warner, ed.), Acad. Press, New York, p. 457.

Chapman, J. N.; Gray, C. C.; Robertson, B. W.; and Nicholson, W. A. P. (1983), X-Ray Spectrometry *12*, 153.

Civan, M. M.; Hall, T. A.; and Gupta, B. L. (1980), J. Membrane Biol. *55*, 187.

Cliff, G., and Lorimer, G. W. (1972), Proc. 5th Euro. Cong. Elec. Mic. (no ed. cited), Inst. of Physics, Bristol, p. 140.

——— (1975), J. Micros. *103*, 203.

Colby, J. W. (1968), "Advances in X-Ray Analysis" (J. B. Newkirk, G. R. Mallett, and H. G. Pfeiffer, ed.), Plenum Press, New York, vol. 11, p. 287.

Dörge, A.; Rick, R.; Gehring, K.; Mason, J.; and Thurau, K. (1975), J. Microscopie et Biologie Cell. *22*, 205.

Dörge, A.; Rick, R.; Gehring, K.; and Thurau, K. (1978), Pflügers Arch. *373*, 85.

Dow, J. A. T.; Gupta, B. L.; and Hall, T. A. (1981), J. Insect Physiol. *27*, 629.

Dubochet, J. (1975), J. Ultrastruc. Res. *52*, 276.

Echlin, P. (ed.) (1979), "Low Temperature Biological Microscopy and Analysis," Blackwell, Oxford. (Articles from J. Micros. *111*, pt. 1, and *112*, pt. 1.)

Echlin, P., and Galle, P. (ed.) (1975), "Biological Microanalysis" (J. Microscopie et Biologie Cell. *22*, No. 2-3), Société Francaise de Microscopie, Paris.

Echlin, P., and Kaufmann, R. (ed.) (1978), "Microprobe Analysis in Biology and Medicine" (Micros. Acta, Suppl. 2), Hirzel Verlag, Stuttgart.

Echlin, P.; Lai, C. E.; and Hayes, T. L. (1982), J. Micros. *126*, 285.

Echlin, P.; Skaer, H. le B.; Gardiner, B. O. C.; Franks, F.; and Asquith, M. H. (1977), J. Micros. *110*, 239.

Egerton, R. F. (1982), J. Micros. *126*, 111.

Erasmus, D. A. (ed.) (1978), "Electron Probe Microanalysis in Biology," Chapman and Hall, London.

Fiori, C. E.; Swyt, C. R.; and Ellis, J. R. (1982), in "Microbeam Analysis 1982" (Heinrich, ed.), San Francisco Press, San Francisco, p. 57.

Forer, A.; Gupta, B. L.; and Hall, T. A. (1979), Experimental Cell. Res. *126*, 217.

Forrest, Q. G., and Marshall, A. T. (1976), Proc. 6th Euro. Cong. Elec. Mic. (Y. Ben-Shaul, ed.), Tal Pub. Co., Jerusalem, vol. II, p. 218.

Franks, F.; Asquith, M. H.; Hammond, C. C.; Skaer, H. leB.; and Echlin, P. (1977), J. Micros. *110*, 223.

Fuchs, W., and Fuchs, H. (1980), Scanning Electron Microscopy/1980/II, 371.

Garland, H. O.; Brown, J. A.; and Henderson, I. W. (1978), "Electron Probe Microanalysis in Biology" (D. A. Erasmus, ed.), Chapman and Hall, London, p. 212.

Grote, M., and Fromme, H. G. (1981), J. Electron Microscopy *30*, 359.

Gupta, B. L. (1979), "Microbeam Analysis in Biology" (C. Lechene and R. Warner, ed.), Acad. Press, New York, p. 375.

Gupta, B. L., and Hall, T. A. (1979), Proc. Fedn. Am. Soc. Exp. Biol. *38*, 144.

——— (1981a), Tissue and Cell *13*, 623.

——— (198lb), "Water Transport Across Epithelia" (H. H. Ussing, N. Bindslev, N. A. Lassen, and O. StenKnudsen, ed.), Munksgaard, Copenhagen, p. 17.

——— (1982), Techniques in Cellular Physiology, P128 (P. F. Baker, ed.), Elsevier/North-Holland, Amsterdam, p. 1.

——— (1983), Am. J. Physiol. *244*, R176.

Gupta, B. L.; Hall, T. A.; and Moreton, R. B. (1977), "Transport of Ions and Water in Animals" (B. L. Gupta, R. B. Moreton, J. L. Oschman, and B. J. Wall, ed.), Acad. Press, London, p. 83.

——— (1980), "X-Ray Optics and Microanalysis" (D. R. Beaman, R. E. Ogilvie, and D. B. Wittry, ed.), Pendell Pub. Co., Midland, Michigan, p. 446.

Gupta, B. L.; Hall, T. A.; and Naftalin, R. J. (1978a), Nature (London), *272*, 70.

Gupta, B. L.; Berridge, M. J.; Hall, T. A.; and Moreton, R. B. (1978b), J. Exp. Biol. *72*, 261.

Hall, T. A. (1971), "Physical Techniques in Biological Research," 2nd ed. (G. Oster, ed.), Acad. Press, New York, vol. IA, p. 157.

——— (1979a), "Microbeam Analysis in Biology" (C. Lechene and R. Warner, ed.), Acad. Press, New York, p. 185.

——— (1979b) J. Micros. *117*, 145.

Hall, T. A., and Gupta, B. L. (1974), J. Micros. *100*, 177

——— (1981), "Microprobe Analysis of Biological Systems" (T. E. Hutchinson and A. P. Somlyo, ed.), Acad. Press, New York, p. 3.

——— (1982a), Proc. 40th Annl. Mtg. EMSA, Claitor's Pub. Div., Baton Rouge, p. 394.

——— (1982b), J. Micros. *126*, 333.

Hall, T. A., and Werba, P. (1971), Proc. 25th Anniv. Mtg. EMAG (no ed. cited), Inst. of Physics, Bristol, p. 146.

Hall, T. A.; Anderson, H. C.; and Appleton, T. C. (1973), J. Micros. *99*, 177.

Hall, T. A.; Echlin, P.; and Kaufmann, R. (ed.) (1974), "Microprobe Analysis as Applied to Cells and Tissues," Acad. Press, London.

Harvey, D. M. R.; Hall, J. L.; and Flowers, T. J. (1976), J. Micros. *107*, 189.

Hayat, M. A. (ed.) (1980), "X-Ray Microanalysis in Biology," Univ. Park Press, Baltimore.

Hirokawa, N., and Heuser, J. E. (1981), J. Cell Biol. *88*, 160.

Hutchinson, T. E., and Somlyo, A. P. (ed.) (1981), "Microprobe Analysis of Biological Systems," Acad. Press, New York.

Ingram, M. J., and Hogben, C. A. M. (1967), Analyt. Biochem. *18*, 54.

James-Kracke, M. R.; Sloane, B. F.; Shuman, H.; Karp, R.; and Somlyo, A. P. (1980), J. Cell Physiol. *103*, 313.

Kramers, H. A. (1923), Phil. Mag. *46*, 836.

Krefting, E. -R.; Lissner, G.; and Höhling, H. J. (1981), Scanning Electron Microscopy/1981/ II, 369.

Läuchli, A.; Spurr, A. R.; and Wittkopp, R. (1970), Planta (Berlin), *95*, 341.

Lechene, C. (1974), "Microprobe Analysis as Applied to Cells and Tissues (T. Hall, P. Echlin, and R. Kaufmann, ed.), Acad. Press, London, p. 351.

Lechene, C. P., and Warner, R. R. (1977), Annl. Rev. of Biophys. and Bioengr. *6*, 57.

——— (1979a), "Microbeam Analysis in Biology" (C. Lechene and R. Warner, ed.), Acad. Press, New York, p. 279.

——— (1979b) (ed.), "Microbeam Analysis in Biology," Acad. Press, New York.

Lechene, C. P.; Bonventre, J. V.; and Warner, R. R. (1979), "Microbeam Analysis in Biology" (C. Lechene and R. Warner, ed.), Acad. Press, New York, p. 409.

Lubbock, R.; Gupta, B. L.; and Hall, T. A. (1981), Proc. Natl. Acad. Sci. USA *78*, 3624.

Marshall, A. T. (1977), Microscopica Acta *79*, 254.

——— (1980), "X-Ray Microanalysis in Biology" (M. A. Hayat, ed.), Univ. Park Press, Baltimore, p. 167.

Morel, F. (1975), J. Microscopie et Biologie Cell. *22*, 479.

Morgan, A. J.; Davies, T. W.; and Erasmus, D. A. (1975), J. Micros. *104*, 271.

Nicholson, W. A. P.; Gray, C. C.; Chapman, J. N.; and Robertson, B. W. (1982), J. Micros. *125*, 25.

Pallaghy, C. K. (1973), Austral. J. Biol. Sci. *26*, 1015.

Quinton, P. M. (1978), Micron *9*, 57.

——— (1979), "Microbeam Analysis in Biology" (C. Lechene and R. Warner, ed.), Acad. Press, New York, p. 327.

Quinton, P. M.; Warner, R. R.; and Lechene, C. (1979), "Microbeam Analysis 1979" (D. Newbury, ed.), San Francisco Press, San Francisco, p. 73.

Ramamurti, K. (1977), Proc. 35th Annl. Mtg. EMSA (G. W. Bailey, ed.), Claitor's Pub. Div., Baton Rouge, p. 560.

Ramamurti, K.; Crewe, A. V.; and Isaacson, M. S. (1975), Ultramicroscopy *1*, 156.

Rick, R.; Dörge, A.; von Arnim, E.; and Thurau, K. (1978), J. Membrane Biol. *39*, 313.

Rick, R.; Dörge, A.; and Thurau, K. (1982), J. Micros. *125*, 239.

Roinel, N. (1977), Proc. 35th Annl. Mtg. EMSA (G. W. Bailey, ed.), Claitor's Pub. Div., Baton Rouge, p. 362.

Roomans, G. M. (1978), Interaction of Cation and Anion Transport in Yeast (doctoral thesis), Stichting Studentenpers, Nijmegen, The Netherlands.

——— (1979), Histochemistry *65*, 49.

——— (1980a), Scanning Electron Microscopy/1980/II, 309.

——— (1980b), Physiol. Plant. *48*, 47.

Roomans, G. M., and Kuypers, G. A. J. (1980), Ultramicroscopy *5*, 81.

Roomans, G. M., and Sevéus, L. (1976), J. Cell Sci. *21*, 119.

——— (1977), J. Submicr. Cytol. *9*, 31.

Roomans, G. M., and Van Gaal, H. L. M. (1977), J. Micros. *109*, 235.

Rowse, J. B.; Jepson, W. B.; Bailey, A. T.; Climpson, N. A.; and Soper, P. M. (1974), J. Physics E: Sci. Instru. *7*, 512.

Russ, J. C. (1974), "Microprobe Analysis as Applied to Cells and Tissues" (T. Hall, P. Echlin, and R. Kaufmann, ed.), Acad. Press, London, p. 269.

Saubermann, A. J., and Echlin, P. (1975), J. Micros. *105*, 155.

Saubermann, A. J.; Beeuwkes III, R.; and Peters, P. D. (1981), J. Cell Biol. *88*, 268.

Saubermann, A. J.; Echlin, P.; Peters, P. D.; and Beeuwkes III, R. (1981), J. Cell Biol. *88*, 257.

Saubermann, A. J.; Riley, W. D.; and Beeuwkes III, R. (1977), J. Micros. *111*, 39.

Shuman, H.; Somlyo, A. V.; and Somlyo, A. P. (1976), Ultramicroscopy *1*, 317.

——— (1977), Scanning Electron Microscopy/1977/I, 663.

Sjöström, M. (1975), J. Micros. *105*, 67.

Smith, N. K. R., and Cameron, I. L. (1981), Scanning Electron Microscopy/1981/II, 395.

Somlyo, A. P.; Somlyo, A. V.; and Shuman, H. (1979), J. Cell Biol. *81*, 316.

Somlyo, A. V.; Shuman, H.; and Somlyo, A. P. (1977), J. Cell Biol. *74*, 828.

——— (1981), "Microprobe Analysis of Biological Systems" (T. E. Hutchinson and A. P. Somlyo, ed.), Acad. Press, New York, p. 103.

Spurr, A. R. (1974), "Microprobe Analysis as Applied to Cells and Tissues" (T. Hall, P. Echlin and R. Kaufmann, ed.), Acad. Press, London, p. 213.

——— (1975), J. Microscopie et Biologie Cell. *22*, 287.

Stenn, K., and Bahr, G. F. (1970), J. Ultrastruc. Res. *31*, 526.

Stewart, M.; Somlyo, A. P.; Somlyo, A. V.; Shuman, H.; Lindsay, J. A.; and Murrell, W. G. (1980), J. Bacteriology *143*, 481.

Van Eekelen, C. A. G.; Boekestein, A.; Stols, A. L. H.; and Stadhouders, A. M. (1980), Micron *11*, 137.

Van Zyl, J.; Forrest, Q. G.; Hocking, C.; and Pallaghy, C. K. (1976), Micron *7*, 213.

Warner, R. R., and Coleman, J. R. (1974), "Microprobe Analysis as Applied to Cells and Tissues" (T. Hall, P. Echlin, and R. Kaufmann, ed.), Acad. Press, London, p. 249.

——— (1975), Micron *6*, 79.

Wilson, A. J., and Robards, A. W. (1982), J. Micros. *125*, 287.

Zs.-Nagy, I.; Lustyik, G.; and Bertoni-Freddari, C. (1982), Tissue and Cell *14*, 47.

Listed below are recommended main literature references by category.

Quantitative x-ray microanalytical microscopy

Hall (1971). General (pre-dates the energy-dispersive era).

Shuman, H.; Somlyo, A. V.; and Somlyo, A. P. (1976, 1977). Continuum method.

Books on x-ray analytical microscopy in biology

Chandler (1977).

Erasmus (1978).

Hayat (1980).

Proceedings of conferences on microprobe analysis in biology

Hall, Echlin, and Kaufmann (1974).

Echlin and Galle (1975).

Echlin and Kaufmann (1978).

Lechene and Warner (1979b).

Hutchinson and Somlyo (1981).

CHAPTER 7

THE BASIC PRINCIPLES OF EELS

D. C. Joy

Bell Laboratories
Murray Hill, New Jersey

I. WHAT IS EELS?

Electron energy-loss spectroscopy (EELS) studies the energy distribution of electrons that have been transmitted through a thin sample. By combining electron spectroscopy and transmission electron microscopy, the analytical power of EELS is coupled with the ability to select, image, and obtain diffraction patterns from small areas. Although the use of EELS as a microanalytical technique was first discussed and demonstrated forty years ago by Hillier and Baker (1944), it is only recently that advances in microscope instrumentation and vacuum technology have made it a practical proposition for routine laboratory use. Like the technique of energy-dispersive x-ray spectroscopy (EDXS) discussed in the previous chapters, electron energy-loss spectroscopy provides a way to identify the elements in the sample and to quantify the amount of each. Since EELS can also detect low-atomic-number elements with high sensitivity, it can offer important advantages over EDS in some applications. In addition, EELS can provide detailed information about the electronic state and chemical bonding of the sample. This chapter describes the principles and practice of obtaining and interpreting EEL spectra, and the following chapters discuss the application of these ideas to problems in materials science and biology.

II. ELECTRON-SOLID INTERACTIONS

The physical and mathematical details of the interactions that occur between electrons and solids have been discussed in the earlier chapters by Newbury and Cowley. The interactions that carry most information about the chemistry of the sample are those in which energy is transferred from the incident electron to the solid through inelastic scattering. Energy-dispersive x-ray spectroscopy observes these events indirectly, but by using electron energy-loss spectroscopy, they can be studied directly by measuring the change in energy, or momentum, of electrons that occurs as a result of the scattering. Not only is this a much more efficient procedure than EDXS, because the primary event is being observed, but it is also more informative because a wider range of interactions can be examined. Figure 1 illustrates the schematic experimental arrangement.

The incident beam of electrons (of energy E_o and intensity I_o) passes through the specimen and into the analyzer, which disperses them so that all those entering with the same energy loss, E, relative to their initial energy, E_o, are focused to the same point. The intensity, I(E), of the transmitted signal as a function of energy-loss E can thus be found by placing a detector at various points in this dispersion plane. If the spectrometer is allowed to collect only those electrons scattered within a narrow angular cone, an energy-loss spectrum could be measured for each scattering angle, θ. This procedure makes it possible to determine the momentum distribution of the transmitted electrons since both their change in energy and direction will have been found. This is an important technique in solid state physics because the resultant information can be interpreted to yield the complex dielectric function of the material.

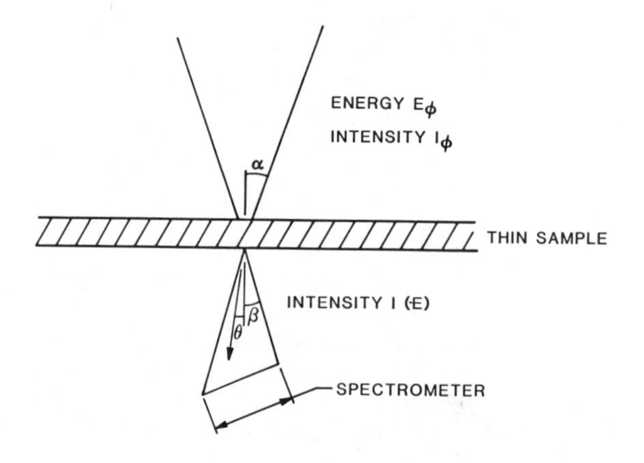

Figure 1 Schematic drawing of the experimental arrangement for energy-loss spectroscopy.

For most purposes, however, it is more efficient to allow the energy analyzer to collect a wide angular cone, so that all transmitted electrons scattered up to some maximum angle, β, are accepted and analyzed. This makes it impossible to determine the momentum distribution, but the resultant energy-loss spectrum can be simply interpreted in terms of the chemical and physical properties of the sample.

The fractional signal, $I(E)/I_0$, collected at energy-loss E and for angular cone β, is directly proportional to the "energy differential cross section," which expresses the probability that an incident electron will be scattered within the appropriate angular range while losing energy E. The constant of proportionality is N, the number of atoms per unit area participating in the scattering event observed, in the volume irradiated by the electron beam. If this cross section can be related to some physical description of the process of inelastic electron scattering, we can obtain microanalytical information about the sample. If the cross section itself can be calculated, we can derive N from the measured EEL spectrum and so obtain quantitative data about the specimen. To see how various inelastic interactions can be related to the properties of the specimen, it is useful to examine the features of a typical energy-loss spectrum, such as that shown in Figure 2, and then identify the physical processes that have given rise to them.

III. THE ENERGY-LOSS SPECTRUM

A. Region 1: The Zero-Loss Peak

The most visible feature in the energy-loss spectrum is that at the incident-beam energy, the so-called "zero-loss peak." Although this feature is of limited use in microanalysis, some discussion of it is required because of its prominence. It is clear that some qualification of the name "zero-loss" is required because this peak has a finite energy width. There are two reasons for this. First, the electron spectrometer has a definite energy resolution, which means that electrons differing in energy by less than this resolution figure (typically 1 to 5 eV) will not be distinguishable. Second, the incident electron beam itself has an inherent energy spread of 0.5 to 1.5 eV, arising both from the natural energy width of the electron source and from instabilities in the

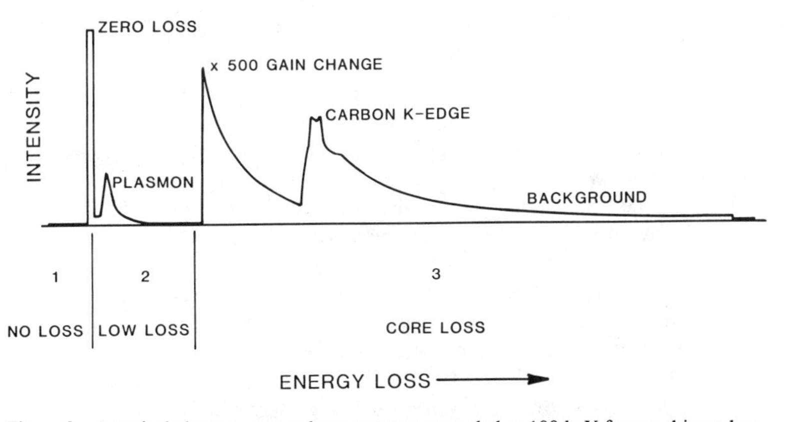

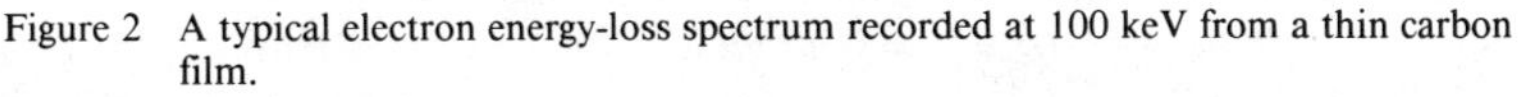

Figure 2 A typical electron energy-loss spectrum recorded at 100 keV from a thin carbon film.

accelerating potential. Together, these two effects give an energy region several volts wide, within which we cannot distinguish electrons that have lost no energy from those that may have lost a small amount.

This zero-loss peak contains contributions from three groups of electrons. First, there are those that are "unscattered," having passed through the sample without being involved in any interaction. Since they have not interacted with the specimen, they can carry no information about it and hence are of no use in microanalysis. The angular distribution of the unscattered beam is, of course, the same as that of the incident beam (Figure 3(a)), i.e., a cone perhaps 10^{-3} rad wide. Second, there are the "elastically scattered electrons" that have been deflected by the nuclear charge of an atom. The cross section for elastic scattering is given by

$$\frac{d\sigma}{d\theta} = \frac{4Z}{[a_o (q^2 + 1/p^2)]^2} \tag{1}$$

where

$q = 4/\lambda \sin(\theta/2)$
θ = scattering angle
λ = electron wavelength
$p = a_o/Z^{1/3}$,

where

a_o = Bohr atomic radius
Z = atomic number of the scattering nucleus.

Note that this cross section contains no energy-dependent term because no energy loss is involved.

The total cross section for elastic scattering into the acceptance angle of the spectrometer, found by integrating Eq. (1) from 0 to β, is $\sim 10^{-18}$ cm^2/atom for a light element such as Si at 100 keV. Another way of expressing this is to define a "mean free path," L, where

$$L = 1/(n\,\sigma) \tag{2}$$

and n is the density of atoms/cm^3. L represents the average distance an incident electron travels between elastic interactions. For example, in silicon, L is ~1000 Å, so in an average TEM specimen of ~1000-Å thickness, most electrons should experience one, or more, such event(s). The angular distribution of these elastically scattered electrons is broad compared to that of the incident beam, typically 30×10^{-3} rad at half height (Figure 3(b)). This fact is significant because it affects our ability to efficiently collect the inelastically scattered electrons in a thick sample since there is a finite probability that the electron will undergo both an elastic and an inelastic event and, because the elastic scattering angle is large compared to typical spectrometer acceptance angles, not be accepted by the analyzer.

The elastically scattered electrons are only of limited use in microanalysis, but it can be shown (e.g., Isaacson, 1978) that if we collect them over a sufficiently wide angular range (by using a large annular detector, for example), the elastic cross section becomes approximately $\sigma_{EL} = k_1 \cdot Z^{3/2}$. This strong dependence on the atomic number makes possible a semiquantitative analysis of samples, after calibration on materials of known composition, by simple measurement of the elastic signal. Although changes in the thickness of the specimen will change the calibration constant, this can be overcome by using the energy analyzer to simultaneously measure the total signal from the inelastically scattered electrons, which can be shown to vary as $Z^{1/2}$. If the ratio of these two signals is taken, the resultant signal is directly proportional to the mean atomic number, Z, of the sampled volume of the specimen but independent of the specimen thickness. After calibration on a suitable standard, this technique can then be used for microanalysis (Egerton *et al.*, 1975). It has also been used for the imaging and identification of single atoms (Isaacson *et al.*, 1979).

The third group of electrons in the zero-loss peak are those that have generated a "phonon excitation." Phonons are vibrations of the atoms in the sample and, although present in all materials, are most significant in crystals. The energy loss associated with phonon scattering is small, typically 0.1 eV, but phonon scattered electrons have an angular distribution that is a broad peak (5 mrad or so wide) centered around each Bragg diffracted beam (Figure 3(c)). They carry no useful information but are of significance because they can lead to a reduction in the number of inelastic electrons entering the analyzer.

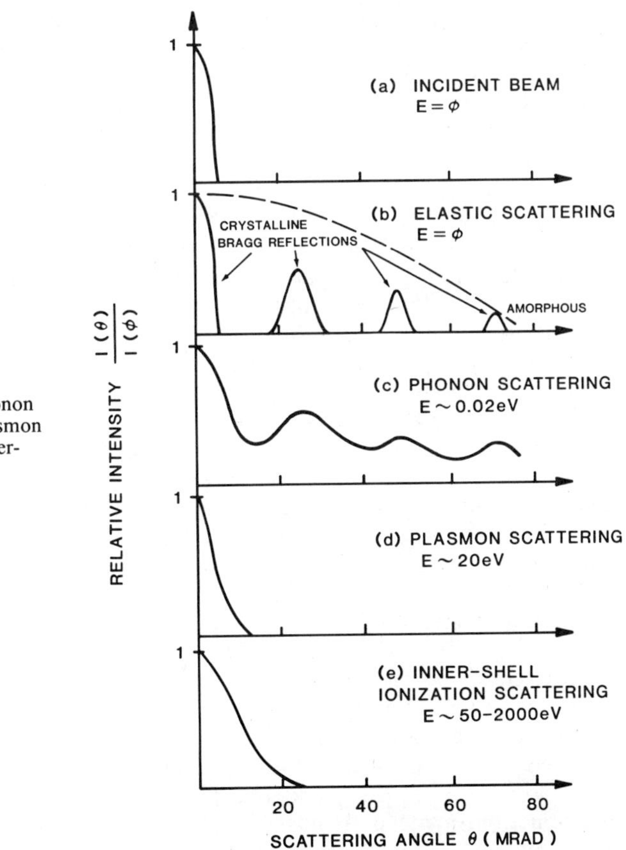

Figure 3 Angular distribution of (a) incident and unscattered electrons, (b) elastically scattered electrons, (c) phonon scattered electrons, (d) plasmon scattered electrons, (e) inner-shell scattered electrons.

Although the microanalytical information carried by the zero-loss peak is small, its importance should not be underestimated. It provides the energy reference for the spectrum, gives a measure of the resolution and accuracy of spectrometer adjustment, and provides the normalization factor required for quantitative microanalytical calculations.

B. Region 2: The Low-Loss Region

The portion of the spectrum extending from the edge of the zero-loss peak out to ~50 eV is usually called the "low-loss" region. Typically, as seen in Figure 2, the signal intensity in this region is 5% to 10% of that in the zero-loss peak. Since ~80% of the electrons passing through a typical sample are either unscattered or elastically scattered, Regions 1 and 2 of the spectrum account between them for all but a few percent of the intensity in the spectrum.

The energy losses in this region are due to direct electrostatic interactions with atomic electrons. In some cases, the interaction involves the transition of a single electron to an unfilled energy level within an atom. The energy-loss structures seen in the low-loss region in these cases are mainly due to the excitation, or ionization, of electrons from various bound states. The lowest energy losses (below 15 eV) come from the excitations of electrons in molecular orbitals. Each molecular type has a characteristic spectrum in this range which, although too complex to be interpreted directly, can be used as a "fingerprint" (Hainfeld and Isaacson, 1978). For example, the graphitic form of carbon shows a peak at 6 eV that is absent in amorphous carbon. Above 15 eV, the spectrum shows structures that are mostly due to valence shell excitations, and it is noticeable in biological samples that there is often a peak, centered around 20 eV, that resembles the plasmon peak found in a metal. One suggested explanation (Isaacson, 1972) is that this peak comes from the excitation of electrons in the π-π^* orbitals. However, more research is required to fully interpret all the data available in this portion of the spectrum.

There are also interactions, known as "plasmon excitations," that occur in metals, such as aluminum, and in alloys that have a large number of free electrons. These electrons behave like a gas, maintaining an equilibrium electron density because of the Coulomb repulsion between them. A fast electron incident on the sample disturbs the equilibrium, and the gas is set into oscillation. Because, in principle, all the valence electrons are involved, this is known as a "collective excitation." The scattering event is not well localized but appears to occur over a volume several tens of angstroms in diameter. The frequency of the plasmon oscillation, ω_P, is proportional to $(n_E)^{1/2}$, where n_E is the number of free electrons/cm^3 in the plasma. To set this gas into oscillation requires an energy, $E_{PL} = h\ w_P$, where h is Planck's constant, so an incident electron that excites a plasmon suffers an energy loss of E_{PL}. For common metals, ω_P is ~10^{16} rad/s, so E_{PL} is ~20 eV.

The angular distribution of the electrons that have undergone either single electron or plasmon excitations is broad, the average scattering angle being 5 to 10 mrad (Figure 3(d)). While the plasmon intensity falls rapidly to zero above 10 mrad, the single electron excitations fall to zero slowly. The plasmon mean free path, L_P, is usually between 500 and 1500 Å at 100 keV, so that in an average specimen, a large fraction of the electrons transmitted will have lost energy by creating a plasmon. If the specimen is thick enough, the electron may excite another plasmon so that its total energy loss becomes $2E_{PL}$, and this process can be repeated. The characteristic EEL spectrum from a metal is thus that of the zero-loss peak followed by several equally spaced plasmon peaks of decreasing amplitude.

Because the plasmon energy loss depends on the free electron density of the sample, it can be used for microanalysis (Williams and Edington, 1976). Unfortunately, all metals and alloys showing good plasmon peaks have approximately the same value of E_{PL}, so the measured result cannot provide a unique identification of an unknown material. However, as discussed in the next chapter, shifts in the plasmon peak position can be used to monitor the change in concentration of one element in an alloy once a calibration is obtained. This technique has been successfully applied to a variety of metallurgical problems (Williams and Edington, 1976). Nevertheless, the limited range of materials to which this technique can be applied makes it less important than the inner-shell losses described in the next section.

Before leaving plasmon excitations, it is worth noting that they provide an accurate way to measure sample thickness. The probability, P(m), of an electron exciting "m" plasmons and so losing $m \cdot E_{PL}$ energy is

$$P(m) = (1/m!)\,(t/L_P)^m \exp(-t/L_P) \tag{3}$$

where t is the sample thickness and L_P is the plasmon mean free path. Thus, the ratio of the probabilities, P(0) and P(1), of exciting no plasmons and one plasmon is

$$P(1)/P(0) = t/L_P\ . \tag{4}$$

On the spectrum this is just the ratio of the intensity of the first plasmon peak to the zero-loss peak, and hence t can be found when L_P is known (e.g., Jouffrey, 1978). This provides a convenient way to assess, directly from the EEL spectrum, whether or not the sample is thin enough since the thickness should be significantly less than L_P for good microanalysis.

C. Region 3: Core Losses

The final portion of the spectrum to be considered is that lying at energy losses above 50 eV. In this region, characteristic features, caused by inelastic interactions with inner atomic shells, are superimposed on a background intensity caused by valence shell excitations. As can be seen in the spectrum of Figure 2, the intensity in this part of the spectrum is very much lower than that in Regions 1 and 2 and falls rapidly as the energy loss increases. Superimposed on the background are the so-called characteristic, or "core-loss" edges, which are the visible evidence that an atom has been ionized. When the incident electron interacts with an atom, a certain minimum amount of energy, equal to the appropriate binding energy E_k, must be transferred from the incident electron to the atom for some shell to be ionized. Below this energy threshold, E_k, no ionization can occur, so the cross section (i.e., probability) is zero. For energy losses greater than the binding energy, however, ionization can occur with a finite cross section. The energy-differential cross section for inner-shell ionizations will therefore change from zero, below the ionization energy, to some finite value above the ionization or "edge onset" energy. The probability of transferring energy $E \gg E_k$ falls rapidly, so the cross section past the edge will fall again. Since the total signal at any energy loss is due to the sum of the background and ionization cross sections, an edge will therefore appear as a discontinuity in the intensity at energy E_k, followed by a gradual decay back toward the signal level that would have been expected in the absence of the edge. Because the energy at which the edge occurs is a unique property of the element concerned, and of the shell ionized, a measurement of the edge energy is sufficient to identify the atom unambiguously.

Since the signal in the edge varies with both the scattering angle, θ, and the energy loss, E, the cross section can be written in a form that is differential with respect to both

$$\frac{d^2\sigma}{d\Omega dE} = \frac{2e^4}{E \cdot m_o v^2 (\theta^2 + \theta_E{}^2)} \times \text{GOS} \quad (5)$$

where

m_o = electron rest mass
v = electron velocity
θ_E = "characteristic inelastic scattering angle" $E/2E_o$.

GOS is the generalized oscillator strength and represents the number of electrons per atom that take part in a particular energy-loss process. For small scattering angles and energy losses near edge energy E_k, the GOS is constant. Therefore, the angular distribution of the inner-shell excitation is of the form $(\theta^2 + \theta_E^2)^{-1}$, with a maximum at $\theta = 0$. At an energy loss, E, of 300 eV and an incident energy, E_o, of 100 keV, θ_E is only 1.5 mrad, so the average scattering angle is ~5 mrad. If the spectrometer collects electrons up to some maximum angle, β, as β is increased, the inner-shell signal will rise, but eventually at an angle of $\sim(2\theta_E)^{1/2}$ (typically 50 mrad), the GOS falls to zero and the signal will saturate. A substantial fraction of the inelastic signal can therefore be collected by an analyzer that subtends 10 or 20 mrad at the specimen. For energy losses much bigger than E_k, the GOS is low at $\theta = 0$ but rises to a maximum at $\theta \sim (E/E_o)^{1/2}$, to produce a peak known as the "Bethe ridge." The scattering is therefore no longer forward peaked.

The total cross section for inner-shell ionization is found by integrating Eq. (5) from $E = E_k$ to E_o, and for all values of θ. The total cross section, σ, for an inner-shell ionization is $\sim 10^{-20}$ cm^2 per atom for a light element (e.g., carbon) at 100 keV. This is two orders of magnitude smaller than the elastic or plasmon cross sections, and consequently the edge signals are weak. Looked at in another way, the mean free path for an inner-shell ionization, L_K, is several micrometers, so that in a typical 500-Å-thick sample, only a few electrons interact in this way. In practice, the spectrometer accepts electrons only up to some maximum angle, β, and over some energy window, Δ. Therefore, the collected signal is actually proportional to a partial cross section, $\sigma(\beta,\Delta)$, which will be some fraction of σ, depending on the angular acceptance and energy window.

In a pure element, the background before the first ionization edge is mostly due to valence excitations and plasmon losses, but past the first edge, the intensity is mainly due to the tail of the edge. In both cases, it has been shown experimentally that the energy differential cross section for the background again has the form $A \cdot E^{-r}$ (Egerton, 1975; Maher *et al.*, 1979). Since r is usually between 3 and 5, the dominant background signal falls rapidly with energy loss, and so the dynamic range in the spectrum will be large. The angular distribution of the inner-shell background contribution (the "tail" of an edge) will follow that of the edge, being forward peaked for energy losses near the edge. The distribution of the valence electron contribution and that of the inner-shell signal far from the edge energy are not, however, forward peaked but also reach a maximum at the Bethe ridge.

IV. COLLECTING THE EEL SPECTRUM

Now that the main features of the EEL spectrum have been described, it is appropriate to see how these data can be collected from the specimen in our microscope. The spectrum is obtained by passing the transmitted signal into an "electron energy analyzer" or spectrometer, which disperses the beam of electrons into its various energy components. This can be done by passing the electrons through a magnetic field, an electrostatic field, or a combination of both. Successful analyzers based on each of these principles have been constructed, but for purposes of microanalysis in an electron microscope, the magnetic spectrometer is the most widely used approach, and the rest of this discussion will concentrate on these devices.

Figure 4 shows the ray paths through such an analyzer, often called a "magnetic prism," both because of its shape and its similarity to the action of a glass prism on white light. The prism behaves like a lens since electrons leaving the object point of the spectrometer are focused back to an image point. However, electrons of different energy are focused to different image points, so electrons leaving the object point with a mixture of energies are dispersed to form a line focus image, each point of which contains electrons of a specific energy. If the energy difference of two rays is δE, and their separation in the image plane is δx, the quantity $D = \delta E/\delta x$ is called the "dispersion" of the spectrometer.

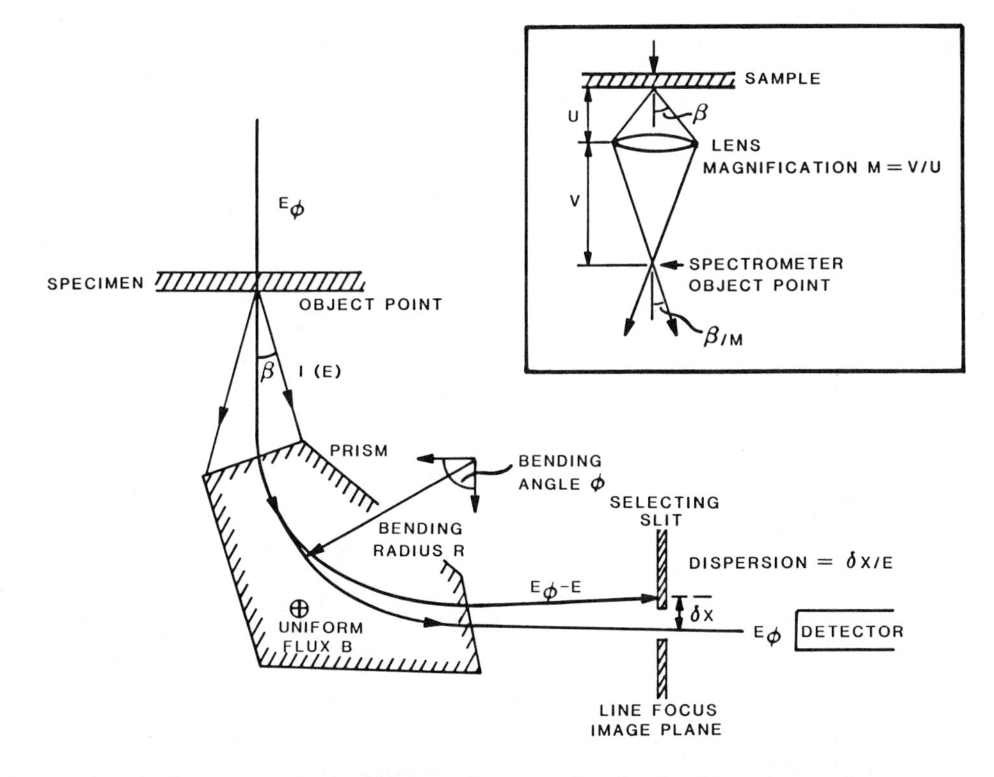

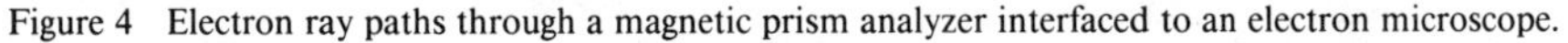
Figure 4 Electron ray paths through a magnetic prism analyzer interfaced to an electron microscope.

For the magnetic prism analyzer

$$D = 2R/E_o \tag{6}$$

where R is the radius of the circle into which the electrons are bent. For 100-keV electrons and a radius of 20 cm, D would therefore be 4 μm/eV. If a slit were placed in the image plane of the prism, only electrons of one specific energy loss would pass through to the detector, and the spectrum could then be recorded by scanning the dispersed electrons over the slit. The minimum energy difference, ΔE, that could be selected with a slit width, S, would be $\Delta E = S/D$. This result would seem to imply that any desired energy "resolution" could be achieved by closing down the slit spacing (subject to the inherent limitation set by the energy spread of the incident beam). However, this is not the case because, like any other lens, the prism has aberrations that set a limit to the performance achieved. It can be shown (e.g., Isaacson, 1978) that for any analyzer

$$\pi\beta^2 (E_o/\Delta E) \sim 1 \tag{7}$$

where

ΔE = effective energy resolution
β = acceptance angle.

To collect a high fraction of the inelastically scattered electrons, we have already seen that β needs to be at least 10 mrad; therefore, for 100-keV electrons, Eq. (7) would estimate a resolution value of 30 eV. Although this may be acceptable for such purposes as energy-filtered imaging, such a resolution would not allow the precise determination of edge onset energies or even the separation of nearby edges; therefore, a higher resolution is usually sought. While this could be achieved by reducing the value of β, this reduction would lead to a fall in the collection efficiency. Fortunately, several other options are available. One is to slow the electrons before dispersing them, using a "retarding field analyzer," since reducing E_o will allow any given spectrometer to achieve a better resolution (as shown by Eq. (7)). This method is often used for very-high-resolution spectrometers, but it can lead to a poor collection efficiency, and the need to feed very high voltages into the analyzer to retard the electrons makes the mechanical and vacuum design difficult. Another solution is to try to reduce the aberrations of the spectrometer by modifying its optics (e.g., Crewe *et al.*, 1971). A gain of as much as an order of magnitude in the energy resolution can be obtained in this way, although such an improvement generally requires high-precision machining of the analyzer.

The most effective procedure, and one that is well suited to the use of the spectrometer in an electron microscope, is to place a lens between the specimen and the spectrometer object point (Figure 4). If the lens has a magnification of M, the entrance pupil into the analyzer becomes β/M for the same collection angle, β, at the specimen, and the constant of unity on the right-hand side of Eq. (7) is increased by a factor of M^2. Since M can be made as large as desired, it becomes easy to combine both high energy resolution and high collection efficiency by using this technique. By correctly choosing the coupling conditions, a resolution of 1 or 2 eV for an acceptance angle of 10 mrad is possible, even with simple analyzer designs (e.g., Egerton, 1978;

Joy and Maher, 1978a). However, there is usually an upper limit to what can be achieved in this way because the coupling lens will also magnify the object area W on the specimen to a size MW. This will be imaged through the analyzer to give a resolution, limited by the dispersion, to MW/D. There may thus be an optimum magnification for this coupling (e.g., Egerton, 1980a; Johnson, 1980). If dimension W is very small, as in STEM where it could be the diameter of the scanned probe (typically 100 Å), M can be made very large before the effect of source size becomes significant. In the TEM, however, where field of view W may be a few micrometers in diameter, only a limited magnification can be used before the resolution is limited by aberrations. Even in the STEM, there will be an upper limit to the field of view over which the beam can be allowed to scan during analysis unless the beam is "descanned" to bring it back onto the axis before it enters the spectrometer because electrons traveling through the analyzer at a high angle to the axis will also be aberrated and worsen the resolution.

To achieve the resolution of which the spectrometer is capable, the analyzer must be "focused" so that its image plane coincides with the selecting slit. Although coarse focusing can be obtained by physically moving the slit, fine focusing is achieved by varying the excitation of the coupling lens or lenses between the analyzer object point and the specimen. The accuracy of focus can be checked by repeatedly scanning the spectrometer through the zero-loss peak while adjusting the lens currents and microscope alignment to give a symmetrical peak with minimum width and maximum height, as shown in Figure 5(a) and (b). If the resolution of the spectrometer is limited by aberrations, as in Eq. (7), the zero-loss peak will have a "Gaussian" shape. If, however, the energy resolution as defined by the slit spacing is greater than the value predicted by Eq. (7), the zero-loss peak will show a "flat-topped" form. In either case, the "energy resolution" of the spectrometer can be defined as the energy width of the zero-loss peak at half maximum height.

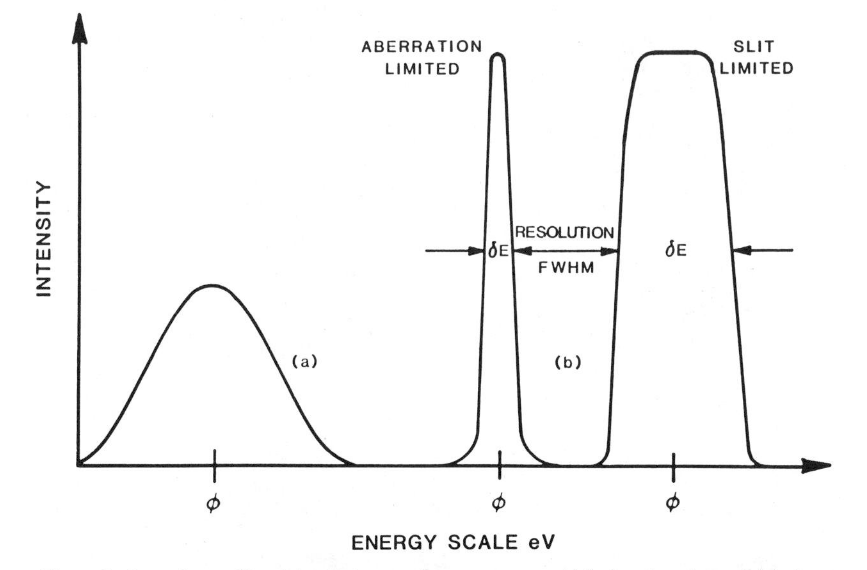

Figure 5 Intensity profile of the slit image of a spectrometer (a) before focusing and (b) after achieving optimum focus.

The choice of energy resolution will depend on the problem. For standard chemical microanalysis, a resolution of 3 to 10 eV is usually enough, the higher value being acceptable for edges occuring at high (>1 keV) energy losses, whereas a resolution of 20 to 30 eV is adequate for energy-selected imaging through the analyzer. But for purposes such as the study of plasmons or the observation of "chemical shifts" in edge energies, a resolution of better than 1 eV may be required. In general, the "best" resolution will be the worst value that will allow the data of interest to be collected because a degraded resolution will increase the partial cross section (by raising ΔE) and so improve the collected signal. Furthermore, if the spectrum is recorded sequentially, the energy increment between successive channels can also be increased as the energy resolution is lowered, thus reducing the time required to scan any given energy range.

The choice of β is also important since it controls two of the most important parameters of the spectrum. Figure 6 shows, schematically, an edge and the definition of two quantities that describe it. "S" is the characteristic edge signal measured above "B," the actual, or extrapolated, background intensity. The ratio $(S + B)/B$ at the edge onset is often called the "jump ratio" and is a measure of the visibility of the edge. The ratio $S/(S + B)^{1/2}$ is a measure of the signal-to-noise, or statistical validity, of the edge. For small values of β, both S and B are small, so the signal-to-noise ratio is bad, but the jump ratio is high.

If the background is due to valence electron scattering, as β is increased, S and B will rise and, because the characteristic edge signal rises faster than the background, give an improved signal-to-noise ratio. At very large values of β, however, S will reach a saturation value, although B will continue rising. The signal-to-noise ratio will therefore show a maximum at some intermediate value of β (Joy and Maher, 1978b), the value typically being $\sim 2\theta_E$. When the background is due to single electron excitations, the background and the edge signals rise together, and the signal-to-noise ratio reaches a plateau. In either case, the largest value of β consistent with maintaining the desired energy resolution is usually the best choice. This is especially true when the sample is thick since a larger value of β will allow more of the electrons that have been both elastically and inelastically scattered to be collected.

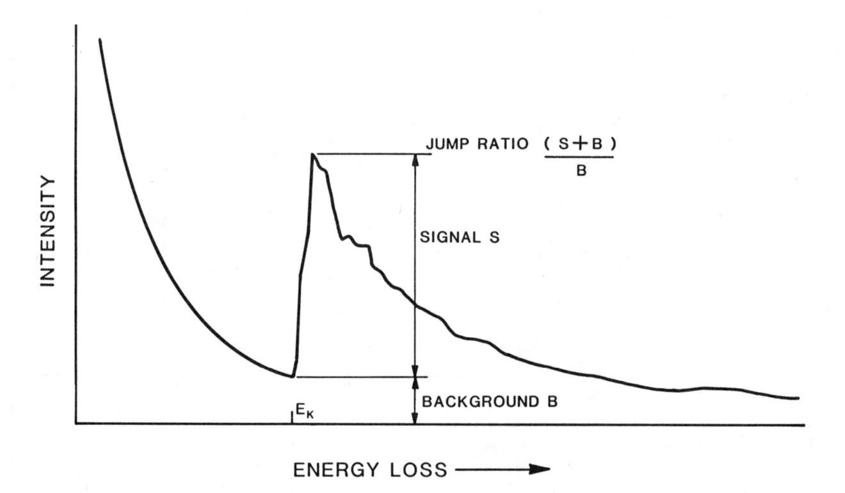

Figure 6 Definition of the parameters that define the quality of an edge.

V. RECORDING THE EEL SPECTRUM

The last step in obtaining an electron energy-loss spectrum is to record and display it in a suitable form for study and subsequent processing. Originally, a photographic plate placed in the plane now occupied by the defining slit acted both to detect and to store the spectrum. Although the photographic process was cheap and efficient, its dynamic range was limited, processing was necessary before the spectrum could be viewed, and microdensitometry was required to obtain numerical data. Consequently, the use of films or plates has been abandoned, and electronic means are used to collect, detect, display, and store the spectra. Figure 7 shows schematically a typical arrangement (Maher *et al.*, 1978). The spectrum is stored in a multichannel analyzer (MCA) operated in the multiscaling mode so that all of the counts arriving during some specified "dwell time" interval, τ, are added to the count total in a particular channel. At the end of the dwell time, the MCA input is addressed to the next channel, and at the same time, the field of the spectrometer is also stepped so as to change the energy passing through the slits by some increment, δE. This procedure is then repeated for the 512 or 1024 channels required to fill the memory. In this way, the spectrum is sequentially built up, the N channels covering a total energy range, $N \cdot \delta E$, in a total recording time, $N \cdot \tau$.

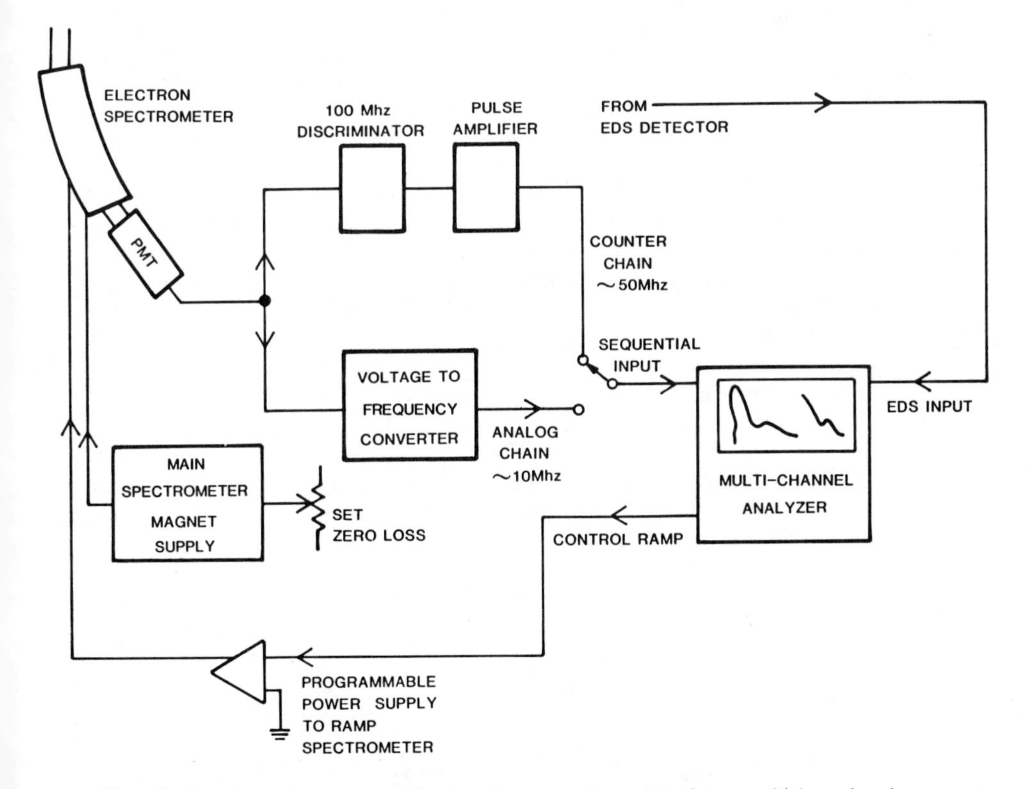

Figure 7 Experimental arrangement for recording energy-loss spectra into a multichannel analyzer.

The signal into the MCA must be in the form of a pulse. Depending on the way that the signal is originally collected, this can be achieved in two ways. If the detector is producing an analogue output (for example, if the signal is taken from the anode of the photomultiplier through a preamplifier), the signal can conveniently be digitized by the use of a "voltage-to-frequency" converter. This produces a pulse train whose instantaneous repetition rate is proportional to the input voltage. A typical device of this type produces a maximum rate of 10 MHz for an input of 10 eV and is linear over more than five orders of magnitude, to give a conversion accuracy equivalent to the performance of an analogue-to-digital converter with a 19-bit resolution.

Alternatively, if the detector has such a fast response that it can generate a pulse for each individual arriving electron, these pulses can be counted directly by the MCA after appropriate amplification and shaping. In principle, the electron counting approach is superior since it provides the highest sensitivity and the stored count is an absolute measure of the collected signal. However, even the fastest photomultipliers cannot count at rates in excess of $\sim$40 MHz; consequently, it is not practical to count in the vicinity of the zero-loss peak except at very low beam currents. The analogue signal and the voltage-to-frequency converter, on the other hand, can easily cover the zero-loss region and produce satisfactory data out to energy losses of 1 keV or more. Care is needed, however, in adjusting V-F converters to achieve the full performance of which they are capable, and the stored count total is only an arbitrary fraction of the true electron count. The ideal arrangement is to have both available and be able to switch from one mode to the other at some appropriate point in the spectrum.

Any of the usual electron detectors can be used in a spectrometer. Semiconductor detectors, such as P-N or Schottky diodes, have been successfully used for analogue detection (Trebbia *et al.*, 1977; Egerton, 1980b), but they are unsuitable for pulse counting operation because of their poor bandwidth. Photomultiplier detectors are more sensitive, have a faster response, and are therefore generally used. However, to obtain the optimum performance, the scintillator must be chosen with care to offer a high count rate, good light output, and freedom from radiation damage effects (Joy, 1983). The very wide dynamic range of the EEL spectrum places stringent requirements on the linearity, noise, speed of recovery from overload, and stability of the detector, its electronics, and those of the MCA. Unless care is taken in setting up the complete system, artifacts can be introduced into the spectrum (Joy and Maher, 1980a), resulting in such effects as the truncation of the low-loss peaks and the distortion of peak and background profiles.

A significant limit in conventional EELS operation comes from the fact that the spectrum is sequentially collected. Thus, in a total recording time of, say, 100 seconds, each channel in the energy-loss range will have been measured for only a few hundred milliseconds. Consequently, the signal-to-noise ratio of the spectrum, and the resultant detection sensitivity, is substantially worse than might be expected from a simple consideration of the operating conditions. An important topic of current research is to circumvent this limit by collecting the spectrum in parallel, as in the EDS case. This can be achieved by using electron detectors, such as arrays of diodes, that are position sensitive (Jones *et al.*, 1977; Chapman *et al.*, 1980; Jenkins *et al.*, 1980). These devices can be placed in the dispersion plane of the analyzer and exposed to the electron beam (Jenkins *et al.*, 1980; Chapman *et al.*, 1980). Alternatively, a phosphor can be used to convert the electrons to light, and the detector array can then be used to image this either directly (Johnson *et al.*, 1981a, b) or after further image intensification (Egerton, 1981a, b; Shuman, 1981). Each of these schemes has both advantages and disadvantages (for a full discussion, see Joy, 1983), but any of them is capable of improving the sensitivity of EELS analysis by typically an order of magnitude. As the technology of these systems is further improved and simplified, it can be expected that they will routinely find their way into systems. In the meantime, however, their use is still experimental, and almost all published data are from the sequential acquisition systems previously described.

VI. MICROANALYSIS USING INNER-SHELL LOSSES

Now that the principles of generating and collecting the EEL spectrum have been outlined, we can consider the practical problems of interpreting an experimental spectrum in order to obtain microanalytical data from it. In principle, the procedure is simple since each element present will produce an edge (or edges), each of which will have a unique energy loss. The task is therefore to identify the features in the spectrum that are edges and to assign to each of them the correct energy-loss value. In EDS analysis, all the characteristic features are Gaussian, so identification of a peak—even in the presence of noise—is relatively straightforward.

The identification of a feature as an edge in an EEL spectrum is less routine because the edges can have a variety of shapes. However, at a resolution of 5 to 10 eV, the most important edge shapes can be schematically represented as shown in Figure 8. K edges (Figure 8(a)) have a characteristic triangular shape, with maximum intensity occuring at onset energy E_k, followed by a steady decay. The L_{23} edges of the elements in the first few rows of the periodic table have relatively little intensity at onset E_k (Figure 8(b)) but rise to a delayed intensity maximum at an energy typically 30 to 60 eV after E_k, the maximum again being followed by a steady decay. For heavier elements, the delay is much less pronounced, and the edges resemble K edges in general form. The other major edges often visible (M_{23} and N_{45}) have a shape similar to that of the L_{23} but with a more delayed maximum and lower intensity at the onset (Figure 8(c)).

At better energy resolution and with a sufficiently good signal-to-noise ratio, fine structures can be seen on edges (Figure 8(d)). In the region around the edge onset are "preionization" or "near-edge" structures. These are present because an electron excited from an inner shell goes first to unoccupied bound states around its atom before going to the vacuum energy level and leaving the atom ionized. The strength of the near-edge structure depends on the transition probability to each available level and the density of unoccupied states at that level; consequently, the form of the structure depends sensitively on the bonding and coordination of the atom. This makes it possible to distinguish between different forms of the same element, for example amorphous carbon, graphite, and diamond (Egerton and Whelan, 1974). Beyond the edge is a weak periodic modulation extending for several hundred electron volts past the edge. This extended energy-loss fine structure (EXELFS) is analogous to the EXAFS effect in x-ray absorption spectra. It arises from a modulation in the ionization cross section caused by interference between the outgoing electromagnetic wave emitted from the ionized atom and components reflected from neighboring atoms (Leapman *et al.*, 1981). The way in which this structure can be analyzed to yield detailed crystallographic information is discussed in the next chapter.

An essential first step in any work on inner-shell edges is to strip the background. This can be done by modeling the experimental data by using the analytical expression of Eq. (5). If logarithms are taken on both sides of Eq. (5), that expression is transformed to a linear relationship

$$\log(I) = \log(A) - r \log(E) . \tag{8}$$

A computer able to read the spectrum stored in the MCA memory can thus find the best-fit values for A and r over some range of energy losses by a standard least-squares fitting procedure (Joy *et al.*, 1979). Alternatively, a procedure based on a graphical analysis can be used (Egerton, 1980b). The calculated values of A and r can then be used to replot a calculated background that can be compared with, or subtracted from, the original data.

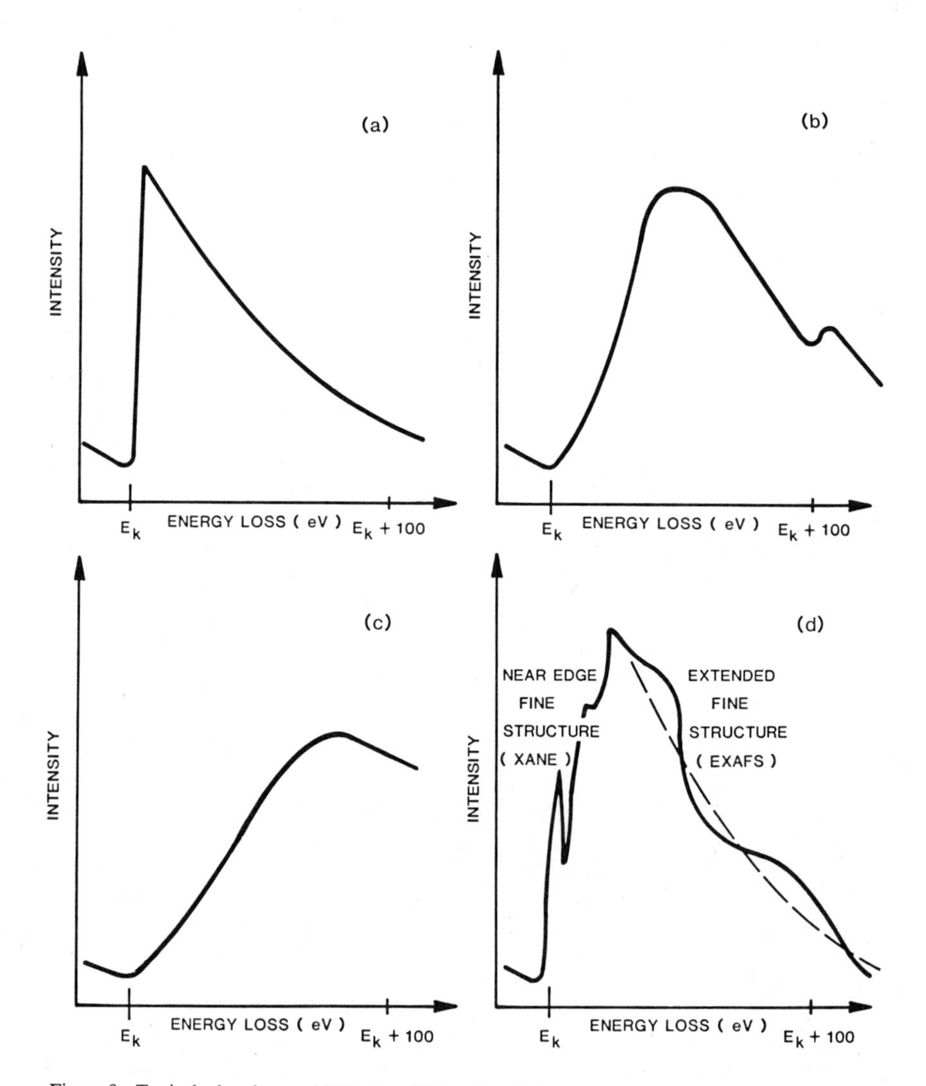

Figure 8 Typical edge shapes: (a) K edge, (b) L_{23} edge, (c) M_{23} edge. (d) illustrates the types of fine structure present on an edge.

The importance of background stripping in EELS microanalysis is demonstrated in Figure 9, which shows an example of applying these procedures to the analysis of precipitates in a silicon device. The original spectrum (Figure 9(a)), taken over a precipitate in the energy-loss range from 200 to 400 eV, shows no obvious edge, although a slight distortion of the background can be seen. However, if a background fit is performed over the range 200 to 260 eV as shown, and the calculated background is then superimposed over the original data and extrapolated, it is clear that the fit is excellent until ~280 eV. At this point, the experimental data diverge from the fitted line, only returning toward it at a loss of ~400 eV. Subtracting away the calculated background reveals what is clearly K edge with an onset energy of ~280 eV, i.e., a carbon edge (Figure 9(b)).

K edges usually have enough intensity to allow their onset energy to be determined without difficulty, but L_{23}, M, and N edges recorded at medium resolutions may have such a low initial intensity that it will usually be impossible to determine the onset energy accurately enough to make an unambiguous identification without background stripping. Note, however, that most edges have relatively intense but narrow line features at their onset energy. This enhances their visibility and the accuracy of energy measurement when the spectrum is recorded by an analyzer of sufficiently high resolution (~1 eV). When several edges are in close proximity, it is necessary to strip the backgrounds sequentially, starting with the lowest energy-loss edge, if good fits are to be obtained. Even then, in some circumstances the simple expression of Eq. (5) will not give an adequate fit, requiring expressions with more adjustable constants (Colliex *et al.*, 1981; Bentley *et al.*, 1981).

The precision of the onset energy determination will depend on the accuracy and linearity of spectrometer calibration. The energy increment (eV/channel) can be found by (a) recording a spectrum containing both the zero-loss peak and an edge of good

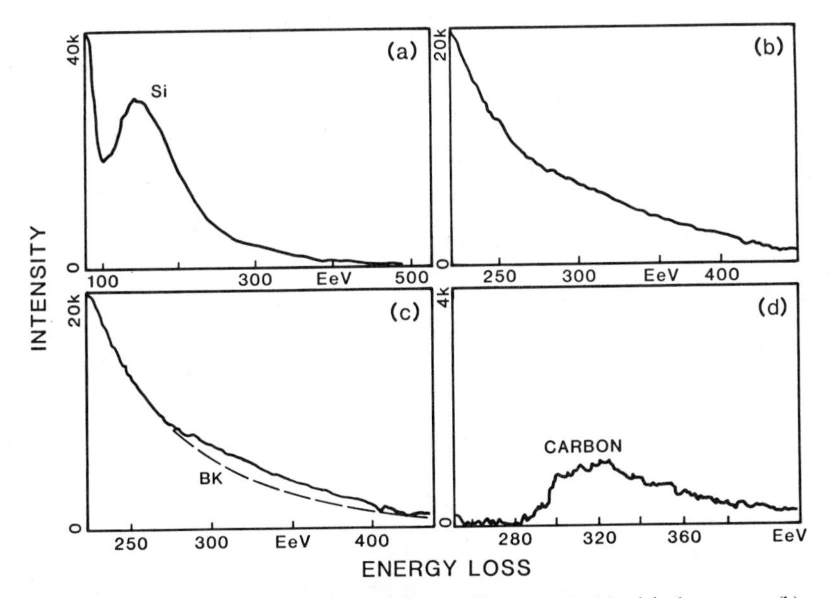

Figure 9 Experimental energy-loss spectra from a silicon sample: (a) original spectrum, (b) same region expanded, (c) spectrum with computed background fit, (d) carbon edge after subtraction of background.

shape and known energy, and (b) measuring the number of channels between these energies. Although identifying the exact zero position is easy, knowing where on the edge profile to assign edge onset energy E_k is less straightforward since the edge has finite width and the spectrometer has limited resolution. Unless there are some identifiable fine-structure features of known energy (Egerton and Whelan, 1974), the onset energy must arbitrarily be placed half way up the edge. Once the spectrometer is calibrated, the energy increment is usually constant enough that subsequent routine calibration only involves locating the zero-loss peak (whose position may vary as the result of drifts in the high voltage or changes of alignment in the microscope). For special purposes, however (measurement of chemical shifts, for example), the assumption that the eV/channel is constant throughout the spectrum may be insufficiently accurate (Egerton, 1982). In that case, the ideal arrangement is to have a known edge or edges near the edge of interest to give additional local calibration points.

When the onset energy has been found, the identity of the edge can usually be found directly with the help of a suitable listing. In some cases of practical importance, however, there may be ambiguity because of the closeness in energy of several edges, each of which may be a possible candidate. In such cases, it is also necessary to use the information derived from the shape of the edge. The schematic edge shapes shown in Figure 8 often are sufficient to allow K, L, M, and N edges to be distinguished; also available are catalogs of experimentally determined edge shapes that can be compared with the unknown edge to give a positive identification (Zaluzec, 1981; Ahn and Krivanek, 1982). Wherever practical, it is also good experimental procedure to record an EDXS spectrum from the sample area at the same time as the EELS data are taken. This will often allow the higher edges (M, N, O) of heavier elements to be identified. After the major edge or edges have been identified, their background contribution should be stripped from the spectrum and the minor edges identified. When edges are very low in intensity, the problem of identification can be severe. One successful approach is to use computed or stored edge profiles and apply statistical techniques to measure the "goodness of fit" between the several possible candidates for the edge and the experimental data (Joy, 1982a).

VII. PROBLEMS IN EELS MICROANALYSIS

The problem most commonly encountered in EELS microanalysis is that a given edge may appear to vary widely in shape, visibility, and even in energy, depending on the specimen and the experimental conditions. These effects are sometimes more apparent than real—for example, the case shown in Figure 10, which compares the Si L_{23} edge in elemental Si and in SiO_2. In the "raw" data form (Figures 10(a) and 10(b)), the edges appear to be different both in shape and in energy, but after background removal (Figures 10(c) and 10(d)), it can be seen that the two edges are, in fact, quite similar. Here the apparent change in edge shape was due to the difference in the magnitude and slope of the background on which the edge was riding. Depending on the other elements present in the compound under examination, the background near the edge of interest can vary widely, and this will interact with the edge profile. Removing the background in such cases will solve the problem.

A more severe problem is the change in edge profile caused by changes in the thickness of the specimen (Figure 11). When the sample is thinner than one mean free path-length for inelastic scattering, on average, most electrons will only be scattered once before leaving the specimen. The low-loss and core-loss regions of the spectrum therefore have the expected appearance shown schematically in Figures 11(a) and (b).

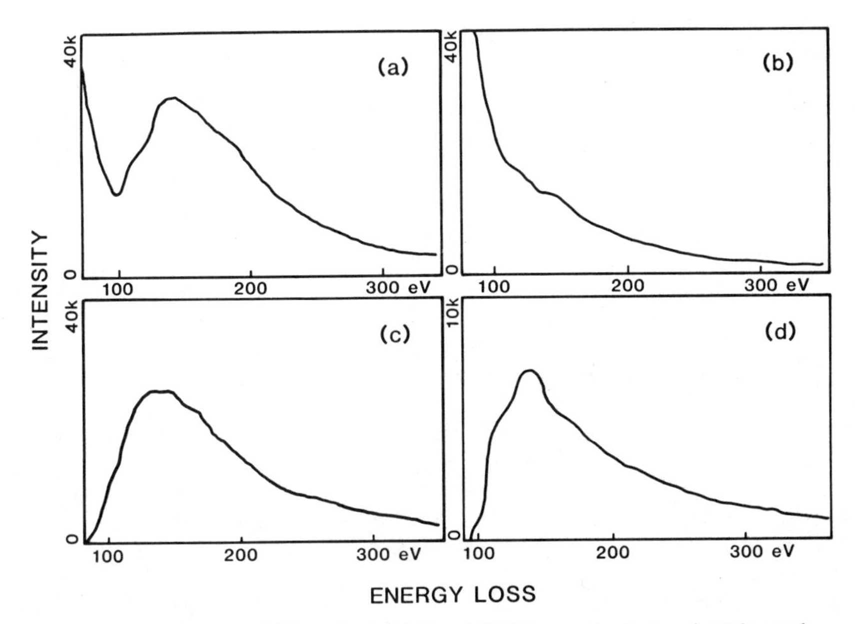

Figure 10 Appearance of Si L_{23} edge in (a) Si and (b) SiO_2 sample of approximately equal thickness. (c) and (d) show the edge appearance in each case after stripping the background.

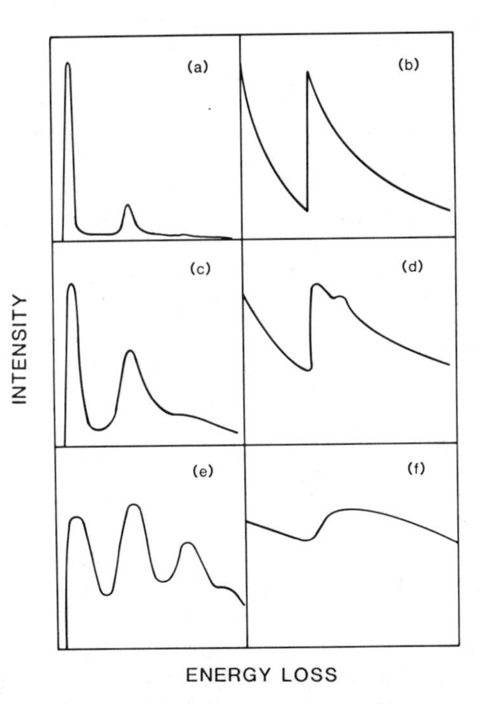

Figure 11 Schematic illustration of the form of the low loss spectrum (a, c, e) and an inner-shell edge (b, d, f) at three different sample thicknesses.

At 100 keV, this would be the EEL spectrum expected from a sample typically 500 Å thick. If the specimen were now made twice as thick, the equivalent spectra would look like those in Figures 11(c) and (d). The low-loss region now shows not just one, but two, plasmon peaks. Measuring the ratio of the heights of these peaks and applying Eq. (3) gives the thickness of the sample in units of the mean free path.

The core-loss edge has also changed. First, a "replica" of the plasmon loss has now appeared behind the edge, indicating that many electrons suffer both a core loss and a plasmon loss before leaving the specimen. Second, the jump ratio of the edge has fallen because the average background level at each energy loss has increased as a result of the increased probability that an electron will be scattered. Doubling the specimen thickness yet again, to perhaps 2000 Å at 100 keV, produces spectra like those in Figures 11(e) and (f). The low-loss spectrum now shows several plasmon-loss peaks, and the zero-loss peak is lower in intensity than the first plasmon, indicating that few electrons now pass through the sample without being scattered. The edge is now substantially changed in form. The effect of the "plural scattering" suffered by each electron is to shift intensity from the edge to the region behind the edge, so the profile becomes more rounded and less distinct. In addition, the jump ratio has fallen still further because the background is still rising more rapidly than the edge signal. A jump ratio that may be as high as 10 in a thin sample may be little more than unity by this thickness.

The interpretation of a spectrum is therefore significantly influenced by sample thickness. When the specimen is thin, the edges are well marked and easy to identify even though the signal-to-noise ratio of the spectrum may be poor. As thickness increases, the edges become less pronounced, and the changes in their profile hamper measurement of the edge onset energy and identification of the edge by matching its shape. But it should be noted that as the thickness increases, the signal strength at any given energy loss also increases. Therefore, the signal-to-noise ratio in the spectrum will improve with thickness, and the best signal-to-noise ratios will be achieved at the thicknesses of about one to two mean free paths (Spence, 1977). Provided that the data are of adequate signal-to-noise, the deleterious effects of plural scattering can be removed from the spectrum by the mathematical process of deconvolution (Leapman and Swyt, 1981; Joy, 1982a, b). Techniques for rapid deconvolution of spectra are now available on minicomputers. Although the procedure is usually not essential, spectra that have been so treated do have the advantage that edge profiles are very reproducible. This is an aid to both the identification and quantification of edges. In general, the best advice is to use the thinnest sample possible as judged from the low-loss spectrum and to work at the highest accelerating voltage since this will increase the inelastic mean free path. The newer microscopes capable of operation at 200 keV, or even 300 keV, will offer substantial benefits in terms of the usable specimen thickness.

Electron energy-loss spectra can also suffer from the problem of the "disappearing" edge, in which an element known to be present will give no visible trace of its existence. There can be several reasons for this disturbing behavior. Excessive specimen thickness, as discussed previously, can result in an edge being low in visibility, particularly when the edge of interest is of low energy loss. Another aspect of this problem occurs in complex samples when many elements are present. Not only may there be cases of edge overlaps, resulting in the obliteration of an edge, but the "tails" of several edges in close proximity produce a large background that may reduce the visibility of a subsequent edge below the necessary threshold.

Finally, it must be remembered that the intensity of an edge falls rapidly with the increase in energy loss at which the edge occurs because of the variation of the partial ionization cross section. For example, in the compound SiO_2, using an energy window of 100 eV and a collection aperture of 7.5 mrad, the silicon L_{23} edge (at 100 eV) has ten times the strength of the oxygen K edge (at 532 eV) even though there are twice as many oxygen as silicon atoms. Consequently, even a major constituent in a compound may produce only a weak edge if its major edge lies at a high loss. The question of EELS microanalysis sensitivity is discussed in the following section. As a general guide, however, it is likely that any element present at a lower level than ~5 at.%, and any constituent represented only by an edge with an energy loss greater than that of the most prominent edge present, is likely to be too low in visibility to be useful.

VIII. THE SENSITIVITY OF EELS MICROANALYSIS

An important question is the sensitivity of EELS microanalysis, both absolutely and in comparison to EDS. As before, the detected signal count rate (electrons/second) in EELS is

$$I_{ELS} = I_o \cdot N \cdot \sigma(\beta,\Delta) \tag{9}$$

where $\sigma(\beta,\Delta)$ is the appropriate partial ionization cross section. The EDS count rate (photons/second) recorded under identical conditions would be

$$E_{EDS} = I_o \cdot N \cdot \sigma \cdot \omega \cdot F \tag{10}$$

where

σ = x-ray cross section
ω = fluorescent yield
F = efficiency of the detector at the energy of the x-ray line examined.

The partial ionization cross section is a function of the experimental parameters of the spectrometer, and by appropriately choosing these, the ratio of $\sigma(\beta,\Delta)$ to $\sigma_{x\text{-ray}}$ can be kept constant at 0.05 to 0.25. On the other hand, the product of the x-ray detector efficiency and fluorescent yield varies widely as the energy of the x-ray photon changes. For an iron K_α line, the fluorescent yield is ~0.3, and the detector efficiency, limited by its geometrical solid angle, is ~0.01. Under these conditions, the EELS and the EDS count rates would be comparable. However, for a light element such as oxygen, the fluorescent yield is only 10^{-3}, and the detector efficiency, limited now by absorption in the beryllium window, is $\sim 10^{-4}$. So in this case, the EELS count rate would exceed the EDS count rate by several orders of magnitude.

When considering the sensitivity of a microanalysis, however, it is the integrated count and the peak-to-background value rather than the count rate that is of significance. In this respect, EELS microanalysis, performed using sequential data storage, fares badly. If in a fixed recording time, T, a spectrum containing C channels is stored, the recording time per channel is T/C for the EEL spectrum but T for the x-ray spectrum. Thus, the integrated count in the x-ray case may still exceed the EELS value even though the count rate is substantially lower. Furthermore, the peak-to-background ratios in the two spectra are quite different since that of a clean EDS spectrum is between 50 and 100 to 1. But the maximum jump ratio in an EEL spectrum is ~10 to 1 for the major element, while for trace elements the ratio may be much less than unity.

These factors must be taken into account when calculating the minimum detectable mass (MDM) of an element in an EELS microanalysis. In the case of most general interest, it is required to calculate the MDM of some element present as a trace constituent in a matrix of some other element. As shown in Figure 2, the edge representing the element of interest is riding on a background with two possible contributions, one from noncharacteristic events such as the tails of plasmon and valence scattering events, and the other from the tail of the matrix edge. Each channel of the MCA contains the signal count recorded at that energy loss, E, for the energy window, δ, defined by the resolution, and for the spectrometer acceptance angle, β. This signal is proportional to the partial ionization cross section for these parameters. It can be shown (Joy and Maher, 1980b) that N, the minimum number of atoms of element A detectable in the matrix, is

$$N = \frac{[(\sigma_o + \sigma_m)M]^{1/2}}{\sigma_A \cdot (J\tau)^{1/2}} \tag{11}$$

where

σ_A, σ_b, and σ_m = partial cross sections for the edge, valence background, and matrix, respectively
J = incident current density (el/cm^2)
M = number of atoms of matrix in the analyzed volume
τ = dwell time (seconds/channel).

This relation indicates that the best sensitivity is obtained for the highest incident current density, longest recording time, and thinnest sample. To a first approximation, all the partial cross sections will change linearly with resolution; consequently, N varies as $\sim 1/\delta^{1/2}$, indicating the benefit of being able to change the energy resolution.

Equation (11) also shows that an element whose detected edge occurs at a lower energy loss than that of the major matrix edge will have a lower MDM than an element that falls after the matrix edge since in the first case, σ_m is zero. This is seen in Figures 12 and 13, which are plots of the MDM (derived by using Eq. (11)) for various elements from Li to Fe in matrices of Si and C, respectively. In both cases, the general rise of the MDM with increasing atomic number is superimposed on the abrupt increase in MDM that occurs at the edge onset. Under the experimental conditions assumed here, which are typical of a conventional TEM/STEM instrument fitted with a tungsten filament, the MDM corresponds to 10^3 to 10^5 atoms of the element of interest, which is of the order of a few atomic percent of that element in the matrix. EELS microanalysis is therefore not a trace detection technique, but it does offer very high absolute sensitivity.

IX. SPATIAL RESOLUTION OF EELS

One of the limitations of EDS analysis is that the analyzed volume is larger in diameter than the incident electron beam. This is because elastically scattered electrons can subsequently generate x-rays from regions far removed from the beam position, which typically limits the spatial resolution to several hundreds of angstroms. In the case of EELS microanalysis, no such effect occurs because only those electrons that leave the specimen within the cone defined by the acceptance angle of the

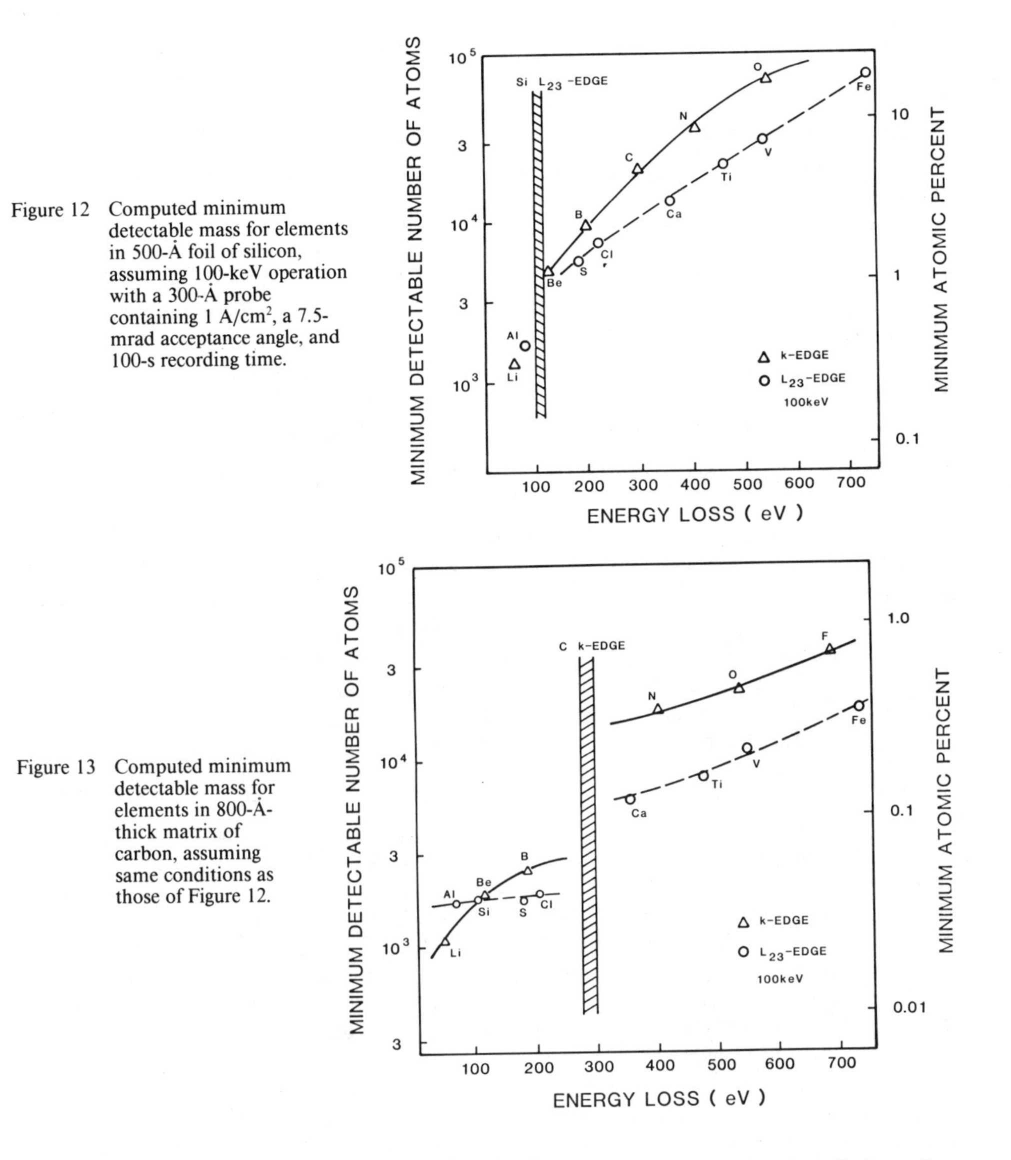

Figure 12 Computed minimum detectable mass for elements in 500-Å foil of silicon, assuming 100-keV operation with a 300-Å probe containing 1 A/cm², a 7.5-mrad acceptance angle, and 100-s recording time.

Figure 13 Computed minimum detectable mass for elements in 800-Å-thick matrix of carbon, assuming same conditions as those of Figure 12.

spectrometer will contribute to the signal. For an acceptance angle of β and a specimen of thickness T, the resolution will be no worse than Tβ. Since β is typically a few milliradians, the spatial resolution defined by the beam broadening is only a few angstroms. A lower limit to the resolution is set by the consideration that the incident electron can ionize an atom even when passing at some distance from it (Howie, 1981).

Calculations show that the silicon L edge (at 100-eV loss) could be ionized by an electron as far as 20 Å away from the atom. In practice, however, the usable spatial resolution is set by the smallest probe size containing sufficient current to achieve an acceptable signal-to-noise ratio in the spectrum. With conventional tungsten filaments, resolutions of ~100 Å are possible, and with a field-emission source, this could be improved to a few tens of angstroms. The actual value will depend on the energy loss and strength of the edge of interest and the recording time. For spectral analysis, the spatial resolution is comparable with that for EDS microanalysis, but much better resolution is possible in energy-loss imaging because the entire analysis time is spent at one energy loss, and images showing detail at resolutions of better than 20 Å have been produced.

X. ENERGY-FILTERED IMAGING

In addition to producing a spectrum by scanning the dispersed electrons across the slit, the energy-loss analyzer field can be held constant so that only electrons that have lost a specified amount of energy are accepted. The spectrometer is then acting as a filter and can be used to form energy selected images. In a STEM system, the filtered image can be obtained by using the usual slit and detector arrangement but using the signal collected to modulate the display screen brightness as the incident beam is scanned. In a conventional TEM system, energy filtering can be achieved with a specially designed filter lens (Andrew *et al.*, 1978) or by using the imaging properties of the magnetic prism.

The imaging mode can be used in several different ways, the simplest of which is as a filter to remove all inelastically scattered electrons from the image. In many cases, the resultant pure elastic image can show better contrast and resolution than the original because of eliminating chromatic aberrations in image and reducing background from plasmon scattering events. This technique is especially effective when applied to convergent-beam and zone-axis diffraction patterns since it sharpens the pattern detail and improves the contrast of weak features (Higgs and Krivanek, 1981).

The spectrometer can also be set to look at any desired energy loss, either an arbitrary value or a specific loss such as a plasmon peak or a characteristic edge. On thick specimens, considerable improvement in penetration can often be obtained by tuning the analyzer to accept the most probable energy loss from the sample rather than the zero-loss peak. The most usual application, however, is for elemental mapping using inner-shell edges. The principle of the technique can be understood from Figure 14(a), which shows an edge and its corresponding background. When the element of interest is present, signal S_E collected in window Δ is S_1, while in an otherwise equivalent region where the element is not present, the signal would be S_2. The variation in the signal level from this window should therefore represent the distribution of the element.

To achieve good-quality images in a reasonable time, energy window Δ, set by the resolution of the spectrometer, is usually set significantly higher (larger) than would be normal for spectral purposes, either by opening the slits or defocusing the spectrometer. In this way, the maximum signal can be collected from the edge. Since energy resolution is not significant, a larger than normal acceptance angle can also be used, again increasing the collected signal. With an optimum choice of conditions, acceptable images can be obtained for recording times of only a few minutes. Because the collection efficiency of the energy-loss spectrometer is so high, the resultant picture looks similar to a regular image rather than to the familiar x-ray “dot” map.

The limitation of this simple approach is that it assumes the sample is completely uniform and that the background near the edge is unaffected by changes in the concentration of any other elements present. If these assumptions are not correct (Figures 14(b) and (c)), the signal level in the window region will not depend

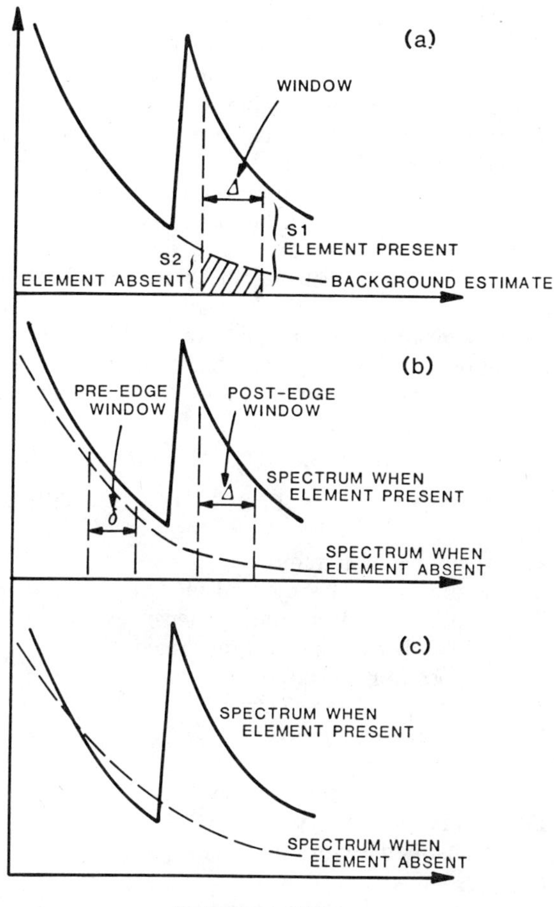

Figure 14 Signal definitions when mapping an element in energy-filtered imaging: (a) for a uniformly thick sample, (b) when the sample thickness changes, and (c) when the background changes because of other changes in the sample chemistry.

systematically on the concentration of the element of interest. Several techniques have been described to overcome this problem. If the background can be assumed to depend only on the mass-thickness of the sample (as in Figure 14(b)), a correction can be made by taking, in addition to the usual image, a second picture using an energy window positioned just before the edge. Since the intensity in this window should vary with sample thickness in approximately the same way as the intensity in the window after the edge, the difference (or the ratio) of these two signals should only be a function of the concentration of the element. Taking the difference between the two images can be done photographically by forming a sandwich of positive and negative transparencies and printing the combination, or electronically by storing the images digitally and processing them in a computer.

Leapman and Swyt (1983) have shown experimentally that in general the background is not only a function of thickness but is also dependent on the local chemistry of the specimen (Figure 14(c)). To compensate for this, it is necessary to model, fit, and strip the background at every point on the image. While this can be done, it requires a substantial amount of computing power and data storage and the ability to scan the spectrometer at high speed. Consequently, for most purposes the assumption of a background that is thickness dependent is the only practical one. In either case, correction for the background requires additional exposure of the specimen. When the sample is sensitive, this extra radiation damage can cause problems because of mass loss or elemental translocation. Considerable care is therefore needed in interpreting the data obtained from such materials as soft tissues since the total exposure dose (typically 1000 el/A^2) is many times the figure of 1 el/A^2 regarded as safe for most biological samples (Wall, 1979).

Because the whole analysis period is devoted to one energy loss, rather than to a scan through the spectrum, the mass sensitivity of elemental mapping is higher than that for a standard EELS microanalysis, and as few as a hundred atoms of an element have been observed in this way. This high sensitivity and the excellent spatial resolution obtainable make this technique an important one in many applications where knowing the composition of the sample is less important than visualizing the distribution of the elements.

TABLE OF CHAPTER VARIABLES

a_o	Bohr atomic radius
A	constant in background modeling equation
B	Background integrated intensity
C	Number of channels integrated
D	Spectrometer dispersion (cm/eV)
e	Charge on the electron
E	Energy loss
E_k	Inner-shell onset energy
E_o	Incident electron energy
E_{PL}	Plasmon energy loss
F	X-ray detector efficiency
h	Planck's constant
I_o	Incident-beam intensity
I(E)	Intensity transmitted at E
J	Incident current density (amp/cm^2)
L	Mean free path
L_K	Inner-shell mean free path
L_{PL}	Plasmon mean free path
m_o	Rest mass of electron
M	Magnification
n	Atoms/unit volume
n_e	Free electrons/unit volume
N	Atoms/unit area
p	$a_o/Z^{1/3}$
q	Scattering vector, $4/\lambda \sin (\theta/2)$
r	Background slope exponent
R	Spectrometer bending radius
s	Post-edge exponent
S	Integrated edge signal
t	Sample thickness
T	Spectrum recording time

v	Electron velocity
W	Object area size
Z	Atomic number
β	Spectrometer acceptance angle
Δ	Energy window
ΔE	Energy resolution
θ	Scattering angle
θ_E	Characteristic scattering angle, $E/2E_o$
λ	Electron wavelength
σ	X-ray cross section
σ_{EL}	Elastic cross section
$\sigma(\beta, \Delta)$	Partial cross section for β, Δ
τ	Dwell time (seconds/channel)
ω	Fluorescent yield

REFERENCES

Ahn, C. C., and Krivanek, O. L. (1982), "A Catalog of Energy Loss Spectra," Arizona State Univ. HREM Facility, Tempe.

Andrew, J. W.; Ottensmeyer, F. P.; and Martell, E. (1978), Proc. 9th Intl. Cong. Elec. Mic. (J. M. Sturgess, ed.), Imperial Press, Toronto, *1*, 40.

Bentley, J.; Lehman, G. L.; and Sklad, P. S. (1981), in "Analytical Electron Microscopy 1981" (R. H. Geiss, ed.), San Francisco Press, San Francisco, p. 161.

Chapman, J. N.; Roberts, P. T. E.; MacLeod, A. M.; and Ferrier, R. P. (1980), Inst. of Physics Conf. Series *52*, 77.

Colliex, C.; Jeanguillaume, C.; and Trebbia, P. (1981), in "Microprobe Analysis of Biological Systems" (J. Hutchinson and A. Somlyo, ed.), Acad. Press, New York, p. 251.

Crewe, A. V.; Isaacson, M.; and Johnson, D. E. (1971), Rev. Sci. Instrum. *42*, 411.

Egerton, R. F. (1975), Phil. Mag. *31*, 199.

——— (1978), Ultramicroscopy *3*, 39.

——— (1980a), in "Scanning Electron Microscopy 1980" (O. Johari, ed.), SEM Inc., Chicago, *1*, 41.

——— (1980b), Proc. 39th Annl. EMSA Mtg. (G. W. Bailey, ed.), Claitor's Pub. Div., Baton Rouge, p. 130.

——— (1981a), in "Analytical Electron Microscopy" (R. H. Geiss, ed.), San Francisco Press, San Francisco, p. 221.

——— (1981b), Proc. 39th Annl. Mtg. EMSA (G. W. Bailey, ed.), Claitor's Pub. Div., Baton Rouge, p. 368.

——— (1982), in "Microbeam Analysis 1982" (K. F. J. Heinrich, ed.), San Francisco Press, San Francisco, p. 43.

Egerton, R. F.; Philips, J.; and Whelan, M. J. (1975), in "Developments in Electron Microscopy and Microanalysis" (J. Venables, ed.), Acad. Press, London, p. 137.

Egerton, R. F., and Whelan, M. J. (1974), J. Electron Spectros. *3*, 232.

Hainfeld, J., and Isaacson, M. (1978), Ultramicroscopy *3*, 87.

Higgs, A., and Krivanek, O. L. (1981), Proc. 39th EMSA Mtg. (G. W. Bailey, ed.), Claitor's Pub. Div., Baton Rouge.

Hillier, J., and Baker, R. F. (1944), J. Appl. Phys. *15*, 663.

Howie, A. (1981), Proc. 39th Annl. Mtg. (G. W. Bailey, ed.), Claitor's Pub. Div., Baton Rouge, p. 186.

Isaacson, M. (1972), J. Chem. Phys. *56*, 1813.

—— (1978), Proc. 11th Annl. SEM Symp. (O. Johari, ed.), SEM Inc., Chicago, *1*, 763.

Isaacson, M.; Ohtsuki, M.; and Utlaut, M. (1979), in "Introduction to Analytical Electron Microscopy" (J.Hren, ed.), Plenum Press, New York, p. 343.

Jenkins, D. G.; Rossouw, C. J.; Booker, G. R.; and Fry, P. W. (1980), Inst. of Physics Conf. Series *52*, 81.

Johnson, D. E. (1980), in "Scanning Electron Microscopy 1980" (O. Johari, ed.), SEM Inc., Chicago, *1*, 33.

Johnson, D. E.; Monson, K. L.; Csillag, S.; and Stern, E. A. (1981a), in "Analytical Electron Microscopy 1981" (R. H. Geiss, ed.), San Francisco Press, San Francisco, p. 205.

Johnson, D. E.; Csillag, S.; Monson, K. L.; and Stern, E. A. (1981b), Proc. 39th Annl. EMSA Mtg. (G. W. Bailey, ed.), Claitor's Pub. Div., Baton Rouge, p. 370.

Jones, B. L.; Jenkins, D. G.; and Booker, G. R. (1977), Inst. of Physics Conf. Series *36*, 73.

Jouffrey, B. (1978), in "Short Wavelength Microscopy," New York Acad. of Sci., New York, p. 29.

Joy, D. C. (1982a), in "Microbeam Analysis 1982" (K. F. J. Heinrich, ed.), San Francisco Press, San Francisco, p. 98.

—— (1982b), Ultramicroscopy *9*, 289.

—— (1983), Proc. NSF Workshop on Electron Interaction, SEM Inc., p. 251.

Joy, D. C.; Egerton, R. F.; and Maher, D. M. (1979), in "Scanning Electron Microscopy 1979" (O. Johari, ed.), SEM Inc., Chicago, *2*, 817.

—— (1978a), J. Micros. *114*, 117.

—— (1978b), Ultramicroscopy *3*, 69.

—— (1980a), in "Scanning Electron Microscopy 1980" (O. Johari, ed.), SEM Inc., Chicago, *1*, 25.

—— (1980b), Ultramicroscopy *5*, 333.

Leapman, R. D.; Grunes, P. L.; Fejes, P. L.; and Silcox, J. (1981), in "EXAFS Spectroscopy" (B. Teo and D. C. Joy, ed.), Plenum Press, New York, p. 217.

Leapman, R. D., and Swyt, C. R. (1981), in "Analytical Electron Microscopy 1981" (R. H. Geiss, ed.), San Francisco Press, San Francisco, p. 164.

—— (1983), in "Microbeam Analysis 1983" (R. Gooley, ed.), San Francisco Press, San Francisco, p. 163.

Maher, D. M.; Joy, D. C.; Egerton, R. F.; and Mochel, P. (1979), J. Appl. Phys. *50*, 5105.

Maher, D. M.; Mochel, P.; and Joy, D. C. (1978), Proc. 13th Annl. Conf. Microbeam Analysis Soc. (D. Kyser, ed.), San Francisco Press, San Francisco, p. 53A-G.

Shuman, H. (1981), Ultramicroscopy *6*, 163.

Spence, J. C. H. (1977), Proc. 35th Annl. Mtg. EMSA (G. W. Bailey, ed.), Claitor's Pub. Div., Baton Rouge, p. 234.

Trebbia, P.; Ballongue, P.; and Colliex, C. (1977), Proc. 35th Annl. Mtg. EMSA (G. W. Bailey, ed.), Claitor's Pub. Div., Baton Rouge, p. 232.

Wall, J. (1979), in "Introduction to Analytical Electron Microscopy" (J. Hren, J. Goldstein, and D. C. Joy, ed.), Plenum Press, New York, p. 333.

Williams, D. B., and Edington, J. W. (1976), J. Micros. *108*, 113.

Zaluzec, N. (1981), in "Analytical Electron Microscopy 1981" (R. H. Geiss, ed.), San Francisco Press, San Francisco, p. 193.

CHAPTER 8

QUANTITATIVE MICROANALYSIS USING EELS

D. C. Joy

Bell Laboratories
Murray Hill, New Jersey

I. INTRODUCTION

An important requirement of any microanalytical technique is that the data it produces be quantifiable. This condition is well met by electron energy-loss spectroscopy since both the low-loss and the inner-shell ionization regions of the loss spectrum can be processed to yield quantitative chemical microanalysis, and the fine structure detail of the core loss edges can be analyzed to provide precise crystallographic data on the sample. In this chapter, the principles of these analyses are described and illustrated by examples.

II. CHEMICAL MICROANALYSIS USING PLASMON LOSSES

It was shown in the previous chapter that the low-loss (below 50 eV) region of the transmitted electron spectrum of metals and semiconductors contained strong features, at characteristic energy losses, called plasmon peaks. In elements such as the light metals, Al or Mg, where the electrons in the conduction band can be considered as being free, the passage of a fast incident electron sets up an oscillation in the free electron gas whose frequency ω_P is proportional to the density of the free electrons.

$$\omega_P = (n\ e^2/\epsilon_o m)^{1/2} \tag{1}$$

where e and m are the electron mass and charge, and ϵ_o is the permittivity of free space. The energy loss corresponding to this oscillation is then $E_{PL} = h\ \omega_P$, where h is Planck's constant, and so should be proportional directly to $n^{1/2}$. In cases where the electrons are not "free," the electron mass must be replaced by an effective value, but the functional relationship

$$E_{PL} = k \cdot n^{1/2} \tag{2}$$

is still obeyed (Raether, 1965).

A simple measurement of a plasmon loss is not sufficient to identify an element unambiguously because all of the materials that give good plasmon peaks have values of E_{PL} lying within a narrow range, say, 15 to 25 eV. However, if element A with plasmon loss E_{PLA} is alloyed with element B, with plasmon loss E_{PLB}, so that the fractional content of B in A is C, we find the resultant alloy has a plasmon loss that, over some range of concentration, follows the relation

$$E_{PLAB} = (1 - C)\ E_{PLA} + C\ E_{PLB} \tag{3}$$

so that changing the concentration leads to a linear variation in the measured plasmon loss. In some systems, such as the Al ($E_{PL} = 15.3$ eV), Mg ($E_{PL} = 10.3$ eV) binary, this relationship is true over the full range of C from 0 to 100%. Thus, for the alloy $Al_{(1-c)}Mg_c$, the plasmon loss E_{PL} is given (Spaulding and Metherell, 1968) as

$$E_{PL} = (1 - C)\ 15.3 + 10.3\ C\ (eV)\ . \tag{4}$$

In other systems, this linearity may only be maintained for a more limited range of concentrations. In either situation, however, if a calibration curve is obtained for several varying compositions, the chemical composition of an unknown alloy of the same basic composition can be determined by simply measuring the plasmon loss and applying Eq. (3) to find C. A similar result is obtained for more complex alloys containing more than two elements, so that its concentration can be determined experimentally from a direct measurement of the plasmon loss, provided only one of the constituents is varied. The element that is being determined does not necessarily itself have to be a metal or semiconductor; for example, increasing the amount of oxygen in silicon shifts the plasmon loss from 17 eV to 22 eV (Joy and Maher, 1981), while the formation of hydrides in materials such as Zr and Ti causes plasmon shifts of ~2 or 3 eV (Zaluzec *et al.*, 1981).

Plasmon loss studies were the first quantitative application of electron energy-loss spectroscopy (Williams and Edington, 1976), and even now the technique has some definite advantages. First, the plasmon loss is the most prominent characteristic feature in the energy-loss spectrum, and it is therefore the one that offers the best chance of producing a good-quality result. Second, the measurement required to obtain the quantitation is straightforward and, at least in principle, requires little data processing. Third, the result is not dependent on the thickness of the sample. In fact, accuracy can improve for samples of approximately one or two inelastic mean free paths in thickness because the multiple plasmon peaks observed permit a more accurate determination of E_{PL}. As discussed in the previous chapter, practical spatial resolutions of $\sim$10 nm are possible although plasmon production is delocalized. Finally, the plasmon method can sometimes be applied to elements (such as H or Li) that are very difficult to study using other analytical techniques.

On the other hand, there are some compelling drawbacks to the technique. First, since the precision of the final result depends on the accuracy with which the plasmon loss can be determined, the technique is readily usable only on those materials, such as Al, Mg, and Si, that naturally give sharp, well-defined plasmon structures. Although it is possible that computer curve-fitting techniques and spectral deconvolution could extend the method's range of usefulness to a few other metals and semiconductors, the plasmon technique is basically limited in its scope. Second, only one elemental variable in the system can be studied at any one time, and a calibration series is necessary for each different experiment. Thus, the method finds its greatest use in such fields as the study of diffusion, where only one constituent need be considered as varying. Finally, the fact that the plasmon loss must be determined to a high degree of accuracy, typically 0.2 eV or better, requires that the stability of the spectrometer and its power supplies be high and that the level of external electromagnetic fields be kept low. The actual energy resolution required of the spectrometer is not very high, typically 1 eV, since the natural width of the plasmon is significant, but instabilities in the analyzer will make measurements irreproducible and thwart attempts to find the true plasmon peak by curve fitting.

III. CORE LOSS QUANTIFICATION

Although the core-loss signals are several orders of magnitude weaker than the plasmon losses discussed previously, they have the advantage that a chemical microanalysis using them can be performed on an unknown system even when it contains several elements. Neither are we restricted to just a few specialized materials. Thus, quantitative analysis relying on inner-shell edges is in much more widespread use than the plasmon technique. From the definitions of the previous chapter, it follows that the detected count rate, R (events/second), of edge-loss events of interest will be

$$R = I \cdot N \cdot \sigma \tag{5}$$

where

I = incident electron flux (electrons/s/m^2)
N = number of atoms contributing to the edge within the volume irradiated by the beam
σ = ionization cross section (m^2/atom).

This equation is perfectly general, but in this form it is of little practical value for several reasons. First, in the form given, we assume that the core-loss electrons can be collected for all possible scattering angles 0 to 2π and for all energy losses from onset E_k to beam energy E_o. Clearly, this is not possible since spectrometers can accept electrons over only a limited angular range, and an edge can be followed for only a limited energy range before it becomes indistinguishable from the background or runs into another edge. Let us assume that the signal is being stored and displayed into a multichannel analyzer set so that each channel represents an energy-loss increment of δ eV, and that the dwell time is t seconds/channel. If a total of m channels is measured after the edge, the edge integral, I_k, is the sum over m channels of the product of the count rate and the recording time

$$I_k = \Sigma_m R(E) \cdot t \tag{6}$$

where R is the appropriate count rate for energy loss E and the data come from an energy "window," $m\delta = \Delta$.

Putting this result into Eq. (5) gives

$$I_k (\beta,\Delta) = (I \cdot m \cdot t) \cdot N \cdot \sigma(\beta,\Delta) \tag{7}$$

since both sides must be treated alike. The quantities β and Δ indicate that this result is now true for a particular range of collected scattering angles and for a specified energy window. $\sigma(\beta,\Delta)$ is thus a partial cross section rather than the original total (or x-ray) ionization cross section. The quantity $(I \cdot m \cdot t)$ is the total number of electrons interacting with the sample. It is not, however, equal to the incident flux since some electrons are, for example, backscattered and so do not participate to produce the measured edge intensity. Following the suggestion of Egerton *et al.* (1976), it is usual instead to replace $(I \cdot m \cdot t)$ with the quantity $I_o(\beta,\Delta)$ which, in a manner exactly analogous to the edge integral, is the total recorded signal in m channels of the spectrum starting at the high (positive "energy loss") side of the zero-loss peak. This again will depend specifically on the chosen values of β and Δ. Thus, finally we get

$$N = \frac{I_k(\beta,\Delta)}{I_o(\beta,\Delta)\, \sigma(\beta,\Delta)} \tag{8}$$

where k defines the type of edge (K, L, etc.).

Although this expression is an approximation (first derived in this form by Egerton *et al.*, 1976), it has been found very satisfactory in use. In the form given, the quantification requires measuring two intensities from the spectrum (Figure 1) and knowing the cross section for the given experimental parameters. The result is then a determination of the number density of the atoms of interest in the irradiated volume. In general, an absolute measure of N is less useful than the relative number of atoms of each constituent present. If two or more elements A, B, etc., are present in the sample, the concentration of each will be proportional to the quantities.

$$N_A = I_{kA}(\beta,\Delta)/[\sigma_A(\beta,\Delta)]_k \tag{9}$$

where A represents the element and k the edge measured, provided that in every case the same experimental conditions β and Δ are used so that the zero-loss integral, $I_o(\beta,\Delta)$, is constant and disappears from the equation (Leapman *et al.*, 1978). When written in this form, the similarity to the k-factor method of quantifying thin-film

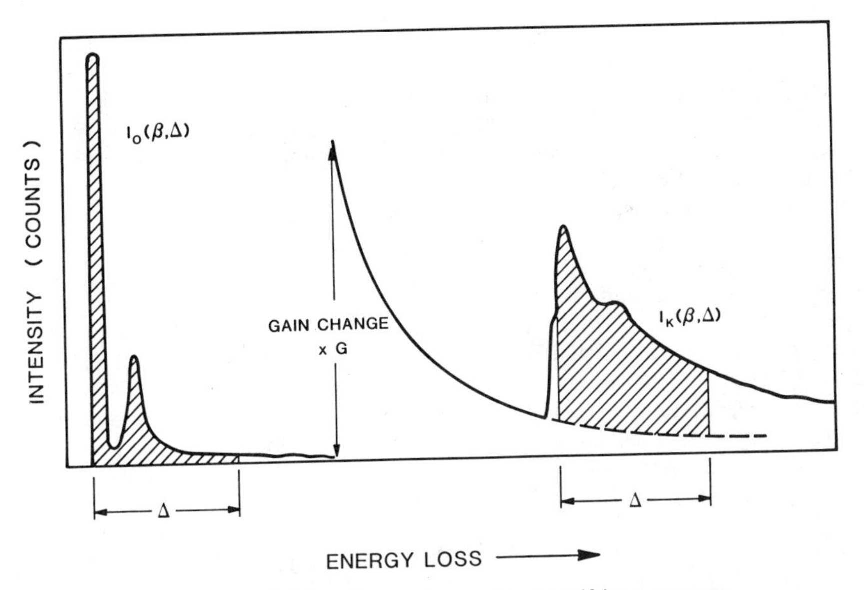

Figure 1 Definition of parameters used in quantifying a spectrum.

x-ray spectra is apparent. Whereas the x-ray k-factor included both physical parameters (cross section, fluorescent yield, partition factor, etc.) and instrumental quantities (detector efficiency), the energy-loss expression on the other hand involves only a single physical variable.

One further and significant difference between the EDS and EELS cases must be noted. In the x-ray case, all of the variables are determined by the experimental geometry and microscope operational conditions; in the EELS case, however, the energy window is chosen only at the time when the data are analyzed. Since any choice is valid, it follows that an important part of the data analysis is arbitrary; therefore, an important part of the data analysis in an energy-loss quantitation is verifying that the result quoted does not depend on a particular choice of energy window. Not only should the effect of varying the window be examined, but in any final result, the values used should be quoted.

IV. APPLYING THE QUANTITATION FORMULA

To apply either Eq. (8) or (9), it is necessary to determine the edge integrals (net edge signal above background), the zero-loss integral (if an absolute quantitation is required), and the relevant partial ionization cross sections.

A. The Edge Integral

It was shown in the previous chapter that the inner-shell edges were riding on large and rapidly changing backgrounds whose intensity I(E) as a function of energy loss E could be represented by the expression (Egerton, 1975; Maher *et al.*, 1979)

$$I(E) = A \cdot E^{-r} \tag{10}$$

where A and r are constants whose particular values depend on the material, its thickness, the beam energy, and the other experimental parameters. Since the edge is usually small in comparison to the background, the contribution to the edge from the background must be removed before the data can be quantified. This can be done in one of several ways. The most usual is to use a minicomputer to fit the expression of Eq. (10) to some portion of the background before the edge of interest and to use a least-squares (regression) analysis to find the best-fit values of A and r.

A program to accomplish this can readily be written in a few lines of either FORTRAN or BASIC code, and the time required for the fit to, say, 100 channels is typically only 1 or 2 seconds on a DEC LSI-11 or similar-size computer. However, this approach does require that the computer be able to read the spectrum on a channel-by-channel basis. When this is not possible, an alternative graphical method suggested by Egerton (1980) is useful. As illustrated in Figure 2, the region over which the fit is desired, starting at energy loss E1 and finishing at E2, is divided into two halves. The integrated intensities in these, I1 and I2, are found (usually possible even on the smallest multichannel analyzers by defining "windows" or "regions of interest" over these segments). Then

$$r = 2 \log (I1/I2)/[\log (E2/E1)] \tag{11}$$

and

$$A = [(I1 + I2)(1 - r)]/[E2^{(1-r)} - E1^{(1-r)}] \,. \tag{12}$$

The two different methods give very similar results. Thus, for the data plotted in Figure 2, where E1 = 180 eV, E2 = 250 eV, I1 = 723392, I2 = 475010, the least-squares method gives r = −2.56 and A = 1.42E10. Applying Eq. (11) and (12) gives r = −2.56 and A = 1.5E10. Either can therefore be used with confidence. It is probable, however, that the second approach is preferable when the spectrum is very noisy because it relies on integrated data rather than on the data in individual channels. Changing the fitting-window start and finish values by even one channel will change the values of both A and r, and this will lead to a statistical error in the value of the peak integral derived. The region over which the background fit is performed must therefore be chosen with care. Its width should be equal to, or greater than, the energy window over which it is proposed to determine the edge integral. However, if the fitting region is too wide, the assumption that parameters A and r are constant may not apply, and the fit will be poor.

As a practical guide, the fitting region should not exceed ~30% of E_k. Thus at carbon (E_k = 284 eV), the fitting region should be <~100 eV. In all cases, the quality of the fit should be visually judged by superimposing the computed background fit over the experimental spectrum rather than by any calculated "goodness of fit" criterion. The modeled background should match the intensity of the experimental background on a channel-by-channel basis (within the expected statistical accuracy) over the whole fitted range, and the extrapolated fit must not intersect the spectrum. Unless there are obviously interfering edges in this range, if a fit of this quality cannot be achieved, it is usually a sign that the spectrum is corrupted by one or more scattering artifacts (Joy and Maher, 1980).

In some circumstances, for example where edges are overlapped, an analytical background fit is not possible. One approach in this situation is to model the first (lower-loss) edge, either by direct calculation (Joy, 1982c) or by using library spectra so

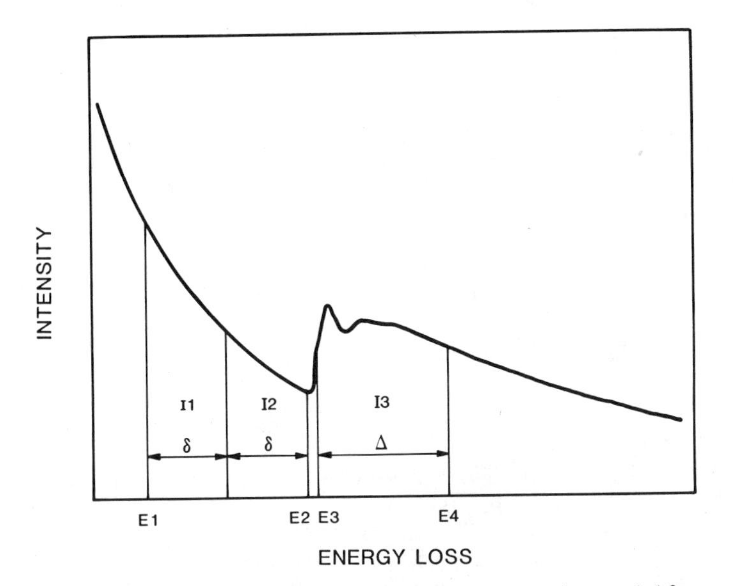

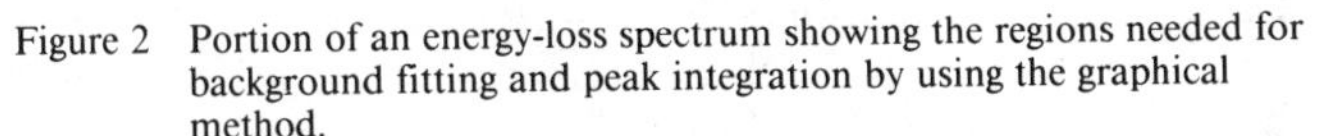
Figure 2 Portion of an energy-loss spectrum showing the regions needed for background fitting and peak integration by using the graphical method.

as to generate a suitable background for subtraction from the second edge. Alternatively, it is sometimes experimentally possible to find a region similar to the one of interest but not containing the element one wishes to quantify. This second region can then be used to provide a model background for the first. The probability of finding a suitable area for comparison depends very much on the nature of the specimen. On specimens such as chemically thinned metal foils, it is difficult to match backgrounds because the sample thickness changes considerably from point to point, but on a sample such as an extraction replica, this technique is very successful. One particular advantage of this approach is that it correctly accounts for any instrumental artifacts (such as scattering, or nonlinearity in the recording system) that might influence the background.

If the modeled, or experimentally determined background can itself be plotted into one of the multichannel analyzer memories, the desired edge integral can be found directly by stripping this from the original spectrum. Even where this is not possible, the integral can readily be calculated, as illustrated in Figure 2. If the gross integral in the edge window $\Delta = E4 - E3$ is I3, the required net integral I_k is

$$I_k = I3 - A\,(E4^{1-r} - E3^{1-r})/(1-r)\ . \tag{13}$$

Thus, again using the data of Figure 2 as an example with E3 = 284 eV, E4 = 354 eV, and I3 = 2349026, and using the values of A and r calculated above, I_k is given from Eq. (13) as 191782. This compares to the value of 1908110 computed by directly stripping the least-squares fitted background from the spectrum. Again, therefore, either computational method can be applied as convenient.

B. The Zero-Loss Integral

Although determining the zero-loss integral is straightforward, since no background has to be subtracted, some amount of caution needs to be exercised. First, spectrometers that are poorly aligned or have bad selecting slits may show a significant amount of scattering in the region of the zero-loss peak. This leads to a long tail of intensity on the high side of the loss peak, and as a result it is difficult to judge the point from which to start measuring the zero-loss window. Of course, while these spectral artifacts should be eliminated, when this scattering is visibly present on a spectrum that must be processed, the correct starting point can be approximated by finding the full width half maximum (FWHM) of the zero-loss peak and starting the integration from the channel that is the FWHM energy above the true zero-loss channel (Figure 3).

An additional problem can arise from the fact that a gain change is often performed between the low-loss and core-loss regions of the electron energy-loss spectrum. To obtain a true absolute quantitation, this gain change must be accurately determined. This can be done by recording a suitable feature, such as a strong edge, at different gain settings and comparing the integrals obtained. It must be noted, however, that because of the large difference in the strength of the zero-loss peak compared to the rest of the spectrum, gain factors obtained on weaker signals may not be correct if the recording system is close to saturation as it passes through the zero-loss peak. If this occurs, not only will the gain factor be wrong, but after the gain is changed, there may be a period (perhaps lasting many seconds) while the system recovers from the overload. This can cause a significant distortion of the spectrum (Joy and Maher, 1980). Such effects are best avoided by sweeping the spectrum from high- to low-energy loss and by not sampling the zero-loss peak at all unless an absolute quantitation is vital. When the zero-loss integral is recorded, the energy window chosen should be at least 50 eV since the approximation of Eq. (8) is valid only if the window includes the majority of the low-loss structure.

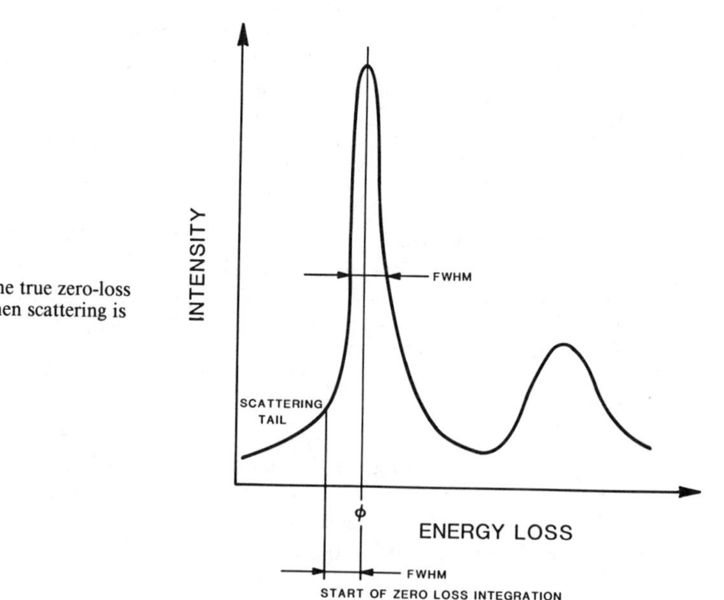

Figure 3 Defining the true zero-loss integral when scattering is present.

C. Calculating the Partial Cross Section

A variety of methods for calculating the required partial cross section, $\sigma(\beta,\Delta)$, from first principles have been described (e.g., Manson, 1972; Leapman *et al.*, 1980; Rez, 1982), and the published results of these studies have shown excellent agreement with experimental data. However, these calculations are complex and require substantial time, even on the largest computers. Consequently, results are available for only a few elements and for a limited range of experimental parameters. Therefore, until considerably more data are produced and made available in suitable tabular form, alternative means of finding the cross section must be resorted to.

The method now in most common use relies on the "hydrogenic" atomic model (Bethe, 1930; Inokuti, 1971), which treats an atom of atomic number Z as if it were a hydrogen atom possessing a nuclear charge of Z, with binding energy equal to that of the K- (or L-) shell of the real atom. Such a model is valuable because its generalized oscillator strength (GOS) can be expressed analytically, and the desired partial ionization cross section can then be found by a simple numerical integration. Two short computer programs, SIGMAK (Egerton, 1979) for K-shells of the elements Li to Si, and SIGMAL (Egerton, 1981) for L-shells of the elements Si to Fe, have been developed that allow the cross sections to be calculated for any set of experimental conditions in a few seconds on a typical mini- or microcomputer. These programs have found wide practical applications because of their simplicity and speed.

It is difficult to judge the absolute accuracy of these hydrogenic models because there are so few reliable theoretical, or experimental, figures with which to compare them. One simple test is to see how the shapes of edges predicted by SIGMAK and SIGMAL compare with experimental ones. Figure 4 compares an experimental carbon K edge (deconvoluted to remove multiple scattering) with the SIGMAK profile calculated for these experimental conditions. The level of agreement is seen to be good, with the exception of pre-ionization and extended fine structure effects, which

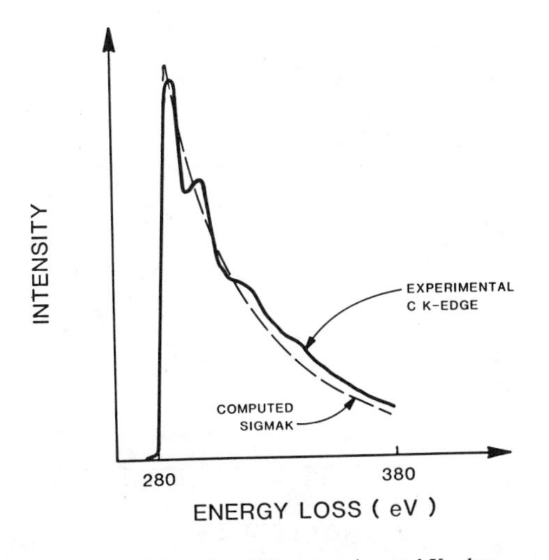

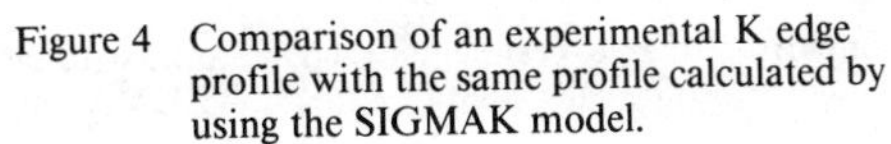
Figure 4 Comparison of an experimental K edge profile with the same profile calculated by using the SIGMAK model.

are, in any case, a minor perturbation of the profile. The same comparison for an L edge (Figure 5) is less convincing because the delayed maximum of this edge is not fully accounted for by the SIGMAL model, although once past this maximum, the agreement is again good. However, in both cases the predicted variation of cross section with, e.g., the integrals beneath the profiles, is well represented by the SIGMAK, SIGMAL data. Comparing x-ray cross sections and values estimated by extrapolation from SIGMAK and SIGMAL shows some substantial discrepancies (Egerton and Joy, 1977), but since the x-ray data are themselves of variable accuracy (Powell, 1976), this is not surprising.

In a few cases, SIGMAK, SIGMAL values can be compared directly with the "first-principles" calculations. Table I compares some selected values, obtained from Leapman *et al.* (1980) by graphical integration, with the corresponding hydrogenic values.

The overall level of agreement is variable but generally encouraging. The best estimates suggest that the accuracy is probably in the range ±20% when relative partial cross sections are compared, as in a quantitation. Much work remains to be done, however, on suitable standard samples to establish these limits properly and to refine the models appropriately.

TABLE I

Element	Edge	β (mrad)	Window (eV)	Leapman *et al.* ($\times 10^{-25}$ m^2/atom)	Hydrogenic	Parametric
C	K	10.0	100	6.5	7.2	6.1
N	K	3.0	100	0.68	0.9	0.86
O	K	3.0	100	0.36	0.37	0.35
Si	L_{23}	10.0	100	128.0	117.0	111.0
V	L_{23}	10.0	100	5.9	5.7	3.7

All data at 80 keV. Leapman *et al.* data from Leapman *et al.* (1980). Parametric data from data of Eq. (14 – 19).

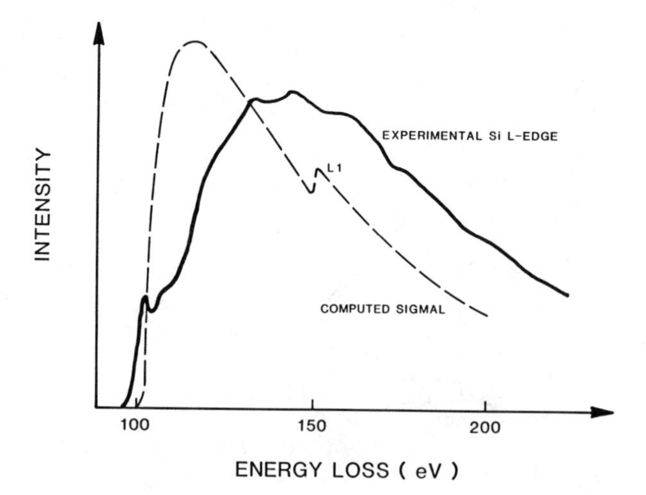

Figure 5 Comparison of an experimental L edge profile with the equivalent profile calculated by using the SIGMAL model.

Although SIGMAK, SIGMAL are readily evaluated on a small computer, there are occasions when no computer is available or when an analytical approximation to the cross section would be useful. This can be obtained by a parametric approximation to the SIGMA values and provides an expression that can be instantly calculated by using no more than a hand calculator. According to Issacson and Johnson (1975), Isaacson (1980), and Joy (1982b, 1984), the partial cross section $\sigma(\beta,\Delta)$ is written as the product

$$\sigma(\beta,\Delta) = \sigma \cdot f(\beta) \cdot g(\Delta) \tag{14}$$

where σ is a saturation cross section given as

$$\sigma = \frac{F \ln (0.055\, E_o/E_k)}{E_o \cdot E_k}\ \mathrm{m^2/atom} \tag{15}$$

and E_o and E_k are the incident-beam energy and the edge-onset energy in eV, and $F = 1.60\ 10^{-17}\ \mathrm{m^2/at.}$ for K edges and $4.51\ 10^{-17}\ \mathrm{m^2/at.}$ for L edges.

The second term, $f(\beta)$, has the form

$$f(\beta) = \frac{\log(1 + \beta^2/\theta^2)}{\log(2/\theta_E)} \tag{16}$$

where β is the spectrometer acceptance angle (in radians) and θ_E is the characteristic inelastic scattering angle for the average energy loss

$$\theta_E = \frac{E_k + \Delta/2}{2E_o} \tag{17}$$

where Δ is the desired energy window ($\Delta < E_k$) above the edge (in eV). The final term has the form

$$g(\Delta) = 1 - [E_k/(E_k + \Delta)]^{s-1} \tag{18}$$

where exponent s is given by the formula

$$s = H - 0.334 \log_e (\beta/\theta_E) \tag{19}$$

and $H = 5.31$ for K edges and 4.92 for L edges.

Equation (14) is easily evaluated on a pocket calculator yet gives results that are in good agreement with the full SIGMAK, SIGMAL computations. Figures 6 and 7 compare some selected K- and L-shell partial cross sections at 100 keV as derived from SIGMAK and SIGMAL, and from the parametric approximation, showing that the level of agreement is good for a wide range of experimental conditions. This agreement is also demonstrated by Table I, which compares the parametric model with the full hydrogenic calculation as well as with "first-principles" computations. The typical accuracy of 20% or better is certainly adequate for a quick analysis or for calculations of detectability limits, etc., where an approximate analytical figure is required.

D. Worked Example

A feel for the sorts of numbers involved in an energy-loss quantitation is best obtained by working through a real example. This is done below by using the graphical background-fitting method (since this can be done "off-line" without a computer) and the parametric cross section formulation so that the interested reader can check the numbers. As a check, however, the original data were also evaluated by using a least-squares background-fitting routine and the full SIGMAK, SIGMAL computations. The two results are in close agreement at all points. Figure 8 shows the silicon L and oxygen K edges recorded from an oxidized silicon foil at $E_0 = 100$ keV, $\beta = 3.10^{-3}$ rad.

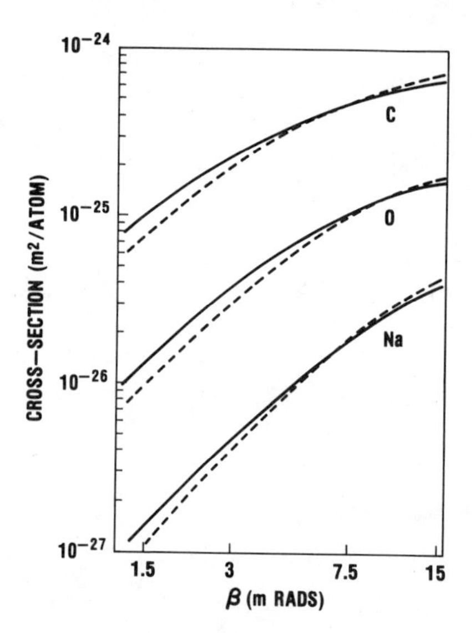

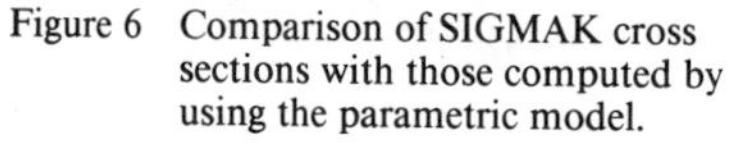

Figure 6 Comparison of SIGMAK cross sections with those computed by using the parametric model.

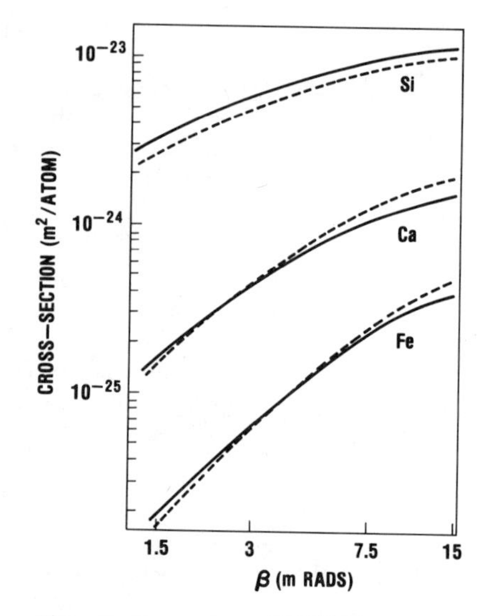

Figure 7 Comparison of SIGMAL cross sections with those computed by using the parametric model.

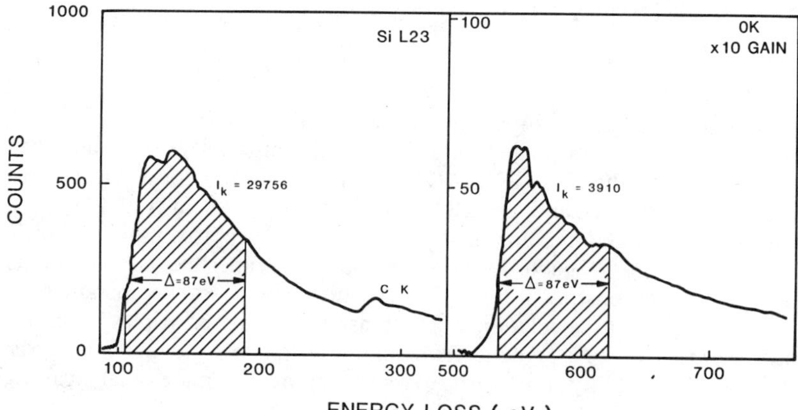

Figure 8 Portion of an energy-loss spectrum showing the silicon L_{23} and oxygen K edges. The oxygen edge was recorded at ten times the gain of the silicon edge.

The oxygen edge was recorded at a gain 10 times higher than the silicon edge. Using the notation of Figure 1, the integrated intensities and start/finish energies for the Si background fit were

$$I1 = 52183;\ I2 = 29149;\ E1 = 72\ \text{eV};\ E2 = 99\ \text{eV}$$

giving, from Eq. (11)

$$r = -3.65;\ A = 3.16\ 10^{10}\ .$$

Choosing a window $\Delta = 87$ eV wide above the edge gives

$$I3 = 74096;\ E3 = 103\ \text{eV};\ E4 = 190\ \text{eV}$$

which, applying Eq. (13) and the A,r values found above, gives the desired edge integral I_k as

$$I_k = 29756\ .$$

Similarly, for the oxygen edge

$$I1 = 5963;\ I2 = 4903;\ E1 = 430\ \text{eV};\ E2 = 508\ \text{eV}$$

giving

$$r = -2.35;\ A = 2.61\ 10^{8}$$

for the background fit. Choosing again an 87-eV-wide window

$$I3 = 11380;\ E3 = 532\ \text{eV};\ E4 = 619\ \text{eV}$$

and using the r,A values found above

$$I_k = 3910\ .$$

Since a gain change of 10 times was involved, the true value of I_k is thus 391.

Finally, the partial cross sections can be evaluated by using the parametric cross section:

For the silicon L edge, $E_k = 99$ eV; $\Delta = 87$ eV; $\beta = 3\ 10^{-3}$ rad;

$$\sigma(\beta,\Delta) = 6.02\ 10^{-24}\ \text{m}^2/\text{atom}\ .$$

Similarly, for the oxygen K edge, $E_k = 532$ eV; $\Delta = 87$ eV; $\beta = 3\ 10^{-3}$ rad;

$$\sigma(\beta,\Delta) = 3.18\ 10^{-26}\ \text{m}^2/\text{atom}\ .$$

Thus, by applying Eq. (9), the concentrations are proportional to the ratio of the quantities $I_k/\sigma(\beta,\Delta)$ thus

$$\text{for oxygen} = 391/381\ 100^{-26}$$

$$\text{for silicon} = 29756/6.02\ 10^{-24}$$

giving as a result

$$\text{O/Si} = 2.07$$

indicating that the film was probably SiO_2. Unlike the situation in EDS analysis, it is evident that this result could not have been guessed from the raw data since, for example, the oxygen edge is only ~1.5% of the intensity of the silicon edge even though there are actually twice as many oxygen as silicon atoms.

V. FACTORS AFFECTING THE ACCURACY OF QUANTITATION

A. Statistics

The accuracy of EELS microanalysis, just as any other analytical technique, will depend on the statistical quality of the data. This is generally expressed by the signal-to-noise ratio

$$S/N = I_k/(I_k + I_b)^{1/2} \tag{20}$$

where I_k is the net integral under the edge and I_b is the corresponding integral under the extrapolated background. The higher the signal-to-noise ratio, the greater the accuracy of I_k. The practical problem is that I_b cannot be measured directly, but only as the result of fitting and extrapolating the background. Egerton (1982) has shown that when the errors associated with these procedures are taken into account, Eq. (20) takes the form

$$S/N = I_k/(I_k + h \cdot I_b)^{1/2} \tag{21}$$

where h is a parameter, greater than unity, whose value depends on such factors as width Γ of the fitting region and energy window Δ beneath the edge. For the best accuracy, h must be as small as possible, and this can be achieved by increasing Γ. However, too large a value of Γ may cause the fitting region to run into the tail of another ionization edge or lead to a systematic error because exponent r is not constant across the fitted region. Γ must therefore be chosen with care to compromise between the systematic and random errors.

If Γ is fixed, as Δ is varied, the signal/noise ratio first increases, then peaks before falling away. The optimum value of Δ is found to be in the range 50 to 100 eV for most conditions. However, the higher, rather than the lower, of these limits should be used where possible because plural scattering and fine structure effects are significant for low values of Δ (typically <20 eV). With $\Gamma \sim 0.3\ E_k$ and $\Delta = 100$ eV, h is typically 4 or 5. Even under the best conditions, such as an edge with a jump ratio of 5 or more, this sets error limits of $\pm 5\%$, while for less prominent edges, the true accuracy may be close to 20%.

B. Electron-Optical Factors

The electron-optical factors are the parameters describing the energy, and entry and exit geometry of the electron beam. In addition to these physical parameters, however, the orientation of the specimen to the beam is important when the sample is crystalline. The partial ionization cross section is modified as a function of the diffraction conditions experienced by the specimen at each point. While this effect can be exploited as a probe of the local environment around the ionized atom (Lehmpfuhl and Tafto, 1980; Krivanek and Tafto, 1982), it is generally necessary to orient a sample so that no strong beams are excited if accurate quantitation is required

(Zaluzec *et al.*, 1980). Even if a condition cannot be found where all diffraction beams are extinguished, the unwanted effects of diffraction can be minimized by using a high value of incident-beam convergence (Maher and Joy, 1976).

Although β is kept constant during any measurement, its value must be accurately determined to ensure acceptable accuracy since cross section $\sigma(\beta,\Delta)$ depends directly on β. The error in the cross section is a function of (β/θ_E), where θ_E is the characteristic inelastic scattering angle $(E_k/2E_o)$. When θ_E is small, the effect of errors in determining β will be small. For example, for the Si L edge with $E_k = 99$ eV, $E_o = 100$ keV, $\theta_E = 0.5$ mrad, then a $\pm 10\%$ error in a nominal value of β of 3 mrad would give about a $\pm 6\%$ error in $\sigma(\beta,\Delta)$, or a $\pm 5\%$ error for a nominal 15 mrad. But at Si_k ($E_k = 1832$ eV, $\theta_E = 9$ mrad), the same $\pm 10\%$ error in the 3-mrad value of β would give a $\pm 18\%$ error in $\sigma(\beta,\Delta)$ or a $\pm 8\%$ error at $\beta = 15$ mrad. Experimentally, β can be measured to an accuracy of 1% or 2% (Maher and Joy, 1976); therefore, the best accuracy will be obtained when β is as large as possible, and under all circumstances β should be kept greater than θ_E.

Incident-beam divergence α does not explicitly appear in the quantitation expression because the assumption has been made that α is insignificant, i.e., the beam is collimated. While this is a sufficiently good approximation for most conditions in the conventional transmission microscope, in the STEM, α can be comparable with β. In this case, the inelastic scattering distribution is convoluted with the incident-beam divergence to give a distribution that is broader than expected; thus, the signal collected will be smaller than predicted. The apparent collection aperture, b, must therefore be replaced by an effective value, β^*. This value can be calculated by numerical integration, and Figure 9 shows how β^* varies for some typical conditions.

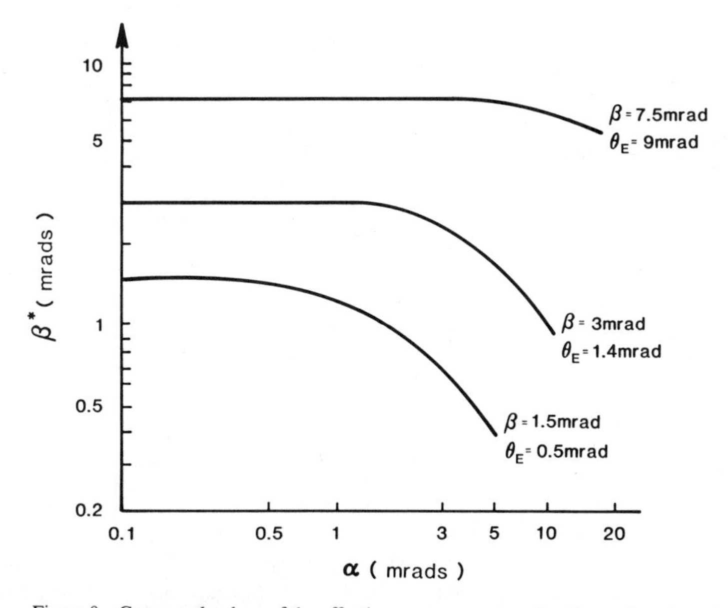

Figure 9 Computed values of the effective spectrometer angle, β^*, as a function of the geometrical value for different experimental conditions.

When inelastic scattering angle θ_E or collection angle β are small, the correction is significant. In many instances, such as the high-resolution STEM $\alpha > \beta$, under this regime it can be shown (Joy, 1982b) that the cross section for collection into the spectrometer is reduced by a factor, R, where

$$R = \frac{\ln(1 + \alpha^2/\theta^2) \cdot \beta^2}{\ln(1 + \beta^2/\theta^2) \cdot \alpha^2} \tag{22}$$

i.e., for $\alpha = 20$ mrad, $\beta = 10$ mrad, R is 0.33, so that only one-third of the expected signal is collected. Although this is an extreme example, significant corrections may still be required even when α and β are comparable. Typically, errors of a few percent in α are reflected by errors of a few percent in the correction.

C. Instrumental Factors

Problems caused by scattering in the spectrometer on the quantification result have already been discussed. However, other instrumental effects can also be important. The quality of the result depends on the assumption that the stored data correctly represent the intensity variation of the original spectrum. Because of the very wide dynamic range in the spectrum, this may not always be the case. If the spectrum is recorded through an analogue conversion as described in the previous chapter, one or more gain changes are usually required to accommodate the data. Each gain change must be measured, ideally from the spectrum itself, so that relative intensities of edges are correctly represented. Since the gain change may be affected by the average signal level, the gain should be calculated by comparing background slopes just before and just after the change has occurred.

In both electron counting and analogue systems, the spectrum eventually enters the multichannel analyzer as a pulse train that must be counted to give the stored spectrum. Although most modern analyzers can recognize pulses at very high rates (50 to 100 MHz), they cannot count them into the memory at this rate. Experimentally, the "dead time" of many units is $\sim$0.5 μs or more (Maher and Joy, unpublished). Since no current units have dead time corrections in the sequential mode, this can produce significant departures from linearity at count rates in excess of a few hundred kilohertz.

Taking into account all the sources of error discussed above, and the probable limits on the accuracy of the cross section models discussed earlier, it seems likely that the accuracy of an EELS quantitation at the present time is $\sim$20% under carefully controlled conditions. Significant improvements in this figure will require the development of well-characterized standard samples as well as advances in instrumentation and operating techniques.

VI. THE EFFECT OF SAMPLE THICKNESS

The theory of quantitative electron energy-loss spectroscopy discussed above is analogous to the "thin-film" approximation for energy-dispersive x-ray analysis since it assumes that each electron is scattered only once as it passes through the sample. This would require, at 100 keV, specimens that are only 30 to 50 nm thick, whereas most samples of practical interest are significantly thicker than this. The thickness of the specimen region being analyzed can conveniently be monitored by the plasmon ratio, I(1)/I(0), where I(0) is the height of the zero-loss peak and I(1) is the height of the first plasmon peak. In a homogeneous material, if the result of an elemental analysis is plotted as a function of this ratio, a result similar to that in Figure 10 is usually obtained (Zaluzec, 1980).

Up to a plasmon ratio of approximately unity, the result is more or less constant because, for moderate thicknesses, the effect of multiple scattering is to redistribute the energy in a direction away from the edge. Provided that the energy window is reasonably large, the value of the integral obtained does not change much. However, edges at differing energies are affected by plural scattering at different rates, with edges at higher energy being less affected than those at lower energy losses. Microanalytical determinations involving edges of widely separated energies are therefore likely to show more variation with thickness than those where the edges are close together. At higher thicknesses, this causes a progressive change in the elemental ratio until eventually the result becomes severely erroneous because the increasing plural scattering raises the background and worsens the statistics.

In principle, this can be overcome by deconvoluting the spectrum to remove the effects of plural scattering. To a good approximation, the profile of an edge in a spectrum that has suffered multiple scattering can be considered (Figure 11) as being the convolution of the ideal single-scattering profile with the low-loss portion of the spectrum (Egerton and Whelan, 1974; Ray, 1980), i.e., as if the ideal profile is distorted by an instrument response whose form is that of the low-loss function. Thus

$$\text{I (experimental)} = \text{I (single scattering)} * \text{I (low loss)} \tag{23}$$

where $*$ represents the convolution operation.

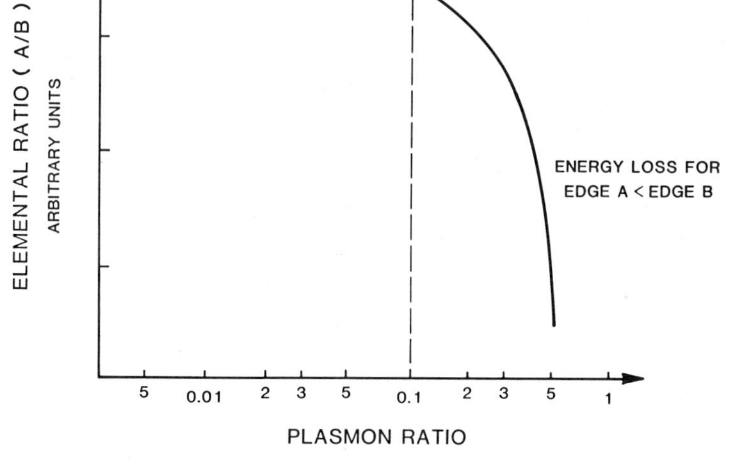

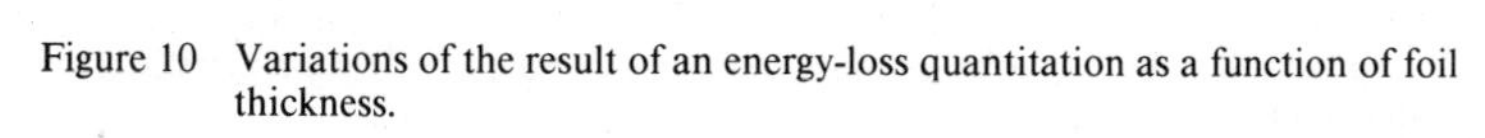

Figure 10 Variations of the result of an energy-loss quantitation as a function of foil thickness.

If the data are Fourier transformed, Eq. (23) becomes

$$\tilde{I}\ (\text{experimental}) = \tilde{I}\ (\text{single scattering}) \cdot \tilde{I}\ (\text{low loss}) \tag{24}$$

where ~ represents the transformed variable. Since the convolution is now a simple product, the expression can be written to give the desired single-scattering profile directly

$$\tilde{I}\ (\text{single scattering}) = \tilde{I}\ (\text{experimental})/\tilde{I}\ (\text{low loss})\ . \tag{25}$$

The transform can now be retransformed to give the actual profile. The transforms can be obtained very rapidly by standard computer algorithms, and the entire process can usually be accomplished in a few seconds. In practice, it is usually necessary to smooth the retransformed data since deconvolution accentuates noise on the spectrum. This can be achieved by multiplying Eq. (25) by an additional term representing the Fourier transform of a symmetrical Gaussian curve equal in width to the resolution of the spectrometer (Joy, 1982a). Alternative strategies have been described by Leapman and Swyt (1981). Figure 12 compares the appearance of an edge before and after deconvolution. It can be seen that the deconvolution has removed the plasmon replica behind the edge and produced a profile that has the expected concave form compared with the convex profile in the untreated spectrum. The jump ratio of the edge and the visibility of the fine structure are also enhanced by deconvolution.

Once processed in this way, the edge can be quantified in the usual manner, except that Eq. (8) now has the form

$$N = I_k(\beta,\Delta)/I_o \cdot \sigma(\beta,\Delta) \tag{26}$$

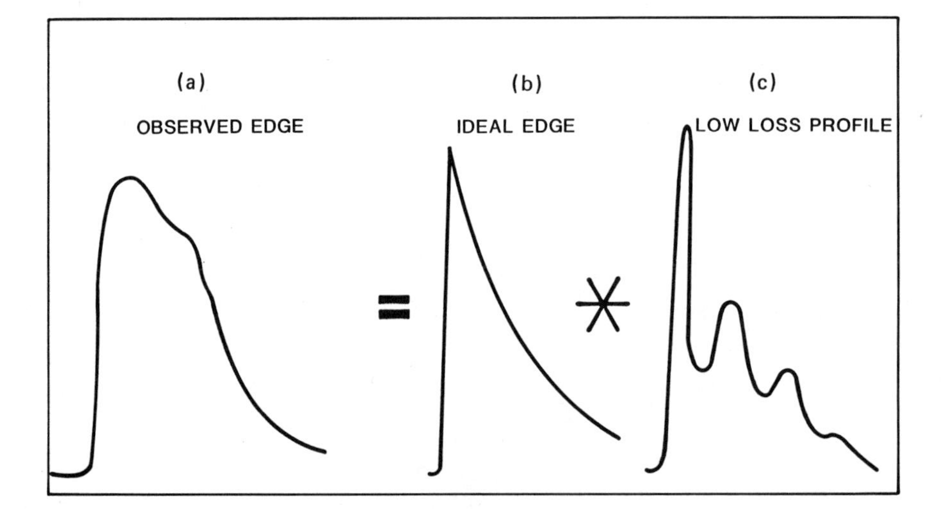

Figure 11 Schematic illustration of the effect of plural scattering. The appearance of the edge in (a) a thick sample is the convolution of (b) the true edge shape and (c) the measured low-loss spectrum.

where I_0 is the incident-beam current. The zero-loss integral, $I_0(\beta,\Delta)$, has been removed because after deconvolution the zero-loss spectrum has become a delta function. In this form, the energy window need not be the same for each edge to be quantified, which allows an optimum choice of window for each edge of interest.

Regrettably, the experimental evidence is that the accuracy of the quantitation result in thick samples is not substantially improved by deconvolution (e.g., Zaluzec, 1983). There are several possible explanations for this. First, deconvolution to be exact requires that all scattering angles be collected. Since this is clearly not possible, the collection angle used needs to be significantly higher than usual (e.g., tens of milliradians) to ensure acceptable accuracy. Second, deconvolution always increases the statistical noise of the spectrum, even after smoothing, so the precision of the integrals is worsened. Finally, the mathematical processing required for the deconvolution can lead to truncation errors unless a large number of significant figures are retained during the calculations.

Despite the marginal benefits to quantitation, deconvolution is an important technique in electron energy-loss spectroscopy. Its most substantial advantage is that it produces a spectrum containing edges whose features are independent of the sample thickness. This makes the identification of these edges much more reliable because libraries of edges, obtained experimentally or by calculation, can be used for comparison. This is of special benefit when overlapping edges are to be quantified since library edges can be used to give a background model in the overlap region (Joy, 1982c).

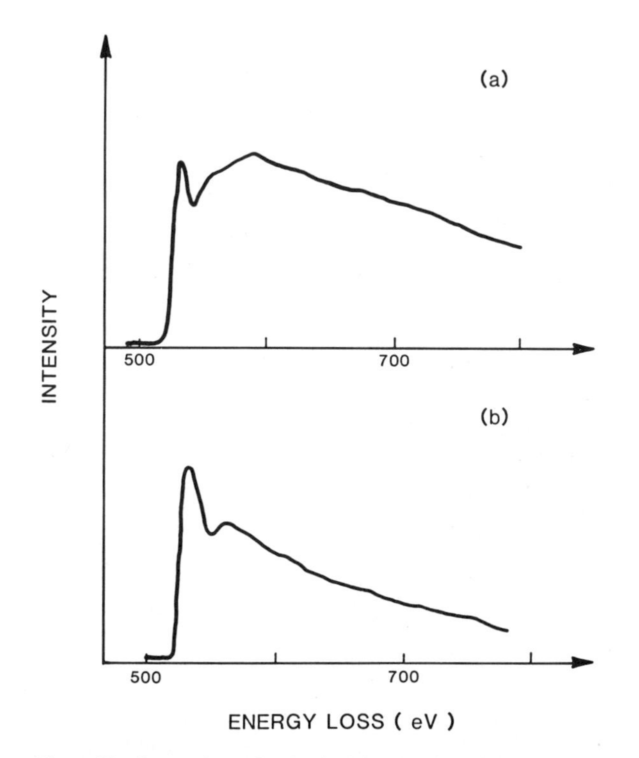

Figure 12 Comparison of an ionization edge (a) before and (b) after deconvolution to remove plural scattering.

VII. EXTENDED FINE STRUCTURE STUDIES

Superimposed on the smooth decay of an ionization edge is a weak oscillatory modulation analogous to the structure seen on the x-ray absorption spectra when a sample is placed in an intense x-ray beam such as that from a synchrotron. That effect is usually termed EXtended Absorption Fine Structure (EXAFS); therefore, the equivalent electron energy-loss effect is often called EXtended Energy-Loss Fine Structure (EXELFS). Unlike the effects discussed earlier that treat the atoms in our sample as if they were free, this fine structure is a consequence of the fact that each atom in the specimen is surrounded by other atoms.

Consider the situation shown in Figure 13, where the atom shown as hatched has been ionized. If the atom's ionization energy is E_k, but the actual energy transferred during the inelastic event was E, there is a surplus energy $(E - E_k)$, and this is radiated as an electromagnetic wave of wavelength

$$\lambda = h/[2m(E - E_k)]^{1/2} . \tag{27}$$

This propagates away from the ionized atom until it meets the ring of nearest-neighbor surrounding atoms. Some fraction of this wave is backscattered from these atoms and travels back toward the central atom.

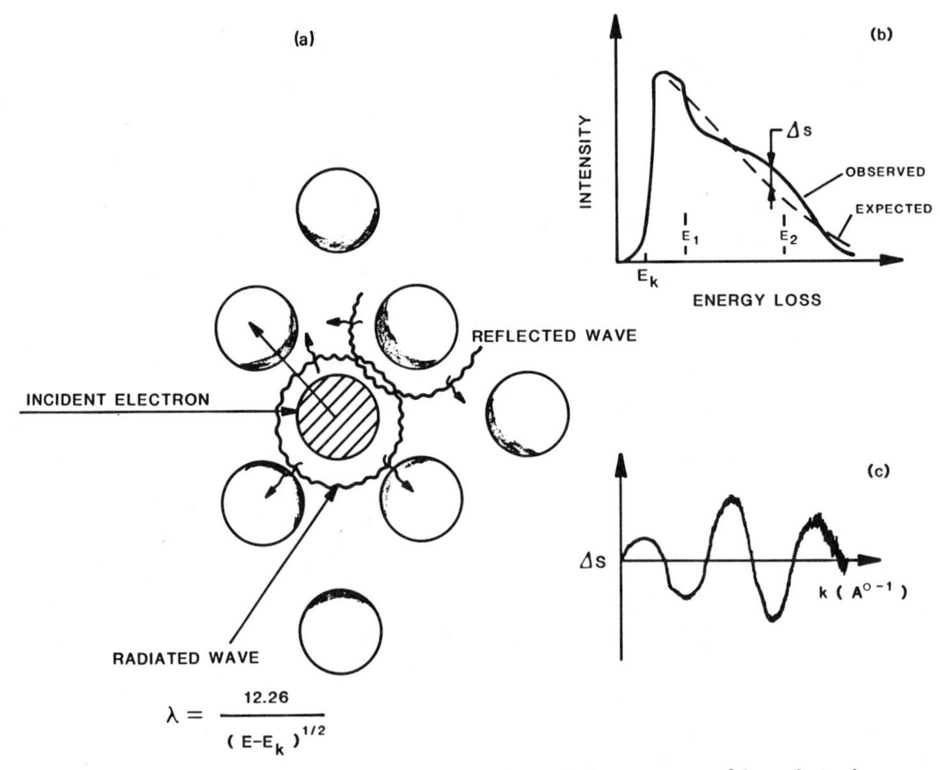

Figure 13 (a) The origin of extended energy-loss fine structure; (b) the appearance of the peaks on the spectrum; (c) the corresponding structure in k-space.

Because the ionization event occupies a time comparable with the time it takes the wave to go out and back, the returning wave can interfere either constructively or destructively with the outgoing wave. This produces the weak modulation (typically 5% or less) in the energy differential ionization cross section. The interference will be constructive when the returning wave is in phase with the outgoing one, thus

$$(2a/\lambda) \cdot 2\pi + \Phi = 2\pi \cdot n \tag{28}$$

where a is the distance to the nearest-neighbor atoms, Φ is the phase shift that occurs on reflection, and n is an integer. This equation shows that the peaks in the modulation are not periodic in energy but in "k-space," where k is the wave number $(1/\lambda)$. If the position of the peaks can be determined, the interatomic spacing can be found from this equation as was done in a pioneering paper by Leapman and Cosslett (1976). If peaks 1,2,3,...j occur at energies E_1, E_2, E_j, a plot of j against $(E_j - E_k)^{1/2}$ gives a straight line with a slope proportional to a, and an intercept equal to phase shift.

In investigations by Leapman *et al.* (1981) and Csillag *et al.* (1981), the modulations are stripped from the edge by using a fitted polynomial to model the profile. The modulations are then replotted in k-space coordinates. Each spacing sampled in the specimen will give a periodic wave in these data, so a Fourier transform of the k-space data will give a spectrum, each point of which represents one of the spacings. Full details can be found in the papers cited above.

EXELFS are important because they offer a way of extracting accurate crystallographic data from samples, such as amorphous materials, which do not give good diffraction data. Under optimum conditions, the precision in finding an interatomic spacing can be very high (~0.1 Å), but practical considerations usually worsen this figure. If the maximum and minimum k-values over which data can be obtained are k_{max} and k_{min}, the spatial accuracy is limited to a figure, R

$$R = \pi/[2(k_{max} - k_{min})] \; . \tag{29}$$

In the usable region of the energy-loss spectrum, edges are spaced by only 100 or 200 eV; therefore, data can usually be obtained over only a limited range. The practical limit is usually 0.2 to 0.3 Å. A second major problem is in obtaining an adequate signal-to-noise ratio since the modulation is weak. Generating data with adequate statistics requires high beam intensities, long exposures, or both; this can produce contamination or radiation damage in the sample. Although these problems have so far limited the quality of EXELFS data produced in the electron microscope, the technique offers considerable promise.

TABLE OF CHAPTER VARIABLES

C	Concentration of A in B
E_k	Onset energy loss for core edge
E_o	Incident-beam energy
E_{PL}	Plasmon energy loss
h	Background statistical weighting factor
I_b	Background integral
I_k	Edge integral
I_o	Incident electron flux

I(E)	Intensity at energy loss E
m	Electron mass
N	Number of atoms in analyzed volume
r	Background slope exponent
s	Post-edge slope exponent
α	Incident-beam convergence angle (rads)
β	Spectrometer acceptance angle (rads)
Γ	Width of fitting region (eV)
δ	Energy increment (eV/channel)
Δ	Energy window (eV)
ϵ_o	Permittivity of free space
θ_E	Inelastic scattering angle
σ	Ionization cross section
ω_P	Plasmon oscillation frequency (rads/s)

REFERENCES

Bethe, H. (1930), Annl. Phys. *5*, 325.

Csillag, S.; Johnson, D. E.; and Stern, E. A. (1981), in "EXAFS Spectroscopy" (B. K. Teo and D. C. Joy, ed.), Plenum Press, New York, p. 241.

Egerton, R. F. (1975), Phil. Mag. *31*, 199.

——— (1979), Ultramicroscopy *4*, 69.

——— (1980), Proc. 38th Annl. Mtg. EMSA (G. W. Bailey, ed.), Claitor's Pub. Div., Baton Rouge, p. 130.

——— (1981), Proc. 39th Annl. Mtg. EMSA (G. W. Bailey, ed.), Claitor's Pub. Div., Baton Rouge, p. 198.

——— (1982), Ultramicroscopy *9*, 387.

Egerton, R. F., and Joy, D. C. (1977), Proc. 35th Annl. Mtg. EMSA (G. W. Bailey, ed.), Claitor's Pub. Div., Baton Rouge, p. 252.

Egerton, R. F., and Whelan, M. J. (1974), Phil. Mag. *30*, 739.

Egerton, R. F.; Rossouw, C. J.; and Whelan, J. J. (1976), in "Developments in Electron Microscopy and Microanalysis," Acad. Press, London, p. 129.

Inokuti, M. (1971), Rev. Mod. Phys. *43*, 297.

Isaacson, M. (1980), Proc. 38th Annl. Mtg. EMSA (G. W. Bailey, ed.), Claitor's Pub. Div., Baton Rouge, p. 110.

Isaacson, M., and Johnson, D. E. (1975), Ultramicroscopy *1*, 33.

Joy, D. C. (1982a), Ultramicroscopy *9*, 289.

——— (1982b), in "Scanning Electron Microscopy 1982" (O. Johari, ed.), SEM Inc., Chicago, *2*, 505.

——— (1982c), in "Microanalysis 1982" (K. J. Heinrich, ed.), San Francisco Press, San Francisco, p. 98.

——— (1984), J. Micros. *134*, 89.

Joy, D. C., and Maher, D. M. (1980), in "Scanning Electron Microscopy 1980" (O. Johari, ed.), SEM Inc., Chicago, *1*, 25.

——— (1981), in "Microscopy of Semiconductors" (A. G. Cullis, ed.), Inst. of Physics (London) Conf. Series 60, p. 229.

Krivanek, O. L., and Tafto, J. (1982), Proc. 40th Annl. Mtg. EMSA (G. W. Bailey, ed.), Claitor's Pub. Div., Baton Rouge, p. 492.

Leapman, R. D., and Cosslett, V. E. (1976), J. Phys. D (Appl. Phys.), L29.

Leapman, R. D., and Swyt, C. R. (1981), in "Analytical Electron Microscopy 1981" (R. H. Geiss, ed.), San Francisco Press, San Francisco, p. 164.

Leapman, R. D.; Grunes, L. A.; Feijes, P. L.; and Silcox, J. (1981), in "EXAFS Spectroscopy" (B. K. Teo and D. C. Joy, ed.), Plenum Press, New York, p. 217.

Leapman, R. D.; Rez, P.; and Mayers, D. F. (1980), J. Chem. Phys. *72*, 1232.

Leapman, R. D.; Sanderson, S. J.; and Whelan, M. J. (1978), Metal Sci. *9*, 215.

Lehmpfuhl, G., and Tafto, J. (1980), in "Electron Microscopy 1980," Proc. 7th Euro. Cong. Elec. Mic., The Hague, *3*, 62.

Maher, D. M., and Joy, D. C. (1976), Ultramicroscopy *1*, 239.

Maher, D. M.; Joy, D. C.; Egerton, R. F.; and Mochel, P. (1979), J. Appl. Phys. *50*, 5105.

Manson, S. T. (1972), Phys. Rev. *A6*, 1013.

Powell, C. J. (1976), Rev. Mod. Phys. *48*, 33.

Raether, H. (1965), Springer Tracts on Modern Physics *38*, 84.

Ray, A. B. (1980), Proc. 38th Annl. Mtg. EMSA (G. W. Bailey, ed.), Claitor's Pub. Div., Baton Rouge, p. 522.

Rez, P. (1982), Ultramicroscopy *9*, 283.

Spaulding, D. R., and Metherell, A. J. F. (1968), Phil. Mag. *18*, 41.

Williams, D. B., and Edington, J. W. (1976), J. Micros. *108*, 113.

Zaluzec, N. J. (1980), Proc. 38th Annl. Mtg. EMSA (G. W. Bailey, ed.), Claitor's Pub. Div., Baton Rouge, p. 112.

——— (1983), Proc. 41st Annl. Mtg. EMSA (G. W. Bailey, ed.), San Francisco Press, San Francisco, p. 388.

Zaluzec, N. J.; Hren, J. J.; and Carpenter, R. W. (1980), Proc. 38th Annl. Mtg. EMSA (G. W. Bailey, ed.), Claitor's Pub. Div., Baton Rouge, p. 114.

Zaluzec, N. J.; Schober, T.; and Westlake, D. G. (1981), Proc. 39th Annl. Mtg. EMSA (G. W. Bailey, ed.), Claitor's Pub. Div., Baton Rouge, p. 194.

CHAPTER 9

ELECTRON MICRODIFFRACTION

J. C. H. Spence and R. W. Carpenter

Center for Solid State Science
Arizona State University, Tempe, Arizona

I. INTRODUCTION

The number and popularity of microdiffraction methods available for the study of matter has increased dramatically during the last decade, since analytical electron microscopes with convergent beam capability, improved vacuum systems and therefore reduced contamination, and a wide variety of beam scanning, rocking, and intensity recording schemes became generally available. Historically, the continuing motivation for the development of microdiffraction techniques has been the growing realization among materials scientists and crystallographers that real materials consist of a great variety of microphases and lattice defects, and that they exhibit various kinds of disorder. These materials can be more fruitfully studied by microbeam methods than by conventional "broad beam" methods that provide structural information statistically averaged over large volumes. Detected count rates from subnanometer-area microdiffraction patterns rival those from synchrotron studies, with the important advantage that a high-resolution electron (real space) image can be obtained from the same area.

These new experimental methods have extended the spatial resolution of electron microdiffraction downward from several micrometers to <1 nm. They are being applied to a steadily increasing range of problems in materials science and solid-state physics, such as the analysis of defects and strain gradients in crystals, point and space group determination, analysis of long- and short-range chemical order, structure of "amorphous" solids, and others. A dramatic example of subnanometer diffraction is shown in Figure 5, demonstrating patterns obtained using a probe smaller than the specimen unit cell. The relationship between "Bragg scattering" and probes smaller than a unit cell is described in Section V.

Probe size is probably the most familiar experimental viewpoint from which to consider microdiffraction methods, but it is not sufficient. Other important considerations are: the incident beam divergence (or convergence), whether the beam is static or rocking, and incident probe coherence. Large divergence is not always synonymous with small probe size. Microdiffraction probes are most often fixed, but may "rock" about a "point" on the specimen, rather than scan an area. Any translational movement of the probe has the same effect as increasing the probe size, producing a diffraction pattern averaged over the specimen volume irradiated during the recording or observation time. Rocking and fixed-beam patterns are related by the reciprocity theorem applied to diffraction, in a manner analogous to TEM and STEM images.

Whether the incident probe is coherent or incoherent has important effects on interpretation of patterns. The diffracted intensity distribution produced by a coherent probe is obtained by an amplitude summation of diffracted beams over the incident radiation cone angle, taking account of the phase changes over irradiation angle caused by the probe-forming lens system aberrations and focus, with the intensity given by the modulus of the amplitude sum. The diffraction intensity distribution produced by an incoherent incident probe is given directly by the intensity sum over incident angle of beams within the irradiation cone, considering only phase changes induced by the specimen. The former is analogous to phase contrast imaging and the latter to amplitude or "diffraction" contrast. Highly coherent convergent probes can only be formed using field emission guns, and coherent microdiffraction theory is an integral part of high-resolution STEM imaging theory. Since instruments with coherent probe capability and useful experimental results requiring coherent wave optics for analysis are beginning to appear, an introduction to the method is given here.

From a more fundamental viewpoint, a general method of structure analysis for crystalline materials by electron diffraction would be required to provide at least the following information:

(a) the dimensions and angles of a unit cell,
(b) the space group,
(c) the atomic number and quantity of each type of atom present,
(d) the atomic coordinates and type of every atom in the asymmetric unit, and
(e) a characterization of the defect structure of the material, since this plays a vital role in controlling mechanical, electronic, magnetic, and thermal properties.

No general solution to (d) exists; however, evidence collected from (a), (b), (c), and (e) will generally limit the possible structures, and an informed guess can often be made, based on natural structures and other experimental observations. High-resolution imaging (HREM), energy dispersive x-ray spectroscopy (EDS), and electron energy-loss spectroscopy (EELS) are all instrumentally compatible with microdiffraction in modern analytical electron microscopes (AEM). EDS and EELS provide useful information for (c) and to some extent for (d), if channelling effects are taken into account. HREM provides information toward (a) and (b) and certainly (e). The difficulties with (d) arise mostly from (i) the absence of a general solution to the inversion problem of coherent multiple scattering, (ii) difficulties in preparing suitably thin specimens of known thickness, (iii) difficulties in determining absorption coefficients, (iv) slow progress in development of quantitative parallel recording systems. The impression that little quantitative electron diffraction has been undertaken because of difficulties with multiple scattering is common and misleading. Section IV.D. of this chapter reviews some of the considerable number of structure factor measurements already performed.

There has been much progress in all the areas mentioned above. We begin our discussion with experimental methods, the area most familiar to the broad group interested in microscopy. Experimental methods are divided into static and rocking probe methods. A discussion of probe coherence is included with static probe methods because the concept will be first encountered when conducting experiments at the limit of spatial resolution with field emission sources. A discussion of reciprocity is included with rocking probe methods because it is necessary to relate them to the more familiar static probe methods and to interpret convergent beam patterns with either method. An introduction to microdiffraction theory follows the description of experimental methods. Descriptions of some current advanced applications of microdiffraction to physical research problems follow the theory section. The last section discusses the relationship between microdiffraction and STEM imaging and EELS. This relationship is discussed here because it is critical, but not often discussed in detail or compared to the TEM case. Extensive references are given for the convenience of the reader. Other recent reviews of microdiffraction with different viewpoints that should be consulted have been given by Reimer (1979), Cowley (1979a), and Steeds (1979).

II. EXPERIMENTAL METHODS

A. Static Beam Microdiffraction

The two basic methods for obtaining a microdiffraction pattern in electron microscopes are the well-known selected-area-diffraction (SAD) method (LePoole, 1947; Boersch, 1936) and convergent-beam diffraction (Kossel and Mollenstedt, 1938, 1942; Hillier and Baker, 1946; Cowley and Rees, 1953; Cockayne *et al.*, 1967; Riecke, 1962). Corresponding geometric ray diagrams are shown in Figure 1. For selected area diffraction, the beam incident on the specimen is ordinarily broad and nearly parallel, focused through the specimen onto a distant detector using the lenses on the electron entrance side of the specimen. For convergent beam, it is focused on the specimen. These two operating modes are the electron analogs to Koehler and critical illumination of optical microscopy.

The terminology used to describe these patterns includes incident beam divergence, angular resolution, and spatial resolution, i.e., the size of the specimen area contributing to the diffraction pattern. The divergence is the half-apex-angle of the cone of incident radiation, α. Several examples are shown in Figure 1(b), (c), and (d). The angular resolution of the diffraction pattern means the angular width of the smallest intensity fluctuation in the pattern that is attributable to the specimen. Divergence can be determined experimentally from the microscope column geometry or directly by measurement from a diffraction pattern. The latter method is preferred because it accounts for changes in divergence caused by the probe-forming lenses exactly.

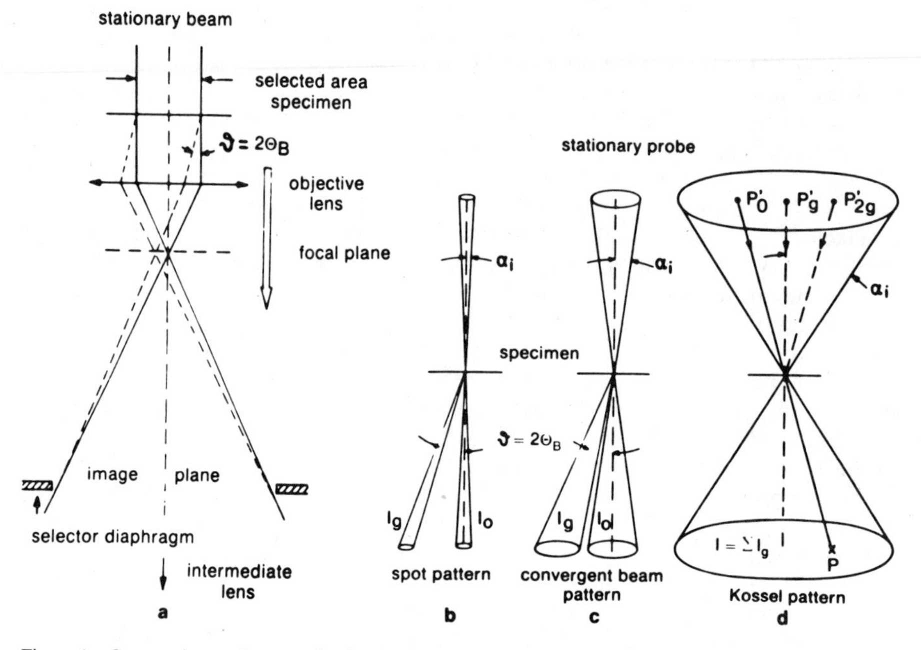

Figure 1 Geometric ray diagrams for fixed-probe microdiffraction. (a) Conventional selected area diffraction (SAD) used in all fixed-beam instruments. The selector diaphragm (aperture) is conjugate to the specimen; (b), (c), and (d) convergent-beam microdiffraction patterns with probe focused on the specimen. The incident probe half-angle, α, determines the angular "diameter" of the Bragg discs. A point, P, in the diffraction pattern (detector or observation plane) is conjugate to a point, P_o, within the cone of incidence. Points P receive forward scattered intensity from P_o and diffracted intensity from points P_g, P_{2g}, . . ., depending on the size of α relative to the Bragg angles, θ_g. See text for discussion.

These parameters are illustrated in Figures 1 and 2. The convergent beam pattern of Figure 2 is from silicon with the 100-kV beam down a <111> direction. The six Bragg discs closest to the origin of the pattern are {220} type. The divergence can be determined directly by measuring the length of the <220> diffraction vectors and noting that the length corresponds to $2\theta_{111} = \lambda/d_{111} = 0.019$ rad. A similar measurement of the radius of the direct beam disc or any of the Bragg discs yields a divergence half-angle of ~7 mrad. This diffraction pattern corresponds approximately to Figure 1(c). When α is increased so that $\alpha = \theta_{220}$, these Bragg discs will just touch; further increases in α lead to the wide-angle CBED or Kossel pattern of Figure 1(d). Figure 1(b) is the same as 1(c), with reduced divergence. The angular resolution limit of the diffraction pattern shown in Figure 2 corresponds to the angular width of the finest lines visible; it is ~0.2 mrad.

It can be seen that the divergence of a nearly parallel incident beam SAD pattern (Figure 1(a)) is very small, and for purposes of discussion, it is often considered zero. However, in the SAD geometry of Figure 1(a), the divergence will be limited by the source size since each point in the "spot" pattern is an image of the source. A typical low value is ~0.05 mrad. The background in SAD patterns is typically much lower than CBED patterns (except for thick specimens), and only sharp Bragg spots are visible. For this reason, divergence and angular resolution are often taken to mean the same thing, particularly for SAD patterns from multiphase specimens with only small differences in lattice parameter and, hence, closely spaced Bragg reflections. The same comment usually applies to small-angle electron scattering (Carpenter *et al.*, 1978).

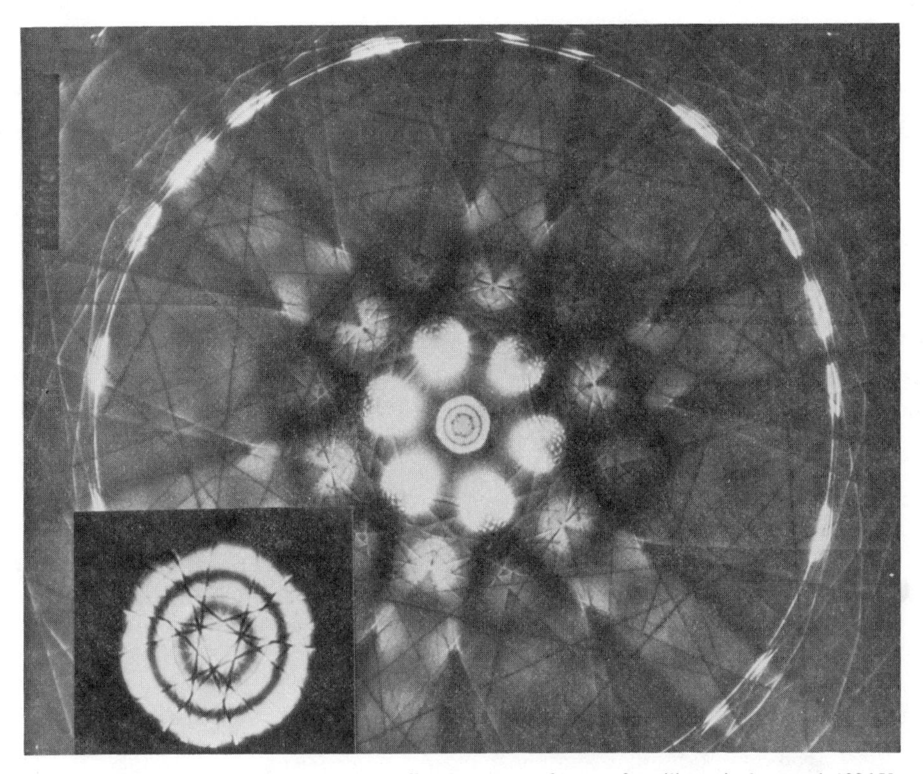

Figure 2 Convergent-beam electron microdiffraction pattern from perfect silicon single crystal, 100 kV. Beam direction is [111] and $\alpha \simeq 7$ mrad, corresponding to Figure 1(c). The first- order Laue zone (FOLZ) ring is visible. The inset shows FOLZ line detail in the central beam disc. The incident probe size was ~10 nm in TEM/CBED mode with field-emission gun.

Riecke (1962) discusses a variation of the SAD method that uses the objective prefield of a TEM/STEM condenser-objective lens as a third condenser to form a highly demagnified source crossover in the front focal plane of the electron-entrance side objective lens (L2 of Figure 3). The crossover is imaged on the object plane (specimen) as a small-diameter disc, on the order of 50 nm in diameter for a thermionic source, conjugate to the source. The selected area size is controlled by the second condenser aperture. The incident divergence is given by r/f, where r is half the diameter of the front focal plane crossover and f is the objective lens entrance side focal length. The divergence will generally be small, but not usually as small as for a well-defocused conventional SAD pattern.

It is important to note that convergent beam diffraction corresponds to irradiation of the specimen over the continuous range of angles of incidence contained within the cone (half-angle α) of convergent electrons focused on the specimen. If the divergence angle α of Figure 1(c) is increased beyond the Bragg angle, it becomes possible for the CBED discs to overlap, producing the Kossel pattern shown in Figure 1(d). Since each ray passing through the sample (assuming a perfect crystal sample in this part of the discussion) can be deflected only by multiples of twice the Bragg angle (or no deflection), it now becomes possible for several source points (such as P_o, P_g, and P_{2g}

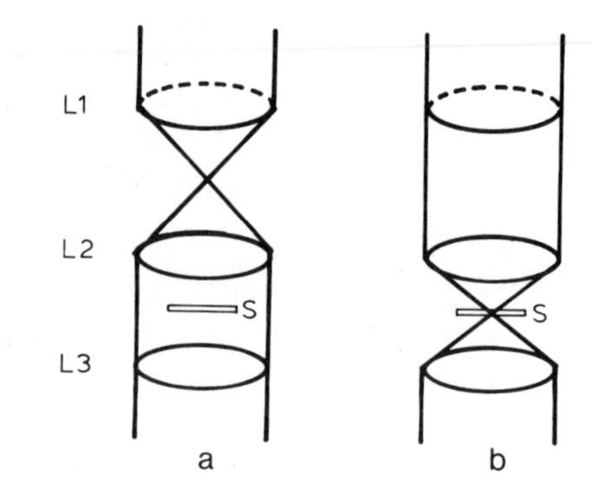

Figure 3 Schematic condenser-objective lens diagrams for a modern analytical electron microscope. The main upper and lower halves of the lens are L2 and L3. L1 is an "auxiliary" lens that interacts with L2. (a) TEM mode, the flux of L1 adds to L2. (b) Convergent beam or SEM mode, the flux of L1 opposes L2. The specimen is indicated by S. This type of lens was first used in Philips EM400 instruments. The objective lens of a D-STEM instrument may be visualized as (b), without lens L3.

in Figure 1(d)) to contribute to a *single* detector point P. This is not possible for $\alpha < \theta_B$. Here, rays from P_o, P_g, and P_{2g} subtend angles that differ by the Bragg angle at the specimen. To calculate the diffracted intensities, one must add either complex amplitudes or intensities from each direction of incidence within the cone, depending on whether the illumination aperture is coherently or incoherently filled.

The final probe-forming lens configurations used for diffraction in most modern analytical instruments may be qualitatively discussed using Figure 3. The objective lens may be considered as three lenses: L1 and L2 on the electron entrance side of the specimen and L3 on the exit side. L2 and L3 are actually the entrance and exit sides of a symmetrical condenser-objective, and L1 corresponds to a prefield lens on the entrance side. Figure 3(a) illustrates near-parallel, large-area beam illumination typical for SAD, obtained by forming a crossover in the front focal plane of L2. Figure 3(b) illustrates convergent beam operation; a near-parallel beam is incident on the front focal plane of L2, producing a convergent probe at the specimen, located in the back focal plane. This is the L3 object plane in a TEM/STEM instrument; for the D-STEM case, L1 and L2 may be considered a single (objective) lens, and L3 need not be present in Figure 3(b). The final probe-defining aperture (not shown) is called the condenser or objective aperture in the TEM/STEM and D-STEM cases, respectively. More detailed discussions of multizone lenses have been given elsewhere (Chapter 2).

For aperture-selected microdiffraction in general, the area of the specimen contributing to the observed diffraction pattern is determined by the "selecting aperture" in some image or object plane conjugate to the specimen, corresponding to a demagnification of the aperture when referred to the specimen. For SAD in ordinary TEM, the aperture is normally in the objective lens image plane. The selected area has nominal diameter ϕ/M, where ϕ is the aperture diameter and M is the lens (or postspecimen image-forming part of a condenser objective) magnification of the objective, usually in the range of 30 to 50 times. For convergent beam diffraction, the specimen area contributing to the diffraction pattern is defined by the diameter of the probe incident on the specimen. The spatial resolution, i.e., the smallest specimen area from which a diffraction pattern can be selected, is limited in both cases by the optics of the particular instrument. The SAD area-selecting aperture is conjugate to the specimen, and the disregistry between the apparent selected area and origin of diffracted beams passed through the selector aperture is given to a good approximation by Hirsch *et al.* (1977):

$$\Delta r = C_s\alpha^3 + (\Delta f)\alpha . \qquad (1)$$

Here,

C_s = spherical aberration coefficient of the postspecimen objective
α = scattering angle of interest
Δf = focusing error of the objective

Typical values of C_s are 2 to 6 mm; for angles α corresponding to low-order reflections (i.e., 111, 200) and small focusing error, the selection error, Δr, can be quite small. For larger angles α and typical focusing errors (~5 μm), the selection error can easily be several micrometers. Because Δf may be positive or negative and $C_s\alpha^3$ is always positive, the selection error can be minimized for some α of interest by proper choice of Δf, but the procedure is tedious and seldom used.

The irradiated area for a probe focused on the specimen (convergent beam) is given approximately by Wells *et al.* (1974):

$$d^2_0 = d_g^2 + d_f^2 + d_s^2 + d_c^2 \quad (2)$$

where d_0 is the total incident probe diameter. The contributions to probe size are the Gaussian beam diameter, d_g, broadening by the spherical aberration for the final probe-forming lens, d_s, and chromatic broadening, d_c, resulting from the energy spread in the beam when it passes through the probe-forming lens. We assume axial astigmatism is fully corrected. The Gaussian beam diameter is equal to the electron source size multiplied by the prespecimen lens demagnification. It is not uncommon to increase the incident convergence angle α somewhat when performing simultaneous microdiffraction and microanalysis experiments, to increase the probe current. The diffraction broadening, d_f, is given by $1.22\lambda/\alpha$, the diameter of the Airy disc, the image of an ideal point source. The spherical aberration broadening, d_s, is $0.3C_s\alpha^3$ based on a Gaussian current distribution in the probe, where this C_s value corresponds to the probe-forming lens, or the electron entrance side prefield of a condenser-objective in a modern TEM/STEM AEM. This C_s value can be determined from shadow images in either TEM or STEM convergent-beam operation (see III.D.). The energy spread in the incident beam contributes to probe broadening through the chromatic aberration term, $d_c = (\Delta E/E)\alpha C_C$, where ΔE is the energy spread in a beam accelerated to energy E, and C_C is the chromatic aberration constant of the probe-forming lens.

For microdiffraction, when d_g^2 is the smallest term in Eq. (2), the aperture diffraction and spherical aberration contributions to probe size become most important; because of their functional dependence on α, there exists an optimum value of the aperture semiangle, α_{opt}, that will produce a minimum incident probe size. For divergence angles larger than α_{opt}, the probe broadening is dominated by the spherical aberration contribution, and the probe is said to be aberration limited. For $\alpha < \alpha_{opt}$, the largest contribution to probe broadening comes from the diffraction term, and the probe is said to be diffraction limited. Here the geometrical demagnification term, d_g, is negligible. It is the case appropriate to modern field-emission (FEG) convergent-beam instruments and "coherent" microdiffraction, since a small value of r_s (the geometrical probe size) will be seen from Eq. (3) to result in a coherently filled illumination aperture. This leads us to the general question of coherence in microdiffraction, which we will now discuss.

The coherence of the illuminating probe focused on the specimen for convergent-beam microdiffraction experiments or STEM imaging has important effects on the diffraction pattern and STEM images. To understand this, it is necessary first to consider the two extreme cases of incoherent and coherent illumination. We consider spatial or transverse (perpendicular to the beam direction) coherence only, since temporal or time coherence is not generally important for microdiffraction. The distinction between coherent and incoherent illumination is made by considering whether the transverse coherence width, X_a, in the plane of the final probe-defining illumination aperture (not at the specimen) is larger or smaller than the aperture diameter, $2R_a$. It can be shown that

$$X_a = \lambda/2\pi\theta_s = \frac{D\lambda}{\pi r_s} \tag{3}$$

where r_s is the diameter of the geometric source image, and θ_s is the semiangle subtended by the *probe* at the illumination aperture. The other parameters are defined in Figure 4.

If $X_a \ll 2R_a$, the illumination aperture can be considered incoherently filled and treated as an ideally incoherent source. The probe will then be partially coherent. If $X_a \gg 2R_a$, then the illumination aperture is coherently filled, and the irradiation can be imagined to have originated from a single point. Thus, the electron source can be replaced by a fictitious point source, and the optical system can be considered to be filled with perfectly coherent radiation. The probe will also be coherent. This is the case for field-emission STEM instruments, with a highly demagnified probe and small illumination half-angles. In this case, the specimen can be thought of as illuminated by an aberrated, converging spherical wave. If the condition for incoherent microdiffraction is fulfilled, the illumination system can be replaced by an incoherent effective source in the plane of the illumination aperture, for which each point is a phase-independent emitter of electrons. This is the usual case for thermionic sources of all types.

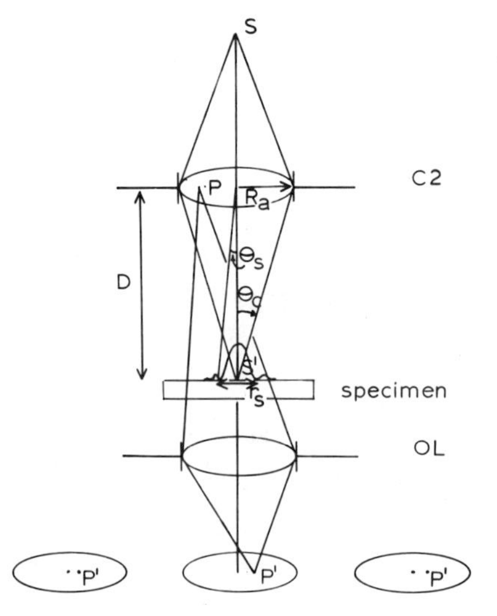

Figure 4 Ray diagram for CBED coherence condition in a TEM/STEM instrument. Here, OL is the objective lens, C2 the probe-forming or condenser lens, and θ_s the angle subtended by the probe at the illumination aperture. The probe "size" is r_s, and P′ is conjugate to P.

An effective source can be defined for all values of X_a; however, the simplifications described above cannot be made in the intermediate case where $X_a \simeq R_a$. These are also based on the commonly made assumption that the exit pupil of the probe-forming lens coincides with the final illumination aperture. For D-STEM field-emission microscopes, the "condenser" lens referred to above is usually called the objective lens; on TEM/STEM instruments, a condenser-objective is used, and the objective lens "prefield" then forms the final "condenser" lens. Note that the size of a coherently diffracting region of the specimen which contributes to a point in the CBED pattern is $\sim\lambda/2\pi\theta_c$ (Figure 4). This is the "coherence patch" size in the object plane (Barnett, 1974) if $X_a \ll 2R_a$. If $X_a \gg 2R_a$, the probe is coherent, and the entire illuminated region of the specimen contributes coherently to the diffraction pattern.

Since each point P in the source is conjugate to a point P′ in the arrangement of Figure 4, there is no reason to expect that the microdiffraction patterns recorded under coherent and incoherent conditions should differ in any way. In each case, the patterns can be thought of as arising from a specimen illuminated by a cone of plane waves, each of which is focused to a set of points (one in each Bragg disc) in the back focal plane of the objective lens. Thus, the phase relationship between component plane waves in the incident cone is unimportant. However, this argument depends on exact focusing conditions (and ignores any diffraction limit imposed by OL), so that in practice, while the in-focus coherent and incoherent patterns are similar, their through-focus appearance differs completely. Computer programs which give "point" diffraction patterns (one numerical intensity value for each Bragg beam) can thus be used for both cases by performing a separate calculation for each incident-beam direction in the illuminating cone of radiation. This is true provided that: (a) the specimen is a perfect, parallel-sided crystal; (b) there is no overlap of Bragg discs; and (c) the "exactly focused" pattern is required (diffraction discs conjugate to effective source). This argument does not apply to WACBED patterns such as Figure 1(d). In that case, an amplitude summation of beams over the irradiation cone must be performed to obtain the diffracted intensity distribution if a coherent source is used; otherwise, a simple intensity sum is permitted. Other focusing conditions are readily obtainable in practice, in which the diffraction discs are conjugate to the electron source (each Bragg spot becomes an electron source image) or to the specimen (each Bragg spot becomes a low-magnification "shadow image" of the specimen).

As a simple example, we consider the coherence conditions in the JEOL 100C top-entry HREM fixed-beam configuration, where the maximum demagnification of the two condenser lenses is 0.03. For a hairpin filament with an electron source size of 15 μm and using D = 251 mm (Figure 4), we then have X_a = 0.327 μm, which is small compared to the 100-μm condenser apertures frequently used. Thus, we would be justified in replacing the illumination system by a fictitious, perfectly incoherent source the same size as the condenser aperture in this case.

In the coherent case ($X_a \gg 2R_a$), there is no simple way to estimate the probe size. The quadrature sum (Eq. 2), which is essentially an incoherent sum of several coherent broadening processes, cannot be used. Computer calculations giving the probe-intensity profile can be readily performed. The results depend strongly on the focus setting of the probe-forming lens, its aberration constants, and the illumination cone size. Note that, in the coherent limit, the entire probe-broadening contribution in the object plane results from diffraction and aberrations. These computer calculations, in which the two-dimensional numerical Fourier transform of the probe-forming lens' transfer function is evaluated, are then the only reliable guide to "probe size" (to the extent that this concept has any useful meaning in the coherent probe case) (Spence,

1978). The results show that the object plane probe-energy distribution is quite broad at most lens focus settings, that the probe can contain strong-intensity oscillations and extended tails, and that it is most compact in the neighborhood of the TEM "dark-field" focus setting (Colliex, 1985)

$$\Delta f = -0.75(C_s\lambda)^{1/2} \tag{4}$$

The minus sign indicates underfocus, a lens weakened from Gaussian focus.

For both the coherent and incoherent cases, the probe size can be measured in principle in TEM/STEM instruments by using the postspecimen lenses at known magnification to form a magnified image of the probe (Bentley, 1980). On "dedicated STEM" instruments, a calculated value of probe size must be used; however, a rough estimate can be obtained by estimating the resolution in the STEM image.

The effect of interference between Bragg diffracted beams when a crystalline specimen is irradiated with a small coherent probe is illustrated in Figure 5 (Cowley, 1981a). When the divergence angle permitted overlap of Bragg diffracted beams at the back focal plane, i.e., recombination of diffracted beams, the intensity distribution in these regions exhibited interference effects and was not the simple sum of the intensities of the individual beams. This result showed that the illumination aperture was coherently filled. When reduced divergence did not permit overlap, these effects were not observable.

Until recently, all fixed-probe convergent-beam microdiffraction methods used a probe-defining aperture together with prespecimen lenses to control incident-beam divergence. This technique leads to complications during convergent-beam examination of large unit cell crystals of relatively low symmetry. In these cases, Bragg discs overlap at small divergence angles because the Bragg spot spacing is small. The result is a wide-angle convergent-beam pattern with overlapping diffraction orders. Symmetry (Eades *et al.*, 1980) and structural information (Fujimoto and Lehmpfuhl, 1974) in these patterns have been discussed, as well as the effects of lattice strain on them (Smith and Cowley, 1971). In general, however, it is desirable to be able to observe the diffracted intensity distribution about a Bragg point out to fairly large angles without the additional complications of overlap. Two methods have been developed to do this: one uses a static probe and the other, a rocking probe. The latter will be discussed in the following section. Both methods are, in principle, the same. A postspecimen limiting aperture is used which, in effect, allows only one diffracted disc to reach the pattern observation screen or detection system. A ray diagram for the static beam method (Tanaka *et al.*, 1980a) is shown in Figure 6. An example of a convergent-beam pattern obtained by this method is shown in Figure 7. No overlap of diffraction orders is present, and the pattern extends to $\sim 4.4°$ in reciprocal space.

The procedure for obtaining a pattern like this (Yamamoto, 1983) for a Philips EM400 with twin lens is:

(a) Set up the microscope in the normal CBED mode. Set up the correct orientation (usually a zone axis setting) of the specimen.
(b) Switch to image mode and focus the electron probe fully on the specimen surface.
(c) Use the specimen height control on the side entry holder to increase the specimen height (i.e., move specimen toward electron gun) until the probe is seen to split into many probe images, one for each Bragg beam.
(d) Isolate the order of interest by inserting the selected area aperture so that the order required passes through the aperture.

(e) Switch to diffraction pattern mode and remove the condenser aperture completely. A wide-angle pattern with no overlap of orders will be seen on the screen.

Note that this is not a small-probe (~20-nm) microdiffraction technique. The size of the irradiated area on the specimen depends on the height increase using the stage and on the probe size with the condenser aperture removed. It is, however, a very useful method, particularly for thin specimens with large unit cells. More details are given in Eades (1984).

Note added in proof: Two recent extensive and useful compilations of CBED patterns and discussions of their applications that should be consulted by researchers are "Convergent Beam Electron Diffraction of Alloy Phases," compiled by J. Mansfield (1984) of the Bristol Group, and "Convergent-Beam Electron Diffraction," by M. Tanaka and M. Terauchi (1985).

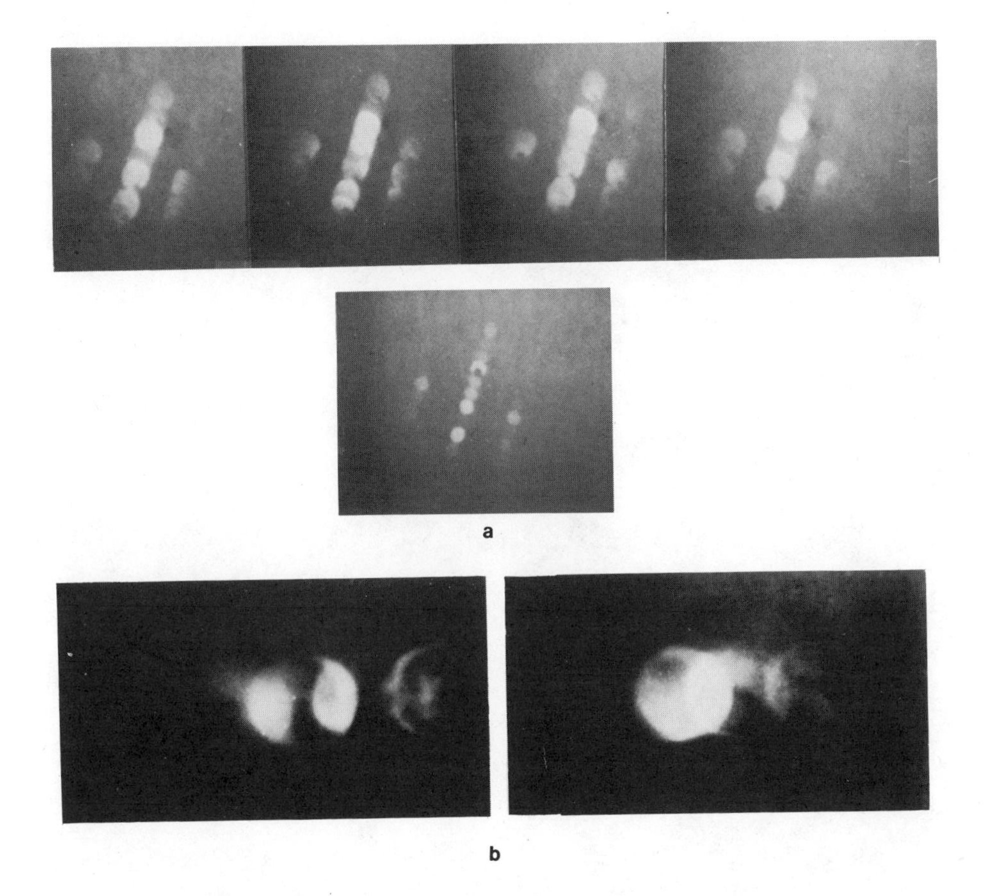

Figure 5 Coherent convergent-beam microdiffraction patterns from a thin crystal of $Ti_2Nb_{10}O_{29}$. (a) The top row of patterns was obtained as the beam was moved from one side of the crystal unit cell to the other; the repeat distance is 2.8 nm. The bottom pattern, with nonoverlapping Bragg discs, shows intensities independent of probe position. (b) Higher magnification of regions within overlapping Bragg disc pattern shown in (a). Regions of coherent interference are visible where the discs overlap.

Figure 6 Ray diagram for the fixed-beam, Tanaka wide-angle CBED method. The source images are I, one for each Bragg disc, and A is conjugate to the selected area aperture. The viewing screen, V, is approximately conjugate to the specimen when inserting the selected area aperture (TEM mode), and to the back focal plane when recording the pattern.

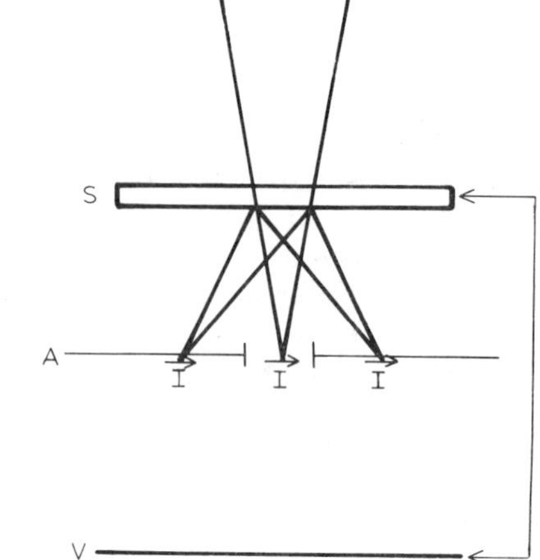

Figure 7 Experimental wide-angle Tanaka pattern showing expanded angular view without disc overlap, for the central disc of silicon with beam direction [111].

B. Rocking Beam Methods

A number of different rocking incident probe-diffraction techniques have been developed for various applications. Among these are SAD methods for use in D-STEM instruments without parallel intensity distribution detection systems, techniques for reducing the aperture-selected specimen area contributing to the diffraction pattern, reducing divergence (in this case considered analogous to angular resolution), electronic intensity measurement, and others.

Ray diagrams corresponding to some of these methods are shown in Figure 8. The diagram of Figure 8(a) depicts SAD in a D-STEM instrument and is reciprocal to Figure 1(a). In the present case, the beam is focused at the front focal plane of the (prespecimen) objective lens, producing approximately parallel illumination of the specimen at the lens image plane. The selected area aperture is located at the lens object plane, conjugate to the specimen. The specimen area contributing to the diffraction pattern is determined by the selected area aperture size and lens demagnification; the effects of lens aberrations and defocus are the same as for the TEM case (Figure 1(a)). The incident-beam divergence half-angle is given by r/f, where r is the radius of the probe crossover in the front focal plane and f is the lens focal length. The scan coils produce a translation of the crossover in the front focal plane,

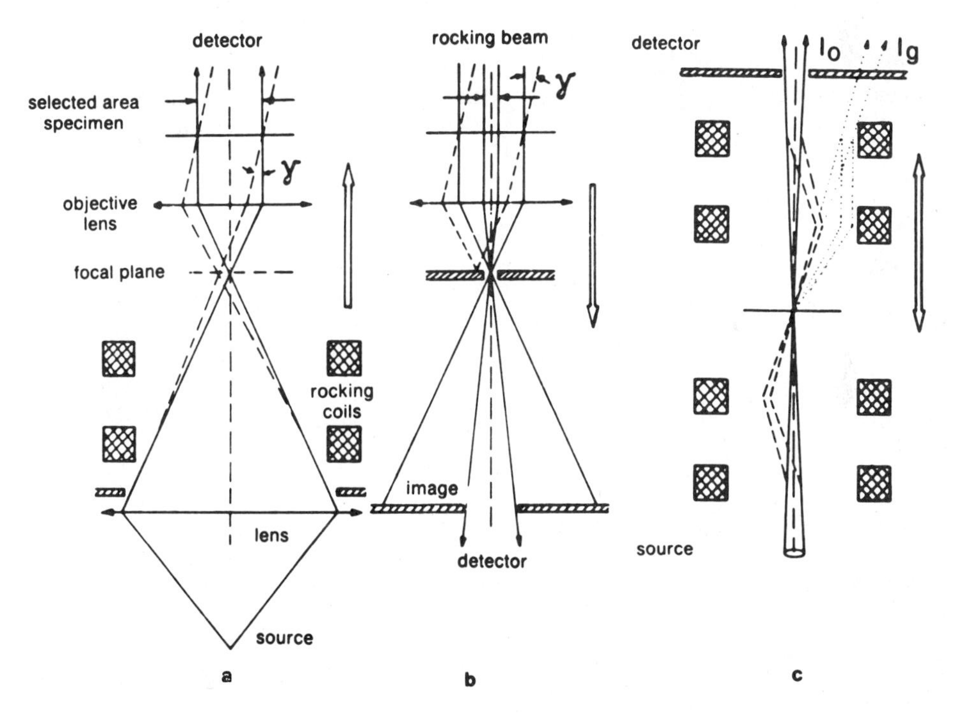

Figure 8 Ray diagrams for rocking beam microdiffraction. (a) Rocking selected area diffraction. This technique is reciprocal to conventional SAD (see Figure 1(a)) and is often used in D-STEM instruments. (b) The rocking beam method of van Oostrum *et al.* (1973). (c) The double-rock (or rock-unrock) method of Eades and earlier investigators. Open arrows show beam direction. The double-rock method can be visualized in either direction, provided reciprocity for detector and collection apertures is considered. Rocking angle is γ.

resulting in a rocking probe at a "point" on the specimen. Concurrently, the Bragg diffraction pattern will move across an apertured electronic detector, and the diffraction pattern will be displayed on a CRT scanned in synchronism with the angular deflection of the incident beam. The "camera length" or magnification of the diffraction pattern can be varied in two ways: by changing the ratio of rocking angle to CRT scan length electrically or by changing the diffraction pattern magnification between the specimen and detector using postspecimen lenses. This results in a change in collection angle when the detector aperture is not changed by the same ratio as the camera length. Large collection angles relative to incident divergence result in Kossel patterns, and small ones yield Bragg "spot" patterns. This effect corresponds to changing the incident-beam divergence in static probe convergent-beam microdiffraction and is one result of the reciprocity principle, which is discussed later.

Several other rocking beam techniques with useful capabilities have been developed by using instruments with postspecimen objective lenses or condenser objectives. A method developed to reduce the size of the selected area (VanOostrum *et al.*, 1973) is illustrated in Figure 8(b). For this technique, the instrument is operated in the image mode; i.e., the specimen is conjugate to the final viewing screen and the detector is near this plane. The active detector diameter, d_{det}, referred back to the specimen determines the selected area diameter, which is d_{det}/M, where M is the total image magnification. Typical M values used are 20,000 to 100,000 times. The convergence angle is determined by the crossover size in the postspecimen aperture relative to the lens focal length.

Further considerations of the method have shown that diffraction patterns can be obtained from ~3-nm specimen areas using a large-diameter, relatively parallel incident probe, which will minimize radiation effects in the specimen and therefore on the resulting diffraction pattern (Geiss, 1976; Chevalier and Craven, 1977). A similar method (Fujimoto *et al.*, 1972) uses a different postspecimen optical scheme; in this case, the transfer lenses focus a magnified image of the scanned diffraction pattern in the back focal plane of the postspecimen objective lens on the final viewing screen/electronic detector. The specimen area contributing to the diffraction pattern is defined by an aperture in the objective lens image plane conjugate to the specimen, just as in the static beam SAD case (Figure 1(a)). Here the diffraction pattern divergence is determined by the projection of the detector aperture (near the final viewing screen) onto the objective lens back focal plane, and the divergence can be reduced to small values by using a large demagnification. These two rocking beam methods are complimentary. The first method noted can be used for microdiffraction from the smallest selected area with moderate divergence and the second for microdiffraction with minimum divergence from a larger selected area. The incident probe in both cases is not focused on the specimen, thus minimizing radiation effects and contamination.

The double-rocking technique (Eades, 1980; and Tanaka *et al.*, 1980b) is the most recently developed rocking diffraction method and the most useful for investigating the symmetry of perfect crystals. A ray diagram for the method is shown in Figure 8(c), and an example of a pattern produced by the method is given in Figure 9. This rocking technique is the analogue of the static-beam Tanaka method shown in Figure 7 and is useful for the same reasons. Note that the simplified ray diagram shows double-tilting coils above and below the specimens, so that the detected scattered rays are on the optic axis of the microscope. This arrangement permits simple filtering or energy-loss analysis of the pattern, whereas single-tilt postspecimen scanning will not; the resulting off-axis rays would severely degrade the spectrum. The pattern of Figure

9 is energy-filtered ($\Delta E \leq 4$ eV) and shows markedly improved contrast relative to an unfiltered pattern (Higgs and Krivanek, 1981).

This discussion of the various static and rocking probe microdiffraction techniques illustrates that essentially the same range of divergence, angular resolution, and wide-angle capabilities are available to the experimenter with either method. Practical differences arise in applications. Film-recorded static-beam microdiffraction methods are often quicker and more generally available; film is a parallel recording system and generally exhibits higher signal-to-noise ratio than commonly available electronic serial (scanning) or parallel (television) systems under the usual current-limited experimental conditions. If, however, quantitative measurements of intensity distributions are required, electronic detection systems offer great advantages. These systems are undergoing rapid development and are discussed further below.

Comparing static and rocking beam microdiffraction methods requires some care in using the reciprocity relationship. The principle of reciprocity, first stated by Helmholtz (Joos, 1950; Born and Wolf, 1980) and frequently used in scattering theory, has been applied to electron microscopy and provides two main useful results. First, this principle has shown that similar contrast is expected from TEM and STEM images (Cowley, 1969), including both phase and diffraction contrast effects, if the

Figure 9 Central portion of an energy-filtered Eades double-rocked, wide-angle CBED pattern from thin silicon, with beam direction [111]. The field of view increases with rocking angle. Compare to Figures 2 and 7. This pattern was recorded serially, through an electron spectrometer that collected only elastically scattered electrons with $\Delta E \leq 4$ eV.

images are recorded under reciprocal conditions. Detailed experiments have confirmed this for diffraction contrast, for example Maher and Joy (1976). The same principles are used to establish the equivalence of various static and rocking microdiffraction techniques. The second application, which gives a relationship between certain diffracted beams, is relevant here since it is used for space-group analysis by microdiffraction. It is discussed in its simplest form below.

The reciprocity theorem states that the complex amplitude at B of a wave originating from a point source at A is equal to the complex amplitude that would be recorded at A if the same point source were moved from B to A. More commonly, one speaks of the "interchange of point source and point detector" leaving the result of a scattering experiment unchanged. The application of this theorem to elastic and small-energy-loss inelastic Bragg scattering in electron diffraction has been discussed in detail by Pogany and Turner (1968). The theorem becomes powerful when it is combined with the symmetry operations of perfect crystals. Note that, as stated, it applies to point sources and detectors, and for the present we consider wave amplitudes (including phases) and elastic scattering. Now a point source at a large distance from a crystal generates a plane wave or beam in the vicinity of the crystal, which in turn excites a set of Bragg beams from the crystal. Thus, as applied to elastic Bragg scattering, the theorem states that the ray-paths for the incident beam and a particular Bragg beam may be reversed. Thus, in Figure 10, the diffracted amplitude, ϕ_g, is due to source S in Figure 10(a) and is equal to the amplitude, ϕ_g, that would be recorded in the direction shown in Figure 10(b) if the same specimen were illuminated from below as shown. But many crystals contain a horizontal central mirror plane of

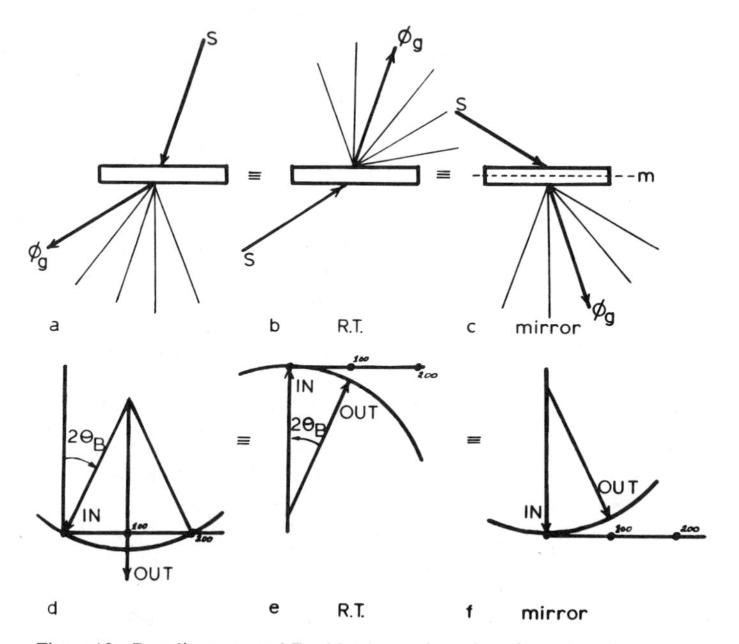

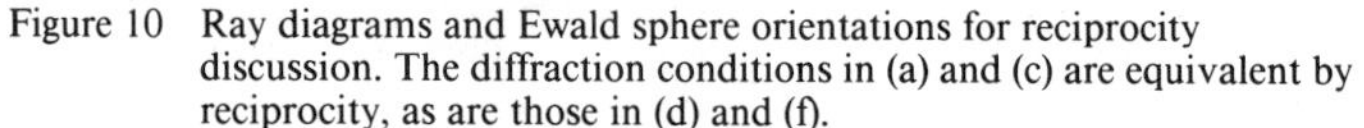
Figure 10 Ray diagrams and Ewald sphere orientations for reciprocity discussion. The diffraction conditions in (a) and (c) are equivalent by reciprocity, as are those in (d) and (f).

symmetry, and thus for these the incident beam in 10(b) can be reflected about this plane as shown in 10(c), and will then give the same scattering amplitude, ϕ_g, in the direction shown. In general, any of the crystal point group operations may be applied to the incident beam, since the crystal "looks" the same from all the equivalent directions related by the crystal symmetry elements (strictly speaking, however, this symmetry element must be a symmetry element of the crystal, its boundaries, and any defects it may contain). In particular, note that in the zero-order Laue zone (ZOLZ) approximation (projection approximation), all crystals have a central mirror plane.

Figure 10 also shows the argument represented using the Ewald sphere construction. As an example, Figure 10(d) shows the particular case of the (100) reflection which we can imagine being used to form a diffraction contrast image with a small objective aperture. We consider the systematics orientation (a single line of reflections), which necessarily allows mirror reflection.

From the geometry of Figure 10(d) (reversing ray paths and reflecting about the horizontal mirror), we see that the amplitude of the (100) beam in an orientation in which the Ewald sphere just passes through the (200) reflection is equal to the amplitude of the (100) reflection in the zone axis (or "symmetrical") orientation. If we write $\psi\,(\mathrm{h},\underline{\mathrm{g}})$ to mean the complex amplitude of beam h in a systematics orientation in which the Ewald sphere just passes through the $\underline{\mathrm{g}}$ reciprocal lattice point, we can write the above result as

$$\psi(100,\underline{200}) = \psi(100,\underline{000}) \,. \tag{5}$$

This surprising result is confirmed exactly by the numerical solutions of the Schroedinger equation for the elastic scattering of 100-kV electrons shown in Table I. Note that all the amplitudes for the two orientations differ, *except* the (100). This reflection must be the same for both orientations for all thicknesses (see Pogany and Turner, 1968, for more detailed calculations) and, since $\underline{\mathrm{g}}$ may be fractional, we may use the column approximation to predict that TEM diffraction contrast images formed in the (100) beam for the two orientations will also be identical.

This result may be generalized, using simplified forms of Pogany and Turner's results. In general, in the systematics orientation, using the above notation

$$\psi(\mathrm{h},\underline{\mathrm{g}}) = \psi(\mathrm{h},\underline{-\mathrm{g}+2\mathrm{h}}) \tag{6}$$

and the intensities $\psi\psi^*$ are of course also equal. In addition, for crystals containing a center of symmetry, it can also be shown that

$$\psi(\mathrm{h},\underline{\mathrm{g}}) = \psi(\bar{\mathrm{h}},\underline{\mathrm{g}-2\mathrm{h}}) \tag{7}$$

in the systematics orientation. The general results for three-dimensional diffraction are

$$\psi[\mathrm{h},\,(\underline{-\mathrm{h}/2+\mathrm{g}})] = \psi[\mathrm{h},(\underline{-\mathrm{h}/2-\mathrm{g}})] \tag{8}$$

for a crystal with a central mirror plane, and

$$\psi(\mathrm{h},\,[\underline{-\mathrm{h}/2+\mathrm{g}})] = \psi[\bar{\mathrm{h}},(\underline{\mathrm{h}/2+\mathrm{g}})] \tag{9}$$

for a crystal with a center of symmetry. In these last two equations, the second, underlined, quantity in each parenthesis must now be interpreted as a vector from the center of the Laue circle (the intersection of the Ewald sphere with the ZOLZ) to the origin of reciprocal space.

TABLE I

Calculated Scattering Amplitudes for Reflection, g, With the Specimen Crystal in Reciprocally Related Orientations. It can be seen that other reflections in the systematics row are not equivalent by reciprocity for the indicated orientations. See text for details.

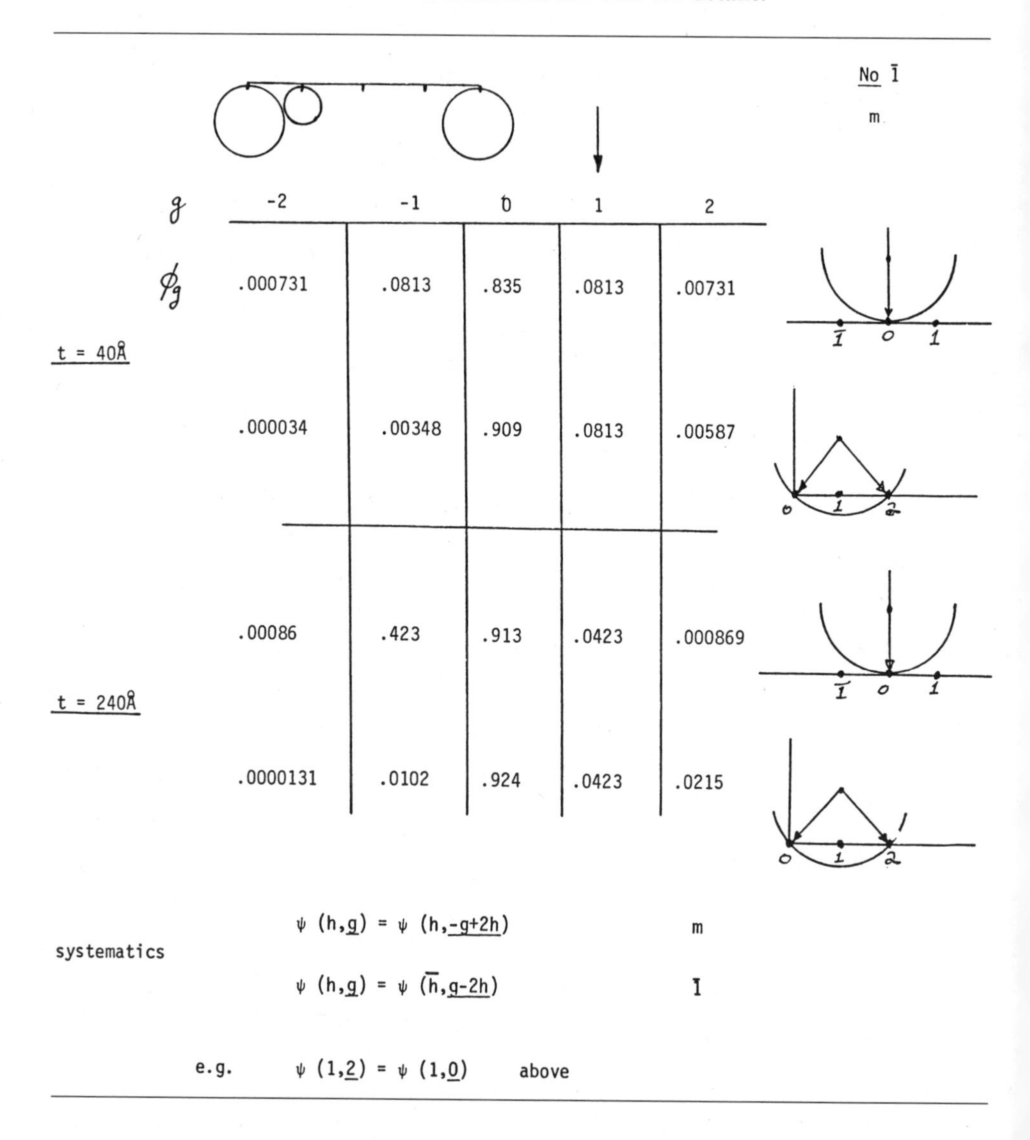

No $\bar{1}$

m

		-2	-1	0	1	2
t = 40Å	ϕ_g	.000731	.0813	.835	.0813	.00731
		.000034	.00348	.909	.0813	.00587
t = 240Å		.00086	.423	.913	.0423	.000869
		.0000131	.0102	.924	.0423	.0215

systematics

$\psi(h,\underline{g}) = \psi(h,\underline{-g+2h})$ m

$\psi(h,\underline{g}) = \psi(\bar{h},\underline{g-2h})$ $\bar{1}$

e.g. $\psi(1,\underline{2}) = \psi(1,\underline{0})$ above

The results (Eq. (6) and (7)) have become the basis of the commonly used method for setting up high-resolution (axis-centered) dark-field images, since it is usually experimentally more convenient to bring a beam conjugate to that required onto the optic axis. Other applications of the reciprocity theorem include certain conclusions about the symmetry of diffraction contrast images of defects and of dark-field convergent-beam discs (i.e., those satisfying the Bragg condition) from perfect crystals.

Finally, Pogany and Turner (1968) were also able to show that there is a reciprocity of intensities for inelastic scattering if the energy loss is small. Experimental evidence for this has recently been obtained (Lehmpfuhl and Tafto, 1980).

C. Intensity Distribution Recording and Measurement

The methods used for recording diffraction patterns can be divided into parallel and serial methods. The oldest method, photographic film, is a parallel method and is by far the most common one used for static-beam microdiffraction patterns. Film is notable for its high sensitivity, simplicity, resolution, and signal-to-noise ratio. The limitations of film, particularly its small dynamic range and the densitometry requirement, make it somewhat difficult to use for quantitative measurements of intensity distribution. The gases in the microscope vacuum system accompanying the presence of film (particularly water vapor) are highly undesirable in modern instruments for a number of reasons. The current solution to the gas problem in all modern microscopes is differential pumping, which is a marked improvement, but it would be desirable to eliminate film completely from UHV instruments if electron detection systems of the same or better resolution and signal-to-noise ratio can be developed. There has been considerable progress toward that objective.

Effort to develop electronic recording systems for diffraction began early. Most of the initial experimentation was done on electron diffraction cameras rather than microscopes. The first experiments used a scanning apertured detector (usually a Faraday cup) moving through a stationary diffraction pattern produced by a fixed probe (Davisson and Germer, 1927; Lennander, 1954). More recently, Grigson and his co-workers developed a number of increasingly sophisticated electron diffraction systems using apertured scintillator/PMT detectors with electrical or magnetic postspecimen serial scanning of the static-probe diffraction pattern; electrostatic filters were used to discriminate inelastically scattered electrons (Grigson and Tillett, 1968). The detection systems from this work form the basis for those used in most scanning transmission microscopes, and postspecimen serial scanning detection methods are now generally called "Grigson scanning" methods. The method has also been used to produce energy-filtered or energy-loss images, diffraction patterns, or selected-area, energy-loss spectra in CTEM instruments.

Serial scanning electronic detection systems are satisfactory for large-probe microdiffraction but are not fast enough for static-beam, small-probe microdiffraction where the time variation of the relative position of the probe and the specimen region of interest and radiation effects in the specimen become limiting. A parallel-recording electronic detection system is desirable for quantitative experiments of this type. Similar systems are useful for fixed-beam HREM imaging. Four main approaches to parallel detection have been tried. These are classified according to the primary detection element, since this invariably controls the overall performance. These are: (a) the use of charged-coupled devices (Chapman, 1981); (b) the use of diode arrays (Booker, 1981); (c) the use of powder phosphor transmission screens (Spence *et al.*, 1982); and (d) the use of thin, single-crystal screens for two-dimensional recording (Joy, 1981). Since the first two methods are currently in the developmental stage, we shall concentrate on the last two.

A parallel-recording system for microdiffraction in the V.G. HB5 STEM instrument has been described in the literature, and the performance of this system has now been proven over several years (Cowley, 1980). A similar system has been developed for TEM instruments (Spence *et al.*, 1982). A thin transmission phosphor is supported on a 2-inch-diameter fiber-optics plate, which forms the detector system vacuum seal on the microscope column. An electrostatic intensifier (gain ~100) is fiber-optics-coupled to the plate to provide an intensified image of the microdiffraction patterns outside the vacuum. In addition to being more compact, this system provides a higher combination of field of view and numerical aperture than can be obtained using coupling lenses. Channel plate intensifiers, while providing much higher gain, are not satisfactory because of their nonlinearity at high intensity. The net quantum efficiency of the above system is close to unity, so that almost every electron arriving at the phosphor is detected.

This work suggests that the P47 powder phosphor is most suitable for applications where TV rate STEM imaging is required, while the slower P22 phosphor gives better results for microdiffraction. Design considerations include the spectral response of the intensifier photocathode, the use of index matching fluid to eliminate Newton's Rings at the fiber-optics interfaces, geometric distortion of the fibers, and electrical interference from the intensifier power supply (15 kV). A modified version of this system using a single-crystal YAG screen is currently available commercially (Swann, 1982). In the system described by Cowley and Spence (1979), the intensifier output is imaged by a system of lenses onto the surface of a Fresnel field lens on which small mirrors and prisms can be placed to direct particular beams or regions of the microdiffraction pattern onto photomultipliers. The entire field is then viewed by a low-light TV system. The sensitivity of this system can be judged from the fact that it allows microdiffraction patterns from subnanometer regions to be recorded directly on video tape at standard TV rates. Thus, rapid changes in specimen structure due to radiation damage can conveniently be studied.

The spatial resolution and sensitivity of powder phosphor systems is generally determined by the thickness of the phosphor or the range of secondary electrons in it. There is an optimum thickness; however, it is difficult to obtain a spatial resolution of better than about five line pairs per millimeter (modulation transfer function = 0.5) at 100 kV. Each 100-kV electron produces ~1000 photons in the phosphor, and the numerical aperture of the fiber-optics substrate must be considered. If fibers could be drawn from YAG and a plate made of this material with a thin doped layer of implanted material near the surface, it is possible that an improved design would result.

The use of single crystals of activated YAG as screens for microdiffraction recording has recently been reported (Joy, 1981). A thorough evaluation and review of these as detectors in electron microscopy can be found in Autreta *et al.* (1983). For two-dimensional recording, the essential problem is that the spatial resolution of a screen is approximately equal to its thickness. Thus, the problem arises of making a self-supporting 2-inch disc of crystalline material <0.1-mm thick of adequate mechanical strength. Where high speed is required, these YAG detectors appear to give excellent results and show good immunity to radiation damage.

Finally, the use of television frame storage and frame integrator devices should be mentioned. Impressive results have been demonstrated of the high-speed integration of many television frames by systems such as the "ARLUNYA" (Swann, 1982), while the continued fall in price of TV rate digitizing frame store devices suggests that it will soon be commonplace for microdiffraction patterns to be recorded directly into computer memory, hard disc, or magnetic tape. An example of a recent system for digital control of the microscope and digital image aquisition can be found in Strahm and Butler (1981). The development of these systems is vital if the quantitative analysis of electron diffraction patterns is ever to prove as useful as the corresponding x-ray techniques.

III. THEORY

A. Perfect Crystals, "Coherent" and "Incoherent" Microdiffraction

We commence with a description of the geometry of diffraction in convergent-beam patterns formed from an extended, ideally incoherent source (Figure 4). Each point on the effective source disc (the probe-defining aperture) defines an independent source of electrons, the direction of a plane-wave incident on the specimen, and an orientation of the Ewald sphere. Each of these source points (such as P) is conjugate to a set of points P′, one in each diffraction disc, whether the discs overlap or not. Thus, in a perfect crystal, elastic scattering into one point in a Bragg disc can only originate from the conjugate source point. This is redrawn in Figure 11, where the line, OX, in the source disc, CA, subtends an angle, α, at the specimen, as shown on the Ewald sphere diagram (Figure 11(b)). Figure 11(c) indicates a view normal to the beam and systematics line, showing the range of Ewald sphere orientations covered by the incident electron cone. If the "shape transform" or "rocking curve" (the variation with incident-beam direction of the intensity of a particular Bragg beam) is drawn normal to the systematics line as shown, then each CBED disc can be thought of as an angular map of the crystal rocking curve. Note the increased number of subsidiary maxima in the outer discs.

Distance across a particular disc is proportional to excitation error. Patterns are commonly recorded either with the axis of the incident cone parallel to a crystal zone axis or at the Bragg angle for a particular "satisfied" reflection. In the first case, geometry shows that intensities for a "spot" pattern at the exact Bragg angle occur at the edges of the discs if these are just touching, as shown by the points X in Figure 11(d). In the second case, the center of the "satisfied" disc will correspond to zero

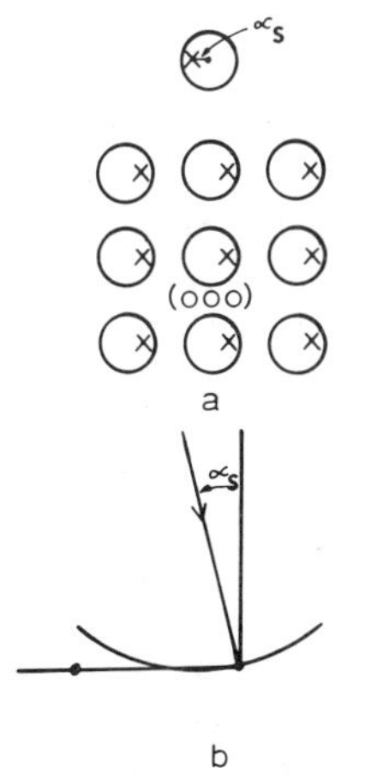

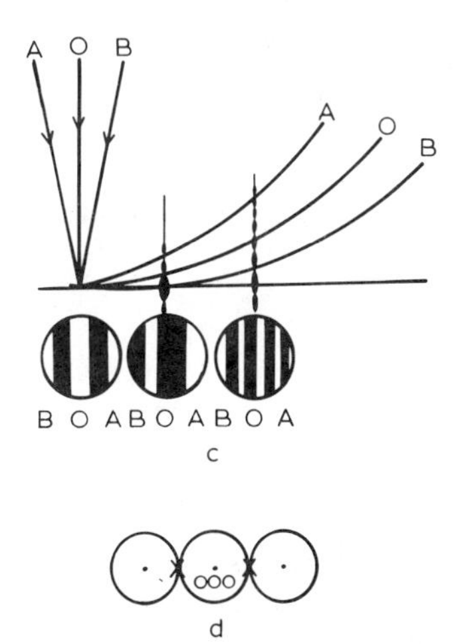

Figure 11 The geometry of CBED patterns. In (a) the condenser aperture is shown looking along the optic axis from the source (above), with the CBED pattern below, for the case of nonoverlapping orders. Conjugate points are marked X. In (b) and (c), the range of Ewald sphere orientation is shown, while (d) indicates at X the positions of the Bragg conditions in a CBED pattern in the symmetrical (zone-axis) orientation.

excitation error for the satisfied reflection. It is sometimes useful in this case to imagine an iris-type condenser aperture being closed down around a particular incident beam direction so that the CBED discs would each shrink to give the corresponding "point" diffraction pattern.

There are various theoretical approximations for the angular variation of intensity across a CBED disc, such as the kinematic approximation, the phase-grating approximation (which gives a rocking curve as a function of direction of projection), and the two-beam approximation. Of these, the two-beam approximation is perhaps the most useful qualitative guide. It can also be used quantitatively in small unit cell crystals, where it forms the basis of perhaps the most flexible and accurate method of local thickness determination (Section IV.B.). This theory gives the intensity variation $I_g(s_g)$ as

$$I_g(s_g) = \frac{1}{1+w^2}\sin^2\left(\frac{\pi t\sqrt{1+w^2}}{\xi_g}\right) \tag{10}$$

for the "satisfied" CBED disc, g, where s_g is the excitation error, $w = \xi_g s_g$, and

$$\xi_g = \frac{\pi}{\sigma V_g}$$

is the extinction distance for the beam, g, with associated Fourier coefficient of crystal potential V_g (in volts). Here, t is the crystal thickness and

$$\sigma = \frac{\pi}{\lambda V_o}$$

(nonrelativistic) where λ is the electron wavelength and V_o is the microscope accelerating voltage. The relationship between excitation error s_g and the angle α_s (Figure 11) is

$$s_g = \frac{2\theta_B\alpha_s}{\lambda} \tag{11}$$

where α_s is the angular deviation from the Bragg angle, θ_B, for the "satisfied" CBED disc, g. (A CBED disc is described as "satisfied" if its center corresponds to the Bragg angle; this is known as a "dark-field disc" or "dark-field image" in the language of the Bristol group (Buxton *et al.*, 1976)).

Equation (10) predicts sinusoidal oscillations in thickness for both I_g and the zero-order beam $I_o = 1 - I_g$. The angular variation is complicated but approaches the kinematic expression for large s_g. The angular width of the central maximum tends to ξ_g^{-1} in thick crystal (the kinematic result is t^{-1}). Thus, the central region of the disc is dominated by "structure factor" or crystal potential information, while the high-order fringes reflect the crystal shape. As observed experimentally, Eq. (10) also predicts that uniform, bright contrast will be seen in CBED patterns from very thin crystals.

The major uses of perfect crystal CBED patterns have been space-group determination (Section IV.A.) and structure-factor refinement (Section IV.D.), where full numerical solutions of the Schroedinger equation are required. For a review of the computing techniques used to do this, see Spence (1981), Metherall (1975), Head *et al.* (1973), and Hirsch *et al.* (1977).

B. Defects: Coherent Microdiffraction

We consider the case of microdiffraction patterns formed from a very small, coherent electron probe in a field emission instrument (Section II). A defect in a crystal scatters a collimated electron beam elastically into all directions, including Bragg directions. For an electron probe a few nanometers in diameter situated over a defect, most of the atoms under the probe may be in disordered positions so that there will be little "Bragg" scattering. In addition, the diffraction limit will require a large objective aperture to form a small probe, resulting in large and blurred CBED "discs" (Figure 5).

There are no simple approximations for estimating the form of this "elastic diffuse" scattering from defects, particularly in a "thick" sample ($t > 100$ Å), due mainly to the presence of coherent multiple scattering and to the fact that in these cases the intensity distribution depends on the precise atomic coordinates of every atom under the probe (unlike the case treated in Section C, below). There is, however, a method of using a standard "many-beam" dynamical electron diffraction computer program to compute the expected form of the elastic diffuse scattering. The method is to define a very large "unit cell" containing all the atoms of the defect and, at the entrance surface of crystal (i.e., in the first "slice" of a multislice calculation), to use the analytic expression for the probe wavefunction. Thus, both the defects in the crystal and the probe are made periodic, with period L. Then the one-dimensional Fourier series for the incident coherent probe wavefunction has Fourier coefficients proportional to

$$H_g = P(U_g) \exp [2\pi i(\Delta f\lambda\, U_g^2/2 + C_s\lambda^3 U_g^4/4)] \tag{12}$$

where $U_g = g/L$, with g an integer, Δf and C_s are the focus setting and spherical aberration constant of the probe-forming (objective) lens, and $P(U_g)$ is equal to unity for $U_g < \alpha/\lambda$ (the objective aperture semiangle is α) and zero elsewhere.

The principle of the method of periodic continuation for calculating coherent small-probe microdiffraction patterns is illustrated in Figure 12. More details are given in Spence (1978), and the result of a typical calculation for a dislocation core is shown in Figure 13 (Cowley and Spence, 1979). Note the increased contrast of the CBED discs as the probe is moved *away* from the core toward more perfect crystal. Note also that in coherent-probe microdiffraction from defects, the entire intensity

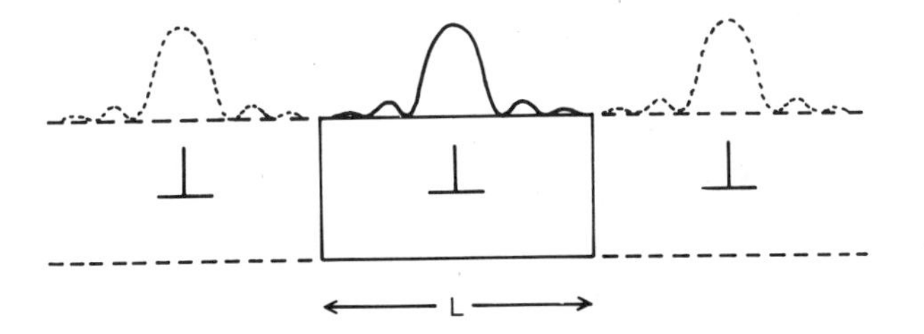

Figure 12 The principle of periodic continuation. The probe is shown incident on a specimen containing a dislocation. If the computer program introduces an artificial periodicity, L, both the defect and the probe will be repeated. The width of a coherent probe is a sensitive function of probe-forming lens focus.

distribution is sensitive to the focus setting, spherical aberration constant, and stigmation of the probe-forming lens, so that accurate values of C_s and Δf are needed to interpret the patterns. For the calculations, the important requirement is that the probe wavefunction fall to negligible amplitude at the boundaries of the cell. Since the energy in coherent probes is widely dispersed except in the neighborhood of the optimum "dark-field" focus (Eq. (4)), these calculations will only be possible over a limited range of focus and will require large amounts of computing space and time for two-dimensional microdiffraction pattern simulation. The exact atom coordinates of every atom under the probe must be entered into the program. Patterns of this type are sensitive to the detailed atomic arrangement under the probe, unlike those in the following section, where the long-range elastic strain field of the defect is sensed. A Bloch wave treatment of coherent microdiffraction has recently been given (Marks, 1985), in which the spatial separation of the wavefield into a "Borrman Fan" is predicted and observed. This approach has much in common with the theory of x-ray topography.

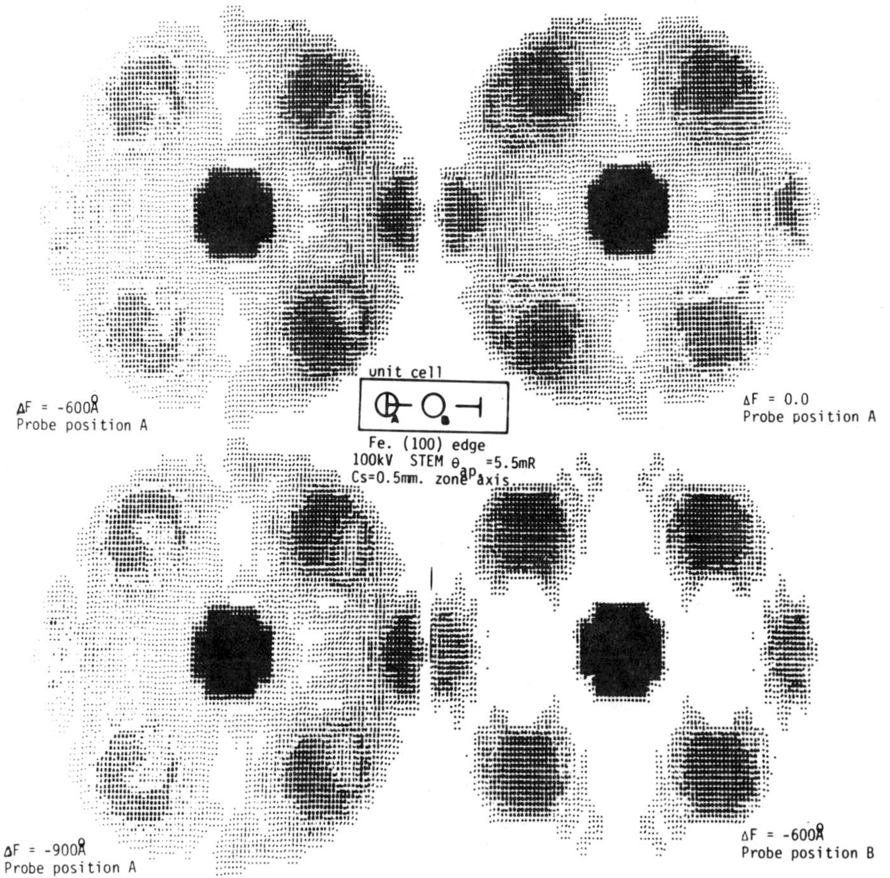

Figure 13 Four computed CBED patterns for an end-on [100] edge dislocation in iron. The two probe positions, A and B, and the core positions in the unit cell are indicated in the inner diagram. Under plane-wave illumination, the four outer discs would sharpen into (110)-type Bragg discs.

C. Defects: Incoherent Microdiffraction

We next consider microdiffraction patterns formed using a probe whose size is determined mainly by the electron source size and the demagnification of the probe-forming lenses rather than by the diffraction limit of the objective lens, and we assume an ideally incoherent effective source.

There are no simple expressions for the electron scattering from defects in "thick" specimens ($t > 200$ Å). There are, however, in many laboratories, computer programs designed to simulate the diffraction contrast (two-beam) TEM images of defects of importance in metallurgy for a specified elastic strain-field. A full description (and listing) of one such program is given in the book by Head *et al.* (1973); the text by Hirsch *et al.* (1977) contains detailed discussion.

These programs (such as those which solve coupled differential equations by Runge-Kutta methods under the column approximation) give an estimate of the intensity, $I_g(x)$, across the exit surface of the crystal due to the contribution of a single beam, g. This is the intensity seen in a diffraction contrast image formed in the electron microscope with a small objective aperture around beam g. These programs do not give the scattering from a defect; they give the image. Nevertheless, it is possible to use programs of this type (without modification) to calculate the intensity variation in CBED discs due to microdiffraction from strained crystal if the principle of reciprocity is invoked, as shown in Figure 14. If lens L1 were perfect and S a point source, we could apply reciprocity between points S and X, so that a program which calculated the TEM image intensity at S due to illumination from X could also be interpreted as giving the intensity at point X in the central CBED disc due to a convergent probe originating from S. Thus, more generally, if the program computes $I_g(x,h)$, the TEM diffraction contrast intensity at point X in the image due to beam g and illumination from direction h, we could also use this result as the CBED intensity in direction h due to illumination from g and a probe centered on X.

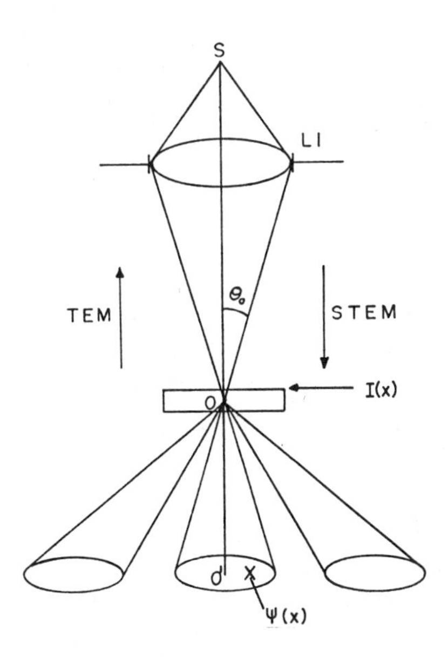

Figure 14 Reciprocity diagram used for column approximation calculation of CBED from defects. The interchange of source (S) and detector (X) allows the CBED intensity at X to be obtained by calculating the TEM image contrast at S due to illumination from X.

However, S is not a point source. In TEM, this is equivalent to imperfect resolution in the detector (e.g., large film grain size). Since an extended detector in TEM is exactly equivalent to an extended source in STEM, the effects of a finite source (and thus finite probe size) in STEM can be obtained from a TEM computer program by using it to give image intensities as they would be recorded on film whose "grain size" is equal to the electron source size in the STEM instrument.

This can be summarized by stating that if $I_g(x,h)$ is the TEM specimen exit-face intensity from a defect imaged in beam g at point X due to illumination from direction h, then if $I_g(x,h)$ is locally smoothed over a distance, d, it will be the intensity in a CBED pattern in direction $-h$ due to a probe of size d_0 (at the specimen) centered on X and illuminated from direction $-g$. This local smoothing can be described by

$$I'_g(x,h) = I_g(x,h) * h_d(x) \tag{13}$$

where the star denotes convolution and $h_d(x)$ is a "top hat" function of width d. For an example of such a calculation for CBED patterns from dislocations in silicon, see Carpenter and Spence (1982). This computational model is not adequate if the probe size becomes smaller than the "column width," c, used in the program, or if an illumination angle θ_c greater than λ/c is used.

We see, therefore, that the information in incoherent zone-axis, small-probe CBED patterns is complementary to two-beam diffraction contrast images of the same defects. The CBED has the advantage that the effects of the displacement field of a single defect on many diffracting vectors can be seen in a single pattern and the defect symmetry determined. The spatial distribution of many defects can be determined from a single image. The two techniques are complementary, and the use of both facilitates defect analysis in crystals.

D. Shadow Images and Ronchigrams

If the probe used to form microdiffraction patterns is focused at a point above or below the specimen, the microdiffraction pattern becomes a convergent-beam shadow image of the specimen, i.e., each Bragg disc contains an out-of-focus, real-space image of the irradiated specimen area. Shadow images may be formed with coherent or incoherent probes. Using geometric ray optics, it is easy to show that the Gaussian magnification of a shadow image is $L_f/\Delta f$, where L_f is the focal plane (i.e., probe) to image observation plane distance and Δf is the probe defocus. For small Δf (large magnification), and particularly for coherent probes, the patterns become more complicated and require wave optics treatment for interpretation. Shadow images have useful and interesting properties. They can be used to align, stigmate, and focus STEM instruments (Cowley, 1979b), to measure C_s and Δf in STEM instruments, and to form a special type of fixed-beam lattice imaging in STEM (Cowley, 1979c), or to measure C_s and Δf for the prefield of a TEM/STEM condenser-objective lens (Carpenter *et al.*, 1982).

For the case of Bragg disc overlap (large α) and small coherent probes, Cowley and Moodie (1960) proposed a form of "lensless" lattice imaging, similar to the Gabor in-line hologram geometry. With modern field emission sources, this proved possible, and an example is shown in Figure 15 (Cowley, 1981a). Recall that the images in Figure 15 were formed in a D-STEM, but with a static probe, using a parallel

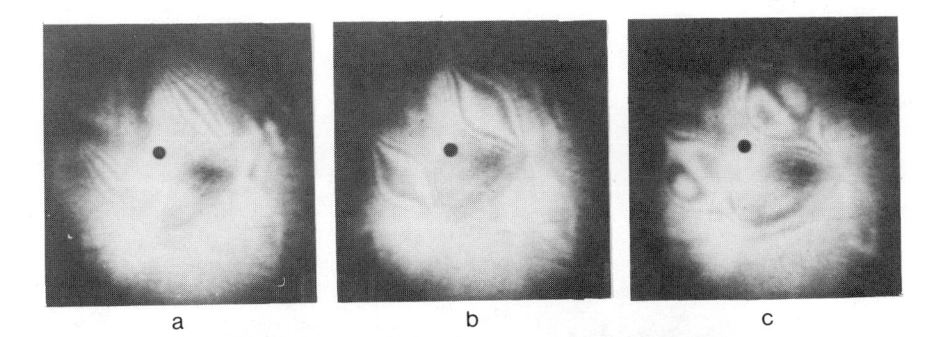

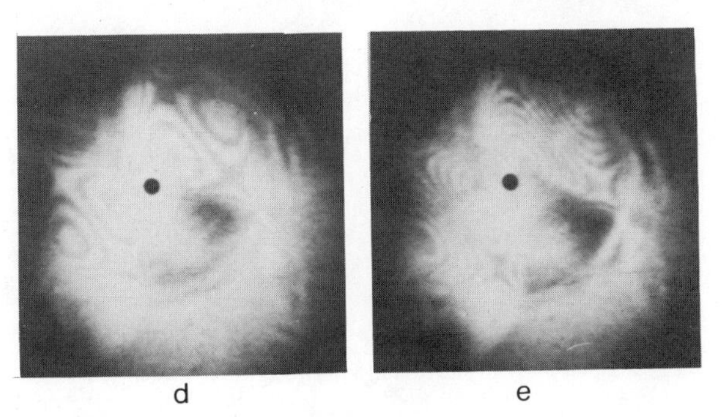

Figure 15 Through-focus series of shadow images of an MgO smoke crystal, showing the Ronchigrams due to the (200) lattice planes (d = 2.1 Å). The black spot is a mirror on the optical system, used to indicate the lens axis: (a) is overfocused; (b) is at the optimum defocus for obtaining (200) fringes in STEM; (c), (d), and (e) are progressively underfocused.

electronic detection system. Images of this type furnish the most direct measurement of system mechanical stability, independent of scan system electronics. For coherent shadow image theory see Cowley (1979c). These images are sometimes known as Ronchigrams, by analogy to a similar optical arrangement for astronomical mirrors.

The Gaussian magnification for a shadow image formed by a perfect lens is $L_f/\Delta f$. When C_s for the probe-forming lens is not zero, two defects appear in the shadow images, illustrated in Figure 16, whose angular positions permit direct measurement of C_s for the probe-forming lens. The defects are circles of infinite magnification. The radial circle results from a crossover on the lens caustic surface of rays with differing angles of incidence, α, and the tangential circle from rays with equal values of α forming a crossover on the optic axis at some distance before reaching the paraxial focal point. The effective shadow image magnification, M, for the real lens is

$$M = \frac{L_f}{(\Delta f + KC_s\alpha^2)} \tag{14}$$

where $K \simeq 1$ for the tangential and ~ 3 for the radial circle. The circles occur where $KC_s\alpha^2 = -\Delta f$, in underfocus shadow images. A crystalline specimen of known structure and spacing can be used to calibrate the angular deflection, and C_s can be

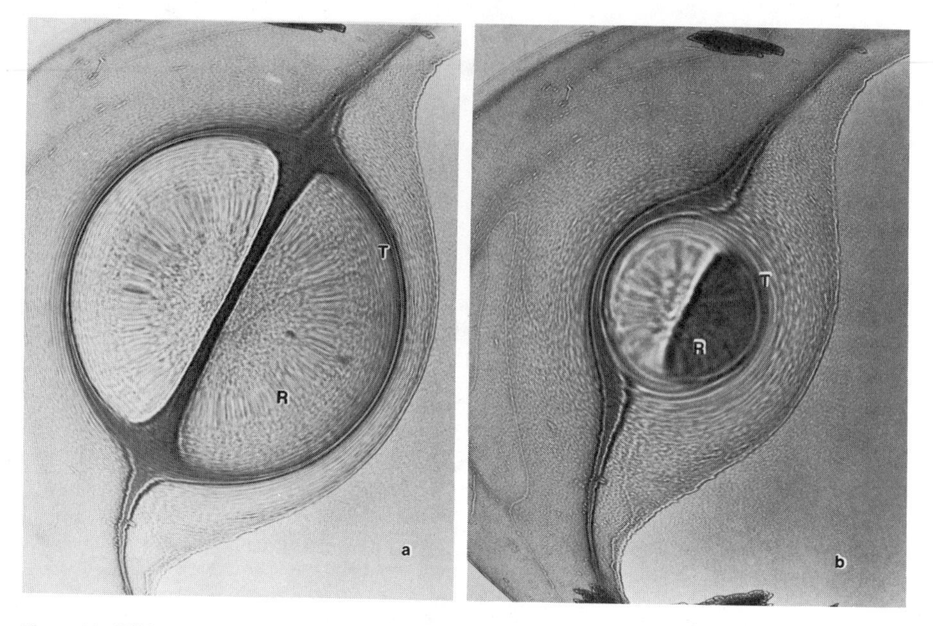

Figure 16 Wide-angle, convergent-beam shadow images from an amorphous carbon film specimen. Taken in a field emission gun TEM/STEM, film recorded. (a) Objective defocus $\simeq -5$ μm, (b) $\Delta f \simeq -1$ μm. The radial circle occurs at smaller angle than the tangential circle. Both decrease in angle as Δf approaches zero.

determined from shadow images of known Δf. For large defocus, the image will appear at low magnification, relatively undistorted. As the probe is brought progressively closer to focus, the magnification and image distortions increase, with the image defects closing to smaller α, nearer the optic axis.

Shadow images are the most useful and reliable method for determining small-probe position on a specimen during various AEM convergent-probe experiments (Carpenter, 1980). At low magnifications, the image distortions are small, and the shadow image is easily correlated with an in-focus TEM or STEM image of the same area (Figure 17(a), (b)). As the magnification is increased ($\Delta f \rightarrow 0$), the field of view decreases, and the probe can be kept centered on the region of interest with beam translators, until at the diffraction point ($\Delta f = 0$) all real-space image detail disappears (Figure 17(e)). This method of probe placement avoids the discontinuous shift in focused probe position accompanying a change from image to diffraction mode, which occurs in most microscopes. The shift results from lens field interaction and is usually small (on the order of a few nanometers) but is critical for quantitative small-probe experiments.

IV. APPLICATIONS

A. Space-Group Analysis

Microdiffraction may be used to determine the crystal space-group of small crystalline regions and particles down to a few nanometers in diameter. It thus provides valuable information for phase identification. This topic is covered in more

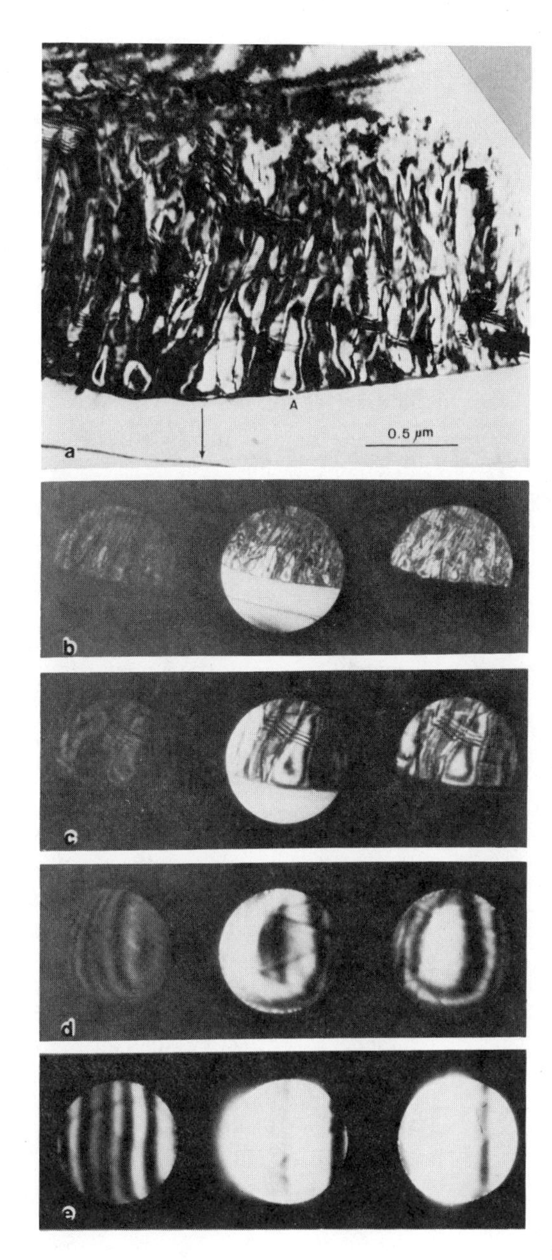

Figure 17 Correspondence between CTEM image and convergent-beam shadow images of varying magnification (varying defocus). The specimen is an Fe-Co-V alloy. (a) Shows a high-angle grain boundary that was migrating in the direction marked on the TEM image. (b) to (e) show shadow images brought to focus on region A by continuous change in condenser lens current, in diffraction mode. Probe is focused in (e). Diffraction discs are $\pm[111]$.

detail in the article by John Steeds (1979) and in references to follow. The broad picture is, however, roughly as follows: Since Friedel's law does not apply to electron diffraction (Goodman and Lehmpfuhl, 1968), microdiffraction patterns express the full crystal point group of the region illuminated. Thus, the three-dimensional diffraction effects from higher order Laue zones (HOLZ) seen in wide-angle microdiffraction patterns may be used to determine the point group elements (mirrors, rotational symmetries, and centers of symmetry) present in the region illuminated (Goodman, 1975; Buxton *et al.*, 1976). In some cases, a second microdiffraction pattern will be needed, and there are some intractable cases.

The determination of the remaining translational symmetry elements (screw and glide) needed to complete a space-group determination may then be made by using the Gjonnes-Moodie rules (Gjonnes and Moodie, 1965). These rules specify that for certain incident-beam directions, a reflection that is kinematically forbidden due to the presence of a translational symmetry element (but which might be thought to occur as a result of double or higher order scattering) in fact remains forbidden due to the cancellation of symmetry-related, multiple scattering paths. The resulting line or cross of zero intensity in the forbidden convergent-beam disc can then be used to indicate the presence of a translational symmetry element. Some features of the convergent-beam method are as follows:

(a) Interfering matrix diffraction effects make the identification of small particles embedded in a host matrix generally very difficult by this method (Titchmarsh, 1978; Steeds, 1980).
(b) The previous discussion refers to a perfect, parallel-sided crystal, a situation that can usually be approached if a sufficiently small probe is available. Field emission sources are very useful in this respect. In general, the symmetry determined by this method is that of the crystal and all its defects, including the effects of the crystal boundary conditions.
(c) In general, it should be remembered that it is strictly possible to prove only the absence of a symmetry element, and not its presence. However, since there appears to be no other technique more sensitive to symmetry changes than CBED, this method may be said to provide the strongest evidence in favor of a particular symmetry classification.
(d) Enantiomorphous pairs may be distinguished by convergent-beam electron diffraction (Goodman and Johnson, 1977).
(e) The previous discussion refers to perfect crystals illuminated with an electron probe much larger than the unit cell. If a probe "smaller" than the cell is used, so that the diffraction discs overlap in the "coherent" case (Section II), the planar point group symmetry of the microdiffraction pattern will be equal to that of the crystal potential, projected in the beam direction, with respect to an origin taken at the probe center (Cowley and Spence, 1979). (Here we have assumed a correctly stigmated probe and negligible upper layer line interactions). Thus, in principle, the site symmetry or local coordination of groups of atoms within a single unit cell could be studied by this method.
(f) A simple method of polarity determination in noncentrosymmetric crystals has been described (Tafto and Spence, 1982). This makes it possible to identify, for example, the gallium and arsenic sites in an HREM lattice image. Two very useful articles that describe the stepwise procedure for space-group determination by CBED are those by Eades (1983) and Steeds and Vincent (1983).

B. Local Thickness Measurement for Crystalline Specimens

Some of the earliest work on electron diffraction (Kossel and Mollenstedt, 1939; MacGillavry, 1940; Ackermann, 1948) showed that the minima in CBED discs (rocking curves) could be used to determine the average thickness of the irradiated specimen volume along the beam direction. With the development of modern analytical microscopes, the need for accurate local thickness information increased, particularly in connection with microspectroscopy experiments. More recent investigations of the method (Kelly *et al.*, 1975; Allen, 1981) showed that local thickness measurements can be made with $\sim\pm 2\%$ accuracy using dynamical two-beam theory, provided that suitable experimental procedures satisfying the assumptions appropriate to the theoretical approximation are used. This research used film-recorded CBED patterns, and whereas the method is straightforward, it is desirable to have thickness data while actually doing other experiments on the instrument, for which a knowledge of thickness is important. Most recently, double-deflection Grigson scanning has been used both to determine the rocking curve parameters necessary to calculate local thickness and to collect the rocking curve intensity distribution (filtered or unfiltered) for on-line computation (Bentley and Lehman, 1982). This procedure enables on-line thickness determination in times of ~ 1 minute within the range of applicability of the method. It is interesting to consider how much faster it might be done with the parallel electronic detection system and digital data acquisition system discussed previously. It has been shown by Blake *et al.* (1978) that two-beam CBED rocking curves, recorded on film with a partial irradiation aperture block (Goodman, 1972), can also be used to determine reasonably accurate extinction distances and anomalous absorption parameters. The previously mentioned Grigson method permits collection of the same data very quickly by sequential collection of filtered and unfiltered rocking curve profiles using a magnetic sector analyzer with the scanning system. Thus, the necessary input data (in addition to diffracting conditions) for diffraction contrast image simulation and defect identification can be obtained within minutes of the experimental micrograph itself. Large errors can, however, be introduced when using the Grigson method if the detector used is not small.

C. Microdiffraction From Defects and Interfaces

Whereas a quantitative analysis of microdiffraction intensities for the structural analysis of defects is possible in principle, most of the work undertaken to date has used certain characteristic and reproducible features of the patterns to make deductions about the specimen. We briefly consider three examples of the early 1980s.

Zhu and Cowley (1982) have used the nanometer-sized electron probe on the Vacuum Generator HB5 instrument to study the nature of antiphase domain boundaries in Cu_3Au. With this instrument it is possible to form microdiffraction patterns from within a single domain. Patterns were recorded by using the previously described parallel recording system and stored on video tape and digital framestore. Earlier work (Cowley and Spence, 1981) had shown that the effect of a crystal edge on coherent microdiffraction patterns was to produce an annular pattern of intensity for each diffraction disc. Similarly, the lattice discontinuity that occurs at an antiphase boundary produces a "splitting" of the superlattice spots, but not the fundamentals.

Figure 18 shows a series of microdiffraction patterns recorded from "good and bad" antiphase domain boundaries, using an electron probe ~ 15 Å in diameter. By analyzing all the possible boundaries that can occur, these authors were able to derive Table II, which allows the type of boundary to be identified from its microdiffraction pattern. (The terms "good" and "bad" refer to the nearest-neighbor coordination of the Au and Cu atoms at the boundary.) It should be noted that, by comparison with $\underline{g}\cdot\underline{b}$ analysis in weak-beam imaging by diffraction contrast, the microdiffraction technique (with nonoverlapping orders) has the advantage that all the spots (those affected by strains or discontinuities and those unaffected) are recorded at once, using a zone-axis

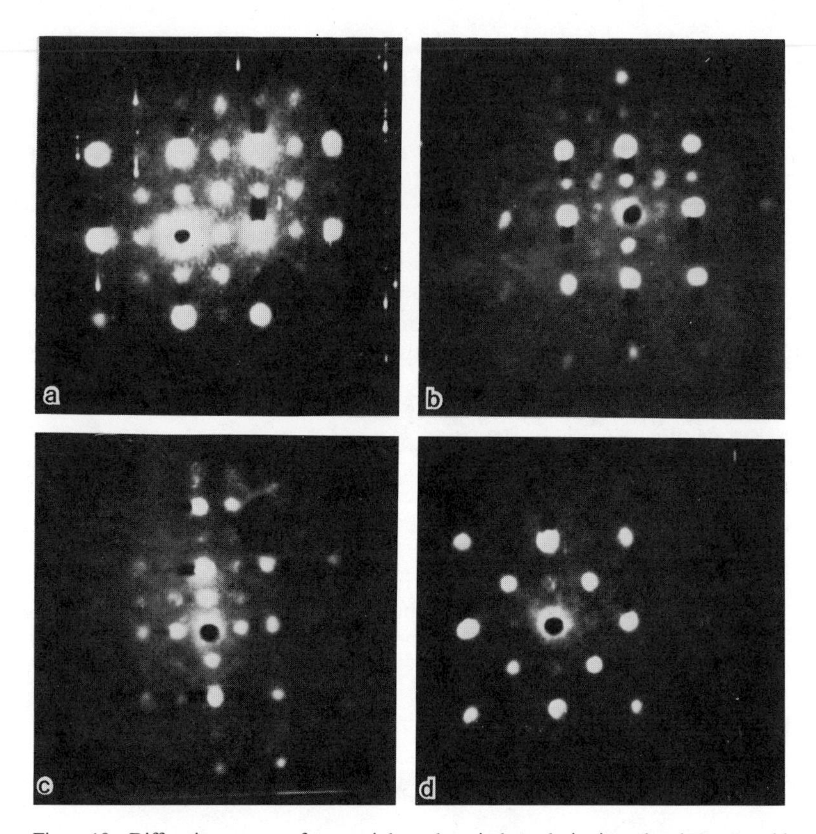

Figure 18 Diffraction patterns from antiphase domain boundaries in ordered copper-gold (Cu_3Au) alloy, obtained with a subnanometer probe. (a) From the crystal, probe not on a boundary. (b) From a "good" boundary. (c) From a "two good" boundary. (d) From a "bad" boundary.

orientation. Similar rules for the effects of stacking faults and twins on spot-splitting have also been derived (Zhu and Cowley, 1983). An attempt to distinguish two current models for the atomic structure of nitrogen platelets in diamond from experimental patterns from individual platelets can be found in Cowley and Osman (1984).

An example of the use of "incoherent" microdiffraction to analyze defects is provided by the work of Carpenter and Spence (1982). These workers found that the effects of strain on Kikuchi lines and HOLZ lines (the fine lines seen to cross the central CBED disc) could be used to determine the three-dimensional Burgers vectors of dislocations. Thus, a single zone-axis microdiffraction pattern often contains enough information to determine an "out-of-zone" fault vector, due to the presence of three-dimensional diffraction effects. Dislocation strain was found to produce a fine splitting of some Kikuchi lines and HOLZ lines, while others remained unaffected. Rules for the type of splitting to be expected in fcc crystals were derived, as shown in Table III. The reciprocity theorem was combined with the column approximation to provide a method of computing "large-probe" (incoherent) electron microdiffraction patterns from defects, thus enabling convergent-beam patterns from strained crystal to be computer-simulated by using the existing, readily available "diffraction contrast"

TABLE II

Diagrams Illustrating "Splitting Rules" for Bragg Spots Resulting From a Coherent Microdiffraction Probe Incident on Several Types of Antiphase Domain Boundaries in Ordered Copper-Gold Alloys. Refer to Section IV.C. for discussion.

Type	Name	Boundary case	Diffraction pattern
Single	'Good' boundary G	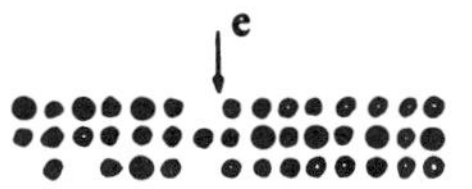 e	
	'Bad' boundary B_1	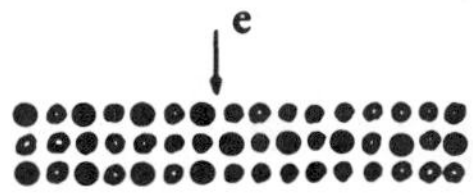 e	
	'Bad' boundary B_2	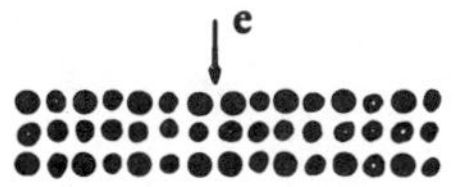 e	
Double	'Good' and 'good' 'Bad$_2$' and 'bad$_2$'	 G, G the area illuminated by incident beam	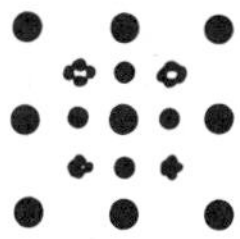
	'Good' and 'bad$_1$' or 'Bad$_2$' and 'bad$_1$'	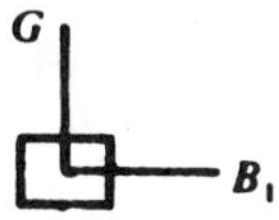 G, B_1	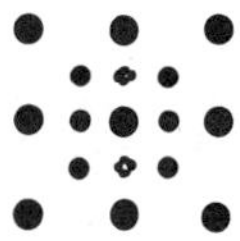
	'Bad$_1$' and 'bad$_2$'	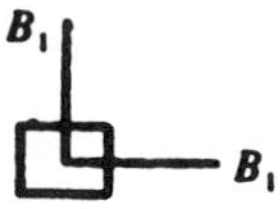 B_1, B_1	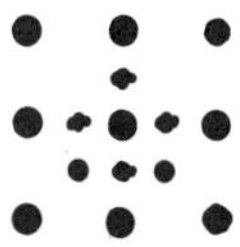
	'Good' and 'bad$_2$'	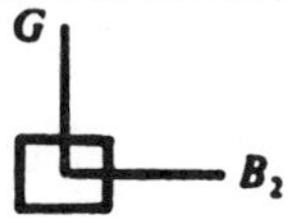 G, B_2	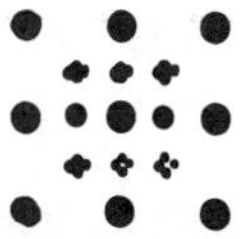

TABLE III

Effect of Dislocations on CBED Patterns From fcc Crystals.

Type of Dislocation	Example (b)	Effect on CBED
In-zone unit dislocation	$\pm\frac{1}{2}[1\bar{1}0]$	$\pm[22\bar{4}]$ in-zone Kikuchi lines unsplit; all FOLZ Kikuchi and FOLZ lines in $[1\bar{1}0]$ zone unsplit
Out-of-zone unit location	$\pm\frac{1}{2}[110]$	$\pm[h\bar{h}0]$ in-zone Kikuchi lines unsplit; $[9\bar{9}1]$ FOLZ line unsplit
In-zone Shockley partial	$\pm\frac{1}{6}[\bar{1}\bar{1}2]$	$\pm[h\bar{h}0]$ in-zone Kikuchi lines unsplit; all FOLZ lines split
Out-of-zone Shockley partial	$\pm\frac{1}{6}[\bar{1}1\bar{2}]$	$\pm[6\bar{2}\bar{4}]$ in-zone Kikuchi lines unsplit
	$\pm\frac{1}{6}[121]$	$\pm[h0\bar{h}]$ in-zone Kikuchi lines unsplit; $[11\bar{1}\bar{9}]$, $[\bar{9}\bar{1}11]$ FOLZ lines unsplit

software (for listings see, for example, Head *et al.*, 1973). In these experiments, with a probe size of ~100 Å, the patterns are more sensitive to the long-range strain-field than the detailed atomic arrangement at the dislocation core, and the situation is similar to diffraction contrast imaging.

It can be seen from Figure 2 that HOLZ lines accurately reflect the three-dimensional symmetry of the specimen crystal along the beam direction. These lines correspond to Bragg reflections from planes of close spacing (e.g., for Si at 100 kV $(11,\bar{5},\bar{5})$ is on the FOLZ ring for B = [111], and $2\theta_B$ = 5.11°); they are, therefore, very sensitive to changes in local lattice parameter or strain (Steeds, 1979). These effects have been applied to examinations of misfit strain and symmetry changes and to measurements of the degree of chemical order in structural alloys.

The properties of nickel-based superalloys containing a distribution of ordered $\gamma'(L1_2)$ precipitate in the γ(fcc) matrix depend on the distribution of γ' in the matrix and its morphology. Since these alloys are most often not in thermodynamic equilibrium, the microstructure is dependent on thermomechanical path. Porter *et al.* (1981) have examined the structure of the γ and γ' phases in several of these alloys after cold work followed by a recrystallization anneal below the γ' solvus. Passage of the recrystallization front takes most of the preexisting γ' into solution, with reprecipitation of γ' by several mechanisms on or behind the moving front. CBED examination of the new γ' morphological forms showed that γ' discontinuously precipitated on the interface was tetragonally distorted rather than cubic; the distortion was attributed to additional solute ordering on the A-sublattice of the A_3B-L1_2 structure. In alloys of high solute content, fine secondary γ' was formed in the matrix on cooling from the recrystallization temperature.

The central discs from two [111] CBED patterns from this "matrix" are shown in Figure 19. The secondary γ' in this case has small misfit and was coherent, but it can be seen that the patterns no longer have three-fold symmetry; only one mirror plane exits in each. Kinematical simulation of these diffraction patterns showed the matrix to be tetragonal, with the difference between a and c axes of ~0.25%. Tetragonal regions with both $c/a > 1$ and $c/a < 1$ were observed. These investigations were extended to similar alloys containing relatively large (~50 nm) and isolated γ' particles, to examine the constrained misfit (Ecob *et al.*, 1982). Local lattice parameters in γ' and γ were determined by comparing the experimental diffraction patterns to computer simulations. Absolute values of misfit from 0.32% to 0.03% were determined. In further work on the same alloy system (Porter *et al.*, 1983), the influence of faceted coherent γ' precipitate on the surrounding γ matrix symmetry was examined. It was found that the matrix surrounding the particles was strained into tetragonal symmetry. CBED patterns similar to Figure 19 were exhibited by the matrix when the probe was closer than ~400 nm to the particles. The location of the tetragonal axis depended on the position of the CBED probe relative to the nonspherical γ' particle symmetry. These results show the usefulness of CBED for measurement of local strain and symmetry in metal alloys. Increasingly wide applications of the method may be expected in metallurgy, and another application of it to the same system can be found in Fraser (1983).

The degree of long-range chemical order in iron, nickel, and vanadium A_3B alloys where A = (Fe, Ni) and B = V has been determined by the CBED/HOLZ method (Braski *et al.*, 1982). The cubic lattice constant of these alloys changes with long-range order parameter. The relationship between these two variables was determined by independent methods. HOLZ line positions are related to local lattice constant a_o and

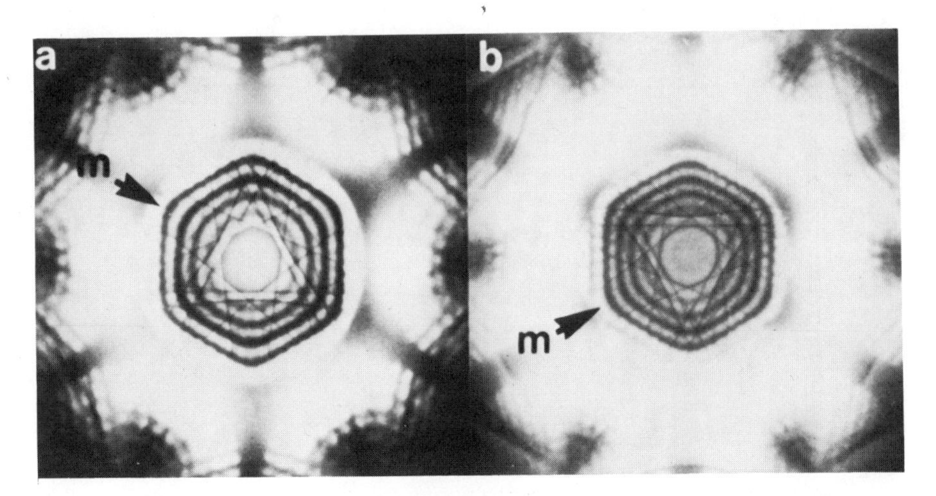

Figure 19 CBED microdiffraction patterns from two matrix regions of a nickel-base γ/γ' superalloy. The matrix contained fine coherent secondary γ', with very small misfit. These patterns were recorded using a STEM probe (cf. Figure 3(b)) at 120 kV. The beam direction was $[111]_{fcc}$. Each pattern contains only a single mirror, indicating a change from cubic to tetragonal symmetry induced by the precipitation reaction. (a) $c/a = 1.0025$; (b) $c/a = 0.9975$. The change in symmetry is most easily determined by observation of the large triangle within the central disc; it originates from $\{5,5,\overline{9}\}$ and $\{5,3,\overline{9}\}$ FOLZ lines.

the microscope accelerating voltage E by $\Delta a/a = \Delta E/2E$ (Steeds, 1979). The microscope used in three experiments was modified to allow continuous control of E, following a design similar to the method developed by the Bristol group.

The lattice constant of a specimen with an unknown degree of order was determined by finding the "crossover voltage" of certain HOLZ lines; the method is illustrated in Figure 20. The crossover voltage corresponds to intersection of the <931> FOLZ lines at a point in the [114] CBED central discs. The lattice constants for ordered and disordered specimens of this alloy are 0.35927 nm and 0.36127 nm, respectively. The accelerating voltage stability of the microscope is 12×10^{-6} over 10 min. Uncertainties from voltage instability introduced in the lattice constant measurements are ~100 times smaller than the change in lattice constant accompanying the ordering reaction; therefore, this source of error is not likely to limit the method here. Since the incident incoherent probe size in this case was ~10 nm, the method is capable of quite high spatial resolution. Ecob *et al.* (1982) found evidence for dispersion effects on absolute lattice parameter determinations by HOLZ methods, which depend on the beam direction in the crystal, i.e., zone axis. Until this effect is more completely investigated, it would be prudent to develop standards for CBED lattice parameter measurement using the same zone axis as that used for HOLZ patterns from specimens of unknown lattice constant. This technique is obviously applicable to segregation research.

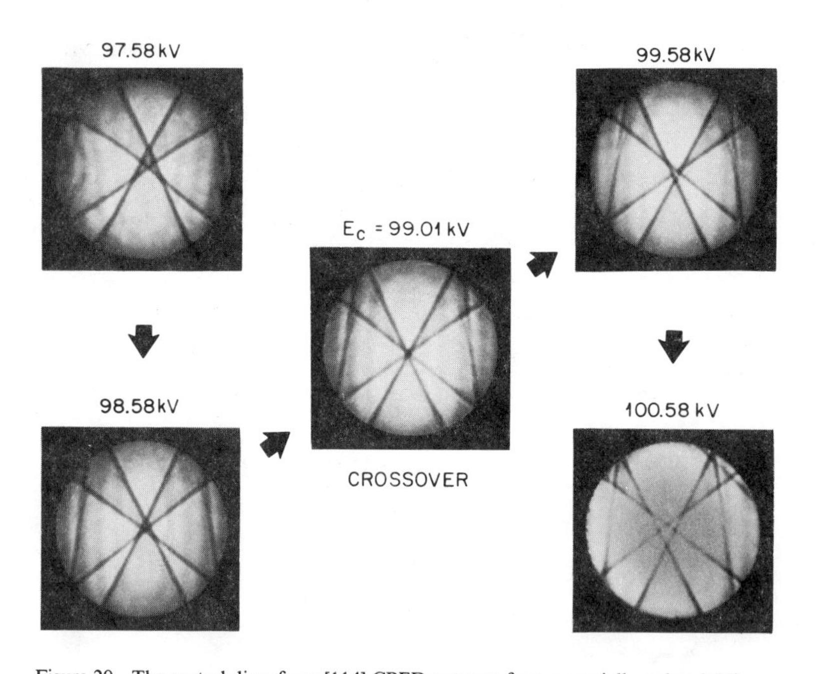

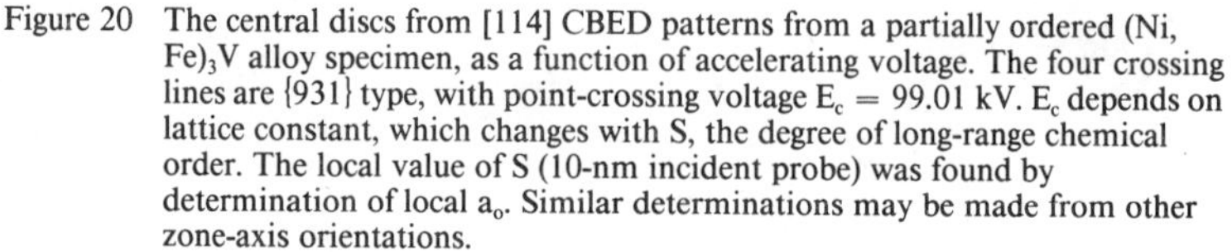

Figure 20 The central discs from [114] CBED patterns from a partially ordered (Ni, Fe)$_3$V alloy specimen, as a function of accelerating voltage. The four crossing lines are {931} type, with point-crossing voltage $E_c = 99.01$ kV. E_c depends on lattice constant, which changes with S, the degree of long-range chemical order. The local value of S (10-nm incident probe) was found by determination of local a_0. Similar determinations may be made from other zone-axis orientations.

Finally, it should be pointed out that methods exist for the convenient computation of small-probe, coherent microdiffraction patterns from defects by the technique of artificial periodic continuation (Spence, 1978). Figure 13 shows the computed patterns (including all multiple scattering effects) that would be expected if a coherent 6-Å probe were placed on, and to one side of, an end-on-edge dislocation in iron. Note the increased diffuse elastic scattering between the Bragg discs as the probe is moved from the more perfect crystal to the dislocation core. Note also the dependence of the intensity on the focus setting of the probe-forming lens, despite the fact that the diffracted orders do not overlap, due to the crystal strain. The deduction of atom coordinates from patterns such as these remains an ambitious project for the future (Cowley and Spence, 1979).

D. Structure Factor Refinement

An important problem in x-ray crystallography is the determination of structure factor phases, which contain information on atom positions. The kinematic intensities in x-ray work are normally independent of the phase of the structure factors. Since the coherent interference between diffracted beams that occurs in dynamical electron diffraction mixes these structure factors, electron intensities *are* highly sensitive to structure factor phases and, therefore, to atom positions, albeit in a very complicated way. Because of the complexity and large number of parameters involved in refinement (e.g., absorption coefficients, thickness, Debye-Waller factors), most studies have concentrated on a few low-order reflections that are difficult to refine by x-ray methods in otherwise "known" structures.

The use of a small probe and a convergent beam allows small, perfect regions of crystal to be examined and so avoids the statistical averaging inherent in the x-ray method. At the same time, it provides a wide angular range across the CBED disc (the crystal "rocking" curve) over which to compare dynamical computed intensities with experimental results. As examples, we cite first the many studies of the (111) and (222) reflections in germanium and silicon (the "forbidden" (222) reflection determines the degree of nonsphericity of the atoms), which have provided the most accurate values of these reflections currently available (Voss, Lehmpfuhl, and Smith, 1980). The use of energy-filtering electron microscopes further increases the accuracy (Pollard, 1972). Second, there is the refinement of the inner reflections of graphite by CBED (Goodman, 1976) in which an accuracy of $<1\%$ was obtained. These are difficult and lengthy studies, requiring great care and large amounts of computer time. As a variant on this method, we also mention the use of the critical-voltage effect on microdiffraction patterns for structure refinement (Sellar, Imeson, and Humphreys, 1980). This method is slightly more accurate than those discussed previously and relies on the extinction of certain reflections at particular accelerating voltages. Accuracies of 0.2% have been reported by using the critical-voltage effect, again from small specimen areas.

E. Microdiffraction From Amorphous Materials

Diffraction patterns from large areas of amorphous material consist of rings of intensity that may be used to determine the radial distribution function for the material (for a review of this work, see Wagner, 1978). This gives the lengths and the weightings of the first few nearest-neighbor interatomic vectors in real space. Is there any new information to be obtained by using diffraction patterns from very small areas (say 1 nm) of amorphous or disordered material? In such patterns, the rings are seen to break up into blobs of intensity (Brown *et al.*, 1976), as might be expected if one considers the extreme case of an isolated cluster of a few atoms illuminated by a coherent cone of radiation. Thus, while the large area patterns contain one-dimensional information on bond lengths (all bond orientations have been averaged over), the small probe patterns contain information on both bond lengths and bond angles for the region illuminated. A statistical characterization of the whole sample would then be possible by collecting a large number of patterns from different areas.

The autocorrelation function of the crystal potential (or Patterson function) provides such a suitable characterization, can be derived from measured intensities, and would in principle enable the two current models of amorphous materials (the "microcrystalline" and the "random network" models) to be distinguished. The main difficulties with this approach are: (a) the effects of multiple scattering, and (b) the projection approximation, which means that only diffraction patterns from the structure projected in the beam direction are normally available. Some progress has, however, been made in relating the Patterson function for amorphous silicon to observed coherent microdiffraction patterns (Cowley, 1981b). Despite the extreme sensitivity of these patterns to the local atomic arrangement and the ready availability of methods for the computer simulation of them, the interpretation of coherent, small-probe microdiffraction patterns from amorphous materials remains an unsolved problem, an outstanding challenge for electron microdiffraction. The use of pattern-recognition techniques in microdiffraction to determine the relative frequency of occurrence of particular atomic arrangements in microcrystalline or near-amorphous materials has been described (Monosmith and Cowley, 1983).

F. Short-Range Order

Since modern field-emission electron microscopes allow microdiffraction patterns to be obtained from nanometer-sized regions of crystal, it is now possible to record diffraction patterns from within a single microdomain in quenched, partially ordered alloys. From the ratio of intensities of the superlattice reflection to the zero order beam, the Bragg-Williams and Cowley-Warren short-range order parameters can be estimated (Chevalier and Craven, 1977). While the values obtained were in good agreement with those obtained by x-ray studies for equiatomic copper-platinum, an important difficulty with this technique is the strong effect of multiple scattering for heavy elements. An analysis of the microdomain structure of $LiFeO_2$ by subnanometer microdiffraction can be found in Chan and Cowley (1981).

G. Small Particles and Catalyst Research

An analytical electron microscope provides the only method for studying diffraction from isolated metal particles a few nanometers in diameter in a nonstatistical way. There have now been many studies of these materials reported in the literature. In all cases a crucial requirement for obtaining microdiffraction patterns from nanometer-sized areas containing fine detail is an efficient, parallel-recording, UHV-compatible electron detector system (Cowley, 1980). It is this fine detail that contains information on the defect structure of the particles. Both experimental microdiffraction patterns and theoretical calculations (Cowley and Spence, 1981) have shown that coherent microdiffraction patterns recorded near an edge or on a particle whose dimensions are comparable with the probe diameter show a characteristic annular appearance in each spot.

Figure 21 shows experimental microdiffraction patterns, while Figure 22 shows the results of dynamical calculations for a similar case. Alternatively, the diffraction discs may break up into smaller spots. Since these spots are smaller than the illumination angle, identification of the reciprocal lattice orientation can be difficult. In general, it is therefore expected that diffracted spots will be split whenever an incident-coherent convergent beam is close to a discontinuity in the lattice potential. The effect was noted previously for microdomain boundaries.

Examples of the analysis of multiple twinning in small gold particles (2 to 5 nm in diameter) can be found in Cowley and Roy (1982) and Roy, Messier, and Cowley (1981). A similar analysis of industrial supported platinum and palladium catalyst particles with a description of a simple parallel detector system can be found in Lynch *et al.* (1981). The identification of microcrystalline oxides on thin films of chromium and iron is described in Watari and Cowley (1981), while an analysis of the Z-contrast method of Crewe as applied to the problem of enhancing the contrast of STEM images of catalyst particles can be found in Treacy, Howie, and Wilson (1978). The general conclusion from the microdiffraction work is that the multiple twinning found in large particles persists in decahedral and icosahedral particles as small as 2 to 3 nm (Cowley and Roy, 1982). The question of localized or distributed strains in multiple twinned particles is discussed in Monosmith and Cowley (1984).

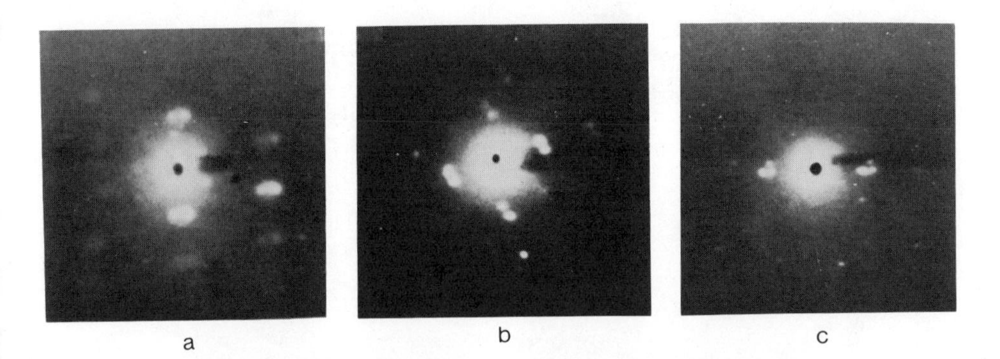

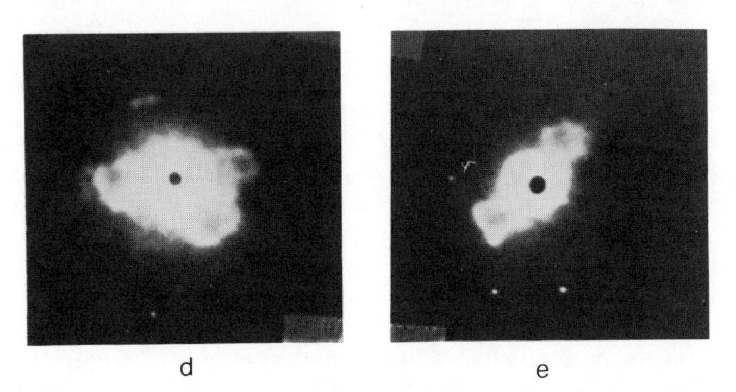

Figure 21 Microdiffraction patterns from small gold crystallites, with the incident beam at the edge of the crystal. The beam convergence angle is $\sim 3 \times 10^{-3}$ rad (beam diameter at specimen, 15 to 20 Å) for (a), (b), and (c), and 2×10^{-2} rad for (d) and (e).

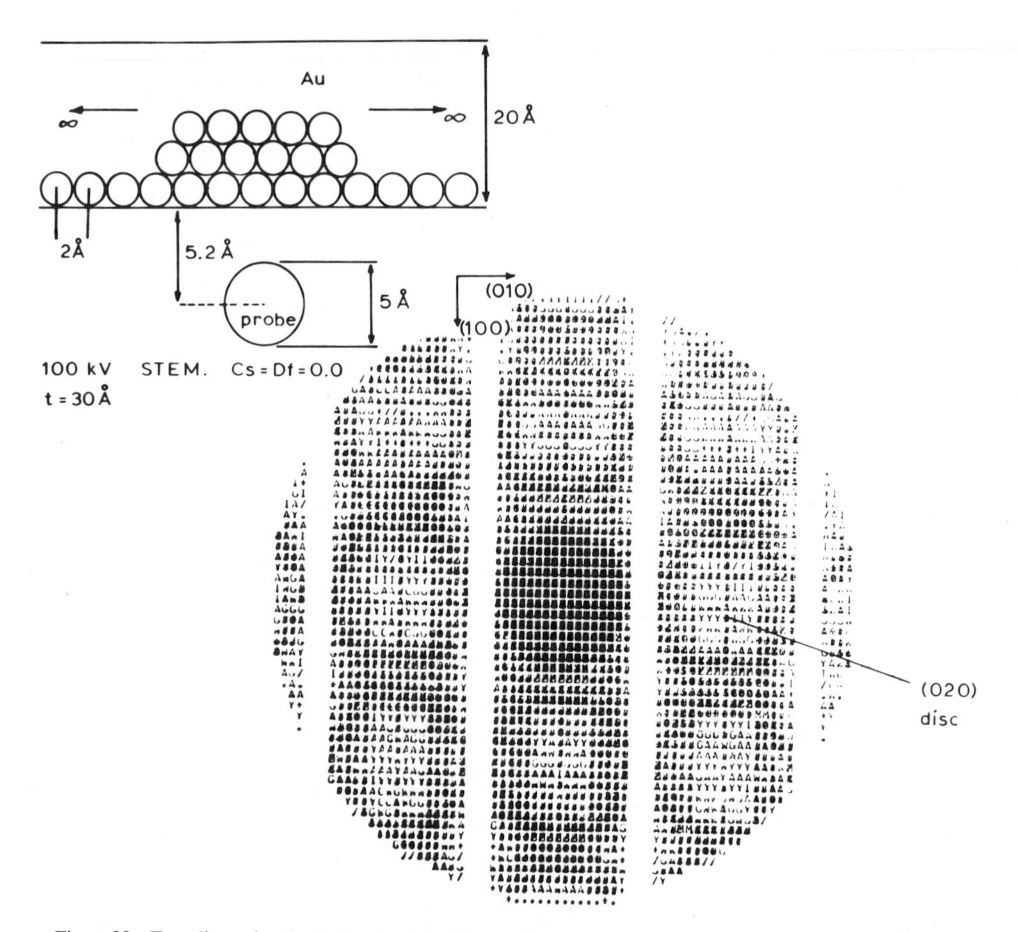

Figure 22 Two-dimensional calculated microdiffraction pattern for an incident probe of diameter 5 Å placed 5.2 Å outside a gold crystal 20 Å wide and 30 Å thick.

H. Microdiffraction From Surfaces

There are two crucial problems in obtaining microdiffraction patterns from surfaces using reflected electrons. First, the small diffraction angles ($\sim$10 mrad) for 100-kV electrons mean that a circular incident probe will intersect the surface in a highly elliptical shape, and thus the diffraction pattern will be obtained from an area $\sim$100 times longer in the beam direction than in the transverse direction. The use of lower voltage to increase these angles unfortunately results in a larger probe because of the increased wavelength, for a lens of constant C_s and focal length. Second, small protrusions and surface roughness will result in an appreciable transmitted contribution to the diffraction patterns formed at these very low angles, making interpretation difficult. A discussion of the optimum choice of voltage for a reflection microdiffraction instrument can be found in Nielsen and Cowley (1976).

The situation has been dramatically altered with the development of the REM technique (Osakabe *et al.*, 1981). For the first time, this technique has made it possible to form images of surfaces prepared in situ with a resolution of a few nanometers. These images have revealed surface steps, emerging dislocations, two-dimensional antiphase boundaries, and surface phase transitions that occur in silicon at high temperature. Similar images can now be obtained by using a coherent small probe in STEM (Cowley, 1982), and if this probe is stopped, a reflection microdiffraction pattern results (Cowley, 1983). By correlating the microdiffraction patterns with the scanning reflection image (SREM), diffraction patterns may be obtained from small regions of atomically flat surface (this is not possible by any other method). These reflection microdiffraction patterns show that the diffraction pattern from an atomically flat surface is *not* streaked, in accordance with theory (Colella, 1972). Streaking in LEED and other reflection patterns arises from surface roughness or phonons and is not (as was sometimes previously supposed) an indication of surface perfection. The problem of sorting out the various transmitted and diffracted beams from small cubes of magnesium oxide is described in detail in Turner and Cowley (1981). For small cubes, it is common to find both Laue (transmitted) and Bragg (reflected) beams contributing to the diffraction pattern, together with the mixed cases (Laue-Bragg and Bragg-Laue) caused by beams entering a side face and exiting a lower surface, etc.

Thus, where suitable heat-treatment facilities exist in situ, the identification of surface superlattices by reflection microdiffraction should be relatively straightforward. The quantitative analysis of reflection microdiffraction intensities, however, in terms of the atomic structure of surface defects such as surface steps, emerging dislocations, and planar faults remains a goal for the future.

V. RELATIONSHIP BETWEEN STEM IMAGING, MICRODIFFRACTION, AND EELS

There is an intimate relationship between STEM images and microdiffraction patterns, since the STEM image is nothing more than the total (integrated) intensity detected within a certain portion of the microdiffraction pattern and displayed on a two-dimensional raster as a function of the probe position. But it is easily shown that, for perfect (unbent) crystals of uniform thickness, the intensity distribution in a CBED pattern does not depend on probe position. How then is it possible to form a lattice image in STEM? In this section, we discuss this question and the relationship between probe "size" and lattice spacing.

The principles of STEM imaging are most easily understood by again using the reciprocity theorem. Figure 23(a) shows a simplified diagram of a single-lens CTEM instrument in which S represents a single point source in the plane of the final condenser aperture, while P is an image point on the film. The reciprocity theorem of Helmholtz states that if a particular intensity, I_p, is recorded at P due to a point source, S, at A, then this same intensity, I_p, would be recorded at A in Figure 23(a) if the source, S, were transferred to the point, P, and the system left otherwise unaltered. The instrument has been redrawn in Figure 23(b), with the source and image points of Figure 23(a) interchanged, and now corresponds to the ray diagram for a STEM instrument. The point source at A in Figure 23(a) illuminates the CTEM specimen with a plane wave (A is a large distance from the specimen), and lens L1 faithfully images all points near the optic axis simultaneously onto the film plane in the neighborhood of P. This "parallel processing" feature of CTEM imaging makes the equivalent STEM arrangement seem very inefficient since, according to the reciprocity theorem, the intensity, I_p, of Figure 23(b) gives the desired CTEM image intensity for only a single point of that image. To obtain a two-dimensional image from the STEM arrangement, it is necessary to scan the focused probe at M across the specimen. The reciprocity argument can then be applied to each scan point in turn. Conceptually, it may be simpler to imagine an equivalent arrangement in which the specimen is moved across a fixed probe to obtain the image signal in serial form (like a television image) from the STEM detector at A. The inefficiency in STEM arises from the fact that, to obtain an image equivalent to the bright-field CTEM image, most of the electrons scattered by the specimen in Figure 23(b) must be rejected by the small detector at A. These scattered electrons carry information on the specimen and can be used by forming a dark-field image.

This reciprocity argument can be extended to cover the case of instruments that use extended sources and detectors. Thus, as shown in Figure 23(c), consider the image formed by a CTEM instrument in which the illuminating aperture, CA, subtends a semiangle, α, the objective aperture subtends a semiangle, θ, and the image is recorded on film, F, with "grain size," d. The illuminating aperture is taken to be perfectly incoherently filled. The reciprocity theorem can be used to show that an identical STEM image can be obtained from a STEM instrument fitted with a finite incoherent source, IS, of size d in which the "illuminating aperture" (called the STEM objective aperture, OA) subtends a semiangle, θ, and in which the STEM detector, DET, subtends semiangle α. The equivalence means that CTEM computed images can be used to assist in the interpretation of STEM images. A computer program that calculated CTEM images will accurately reproduce STEM images of the same specimen so long as the incoherent sum over CTEM illumination angles is reinterpreted as an incoherent sum over the STEM detector aperture.

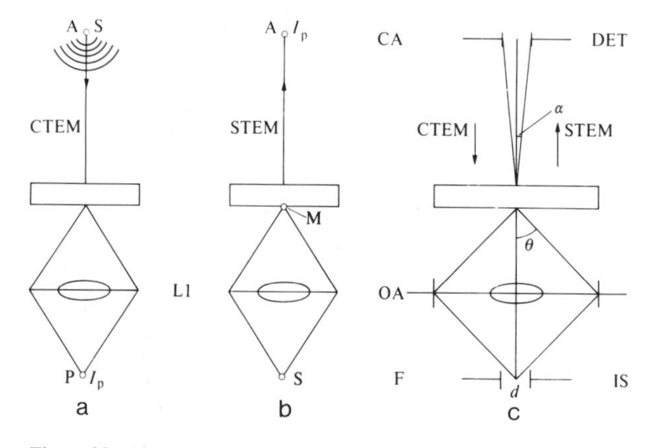

Figure 23 In (a) the ray diagram for a simplified CTEM instrument is shown. Here, S is a point within the final condenser aperture (strictly the "effective source"), which emits a spherical wave to illuminate the specimen. All points on the lower side of the specimen are imaged to conjugate points, such as P in the film plane by the single lens, L1. The intensity at P is I_p. In (b) the source, S, and "detector" (the image point, P) have been interchanged, producing the ray diagram for STEM. In (c) the condenser aperture (CA), objective aperture (OA), film plane (F), and film grain size (d) are defined for CTEM on the left, with the corresponding quantities for STEM shown on the right. Here, DET is the STEM detector and IS refers to the incoherent STEM source plane, taken to be an electron emitter of size d_0.

These results can be used to find STEM analogues both for incoherent CTEM imaging and for the use of conical illumination in CTEM. We now consider the STEM analogue for many-beam lattice imaging in the symmetrical zone-axis orientation using the three-beam case as a simple example. Figure 24, reading down the page, shows the ray paths for CTEM three-beam lattice imaging. A spherical wave diverging from D illuminates the specimen, becoming a plane wave at O, which is diffracted into three Bragg beams, passing through the objective aperture at A, B, and C to be synthesized by the lens into an image point at P. If we assume the use of a "point" field-emission source in STEM (and no resolution limit due to the film used in CTEM), the reciprocity theorem indicates that an identical STEM image can be formed (reading up the page) if a small STEM detector, D, is used together with a STEM objective aperture of semiangle $2\theta_B$. This angle must be sufficiently large to accept the three Bragg beams of interest as diffracted from an imaginary point source placed at the STEM detector, D. Thus, when setting up a STEM instrument in which the electron gun is below the specimen, the choice of objective aperture for lattice imaging with a small central detector can be made by imagining the specimen to be illuminated from above by an axial plane wave and using the same criteria that apply to the choice of objective aperture in CTEM. For a STEM instrument in which the electron gun is mounted above the specimen, the objective aperture is again selected by imagining the specimen to be illuminated from a point at the STEM detector (now below the specimen); however, the STEM image is now the same as that which would be obtained by placing the inverted specimen in a CTEM instrument. This inversion is unimportant in crystals containing a horizontal mirror plane of symmetry, as described earlier in the discussion of reciprocity.

The choice of STEM detector size for lattice imaging can be understood by drawing out Figure 24 more fully in the STEM case. This is done in Figure 25(a). The specimen is illuminated from below by a cone of radiation (semiangle $2\theta_B$) which, for a point source at P, forms an aberrated spherical wave (aberrations from the probe-forming lens) converging to the focused electron probe, M, on the specimen. Thus, around each scattered Bragg direction (for the illumination direction, PM) on the specimen exit side, we must draw a cone of semiangle θ_B. The result is a set of overlapping "convergent beam" diffraction discs as shown in Figure 25(a). Three of these overlap at the detector, D, in the three-beam case shown. It is the coherent interference between these three discs at D that produces the lattice image as the

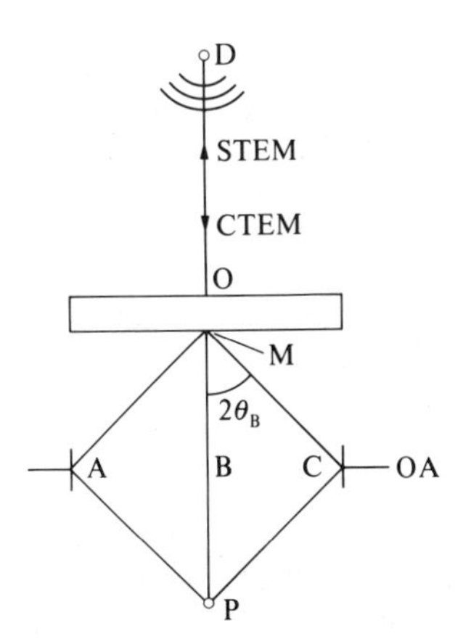

Figure 24 Ray diagram for symmetrical three-beam lattice imaging in CTEM (reading down the page) and STEM (reading up the page). In CTEM, the optical paths MA, MB, and MC represent the directions in which three Bragg beams are scattered by the specimen, which is illuminated by a plane wave in the direction DO. These are focused by the lens to interfere at the image point, P. In STEM the specimen is illuminated from below simultaneously by three coherent waves in directions AM, BM, and CM. The Bragg condition permits scattering only through angle $2\theta_B$ (or zero), so that each of these incident plane waves can scatter in the direction OD of the detector, where they interfere to produce the STEM image signal. Note that for Bragg scattering, points A, B, and C are the only source points that can scatter into the detector, D.

probe is scanned across the specimen. Figure 26 shows the arrangement that would be used to form the analogue of tilted illumination, two-beam fringes in CTEM. A detailed analysis of these arrangements (Spence and Cowley, 1978) has established the following general results, which take account of all multiple scattering effects:

(a) For crystalline specimens devoid of any imperfections illuminated in a STEM instrument by an objective aperture of semiangle $\theta < \theta_B$ (θ_B is the Bragg angle), the set of convergent-beam diffraction discs observed in the detector plane show an intensity distribution that is independent of the aberrations of the probe-forming lens and of the probe position. In this case, the "size" of the electron probe is much greater than that of the crystal unit cell.
(b) The intensity within regions of disc overlap depends both on the probe position and the aberrations and focus setting of the probe-forming lens. The intensity at D of Figure 26 varies sinusoidally as the probe moves across the specimen and may be estimated by two-beam theory (for a small detector). The effect of enlarging the detector is exactly analogous to the effect of enlarging the final condenser aperture in CTEM. Outside the regions of Bragg disc overlap, the diffracted intensity does not depend on the probe position or lens aberrations.
(c) The intensity at the midpoint between overlapping discs is a special case. Here, the intensity depends on the probe position but not on those lens aberrations that are an even function of angle, such as focus setting and spherical aberration.

A brief note concerning the frequently discussed question of probe size can be added here. In general, the intensity distribution of the STEM probe incident on a specimen is a complicated and not particularly useful function. Near the "dark-field" focus setting for the probe-forming lens it does, however, form a well-defined peak, but with rather extensive "tails." The width of the central peak, "the probe size," is given very approximately in the absence of spherical aberration by $d = 0.6\lambda/\theta_R$ if we use the Rayleigh criterion and imagine the STEM lens to be imaging an ideal point field-emission source (Figure 26). If we set this probe size equal to the lattice spacing, a, then the illumination angle (STEM objective aperture semiangle) needed to match the probe "size" to the lattice spacing is $\theta_R = 0.6\lambda/a$. But the Bragg angle is $\theta_B = 0.5\lambda/a$, and this is the condition that adjacent diffraction discs just begin to overlap. In this rather loose sense, lattice imaging becomes possible as the electron probe becomes "smaller" than the crystal lattice spacing, and this is possible only if the diffraction discs are allowed to overlap. In fact, the important condition for lattice imaging in STEM is that the exit pupil of the probe-forming lens be coherent over an angular range equal to that covered by the Bragg beams one wishes to image. As the probe becomes very much smaller than the unit cell, information on the translational symmetry of the crystal is progressively lost (it becomes difficult to distinguish the reciprocal lattice using the overlapping convergent-beam pattern), leaving only information on the point group symmetry of the unit cell contents and the probe position. Figures 27(a) and 27(b) show bright and dark-field STEM lattice images of niobium oxide recorded by J. M. Cowley on the V.G. HB5 STEM instrument. The dark-field image was recorded using an annular detector.

Finally, the relationship between energy-filtered microdiffraction patterns and energy-filtered STEM lattice images should be mentioned. In a crystal containing many atomic species, it might be thought that a lattice image can be formed showing only one of these if an energy filter were used so that only those transmitted electrons were

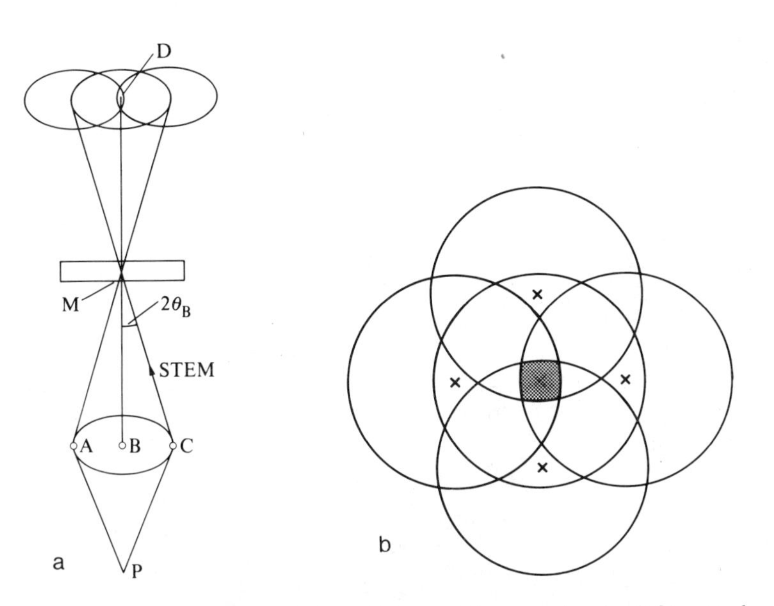

Figure 25 (a) The ray diagram of Figure 24 drawn with a cone of semiangle $2\theta_B$ around each Bragg direction. "Bragg direction" means the scattered directions resulting from axial plane-wave illumination, and these fall at the disc centers. In (b) the pattern is drawn out as it would appear in two dimensions for five-beam imaging; here, the crosses denote Bragg directions, while the hatched area indicates a possible STEM detector outline. Within this region, the intensity is sensitive to the probe position (scan coordinate). For thick crystals, these discs will contain intensity variations characteristic of the crystal rocking curve.

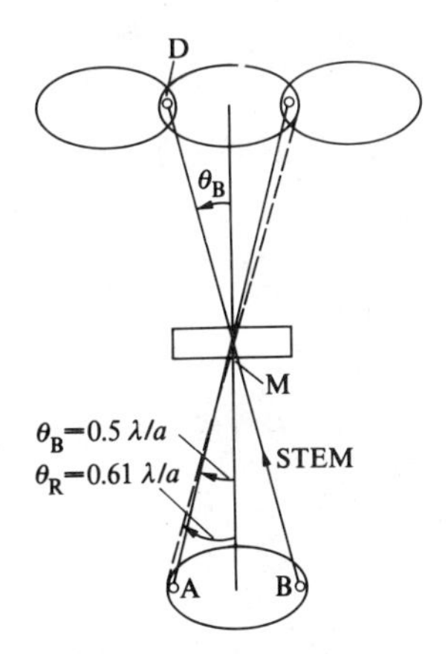

Figure 26 The STEM analogue of tilted illumination, two-beam CTEM lattice imaging. CTEM conditions are obtained by placing the electron source at D, so that MA and MB become the two Bragg beams used for imaging. The figure also shows the illumination angle, θ_R, needed to make the electron probe "width" approximately equal to the crystal lattice spacing, a. As drawn, the diffraction discs subtend a semiangle slightly larger than the Bragg angle and therefore overlap.

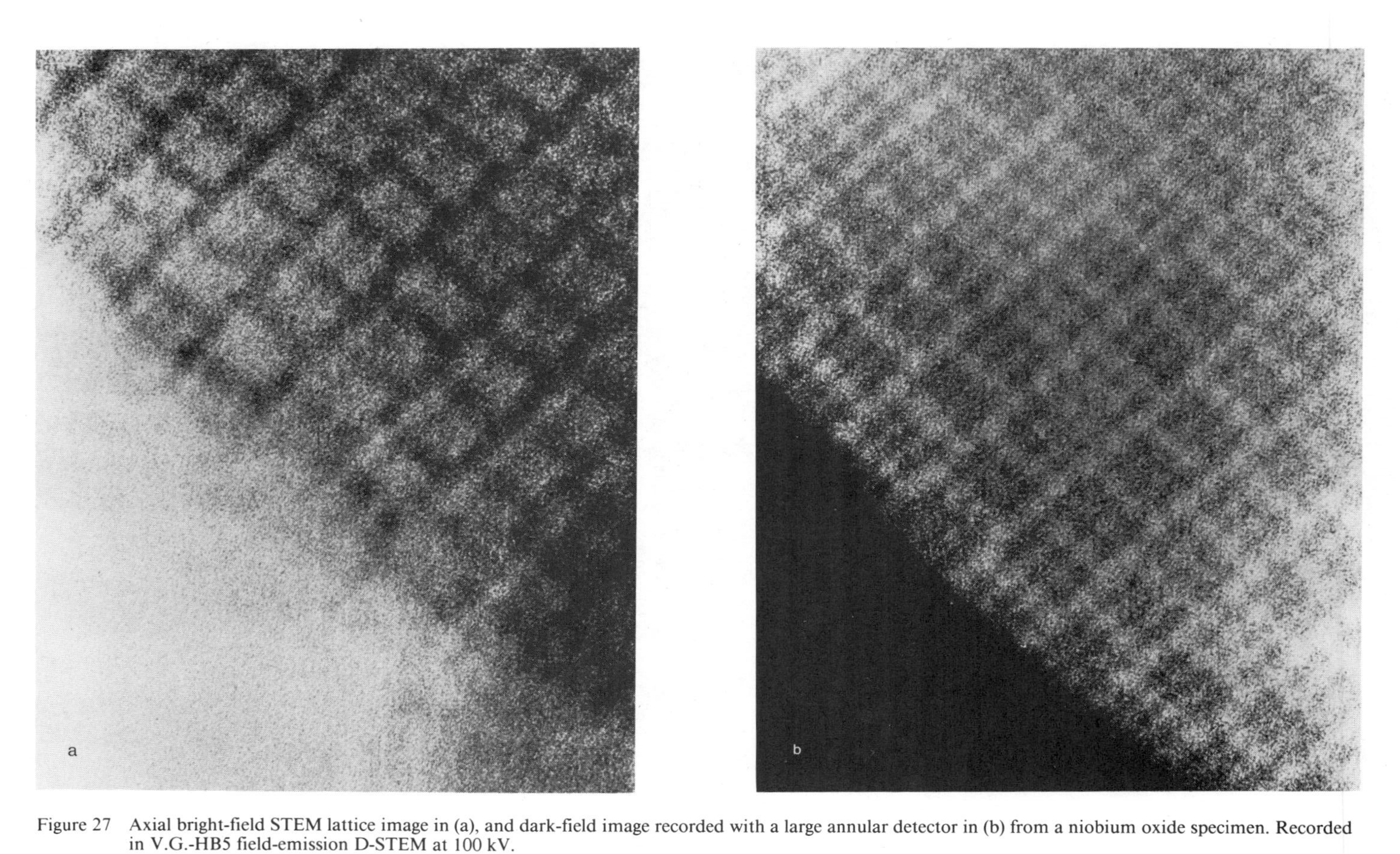

Figure 27 Axial bright-field STEM lattice image in (a), and dark-field image recorded with a large annular detector in (b) from a niobium oxide specimen. Recorded in V.G.-HB5 field-emission D-STEM at 100 kV.

selected to form the image that have lost a characteristic amount of energy in exciting, say, the K-shell ionization process for a particular species. In general, this is not so, because these inelastically scattered energy-loss electrons will "subsequently" be elastically scattered in the same specimens, particularly thicker specimens, by other species, whose presence will thus be revealed in the image. (For an example, see Figure 10 of Spence and Lynch, 1982). A similar comment applies to energy-filtered microdiffraction patterns. A related problem concerns the difficulty in absolutely locating a subnanometer probe with respect to a crystal lattice, in view of the contrast reversals possible in lattice images. These questions have been discussed in detail by Spence and Lynch (1982). They find that:

(a) If facilities exist for real-time observation of the elastic microdiffraction pattern, the position of a subnanometer probe within the unit cell can be determined from the symmetry of the patterns. This allows the probe to be situated over particular species that fall on known symmetry elements within the unit cell.
(b) For localized inelastic scattering, changes in the core losses can be detected as the probe is moved within a single unit cell.
(c) Interpretable core-loss filtered STEM lattice images can thus be formed under certain severely restricted conditions of specimen thickness, energy loss, and focus.

Since the characteristic scattering angle, θ_E, for most energy-loss processes is proportional to energy loss, the form of energy-filtered microdiffraction patterns (for a small illumination angle) goes roughly as follows with increasing energy loss. At small losses, a sharp "spot" pattern will be seen, formed from delocalized excitations such as plasmons. The region in real space contributing to this pattern has a (large) characteristic dimension, $L \approx \lambda/\theta_E$. As the energy loss is increased and θ_E increases (to become comparable with the Bragg angle), the Bragg spots broaden out and overlap (as the corresponding localization, L, in real space shrinks to become comparable with the unit cell dimension). At large losses, this inelastic pattern will be crossed by sharp Kikuchi lines due to elastic scattering of the inelastically scattered electrons (blocking). This simplified account of a highly complex process ignores channeling and absorption effects; for more details the reader is referred to Gjonnes and Hoier (1971).

TABLE OF CHAPTER VARIABLES

a,c	Lattice constants of crystals
$\underset{\sim}{B}$	Incident beam direction, normally given as a direction in the target crystal
c	Column width used in column approximation calculation
C_c	Chromatic aberration coefficient
C_s	Spherical aberration coefficient
d	Electron emitter size, or "grain size" of a recording system in discussion of reciprocity
d_c	Chromatic aberration broadening of probe diameter
d_{det}	Detector diameter
d_f	Diffraction broadening of probe diameter
d_g	Gaussian probe diameter in the object plane
d_s	Spherical aberration broadening of probe diameter
d_0	Total diameter of focused probe in the object plane

D	Distance from plane of probe-forming lens to object plane
f	A lens focal length
g,h	Indices for reciprocal lattice vectors
$h_d(x)$	Top-hat aperture function of width d and unit amplitude
I_g	g^{th} Bragg scattered-beam intensity
I_o	Forward scattered-beam intensity
K	A numerical constant
L	Lateral period of a scattering object
L_f	Illuminating probe-to-image distance in a convergent-beam shadow image
M	A lens magnification
r	Half-diameter of probe crossover in some plane of interest
r_s	Geometrical Gaussian probe diameter
R_a	Probe-defining aperture radius
s_g	Bragg diffraction error for reflection g
t	Specimen thickness
v_g	g^{th} Fourier coefficient in crystal potential expansion, in volts
V_o	Microscope accelerating voltage
w	Dynamical parameter, equal to $s_g\xi_g$
X_a	Transverse (spatial) coherence width
α	Divergence or convergence half-angle
α_{opt}	Optimum divergence half-angle for minimum probe size
α_s	Scattering angle
γ	Beam rocking angle
ΔE	Energy spread in a beam of energy E
Δf	Objective lens focusing error
Δr	Disregistry between selected area and origin of some diffracted beam in SAD
θ	Bragg scattering angle
θ_c	A divergence half-angle usually equal to α
θ_E	Characteristic inelastic scattering angle equal to $\Delta E/2E$, where ΔE is the energy loss and E the incident-beam energy
θ_R	Illumination semiangle satisfying Rayleigh criteria
θ_s	Half-angle subtended by probe in object plane at the probe-defining aperature
λ	Electron wavelength
ξ_g	Elastic extinction distance for reflection g
σ	Interaction parameter, equal to $\pi/\lambda V_o$ (nonrelativistic)
ϕ	An aperture diameter
ϕ_g	Diffracted amplitude
ψ	Electron wavefunction

Abbreviations

AEM	Analytical electron microscopy
CBED	Convergent-beam electron diffraction
D-STEM	Dedicated scanning transmission electron microscope
EDS	Energy-dispersive x-ray spectroscopy
EELS	Electron energy-loss spectroscopy
FOLZ	First-order Laue zone
HOLZ	Higher-order Laue zone

HREM	High-resolution electron microscopy
REM	Reflection electron microscopy
SAD	Selected area diffraction
SREM	Scanning reflection electron microscopy
STEM	Scanning transmission electron microscopy
TEM	Transmission electron microscopy
YAG	Yttrium-aluminum-garnet, an inorganic crystal
ZOLZ	Zero-order Laue zone

ACKNOWLEDGMENTS

The authors are indebted to many colleagues for informative discussions on various aspects of this subject and to the following researchers for figures illustrating their results: J. M. Cowley (Figures 4, 15, 18, and 27), L. Reimer (Figures 1 and 8), D. Braski (Figure 20), N. Yamamoto (Figure 7), A. Higgs (Figure 9), and B. Ralph (Figure 19).

This work was supported by NSF Grants DMR8002108 and DMR8221443, and by the NSF/ASU National Center for High Resolution Electron Microscopy.

REFERENCES

Ackermann, I. (1948), Annl. Physik *2*, 19.

Allen, S. (1981), Phil. Mag. *43*, 325.

Autreta, R. P.; Schaner, S. T.; and Krapil, J. (1983), SEM/1983, (O. Johari, ed.), SEM Inc., AMF O'Hare, Chicago, p. 36.

Barnett, M. E. (1974), J. Micros. *102*, Pt 1, 1.

Bentley, J. (1980), Proc. 38th Annl. Mtg. EMSA (G. W. Bailey, ed.), Claitor's Pub. Div., Baton Rouge, p. 72.

Bentley, J., and Lehman, J. (1982) Proc. 40th Annl. Mtg. EMSA, (G. W. Bailey, ed.), Claitor's Pub. Div., Baton Rouge, p. 694.

Bentley, J.; Goringe, M. J.; and Carpenter, R. W. (1981), SEM/1981/I (O. Johari, ed.), SEM Inc., AMF O'Hare, Chicago, p. 153.

Blake, R.; Jostons, A.; Kelly, P.; and Napier, J. (1978), Phil. Mag. *37*, 1.

Boersch, H. (1936), Annl. Physik *27*, 75.

Booker, G. R. (1981), Inst. Phys. Conf. Series Vol. 61, Brit. Inst. Phys., London (Bristol, Proc. EMAG), p. 135.

Born, M., and Wolf, E. (1980), "Principles of Optics," 6th ed., Pergamon Press, New York, p. 381.

Braski, D.; Bentley, J.; and Cable, J. (1982), Proc. 40th Annl. Mtg. EMSA (G. W. Bailey, ed.), Claitor's Pub. Div., Baton Rouge, p. 692.

Brown, L. M.; Craven, A. J.; Jones, L. G. P.; Griffith, A.; Stobbs, W. M.; and Wilson, C. J. (1976), SEM/1976/I (O. Johari, ed.), IITRI, Chicago, p. 353.

Buxton, B. F.; Eades, J. A.; Steeds, J.; and Rackham, G. (1976), Phil. Trans. *281*, 15.

Carpenter, R. W. (1980), Microbeam Analysis – 1980, Proc. 15th Annl. Conf. Microbeam Anal. Soc. (D. B. Wittry, ed.), San Francisco Press, p. 1.

Carpenter R. W., and Spence, J. C. H. (1982), Acta Cryst. *A38*, 55.

Carpenter, R. W.; Bentley, J.; and Kenik, E. A. (1978), J. Appl. Cryst. *11*, 564.

Carpenter, R. W.; Chan, I.; and Cowley, J. M. (1982), Proc. 40th Annl. Mtg. EMSA (G. W. Bailey, ed.), Claitor's Pub. Div., Baton Rouge, p. 696.

Chan, I. Y. T., and Cowley, J. M. (1981), Proc. 39th Annl. Mtg. EMSA (G. W. Bailey, ed.), Claitor's Pub. Div., Baton Rouge, p. 350.

Chapman, J. N. (1981), Inst. Phys. Conf. Series Vol. 61, Brit. Inst. Phys. London (Bristol, EMAG), p. 131.

Chevalier, J. P., and Craven, A. J. (1977), Phil. Mag. *36*, 67.

Cockayne, D.; Goodman, P.; Mills, J.; and Moodie, A. F. (1967), Rev. Sci. Instr. *38*, 1093.

Colella, R. (1972), Acta Cryst. *A28*, 11.

Colliex, C. (1985), personal communication.

Cowley, J. M. (1969), Appl. Phys. Lett. *15*, 58.

——— (1973), Acta Cryst. *A29*, 529.

——— (1979a), Adv. Elec. and Elec. Phys. *46*, 1.

——— (1979b), Ultramicroscopy *4*, 413.

——— (1979c), Ultramicroscopy *4*, 435.

——— (1980), SEM/1980/I (O. Johari, ed.), SEM Inc., AMF O'Hare, Chicago, p. 61.

——— (1981a), Ultramicroscopy *7*, 19.

——— (1981b), "Electron Microdiffraction and Microscopy of Amorphous Solids" (I. Hargittan and W. J. Thomas, ed.), Orville, Akademiai Kiado, Budapest.

——— (1982), SEM/1982/I (O. Johari, ed.), SEM Inc., AMF O'Hare, Chicago, p. 51.

——— (1983), J. Micros. *129*, 253.

Cowley, J. M, and Moodie, A. (1960), Proc. Phys. Soc. (London) *76*, 378.

Cowley, J. M., and Osman, M. A. (1984), Ultramicroscopy *15*, 311.

Cowley, J. M., and Rees, A. L. G. (1953), J. Sci. Instr. *30*, 33.

Cowley, J. M., and Roy, R. A. (1982), SEM/1981 (O. Johari, ed.), SEM Inc., AMF O'Hare, Chicago, p. 143.

Cowley, J. M., and Spence, J. C. H. (1979), Ultramicroscopy *3*, 433.

——— (1981), Ultramicroscopy *6*, 359.

Davisson, C., and Germer, L. H. (1927), Phys. Rev. *30*, 705.

Eades, J. A. (1980), Ultramicroscopy *5*, 71.

——— (1983), SEM/1983/III (O. Johari, ed.), SEM Inc., AMF O'Hare, Chicago, p. 1051.

——— (1984), J. Elec. Micros. Tech. *1*, 279.

Eades, J. A.; Shannon, M. D.; and Buxton, B. F. (1980), J. Appl. Cryst. *13*, 368.

Ecob, R.; Ricks, R.; and Porter, A. (1982), Script. Met. *16*, 1085.

Fraser, H. (1983), J. Micros. Spectros. Electron *8*, 431.

Fujimoto, F., and Lehmpfuhl, G. (1974), Z. Naturforsch. *29a*, 1929.

Fujimoto, F.; Komaki, K.; Takagi, S.; and Koike, H. (1972), Z. Naturforsch. *27a*, 441.

Geiss, R. H. (1976), SEM/1976/I (O. Johari, ed.), IITRI, Chicago, p. 337.

——— (1979), in "Introduction to Analytical Electron Microscopy" (J. Hren, J. Goldstein, and D. Joy, ed.), Plenum Press, New York, Ch. 2..

Gjonnes, J., and Hoier, R. (1971), Acta Cryst. *19*, 65.

Gjonnes, J., and Moodie, A. F. (1965), Acta Cryst. *19*, 65.

Goodman, P. (1972), Acta Cryst. *A28*, 92.

——— (1975), Acta Cryst. *A31*, 804.

——— (1976), Acta Cryst. *A32*, 793.

Goodman, P., and Johnson, A. W. S. (1977), Acta Cryst. *A33*, 997.

Goodman, P., and Lehmpfuhl, G. (1968), Acta Cryst. *A24*, 339.

Grigson, C. W. B., and Tillett, P. I. (1968), Int. J. Electronics *24,* 101.

Head, A.; Humble, P.; Clareborough, L.; Morton, A.; and Forwood, C. (1973), "Computed Electron Micrographs and Defect Identification," North-Holland Pub. Co., Amsterdam.

Higgs, A., and Krivanek, O. L. (1981), Proc. 39th Annl. Mtg. EMSA (G. W. Bailey, ed.), Claitor's Pub. Div., Baton Rouge, p. 346.

Hillier, J., and Baker, R. F. (1946), J. Appl. Phys. *17*, 12.

Hirsch, P. B.; Howie, A.; Nicholson, R. B.; Pashley, D. W.; and Whelan, M. J. (1977), "Electron Microscopy of Thin Crystals," R. E. Krieger Pub. Co., New York.

Joos, G. (1950), "Theoretical Physics," 3rd ed., Hafner Pub. Co., New York, p. 379.

Joy, D. (1981), private communication.

Kelly, P.; Jostons, A.; Blake, R.; and Napier, J. (1975), Phys. Stat. Sol. *31*, 771.

Kossel, W., and Mollenstedt, G. (1938), Naturwis. *26*, 660.

——— (1939), Annl. Phys. (Leipzig) *36*, 113.

——— (1942), Annl. Phys. (Leipzig) *42*, 287.

Lehmpfuhl, G., and Tafto, J. (1980), Proc. 7th Euro. Conf. Elec. Mic. *3*, 62 (P. Brederoo, ed.), Elec. Mic. Fdn., Leyden.

Lennander, S. (1954), Arkiv Fysik *8*, 551.

LePoole, J. P. (1947), Philips Tech. Rev. *9*, 33.

Lynch, J. R.; Lesage, E.; Dexpert, H.; and Freund, E. (1981), Brit. Inst. Phys. Ser. No. 61, 67 (I.O.P. Bristol).

MacGillavry, C. H. (1940), Physica (Utrecht) *7*, 329.

Maher, D., and Joy, D. (1976), Ultramicroscopy *1*, 239.

Mansfield, J. (1984), "Convergent Beam Electron Diffraction of Alloy Phases," A. Hilger, Ltd., Bristol, UK.

Marks, L. (1985), Proc. EMSA (G. W. Bailey, ed.), San Francisco Press, in press.

Metherall, A. J. F. (1975), in "Electron Microscopy and Materials Science, Third Course of the International School of Electron Microscope" (U. Valdrè and E. Ruedl, ed.), Commiss. of Euro. Communities, Directorate General, "Scientific and Technical Information," Luxembourg, vol. II p. 397.

Monosmith, W. B., and Cowley, J. M (1983), Ultramicroscopy *12*, 51.

——— (1984), Ultramicroscopy *12,* 177.

Nielsen, P. E., and Cowley, J. M. (1976), Surf. Sci. *54*, 340.

Osakabe, N.; Tanishiro, Y.; Yagi, K.; and Honjo, G. (1981), Surf. Sci. *109*, 353.

Pogany, A., and Turner, P. (1968), Acta Cryst. *A24*, 103.

Pollard, I. (1972), PhD. Thesis, Univ. Melbourne.

Porter, A.; Ecob, R.; and Ricks, R. (1983), J. Micros. *129*, 327.

Porter, A.; Shaw, M.; Ecob, R.; and Ralph, B. (1981), Phil. Mag. *A44*, 1135.

Reimer, L. (1979), Scanning *2*, 3.

Riecke, W. D. (1962), Optik *19*, 273.

Roy, R. A.; Messier, R.; and Cowley, J. M. (1981), Thin Solid Films *79*, 207.

Sellar, J.; Imeson, D.; and Humphreys, C. (1980), Acta Cryst. *A36*, 686.

Smith, C. and Cowley, J. M. (1971), J. Appl. Cryst. *4*, 482.

Spence, J. C. H. (1978), Acta Cryst. *A34*, 112.

——— (1981), "Experimental High Resolution Electron Microscopy," Oxford Univ. Press, Oxford.

Spence, J. C. H., and Cowley, J. M. (1978), Optik *50*, 129.

Spence, J. C. H., and Lynch, J. (1982), Ultramicroscopy *9*, 267.

Spence, J. C. H.; Higgs, A.; Disko, M.; Wheatley, J.; and Hashimoto, H. (1982), Proc. Intl. Cong. Elec. Mic. Hamburg, p. 43.

Steeds, J. W. (1979), "Introduction to Analytical Electron Microscopy" (J. Hren, J. Goldstein, and D. Joy, ed.), Plenum Press, New York, p. 387.

——— (1980), Inst. Phys. Conf. Series No. 52 (I.O.P., London).

Steeds, J. W., and Vincent, R. (1983), J. Appl. Cryst. *16*, 317.

Strahm, M., and Butler, J. (1981), Rev. Sci. Inst. *52*, 840.

Swann, P. (1982), personal communication, GATAN Inc., Pittsburgh.

Tafto, J., and Spence, J. C. H. (1982), J. Appl. Cryst. *15*, 60.

Tanaka, M., and Terauchi, M. (1985), "Convergent-Beam Electron Diffraction," JEOL, Ltd., Tokyo.

Tanaka, M.; Saito, R.; Ueno, K.; and Harada, Y. (1980a), J. Elec. Mic. *29*, 408.

Tanaka, M.; Ueno, K.; and Harada, Y. (1980b), Japan J. Appl. Phys. *19*, L201.

Titchmarsh, J. M. (1978), Proc. Intl. Cong. Elec. Micros., Toronto, vol. 1, p. 318.

Treacy, M.; Howie, A.; and Wilson, A. (1978), Phil. Mag. *A38*, 569.

Turner, P. S., and Cowley, J. M. (1981), Ultramicroscopy *6*, 125.

Van Oostrum, K. J.; Leenhouts, A.; and Bre, A. (1973), Appl. Phys. Lett. *23*, 283.

Voss, R.; Lehmpfuhl, G.; and Smith, P. J. (1980), Z. Naturforsch. *35a*, 973.

Wagner, C. (1978), J. Non-Cryst. Sol. *31*, 1.

Watari, R., and Cowley, J. M. (1981), Surf. Sci. *105*, 240.

Wells, O. C.; Boyde, A.; Lifshin, E.; and Rezanowich, A. (1974), "Scanning Electron Microscopy," McGraw-Hill, New York.

Yamamoto, N. (1983), private communication.

Zhu, J., and Cowley, J. M. (1982), Acta Cryst. *A38*, 718.

——— (1983), J. Appl. Cryst. *16*, 171.

CHAPTER 10

BARRIERS TO AEM: CONTAMINATION AND ETCHING

J. J. Hren

Department of Materials Science and Engineering
North Carolina State University
Raleigh, North Carolina

I. INTRODUCTION

Contamination and etching are terms used quite negatively by the analytical electron microscopist. The mental "images" evoked are of a loss of resolution, a decrease in signal/noise ratio, destruction of the microstructure, etc. Yet more propitious results may be achieved by using our empirical understanding of the phenomena to manufacture microcircuits, measure foil thickness, or to preferentially etch microstructures. For the most part, however, the effects of rapid contamination buildup or the local loss in specimen mass are viewed as worse than annoying. They are, in fact, significant barriers to continued advances in true microchemical and high-resolution microstructural analyses. This chapter attempts to summarize our understanding of the physical processes contributing to each phenomenon, to illustrate their effects on AEM, to describe some working cures (and even a few applications), and to project some avenues for further improvement.

The description of the present state of our understanding will follow a largely chronological course. Although this is the general approach to a review, we will not try to be comprehensive but rather place our emphasis here on circumscribing our present bounds of understanding. That is, some of the intermediate stages of understanding will be minimized, and those results that are still debatable will be left inconclusive. On the other hand, working solutions that have proven successful will be described, and still others that appear useful but remain largely untested will be suggested.

II. SOME DEFINITIONS

There are some uncertainties in the precise meaning of the terms *contamination* and *etching* and other terms such as *mass loss*, *radiation damage*, *bulk* and *surface impurities*, etc, all commonly used in the literature of electron microscopy. Some confusion may be avoided by specifying a working definition. Simply put, by *contamination*, we mean the unintentional act of adding mass to the surface(s) of a thin specimen during observation or analysis by an electron beam.

In our definition, *etching* is just the reverse process of contamination, qualified only to draw a reasonable line between etching and radiation damage. We therefore restrict our working meaning to that given by Glaeser (1979): the removal of material by " . . . the synergistic action of electron irradiation and . . . certain residual gases, such as oxygen or water vapor . . . ." Of course, to the practicing microscopist, local mass loss is local mass loss, whatever the mechanism; a careful reading of both Glaeser (1979) and Hobbs (1979) is thus mandatory. One result of this somewhat arbitrary division is that the present discussion of etching will be largely restricted to organic, carbonaceous, and polymeric materials. However, contamination effects, as here defined, extend to all known materials.

III. EARLY OBSERVATIONS OF CONTAMINATION

Historically, all of the early observations of contamination and etching were observed with large electron beams (>1 μm diameter) varying in energy from a few hundred volts to one hundred kilovolts. As we shall see, both effects become significantly more severe as the beam size decreases. Etching was, in effect, "discovered" by the attempts to control the buildup of contaminating layers.

The earliest reference to the subject of contamination deposits induced by electron beams is probably untraceable. However, Stewart (1934) is a suitable starting point. He points out: "In an evacuated tube in which the slightest traces of organic vapors may occur, . . . insulating layers are formed on surfaces subject to electron . . . bombardment. These layers may be attributed to carbon compounds, and their formation is related to the polymerization of organic vapors . . . ." Stewart's analysis, as far as it went, remains unimpeachable today and has been reduced many times since.

Stewart was not an electron microscopist, and his measurements were made at very low electron energies ($\sim$200 eV). Contamination effects were soon encountered and studied by a number of practicing electron microscopists. Watson (1947) was the first to report a significant change in the mean particle size and shape of carbon black while under examination in the electron microscope. He concluded that the contaminants were condensed organic vapors polymerized by the electron beam. Watson's observations were incomplete and even somewhat contradictory. He recognized the need for further studies and called for an exchange of experience. About the same time, Kinder (1947) noted that a transparent skin of contamination was left around crystals after prolonged observation. The crystals themselves had often been completely removed under the influence of the beam, presumably by evaporation. Specimen degradation in a number of substances, mainly ionic crystals, was soon observed by Burton *et al.* (1947), who reported both increased electron transparency and debris tracks as a consequence of electron irradiation. Although they had no conclusive explanation of the phenomena, Burton *et al.* thought the debris to be the same as that observed by Stewart (1934) and Watson (1947). Sublimation and redeposition because of beam heating was ruled out as an explanation. Cosslett (1947) responded to Watson's call for further observation of contamination. His observation on zinc oxide and magnesium oxide crystals complicated the picture still further. Cosslett's observation tended to minimize the influence of organic vapors and emphasize the importance of ejected particles from the grid or specimen itself. He suggested that the contaminants observed by Watson came from the carbon black.

IV. THE NATURE OF THE CONTAMINANT

Hillier (1948) and König (1948) also responded to the now-obvious need for more controlled studies. Some notable observations of the nature of the contaminant itself were made by Hillier. He observed that: (a) the scattering power was slightly less than carbon and the deposit was amorphous as deduced from diffraction; (b) only carbon was found when analyzed with his electron microanalyzer (a very early version of an electron energy loss spectrometer); (c) surface migration of the contaminant occurred; (d) shielding the specimen (at room temperature) reduced the rate of deposition only slightly. Hillier concluded that the contaminant product was formed by the polymerization of hydrocarbon molecules as originally proposed by Stewart and Watson and that material in the gas phase and from the instrument surfaces were both important sources. The responsible vapors were not just from the diffusion pumps. Hillier proposed several solutions, one based upon a low-energy electron spray of the chamber (Hillier and Davidson, 1947), a method since adopted by others. König (1948) confirmed the hypothesis that the contaminants were primarily hydrocarbons and were vacuum-system dependent. He pointed out that improvements in the general vacuum of an electron microscope could and should be made and that these were well within the state of the art at that time. Subsequently, König (1951) showed by electron diffraction that organic material such as collodian could indeed be converted into amorphous carbon by intensive electron bombardment. König and Helwig (1951) also showed that a thin coating of polymerized hydrocarbon was formed on top of underlying thin films in the electron microscope. Two other researchers were investigating the contamination phenomenon at about the same time and soon reported further results.

Ellis (1951) suggested that the principal source of organic molecules was condensate on the specimen supports which then migrated to the irradiated area of the specimen. Extensive studies by Ennos (1953, 1954) concluded that surface migration was not an important supply mechanism, but that organic molecules were continuously absorbed onto the irradiated surfaces from the vapor phase. The relative importance of various vacuum-based contaminant sources in descending order was found by Ennos to be: diffusion pump oil, vacuum grease, and rubber gasket materials, followed by other vacuum system components. Uncleaned metal surfaces were also recognized to be substantial sources of the contaminating vapor. In short, the application of good high-vacuum practices was strongly urged by Ennos. Two expedients were demonstrated that considerably reduced contamination: specimen heating (to ~200°C, if possible) and cryoshielding (i.e., a cold finger). Rates of buildup were monitored by measuring the step height of the contaminating layer using optical interference methods.

V. RELATIONSHIP BETWEEN CONTAMINATION AND ETCHING

Following the publications of Ennos, cryoshielding was commonly used to decrease the rate of contamination buildup. However, in cooling the area around the specimen, the specimen itself was also often cooled.* For example, in using a Siemens Elmiskop I, Leisegang (1954) noticed that below ~−80°C (for both the cryoshield and the specimen), not only was carbon contamination stopped, but the structure of the organic specimens under observation was destroyed. At about the same time, a number of other investigators noted that the introduction of various gases (e.g., air, hydrogen, nitrogen, and even inert gases) resulted in the reduction of carbonaceous deposits. For example, Castaing and Descamps (1954) discovered that the introduction of air would prevent the buildup of a contaminating layer during specimen analysis in their early electron microprobe. That is, contamination buildup could be made to slow to a stop and even occasionally to reverse. In fact, similar observations were originally reported by Ruska (1942) in an electron microscope but never investigated further.

The combined effects of low specimen and cryoshield temperatures and the introduction of certain gases were studied in detail by Heide (1962, 1963) in an attempt to find the optimum conditions to prevent specimen contamination and damage. Although his work was concerned mainly with carbon removal (C-Abbau), it was also the forerunner of subsequent studies on specimen etching. The link between specimen contamination and specimen etching was now forged. (Carbon removal, after all, is not always desirable if it is an important constituent of the specimen under study.)

Heide's view of the contamination process was in the following mechanistic sequence: condensation of hydrocarbon molecules onto the surface – ionization – polymerization and reduction to an immobile carbon deposit. In all previous attempts to prevent contamination, both the specimen and its surroundings were cooled (with or without the introduction of gases into the specimen chamber); therefore, these attempts almost invariably led to specimen damage by carbon removal, particularly if the specimens were biological in nature. Naturally, one preferred to observe a specimen with neither contamination buildup nor carbon removal (unless only the contaminant was removed). Heide thus built an independently regulated cryoshield in

*There are other reasons why specimen cooling may be useful to prevent radiation damage (Glaeser, 1979; Hobbs, 1979), but we concentrate here only on the effects of specimen contamination and etching as defined in the introduction. Also see Chapter 11 in this volume, by D. G. Howitt.

an electron microscope. That is, he separately controlled the temperature of the shield surrounding the specimen and the specimen itself.

The results were dramatic and are best illustrated by Figures 1(a) and (b). Heide found that keeping the specimen at room temperature while cooling the surrounding cryoshield below $\sim -130°C$ seemed to achieve the simultaneous goals of contamination prevention without carbon removal. Heide's interpretation of these data involved a mechanism that required the presence of vapors other than those of hydrocarbons (Σ CH). He assumed that under normal operating conditions (in the Elmiskop I), the partial pressures of various gases were as follows: H_2O, 2×10^5 Torr; Σ CH, 5×10^6 Torr; Co, N_2, CO_2, and H_2, 1×10^6 Torr each; O_2, 1×10^7 Torr. He then concluded that when the temperature was lowered, the partial pressure of the hydrocarbons also lowered to the extent that at temperatures below $-60°C$ their partial pressures were negligible. No further specimen contamination caused by the deposition of hydrocarbons thus took place. The partial pressures of the other gases, however, were very little affected until the temperature dropped to $-100°C$ to $-130°C$. At these temperatures the partial pressure of water vapor dropped dramatically from $\sim 1 \times 10^5$ Torr to $\sim 1 \times 10^8$ Torr. The process of carbon removal was then postulated to take place through reactions of the type

$$H_2O \rightarrow H_2O^+ + e \tag{1}$$

and

$$H_2O^+ + C \rightarrow H + H^+ + CO \tag{2}$$

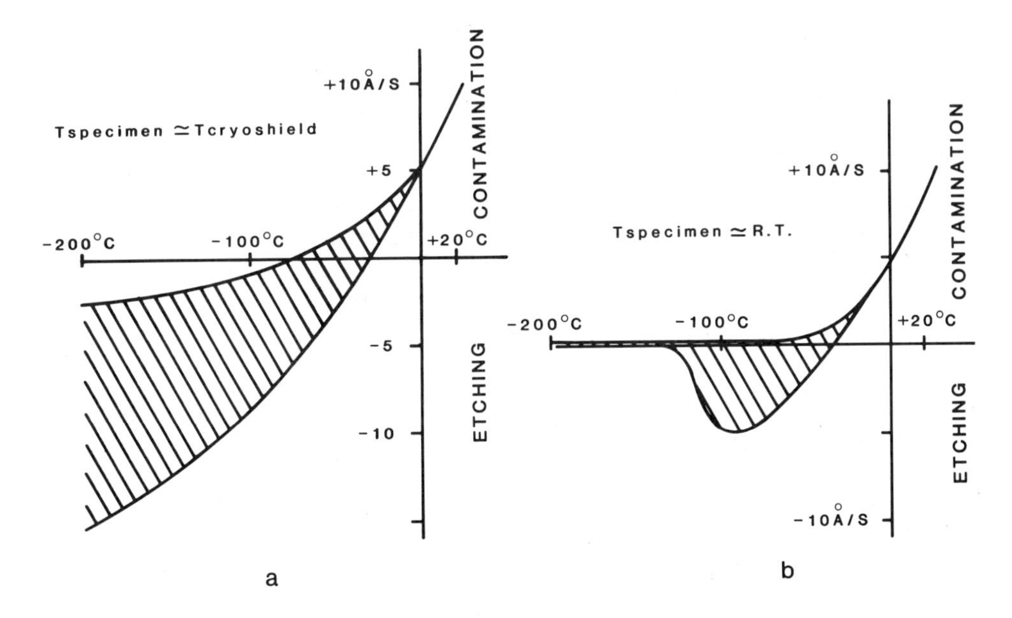

Figure 1 Rate of contamination buildup or carbon removal as a function of the temperature of the specimen and cryoshield: Beam diameter $\sim 2\ \mu m$; microscope vacuum, $\sim 5 \times 10^5$ Torr. Shaded areas indicate data scatter (after Heide, 1962). (a) Specimen and cryoshield cooled together; (b) cryoshield cooled and specimen at room temperature.

under the action of the electron beam. If the specimen was cooled along with the cryoshield, water molecules would absorb onto its surface, thereby providing a plentiful supply of reactants to remove carbon. If the specimen was not cooled, but the cryoshield was, the water vapor would preferentially absorb onto the cryoshield and protect the specimen.

It seemed that, with the conclusion of Heide's work, the way was open to high-resolution electron microscopy, and to a considerable extent this was true. Unfortunately, as progress toward better and better resolution continued, further difficulties appeared. In particular, the use of smaller and smaller electron probes for both imaging and local elemental analysis aggravated the processes of contamination and etching in a surprisingly intense manner.

VI. SURFACE DIFFUSION AND BEAM SIZE EFFECTS

Vacuum technology has improved substantially in parallel with improvements in electron microscopes. For example, microscope column pressures of $<10^7$ Torr in a modified electron microscope were achieved by Hartman and Hartman (1965) using ion pumps. Differential pumping of the specimen chamber was demonstrated by several investigators to reduce pressure to the region of 10^{-8} Torr (Grigson *et al.*, 1966; Hart *et al.*, 1966; Valdré, 1966). As a result of these improvements, more subtle effects of contamination and etching were detected, and the earlier explanations were found to be wanting. In contrast to earlier conclusions, Hart *et al.* (1966), for example, determined that surface diffusion of the contaminant to the irradiated area was required to account for the observed growth rate. Differential pumping, mass spectrometric analysis, and some regulation of the partial pressure of the background gases were also features of this work. This more extensive control of the environment led to attempts to measure contamination rates quantitatively.

Confirmation of the importance of surface diffusion came from a somewhat different direction. It was found that controlled local growth of contamination could be used to write patterns of incredibly small dimensions on a surface (Broers, 1965; Oatley *et al.*, 1965; Thornhill and Mackintosh, 1965). Literally, the Bible could be put on the head of a pin! The carbonaceous deposits formed by the electron beam resisted subsequent removal by chemical etching or ion bombardment in the same way that the photoresists used in solid-state microfabrication did. Thus, the size scale of the patterns could be made in the range of several hundred angstroms rather than a few micrometers as limited by the wavelength of light. The control of such "electron-beam writings" required at least an empirical knowledge of contamination rates since these rates determined the total writing time for the circuits. A search for such quantitative control led Müller (1971a, b) to an extensive study of the dependence of the contamination rate on the electron-beam diameter and current density. He used a thin carbon film as the "recording" medium. Halftone pictures 5 μm by 5 μm could be produced in this way by varying the electron-beam scan rate! Müller's derived contamination growth rates varied with the beam diameter, but not with the current density. His model also required surface migration of the precursive material from the immediate surroundings of the illuminated region to account for the observed rates. Experimental results, fitting his theory, were reported for probe diameters of 15 nm to 3 μm.

VII. RECENT STUDIES OF CONTAMINATION AND ETCHING

Substantial differences in the kinetics of contamination associated with beam size effects were encountered in electron microscope applications at about the same time as the early beam writing studies. For example, the possibilities of STEM imaging using electron probes of nearly atomic dimensions were just beginning (e.g., Crewe and Wall, 1970). Convergent beam diffraction, using stationary probes of 30 nm, was also more and more often employed (e.g., Glaeser *et al.*, 1965; Mills and Moodie, 1968; Riecke, 1969). Extreme measures had to be taken to avoid contamination (e.g., through the use of ultrahigh vacuums, by specimen heating, etc). Even so, many of the cures found were impractical for the vast majority of practicing microscopists. Still further investigations of the detailed mechanisms of contamination and etching were obviously required.

The work of Isaacson *et al.* (1974) was particularly informative since they could combine very high image resolution with an analysis of the electron energy-loss signal (EELS). Their specimens were placed on very thin (~2-nm-thick) carbon support films and investigated under ultrahigh vacuum (UHV) conditions. Many contaminant molecules still remained absorbed to the films from prior handling unless extraordinary means to clean the specimens were taken beforehand. Isaacson *et al.* could observe the effects of absorbed contaminant molecules by changes in the electron scattering power. It was clear to them that the adsorbates were responsible for both etching and contamination in the beam. Etching was observed to proceed rapidly at first and was highly specimen dependent, being especially rapid when organic salts were placed on the film. The measured characteristics of the contaminants were indistinguishable from those of the film, and the rate of buildup depended highly on the type of specimen and the gases to which the film was exposed. All of the data of Isaacson *et al.* were consistent with the hypothesis that surface diffusion was the principal source of the molecules that formed their contaminating layers. They found that contamination could be virtually eliminated by gently heating the specimen to ~50°C in the ultrahigh vacuum before observation.

Further semiquantitative studies of contamination were conducted by Egerton and Rossouw (1976) using a conventional TEM equipped with an electron energy-loss analyzer. The relative thickness, t, of the contaminating film could be deduced directly from the heights of the carbon peaks for zero energy loss, h_o (elastically scattered), and the plasmon loss maximum, h_i. The measurement could be made approximately quantitative by determining a proportionality constant, C, calibrated with a carbon specimen of known thickness. The relationship then became simply

$$t = C\frac{h_i}{h_o}. \tag{3}$$

Egerton and Rossouw proceeded to monitor specimen thickness as a function of irradiation time for a series of specimen temperatures. Their results are summarized in Figure 2. The thinning effect (etching) was found to be confined to the area illuminated by the electron beam. The authors evaluated a number of possible mechanisms: (1) ionization of the molecules in the vacuum system in the gas phase that could then react with the specimen upon being adsorbed; (2) ionization of gas absorbed on the specimen surface (e.g., H_2O) that could then react with carbonaceous material; (3) ionization of the atoms of the specimen surface itself or any carbonaceous deposits that could subsequently react with the absorbed gases; (4) direct sputtering of the surface atoms by the highly energetic incident electrons.

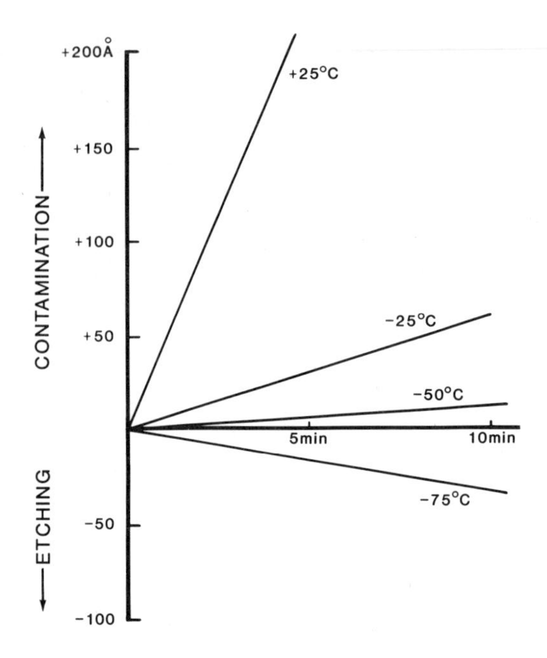

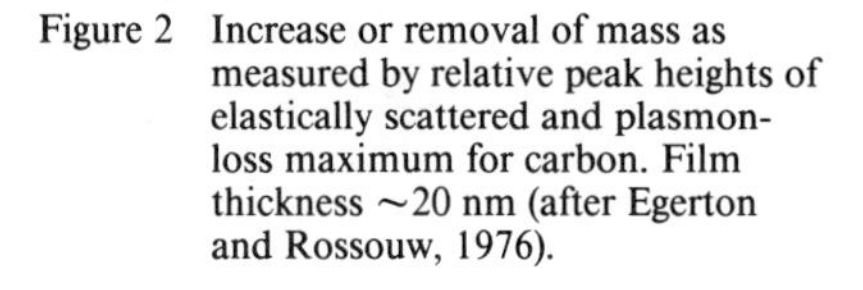
Figure 2 Increase or removal of mass as measured by relative peak heights of elastically scattered and plasmon-loss maximum for carbon. Film thickness ~20 nm (after Egerton and Rossouw, 1976).

After careful consideration, none of these mechanisms could be ruled out completely. Mechanisms (1) and (2) were those proposed originally by Heide (1963), with the latter likely to predominate. However, mechanism (3) could also reasonably account for the observed thinning rates (~5 Å/min) if one assumed a surface coverage of ~0.5% absorbed gas (say H_2O) that provided reactants for the ionized carbon atoms. In addition, Egerton and Rossouw noted that areas that had been thinned by the electron beam appeared rougher than their surroundings. They concluded that impurity metal atoms might be acting as localized catalysts for the oxidation of carbon in agreement (for example) with the observations of Thomas and Walker (1965) on the etching of graphite.

A series of papers by Hartman *et al.* (1968), Calbick and Hartman (1969), Hartman *et al.* (1969), and Hartman and Hartman (1971) was specifically directed toward defining the mechanisms of residual gas reactions during irradiation under controlled conditions in the electron microscope. These authors concluded that inelastic scattering can result in the transfer of sufficient energy to the surface atoms of the specimen that the activated molecule which results can decay by a large number of possible mechanisms. In many cases, the most probable decay mechanism is simply that the activated molecule returns to its original state with no permanent damage. However, quite often damaging etching reactions (for organic specimens) such as the following could occur

$$C_nH_{2n} + 2H_2O \xrightarrow{e} C_{n-1}H_{2n} + CO + 2H_2 \ . \qquad (4)$$

Hartman *et al.* showed that such reactions could indeed have a negative temperature dependence in agreement with the findings of Heide (1962).

In more recent work, etching reactions in biological molecules were studied using the loss of ^{14}C caused by electron irradiation on labeled T4 bacteriophages and E. coli bacteria (Dubochet, 1975). He used a variety of commercial electron microscopes with both fixed and scanned beams. The irradiation doses were those which approximated practical observational conditions. Radiographic methods by high-resolution photographic recording were used to determine the ^{14}C distribution. Dubochet concluded that: (1) the sensitivity to carbon loss decreased with exposure; (2) surface migration of molecular fragments and absorbed molecules is involved in the mechanism of beam damage; (3) there is no perceptible carbon loss when irradiation takes place at liquid helium specimen temperatures.

In an extensive series of studies, Fourie (1975, 1976a, 1976b, 1978a, 1978b, 1979) attempts to develop a quantitative theory capable of explaining the drastic difference in the kinetics of contamination between large and small electron-beam diameters and to develop a satisfactory cure. Fourie states flatly that surface diffusion alone is the source of contaminant provided that the specimen is suitably cryoshielded or that a UHV microscope is used. He also argues that the influence of an induced electric field on the specimen surface (arising from the ejected secondary electrons) dramatically enhances surface diffusion as the beam size decreases.

The general form of the contamination buildup shifts with decreasing beam size, as indicated in Figure 3. Surface charging effects become dominant, according to Fourie, as the beam size decreases to well below the lateral range of the "high-energy secondaries." He shows, for example, that charging effects are most dramatic with dielectric samples that are irradiated distant from conducting pathways (such as grid bars, coatings, substrates, etc). A further observation by Fourie is that *all* pretreatments of the specimen before it is in position for analysis are critical. Pretreatment includes specimen preparation (including handling), cleaning or rinsing before insertion, prepumping, exposure to hydrocarbon vapors within the microscope before cryoshielding or viewing, etc. Only by carefully controlling specimen pretreatments (e.g., prepump times) could Fourie obtain reproducible kinetic results of the growth of contaminant deposits. Apparently, many of the inconsistencies in earlier attempts to develop a quantitative theory can be traced to a lack of control of the specimen before analysis.

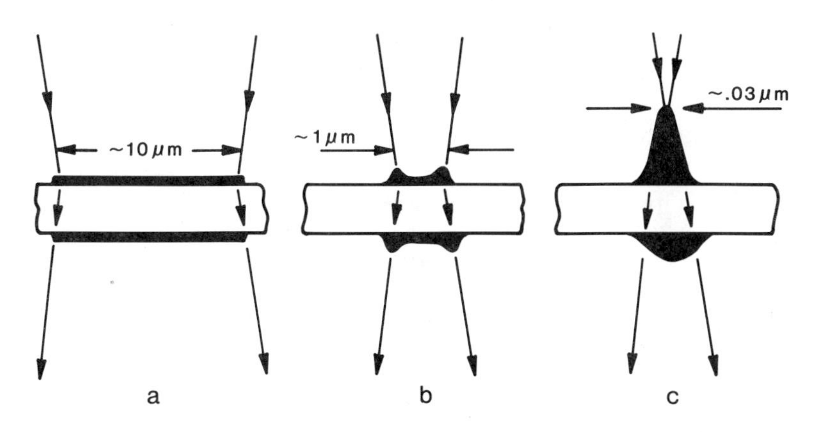

Figure 3 Schematic illustration of the change in appearance of the contamination deposit with different beam sizes. Note the change in scale from (a) to (c).

VIII. SUMMARY OF PHENOMENOLOGICAL OBSERVATIONS

This chronological review of the major literature on contamination and etching has given us a reasonably complete description of the qualitative conditions leading to specimen contamination and a somewhat less complete idea of the conditions leading to specimen etching. Although there have been numerous attempts to provide quantitative contamination rates, the variety of experimental parameters encountered makes such attempts necessarily incomplete or only narrowly applicable. In any case, a full, quantitative description is probably needed only in very special cases (perhaps for electron-beam writing). Actually, the practicing microscopist only wants to know how to avoid deleterious effects during analysis or how to buy sufficient time to obtain the results of interest.

As a first step to evolving practical solutions, we simply summarize the most important observations (Table I). The properties selected, those left out, and the brief descriptions of each are purely subjective judgments by the author. Several conclusions are evident immediately. Our qualitative understanding of contamination is significantly better than that of etching. On the other hand, much is still unclear or unstudied in regard to both. Of course, many of the control parameters are specimen related and cannot be arbitrarily altered. Conversely, there is much that can be controlled and yet often is not. If we strip the empirical observations down further, we are left with the following obvious parameters: the operator, the specimen, and the microscope! Of course, this should have been clear to all from the beginning. The interaction of these three and their principal effects are described for emphasis in Figure 4. Although the operator cannot affect the nature of the specimen, he (she) can affect its pretreatment and can control the mode of operation of the microscope. He (she) can also use (or add) accessories, and this very act influences manufacturers' designs of future microscopes.

TABLE I

Parameter or Condition	Contamination	Etching
Residual Gases	H_xC_y, other large molecules	$p(H_2O)$, $p(O_2)$, $p(H_2)$ unclear
Initial Surface Composition	Critical for small beam diameters	Effect unclear, but probably more severe with small beam size
Electrical Conductivity	Important for insulators	Effect unclear, probably only indirectly involved
Thermal Conductivity	Probably of some importance, but highly variable	Effect unclear, may be secondarily important
Beam Size and Current Density	Critical for small beam sizes	Effect unclear, probably more severe with smaller beam size
Secondary and Back-scattered Electrons	May be important, especially for insulators	Effect unclear, probably only indirect
Specimen Temperature	Important, but can be kept low	Important; determines reaction rate; can be kept low

IX. THE MECHANISMS OF CONTAMINATION

As we have shown, contamination and etching effects are often related; however, it will be less confusing to consider their mechanisms separately. The plural of mechanism has been used with forethought. To speak of a singular mechanism is a gross oversimplification in either case. We already know that working cures such as cold fingers (or cryoshields) appear to be perfectly satisfactory for some specimens under specific working conditions. However, changing just one variable—for example, decreasing the beam size or changing the mode of operation (to stationary)—is known to alter the rate of contamination buildup drastically.

What then are the fundamental processes whose relative importance affects the rate kinetics and the form of the buildup so dramatically? Figure 5 illustrates schematically all those mechanisms that are now known. We consider these processes in order.

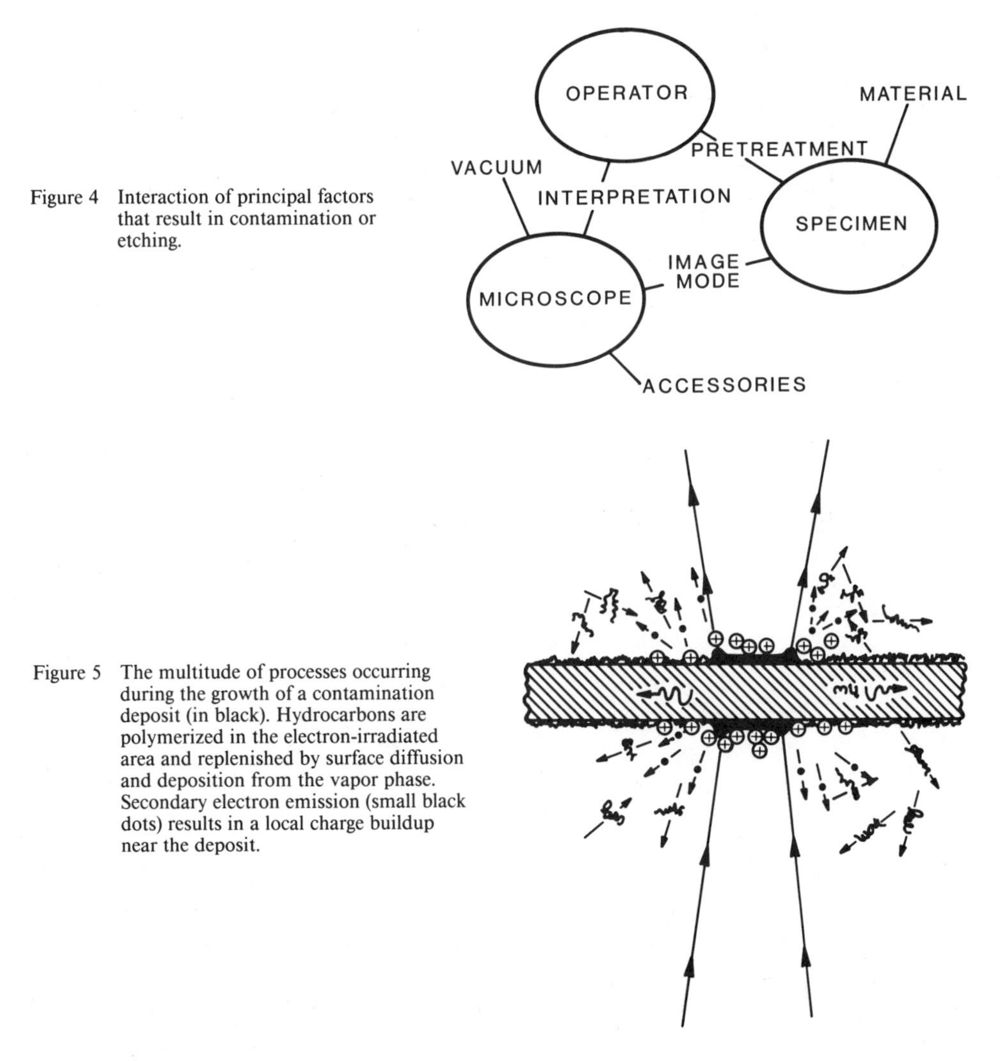

Figure 4 Interaction of principal factors that result in contamination or etching.

Figure 5 The multitude of processes occurring during the growth of a contamination deposit (in black). Hydrocarbons are polymerized in the electron-irradiated area and replenished by surface diffusion and deposition from the vapor phase. Secondary electron emission (small black dots) results in a local charge buildup near the deposit.

A. Physisorption of Hydrocarbon Molecules

Any low-vapor-pressure molecule deposited before or during the observational or analytical step will probably contribute to the contamination process. The specimen itself could be such a source, but evidence so far points overwhelmingly toward other external sources. The key property seems to be weak absorptive bonds (physisorbed, not chemisorbed). The large, adsorbed molecules come either from sources in the microscope vacuum (or prepump chamber) or from an earlier preparation step. The incoming molecular flux, f, is controlled by a physical law of the following form

$$f \propto \sqrt{m/(TP)} \tag{5}$$

where m is the mass of the hydrocarbon molecules, P is their partial pressure, and T is the absolute temperature.

B. Surface Diffusion of Hydrocarbon Molecules

Because adsorbed hydrocarbons are not tightly bound to the surface, they may diffuse rapidly under the influence of any suitable gradient (chemical, thermal, or electrical). Such gradients may be created by several mechanisms that vary strongly with the nature of the specimen and the imaging conditions. For example, the diffusivity, D, is determined by an equation of the kind

$$D_s \propto \exp\,[-Q_s/kT] \tag{6}$$

where Q_s is the activation energy for surface diffusion (and thereby strongly related to the strength of the molecule surface bond), k is Boltzmann's constant, and T is the absolute temperature.

C. Polymerization and Fragmentation of Hydrocarbon Molecules in the Electron Beam

The specimen surfaces (top and bottom) are coated with hydrocarbons. Passing an intense, energetic electron beam through the specimen is known to polymerize and fragment the hydrocarbon molecules. In turn, polymerization immobilizes them and produces a carbonaceous layer (blackened in Figure 5), which then creates a surface concentration (chemical) gradient. Hydrocarbons from outside the irradiated area fill the void by random-walk surface diffusive processes. The local concentration changes with time in a gradient $\partial C/\partial x$ according to Fick's second law

$$\frac{\partial C}{\partial t} = D_s \frac{\partial^2 C}{\partial x^2} \tag{7}$$

where D_s is the surface diffusivity as defined above.

D. Beam-Induced Thermal Gradients

No controlled study of the effects of thermal gradients on contamination in thin films has as yet been reported, probably because temperature measurement and control at the spatial levels required is very difficult. However, many theoretical calculations of beam heating have been reported (Glaeser, 1979; Hobbs, 1979). In addition, heat flow is governed by (a different form, but) the same Fick's laws as those for molecular diffusion; therefore, boundary conditions such as the cross-sectional area of the specimen, the distance to a thermal sink, the thermal conductivity of the specimen, etc, are all important variables. For example, even a thin gold foil can be melted under suitable conditions in an electron microscope (the author can attest to this). On the other hand, under the right conditions polymers and biological thin sections can be observed for hours without obvious image deterioration. In short, although local temperature (or a temperature gradient) can drastically affect the rate of surface diffusion and adsorption (for example, Eq. (5), (6), and (7)), its quantitative importance is still subject to extreme variation.

E. Electrical Gradients in the Surface

As illustrated in Figure 5 (by the escaping secondary electrons and their residual holes), a positive potential is being constantly created under and near the electron-irradiated area. The electric field so created may be quickly neutralized (e.g., in a metallic specimen with little or no oxide on its surfaces) or it may build up until local dielectric breakdown takes place (e.g., in a thin, uncoated insulating film). In very large electrical gradients and small beam sizes, the physisorbed hydrocarbons may become polarized by the field and drain the area surrounding that being irradiated at a surprisingly rapid rate (Fourie, 1979). For large beams (>1 μm) and insulating specimens, contamination may actually be suppressed by the electric field gradient (Fourie, 1975).

Although the precise electrical field and its effects on contamination are as widely variable as the thermal gradients cited above, their origins seem well defined. Secondary electrons must be emitted by the incident beam (and its elastically and inelastically scattered successors) at a sufficient rate that charge neutralization does not take place instantaneously. Such charging effects are commonly observed in SEM imaging and may be overcome by coating with a conducting layer or intermittent flooding by an electron source.

The five processes just described seem to be at the heart of the mechanisms creating contamination as we encounter it in analytical electron microscopy. If this qualitative understanding is correct, we should be able to regulate, eliminate, or at least minimize the undesirable effects of contamination. We now seem to be approaching that condition and will describe in the following section the methods that seem to be most successful to date and the additional improvements that can be hoped for.

X. THE MECHANISMS OF ETCHING

We will try to follow the same format in explaining etching mechanisms as we did for contamination. Unfortunately, we find that there have been fewer explicit studies of etching than of contamination. Further, the etching phenomenon is a form of beam-induced radiation damage and hence overlaps the subject matter of Chapters 6 and 11. By restricting ourselves to Glaeser's (1979) definition, however, we can present a somewhat limited picture of our current understanding.

In place of the mechanisms depicted in Figure 5, we present a somewhat different model, Figure 6. Although water vapor is not the only possible reacting gas (O_2, H_2, etc. are possible), it is a proven culprit. Furthermore, water vapor is a common residual gas in all vacuum systems—certainly in most electron microscopes. It follows that a certain fraction of water molecules will always be adsorbed on the specimen surface along with the hydrocarbons responsible for contamination. In Figure 6, we have described only the hypothetical situation where carbon-containing molecules (comprising the specimen) react with activated water molecules adsorbed to the specimen surface. We could also have added a contaminant layer in this model with similar results. Furthermore, the mechanisms shown need not be only hydrocarbons. What generalization can we make with regard to possible mechanisms?

A. Physisorption of a Potentially Reactive Gas

Water vapor (or other gaseous molecules) can be made to react with either the contaminant or the specimen directly under the conditions of electron irradiation. Most observations of etching to date have been on organic specimens or on carbon-bearing contaminants themselves. Other substrate materials can certainly be susceptible to etching. However, in order that there be a sufficient physisorbed gas, the specimen temperature must be lower than ambient (generally much lower) or the partial pressure of the reactive gases must be high. Of course, a specimen at low temperatures will act as a cryopumping surface and thereby increase its surface concentration of physisorbed gas.

B. Activation of the Reactive Gas by Electrons

It is difficult to separate this step from that of physisorption. What is required is that either a sufficient number of gaseous molecules be available or that their activation cross sections be high enough that a measurable reaction rate between the gas and substrate is attained. The electron energies available to create such reactive ions are obviously more than large enough (Glaeser, 1979).

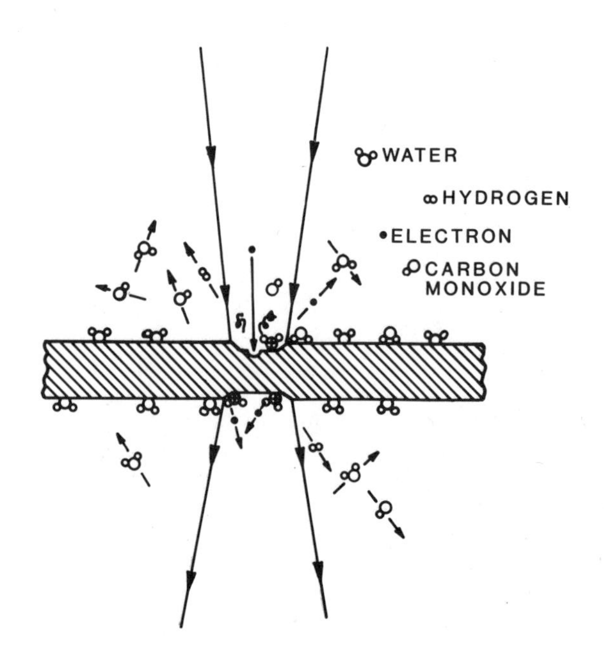

Figure 6 Schematic illustration of the etching process for reactions of the types that produce CO by interaction with H_2O ions. The specimen is assumed to be carbon-containing.

C. Specimen or Contaminant Molecules That Will React With the Excited Physisorbed Gas

Mass is removed by a chemical reaction between the excited gas molecule and ionized fragments of either the specimen or the contaminant layer. The prototype reactions of Hartman *et al.* (1968) are typical (Eq. 4). Gaseous products are released (e.g., H_2 and CO_2) and not readsorbed. The reacting species may be provided by fragmentation of larger molecules, but because the lifetime of such activated species is short, the supply of physisorbed gas must be large enough to provide a high reaction probability.

D. The Reactant Molecules Must be Volatile

Mass can be removed from the specimen only if one or more of its atomic species forms a volatile gas upon reaction with adsorbates and is not readsorbed upon reacting (e.g., Eq. (1) and (2)). Naturally, this requirement is linked to the other thermodynamic variables, pressure and temperature, so that a limited class of potential etching conditions appears possible given the usual constraints of electron microscopy. For example, metallic specimens and most ceramics seem unlikely to be susceptible to etching under most AEM conditions, but their contaminating layer will. Further (hypothetical) steps could be added. The effects of local heating and charging could alter conditions sufficiently to affect the reaction rate as well. There is little point in speculating further without reasonably secure verifications from experiment, and these are still lacking.

It can be safely concluded that mass can be removed by beam-induced chemical reactions. The most certain combination leading to etching involves mass removal of cooled organic specimens where there is also a sufficiently high partial pressure of water vapor. Since such conditions are often met, this is an important practical case. Does this imply that specimen cooling will not help prevent its destruction during observation? Fortunately, it does not, because: (a) the partial pressure of water vapor and other culprit gases can be reduced below harmful levels with sufficient care, and (b) specimen temperatures that are low enough can also slow surface diffusion and chemical reaction rates to a snail's pace. On the other hand, etching may still be useful as an in situ contaminant-removal process. This still remains to be reliably demonstrated.

XI. WORKING SOLUTIONS: PROVEN AND POTENTIAL

Contamination and etching have been with us since the infancy of microscopy, and so have many cures. We review here some of these proven solutions to reduce contamination and suggest why they are or are not adequate.

Shielding or cryoshielding the specimen has been used in numerous forms (e.g., Ennos, 1954; Heide, 1963; Rackham, 1975; Hren *et al.*, 1977; Tomita *et al.*, 1978). The philosophy behind its use is simple and eminently practical. Cooling the region surrounding the specimen to a temperature as low as possible achieves several desirable objectives. First, contaminant (or etchant) molecules are shielded by line of sight from the specimen. Second, the undesirable molecules are captured (adsorbed) and held by the cryoshield. Third, the cryoshield will act as a local pump toward the specimen, grid, and holder and clean them up with time. Fourth, a cryoshield can also help eliminate system-generated background x-rays and stray electrons. For all of these reasons a cryoshield is a highly desirable addition for nearly any analytical electron microscope.

In its idealized configuration, the cryoshield should completely surround the specimen. Since this presents some practical difficulties, openings through it should be made as small as possible (~1-mm openings top and bottom). Clearly, an unworkable condition can be made marginally useful by good cryoshielding design. For example, the volume of the contaminant formed by a 10-minute irradiation using a cryoshield at liquid nitrogen temperature is about the same as that for a 1-minute irradiation with the cryoshield at room temperature. Cryoshielding is, therefore, a first-line cure, and it is useful in nearly all cases.

Better vacuum systems are, of course, *nearly* a cure-all, but there are two remaining barriers to analysis: the specimen and the operator (see especially Sutfin, 1977). Experience with ultrahigh vacuum conditions is now very extensive. It stands to reason that hydrocarbon-based contaminants can be eliminated by simple exclusion from the microscope (provided that they are also excluded from the specimen beforehand). Furthermore, good UHV techniques should result, as well, in substantial decreases of residual gases such as H_2O, O_2, etc., and therefore in a decreased etching rate. Although a better vacuum is an obvious cure, it is placed second among the solutions described here because many years will be required to upgrade the current generation of microscopes to UHV capabilities. However, certain measures can be taken now that will help the average operator. Vacuum fluids and sealants (greases) can be substantially improved by replacement with new low-vapor-pressure fluids and materials (Ambrose *et al.*, 1972; Holland *et al.*, 1973; Elsey, 1975a, b). Electron microscope manufacturers can help immensely by taking advantage of well-developed vacuum procedures commonly used in other instruments (mass spectrographs, surface analytical instruments, etc.). In fact, this upgrading process is now being started.

The other major area for immediate improvement is in specimen preparation and pretreatment. Every handling and preparation step is important up to the time when the specimen is actually in place and being analyzed. This includes the step of insertion into the analyzing chamber. Residues from electropolishing, hydrocarbons deposited during freeze drying, fingerprints, etc.—all are sources of contamination. It should not be surprising that the insertion of a dirty specimen into a sophisticated STEM (even one with a UHV pumping system) will still yield a cone of contaminant when examined with a fine electron beam. On the other hand, even with a dirty specimen, a UHV system will drastically slow if not eliminate etching effects. This will be true even for low specimen temperatures and no cryoshield (Wall *et al.*, 1977; Wall, 1979).

A number of other workable solutions exist as well. Most are restricted to specific analytical techniques or to certain classes of specimens. One widely used approach involves flooding an area much larger than that to be analyzed with electrons or ultraviolet radiation and thus polymerizing (immobilizing) the physisorbed supply of hydrocarbons. The diffusion length for a fresh supply of contaminant now becomes approximately the radius of the irradiated area. An analysis using Eq. (6) and (7) is consistent with the experimental observation that useful working times up to 10 minutes in the stationary spot mode may thus be obtained (see Rackham, 1975, for a review). Microdiffraction information from areas of diameter as small as 3 nm were achieved by Geiss (1975) while continuously flooding a much larger area (~1 μm) with electrons. He thus prevented contamination buildup while collecting the analytical equivalent of stationary microbeam data. The method is described by Warren (1979).

Long diffusion times, and thus long working times, may be obtained by specimen cooling provided that specimen etching effects are avoided by using suitable vacuum pumping, cryoshielding, or both. Desorption of the physisorbed contaminants before analysis provides another approach. One way to achieve desorption is by heating the specimen in a prepump chamber free of contaminating molecules (Isaacson *et al.*, 1974) or in situ. A simple but time-consuming alternative is to pump on the specimen in the analyzing position for a long period (hours) before the actual analysis. In either case, the specimen must be free from further contaminating adsorbants when in the microscope. That is, either the vacuum system must be intrinsically contaminant free or a good cryoshield must be in place and cooled during the entire preanalytical period. Note that the cryoshield acts as a pump toward the specimen and, in effect, attracts contaminants to it and from the specimen.

Etching effects may also be reduced by such procedures. Alternative methods such as glow discharge have been proposed to achieve this same end (Bauer and Speidel, 1977). Still another avenue is to substitute a high-vapor-pressure contaminant (e.g., a volatile solvent such as ethyl alcohol, which is commonly used for a final rinse) for the low-vapor-pressure hydrocarbons (Tomita *et al.*, 1978). The specimen is then inserted while still wet with the solvent, and the volatile molecules are quickly desorbed. Provided that further contamination in the microscope is avoided, the area analyzed will contaminate at a much lower rate than before. This procedure is not really good high-vacuum practice since the solvent may eventually cause difficulties in other parts of the vacuum system, affecting seals, greases, etc.

Fourie (1979) suggests that the rate of buildup may be largely suppressed by irradiation in situ with an auxiliary source of low-energy electrons. He proposes that this procedure prevents charging effects, thereby drastically cutting the kinetics of the buildup of a cone for small spot sizes. Such procedures have been suggested before (Hillier and Davidson, 1947; LePoole, 1976). These same low-energy electrons will simultaneously polymerize a large area surrounding the beam and help in this way as well (Stewart, 1934).

Finally, intermittent analysis (e.g., by beam pulsing) appears to be an attractive means to achieve a number of desirable results in AEM without requiring extensive vacuum modifications (Smith and Cleaver, 1976; Hren, 1978). If the purpose of a spot analysis is to gain spectroscopic (compositional) data, then suitably controlled pulsing intervals will actually increase the useful information per unit real time (Statham *et al.*, 1974). A reduced irradiation time also reduces the contamination and etching rates and any charging effects in at least linear proportion to the off-time. In all probability, the gains will be considerably better than linear since the kinetic processes involved are higher order or exponential.

XII. SOME EFFECTS OF CONTAMINATION AND ETCHING ON AEM

This section could have been placed at the beginning of this chapter as a means of convincing the reader that the subject is of importance to AEM. Such emphasis was considered redundant since earlier authors in this book have already achieved this result, and nearly every reference cited describes such effects. The proceedings of an AEM workshop in 1978 (Silcox, 1978) also provide an extensive summary of many specialist experimental limitations resulting from contamination and etching effects, especially the former. It seems fair to assume that those contemplating research using the AEM are forewarned. The present section will therefore present only a few selected encounters with these barriers to AEM.

One of the most serious and directly adverse consequences of contamination is a loss of spatial resolution. Historically, even the very first observations (e.g., Watson, 1947) were most concerned with a loss of image detail (edges lost, contrast reduced, etc). The effective loss of resolution can be made more quantitative, especially in the extreme case of a stationary probe. Chapters 4 and 5 provide means to calculate the extent of spreading of the electron beam as it passes through a contaminating film. Another way is from the information contained in an electron-diffraction pattern of a film of amorphous carbon (or any light element or combination thereof). An easily measurable intensity maximum is recorded in the photographic film of such patterns at 1 Å (Heidenreich, 1964). From a simple Bragg's law calculation, this corresponds to $\theta \cong 0.02$ rad. Thus, for a point electron source impinging at the top of a 100-nm-thick film of carbonaceous contaminant, one would expect a circle of *at least* 4-nm diameter striking the surface of the specimen. Since typical probe sizes are ≤ 10 nm and contaminant buildup for stationary probes is typically >100 nm, the accompanying loss in resolution is quite substantial (Fraser, 1978). There is also some evidence that a stationary electron probe may be deflected by charging effects inherent to the poorly conducting contaminant (Zaluzec and Fraser, 1977; Fourie, 1976a; Tonomura, 1974). Obviously, this effect will seriously degrade the expected resolution even more. An indirect effect, during convergent beam diffraction experiments, is the local specimen bending induced in crystals during their examination (Kambe and Lehmpfuhl, 1975). The lack of local stability not only affects the orientation determination, but interferes directly with observations of atomic-sized step heights on the surfaces. The effects of etching on resolution are considerably more subtle and complex. Examples clearly fall into the regime of radiation damage, and the reader is referred especially to Glaeser (1979).

The measurement of local composition provides what may be the strongest driving force toward the minimization of contamination and etching. If spires of contaminant are built up on the entrance surface to the specimen, then clearly there are limitations to the minimum volume that can be analyzed (see above). This limitation affects both energy-dispersive spectroscopy (EDS) and electron energy-loss spectroscopy (EELS), particularly the latter, since it is often used for its sensitivity to light elements. Etching, on the other hand, not only decreases the volume analyzed, but it may significantly alter the apparent (and even real) local composition (Saubermann, 1977).

Contamination deposits will also affect the apparent local concentration measurement in a number of other ways. In the case of EELS measurements, peak-to-background (P/B) ratios will be altered drastically for specimens containing light elements. Even in EDS systems, characteristic peaks may be built up if the contaminating molecules contain elements ≥ 10. For example, Figure 7 shows an expanded region of an EDS spectrum recorded from a thin sample of homogeneous β-NiAl (Zaluzec, 1978).

Three peaks are resolved: NiL_{α}, AlK_{α}, and SiK_{α}. The silicon came from the fluids and greases in the vacuum components. Even if such characteristic peaks from the contaminant are not obtained, the effective mass sensitivity for x-ray analysis can be substantially reduced by the presence of contaminants because of the increased bremsstrahlung generated (Zaluzec and Fraser, 1977; Shuman *et al.*, 1976). Absorption effects from the contaminant affect not only EELS, but can substantially alter characteristic x-ray intensity ratios (Zaluzec and Fraser, 1977). For example, Figure 8 shows the intensity ratio of NiK_{α} to AlK_{α} plotted as a function of measurement time (hence increasing contaminant thickness).

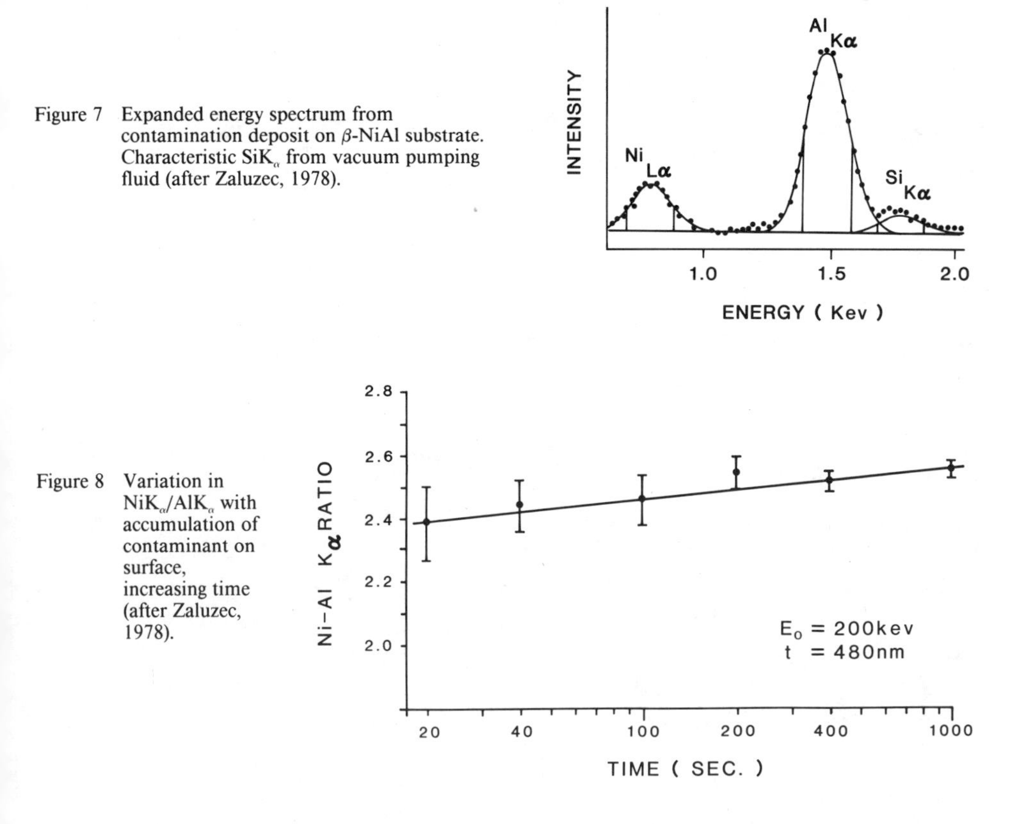

Figure 7 Expanded energy spectrum from contamination deposit on β-NiAl substrate. Characteristic SiK_α from vacuum pumping fluid (after Zaluzec, 1978).

Figure 8 Variation in NiK_α/AlK_α with accumulation of contaminant on surface, increasing time (after Zaluzec, 1978).

There is no need to belabor the point of this section. Clearly, contamination and etching effects have numerous deleterious influences on AEM. After all, the rates of contamination buildup have been measured by observing image or diffraction pattern degradation. Etching effects can be seen (imaged) or measured as a mass loss. These obvious results are sufficient in themselves to indicate the seriousness and magnitude of the problem. Yet we can close on a less negative note by remembering that etching can possibly be used to remove contamination (two negatives would indeed produce a positive!) and that contaminants can be used to mark areas, to measure specimen thickness and variations in thickness, and to detect drift (or its lack). These can be useful phenomena occasionally, and we would like to turn them on and off at will. Perhaps we shall be able to achieve just that goal, soon.

TABLE OF CHAPTER VARIABLES

C	Proportionality constant
D_s	Surface diffusivity
f	Molecular flux
h_i	Plasmon loss maximum

h_o Intensity of carbon zero-loss peak
k Boltzmann's constant (8.62×10^{-5} eV/K)
m Mass of hydrocarbon molecules
P Hydrocarbon gas partial pressure
Q_s Activation energy for surface diffusion
t Thickness
T Absolute temperature

REFERENCES

Ambrose, B. K.; Holland, L.; and Laurenson, L. (1972), J. Micros. *96*, 389.

Bauer, B., and Speidel, R. (1977), Optik *48*, 237.

Broers, A. N. (1965), Microelec. & Reliability *4*, 103.

Burton, E. F.; Sennett, R. S.; and Ellis, S. G. (1947), Nature *160*, 565.

Calbick, C. J., and Hartman, R. E. (1969), Proc. 27th Annl. EMSA Mtg., Claitor's Pub. Div., Baton Rouge, p. 82.

Castaing, R., and Descamps, J. (1954), C. R. Acad. Sci. (Paris) *238*, 1506.

Cosslett, V. E. (1947), J. Appl. Phys. *18*, 844.

Crewe, A. V., and Wall, J. (1970), J. Mol. Biol. *48*, 375.

Dubochet, J. (1975), J. Ultrastr. Res. *52*, 276.

Egerton, R. F., and Rossouw, C. J. (1976), J. Phys. D.: Appl. Phys. *9*, 659.

Ellis, S. G. (1951), Paper read to EMSA, Washington, DC, Nov. 1951.

Elsey, R. J. (1975a), Vacuum *25*, 299.

——— (1975b), Vacuum *25*, 347.

Ennos, A. E. (1953), Brit. J. Appl. Phys. *4*, 101.

——— (1954), Brit. J. Appl. Phys. *5*, 27.

Fourie, J. T. (1975), Optik *44*, 111.

——— (1976a), Proc. 9th Annl. SEM Symp., IITRI, Chicago, p. 53.

——— (1976b), Proc. 6th Euro. Cong. on E. M., Jerusalem, vol. 1, p. 396

——— (1978a), Proc. 9th Intl. Cong. on E. M., Toronto, vol. 1, p. 116.

——— (1978b), Optik *52*, 91.

——— (1979), Proc. 12th Annl. SEM Symp., Washington, DC.

Fraser, H. (1978), Scanning Electr. Micros. *1*, 627.

Geiss, R. H. (1975), Proc. 33rd EMSA Annl. Mtg., Claitor's Pub. Div., Baton Rouge, p. 218.

Glaeser, R. M. (1979), "Introduction to Analytical Electron Microscopy" (J. J. Hren, J. I. Goldstein, and D. C. Joy, ed.), Plenum Press, New York, p. 423.

Glaeser R. M.; Goodman, P.; and Lehmpfuhl, G. (1965), Z. Naturforsch. *20a*, 110.

Grigson, C. W. B.; Nixon, W. C.; and Tothill, F. (1966), Proc. 6th Intl. Conf. on E. M., Kyoto, vol. 1, p. 157.

Hart, R. K.; Kassner, T. F.; and Maurin, J. K. (1966), Proc. 6th Intl. Conf. on E. M., Kyoto, vol. 1, p. 161.

Hartman, R. E., and Hartman, R. S. (1965), Lab. Invest. *14*, 409.

——— (1971), Proc. 29th Annl. EMSA Mtg., Claitor's Pub. Div., Baton Rouge, p. 74.

Hartman, R. E.; Akahori, H.; Garrett, C.; Hartman, R. S.; and Ramos, P. L. (1969), Proc. 27th Annl. EMSA Mtg., Claitor's Pub. Div., Baton Rouge, p. 82.

Hartman, R. E.; Hartman, R. S.; and Ramos, P. L. (1968), Proc. 26th Annl. EMSA Mtg., Claitor's Pub. Div., Baton Rouge, p. 292.

Heide, H. G. (1962), Proc. 5th Intl. Conf. on E. M., Philadelphia, p. A-4.

——— (1963), Zeit. Angew. Phys. *XV*, 117.

Heidenreich, R. D. (1964), "Fundamentals of Transmission Electron Microscopy," Interscience Pub., New York.

Heidenreich, R. D., Hillier, J.; and Davidson, N. (1947), J. Appl. Phys. *18*, 499.

Hillier, J. (1948), J. Appl. Phys. *19*, 226.

Hobbs, L. W. (1979), "Introduction to Analytical Electron Microscopy" (J. J. Hren, J. I. Goldstein, and D. C. Joy, ed.), Plenum Press, New York, p. 437.

Hobbs, L. W.; Holland, L.; Laurenson, L.; and Fulkner, M. J. (1973), Japan J. Appl. Phys. *12*, 1468.

Hren, J. J. (1978), Proc. Workshop on AEM, Cornell Univ., p. 62; Ultramicroscopy *3*, 375.

Hren, J. J.; Jenkins, E. J.; and Aigeltinger, E. (1977), Proc. 35th Annl. EMSA Mtg., Claitor's Pub. Div., Baton Rouge, p. 66.

Isaacson, M.; Langmore, J.; and Wall, J. (1974), Proc. Annl. SEM Symp., Chicago, p. 19.

Kambe, K., and Lehmpfuhl, G. (1975), Optik *42*, 187.

Kinder, E. (1947), Naturwiss *34*, 23.

König, H. (1948), Naturwiss *35*, 261.

——— (1951), Zeit. für Phys. *129*, 483.

König, H., and Helwig, G. (1951), Zeit. für Phys. *129*, 491.

Leisegang, S. (1954), Proc. Intl. Conf. on E. M., London, p. 184.

Le Poole, J. B. (1976), "Developments in Electron Microscopy and Analysis" (J. A. Venables, ed.), Acad. Press, New York, p. 79.

Mills, J. C., and Moodie, A. F. (1968), Rev. Sci. Instru. *39*, 962.

Müller, K. H. (1971a), Optik *33*, 296.

——— (1971b), Optik *33*, 331.

Oatley, C. W.; Nixon, W. C.; and Pease, R. F. N. (1965), Adv. Elect. Electron Phys. *21*, 181.

Rackham, G. M. (1975), Ph.D. Dissertation, Univ. of Bristol, p. 39.

Riecke, W. D. (1969), Zeit. Angew. Phys. *27*, 155.

Ruska, E. (1942), Kolloid. Z. *100*, 212.

Saubermann, A. J. (1977), Proc. 35th Annl. EMSA Mtg., Claitor's Pub. Div., Baton Rouge, p. 366.

Shuman, H., and Silcox, J. (1978), "Analytical Electron Microscopy: Report of a Specialist Workshop," Cornell Univ., pp. 62-129.

Shuman, H.; Somlyo, A. V.; and Somlyo, A. P. (1976), Ultramicroscopy *6*, 317-39.

Smith, K. C. A., and Cleaver, J. R. A. (1976), "Developments in Electron Microscopy and Analysis" (J. A. Venables, ed.), Acad. Press, New York, p. 75.

Statham, P. J.; Long, J. V. P.; Waite, G.; and Kandiah, K. (1974), X-Ray Spectrometry *3*, 153.

Stewart, R. L. (1934), Phys. Rev. *45*, 488.

Sutfin, L. V. (1977), Proc. IXCOM, Boston, p. 66A.

Thomas, J. M., and Walker, P. L., Jr. (1965), Carbon *2*, 434.

Thornhill, J. W., and Mackintosh, I. M. (1965), Microelec. & Reliability *4*, 97.

Tomita, T.; Harada, Y.; Watanabe, H.; and Etoh, T. (1978), Proc. 9th Intl. Cong. on E. M., Toronto, vol. 1, p. 114.

Tonomura, A. (1974), Optik *39*, 386.

Valdré, U. (1966), Proc. 6th Intl. Conf. on E. M., Kyoto, vol. 1, p. 157.

Wall, J. (1979), "Introduction to Analytical Electron Microscopy" (J. J. Hren, J. I. Goldstein, and D. C. Joy, ed.), Plenum Press, New York, p. 333.

Wall, J.; Bittner, J.; and Hainfeld, J. (1977), Proc. 35th Annl. EMSA Mtg., Claitor's Pub. Div., Baton Rouge, p. 558.

Warren, J. B. (1979), "Introduction to Analytical Electron Microscopy" (J. J. Hren, J. I. Goldstein, and D. C. Joy, ed.), Plenum Press, New York, p. 369.

Watson, J. H. L. (1947), J. Appl. Phys. *18*, 153.

Zaluzec, N. (1978), PhD. Thesis, Univ. of Illinois; Oak Ridge National Laboratory Report ORNL/TM-6705.

Zaluzec, N., and Fraser, H. (1977), Proc. 8th Intl. Conf. on X-ray Optics and Microanalysis, Science Press, Princeton, New Jersey.

CHAPTER 11

RADIATION EFFECTS ENCOUNTERED BY INORGANIC MATERIALS IN ANALYTICAL ELECTRON MICROSCOPY

D. G. Howitt

Department of Mechanical Engineering
University of California, Davis, California

I. INTRODUCTION

In analytical electron microscopy (AEM), radiation damage will be of only structural consequence unless it occurs during specimen preparation or the elements that are the subject of analysis depart from the vicinity of the electron probe. At electron intensities between 10 and 10^3 A/cm^{-2}, the mechanical and thermal responses of some specimens can often be spectacular, but the concomitant variations in chemical composition—averaged over the volumes exposed to the electron beam—need not be. The processes responsible for radiation damage are the same electron-atom (elastic) and electron-electron (inelastic) collisions that provide diffraction contrast and chemical information in electron microscopy; the effects of radiation damage upon the image contrast of crystalline materials in transmission electron microscopy have already been extensively reviewed (Hobbs, 1979).

It is also appropriate in AEM to examine the effects of radiation damage upon the chemical information obtained from a thin specimen, and ideally one would like to compare the rates at which atoms are lost from the vicinity of the electron probe to the rates at which characteristic x-rays or energy losses can be collected. It can be shown that the atomic rearrangement brought about by the electron scattering events themselves is generally of lesser importance to the composition stability within the probe than the subsequent atom migration caused by diffusional processes. Radiation damage in AEM is therefore dependent upon beam flux and specimen temperature in addition to the magnitude of the elastic and inelastic cross sections.

An outline of the mechanisms by which elastic and inelastic collisions produce radiation damage in the electron microscope and a brief methodology for making quantitative estimates of each process are given in the first section of this chapter. The other sections are devoted to the estimation of the collection efficiencies for electron energy loss and energy dispersive x-ray spectroscopy (EELS/EDXS) and the manner in which radiation damage will affect their reliability. Some quantitative estimates of the significance of radiation damage to AEM are also presented, as well as a brief description of the radiation damage encountered from ion-milling specimen-preparation techniques, which have been treated in detail elsewhere (Howitt, 1984).

A rigorous justification for the calculations of many of the quantities used here has been omitted, and some of the calculations have been deliberately oversimplified, particularly with respect to the estimation of background in EDXS and EELS. It is hoped that these measures will help to preserve clarity and permit the straightforward incorporation of more appropriate cross sections and efficiencies as they become available.

II. DIRECT DISPLACEMENT AND IONIZATION PROCESSES

The radiation damage associated with electron-atom interactions is the rather unlikely process of direct displacement and, as the epithet suggests, involves the relocation of an atom in the structure. Relocation can occur only when the energy of the electron beam exceeds a threshold that is defined by the angle between the direct and off-axis trajectories (θ), the rest energy of the electron (m_oc^2), the mass of the displaced atom (M), and the energy necessary to displace it (E_d). The threshold is given by

$$E_{th} = m_oc^2\left[\left(\frac{M\,E_d}{2m_o^2c^2\cos^2\theta}+1\right)^{1/2}-1\right] \tag{1}$$

and even the minimum threshold ($\cos\theta = 1$) is substantial (Table I). It is substantial because only a very small proportion of the energy of a moving body can be transferred in a collision to something many thousand times heavier. In multicomponent solids, the thresholds vary among the constituent atoms because of their different masses and because their displacement energies reflect their different binding energies.

In most metals and ceramics, the displacement energies are high enough that the direct displacement of atoms in the interior of the specimen will not occur at 100 keV. In organic specimens this is not the case, but displacement damage represents only a small contribution to the radiation damage observed in these materials (Howitt and Thomas, 1977). In higher-voltage electron microscopes, the immunity to displacement damage is shared by fewer materials, and these processes will become important.

TABLE I

Minimum Displacement Energies for Several Solids at Room Temperature Compared to Maximum Energy Transfer (E_{max}), Calculated From Eq. (1).

Materials	E_D (eV)	E_{th} (keV)	References
Mg	10	101	Faust, O'Neal, and Chaplin (1969)
Al	16	169	Iseler *et al.* (1966)
Mo	34	822	Jung and Lucki (1975); Rizk *et al.* (1973)
Fe	17	327	Maury *et al.* (1976)
Fe (FeB)	22	402	Audouard *et al.*
amorphous	22	402	(1982)
Zn	14	318	Maury *et al.* (1974)
ZnO Zn	40	704	Yoshiie *et al.* (1979)
O	40	326	
Al_2O_3 Al	18	187	Pells and Phillips
O	75	394	(1979)
GaAs Ga	9	233	Hobbs (1979)
As	9.4	256	

The ionization process, which is the electron interaction with the atomic electrons, is notably different because the colliding particles are of the same size, and the restriction upon the proportion of energy that can be transferred from the incident electron no longer applies. Nevertheless, only half the energy of the incident electron can in fact be transferred because the electrons are synonymous during the collision. These observations are reflected in the expression for the threshold, which is identical to Eq. (1) except for the factor of one-half and the equality of M and m_o. It is therefore only the magnitude of the binding energy of the target electron, which is substantial for an inner-shell electron, that will prevent ionization.

The distribution of the electron interactions in a solid is along the primary track of the incident electron and the subsidiary tracks caused by the scattered electrons with energies above ~100 eV (δ-rays). The vast majority of the scattered electrons produced from even a 1-MeV primary electron are incapable of producing direct displacement; consequently, displacement processes are confined to single events along the primary track. Ionization events are prolifically produced by the scattered electrons; in the case of complete absorption of the primary electron, the interactions of the δ-rays account for almost half of the total ionization, even though only ~5% of the primary electron interactions generate them (Newton, 1963). In a thin foil, however, it is the primary electron interaction that produces most of the low-energy secondaries. Using Monte Carlo techniques (e.g., Kyser, 1979), it is straightforward to calculate, for a particular beam diameter and specimen thickness, the effective beam diameter throughout which ionization processes will occur (Figure 1).

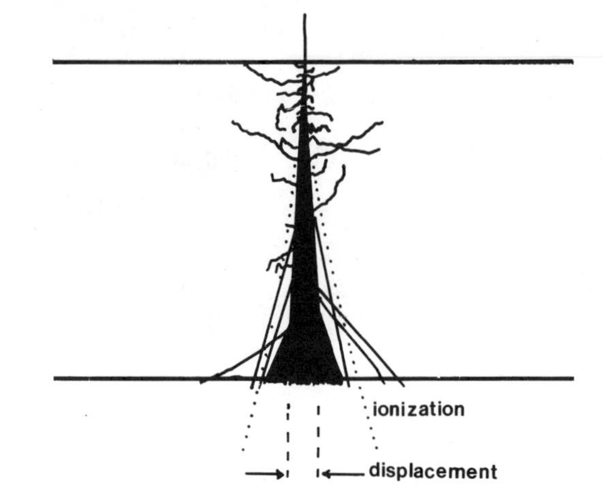

Figure 1 Monte Carlo simulation of the electron trajectories down to 10 eV in an aluminum foil irradiated with 300-keV electrons. The spatial extent of ionization and displacement events for a 50-Å probe are also shown for the 300-Å-thick specimen.

The probability that a primary displacement or ionization event will occur in a specimen can be evaluated from the cross sections for these processes. Oen (1973) has numerically evaluated the expression for the total displacement cross section

$$\sigma_T = \frac{\pi Z^2 e^4 (1 - \beta^2)}{m_o^2 c^4 \beta^4} \int_{E_d/E_{max}}^{1} \frac{dy}{y^2} M(y,E) \tag{2}$$

at 4-eV intervals of the displacement threshold (E_d),

where

E = kinetic energy of the electron
β = ratio of the electron velocity to that of light
Z = atomic number
e = electronic charge
M(y,e) = ratio of Mott to Rutherford cross section (Mott, 1929; Mott and Massey, 1965).

The calculation of the ionization-damage cross section along these lines requires some knowledge of the efficiency of the damage process, which is not usually expressed appropriately for coupling with the inelastic cross section. The destruction of a material by ionization damage (radiolytic decomposition) is most often reported in terms of specific and total G values, which are the numbers of radiation damage events that follow the absorption of 100 eV of energy. The ionization-damage cross section can be conveniently expressed in this way as

$$\sigma_D = \frac{G}{N \times 100} \frac{dE}{dx} \tag{3}$$

where N is the number density, and the energy loss rate

$$\frac{dE}{dx} = -\frac{2\pi Z e^4 N}{m_o V^2}\left[\ell n \frac{m_o V^2 E}{169 Z^2 (1-\beta^2)^2} + 1 - \beta^2 + \left(\beta^2 - 1 - 2\sqrt{1-\beta^2}\right) \ell n\, 2 + \tfrac{1}{8}\left(1 - \sqrt{1-\beta^2}\right)^2\right] \tag{4}$$

is the relativistic stopping power for an electron (Bethe, 1933).

The combination of Eq. (3) and (4) yields a particularly simple expression for the ionization-damage cross section for 100-keV electrons of

$$\sigma_D \doteq 2GZ \times 10^{-20} \text{ cm}^2 \ . \tag{5}$$

The reciprocal of this cross section, when G is the total radiolytic yield, is known as the D_{37} dose—the dose that will induce an average of one event per target. In reality, 37% of the targets (either atoms or molecules, depending upon the interaction processes involved) are intact at this dose because some 63% have suffered more than one interaction.

The mechanisms by which individual materials dissipate the energy they retain from electron interactions are diverse in both organic and inorganic materials (organic: Glaeser, 1979; Reimer, 1975; Pacansky, 1982; inorganic: Corbett and Watkins, 1971; Makin, 1971; Sonder and Sibley, 1972; Corbett and Bourgoin, 1975; Hobbs, 1975; Norris, 1975; Peterson and Harkness, 1976; Pooley, 1975; Hobbs, Howitt, and Mitchell, 1978). These mechanisms are evidenced in the electron microscope as the transformation of the structure to a new, steady-state configuration. Some spectacular examples in inorganic materials include quartz, which transforms to an amorphous state ($G \sim 0.002$) (Das and Mitchell, 1974; Howitt and DeNatale, 1983); alkali halides, which produce halogen dislocation loops and metal colloids ($G \sim 0.3$) (Hobbs, Hughes, and Pooley, 1973); and silicate glasses, which phase separate ($G \sim 0.1$). These transformations are often as easily distinguished in diffraction as they are in the image and can be both rapid and devastating to microstructural analysis.

The aggregation of vacancies and interstitials to form contributing defects to these steady-state structures is not a good measure of the extent of radiation damage because their aggregation suffers from the encumbrance that the bias is small for interstitial defects and nonexistent for voids. The vast majority of point-defect couplings occur, therefore, between vacancies and interstitials, with the result that the defect aggregates contain only a very small proportion of the point-defect population that has been created. Since the mobility of self-interstitials is always high, interstitial dislocation loops will always form and disrupt the image contrast except at very low temperatures. Vacancies, on the other hand, will form aggregates in a material only at about one-fifth to three-fifths of the melting temperature; these temperatures mark the extent of low mobility and high equilibrium concentration, respectively. Since surfaces and grain boundaries are unbiased sinks, conducive sinks such as dislocations or aggregated interstitials, as well as an appropriate temperature, are prerequisites for the formation of voids.

The materials that undergo radiation damage from electrons at 100 keV are invariably displaying susceptibility to ionization rather than displacement damage; it is therefore the probability of a chemical interaction, the appropriate G value, that determines the sensitivity to radiation damage in the voltage ranges appropriate to current analytical electron microscopy. In metallic systems there is no permanent mechanical response of the solid to such electron-electron interactions, and the radiation effect at 100 keV is usually confined to a rise in temperature.

The anticipated rise in specimen temperature from an electron probe can be estimated from the differential equations for heat conduction in an isotropic solid (Gale and Hale, 1961; Fisher, 1970; Thornburg and Wayman, 1973). Assuming steady-state conditions, the temperature increase caused by a heat source of radius r_o about the origin is, at distance r, given by

$$\Delta T = \frac{Q \,\ell n\,(r/r_o)}{2\pi\, k\, x} \tag{6}$$

where

Q = heat flux (a proportion of the Bethe stopping by the specimen)
k = thermal conductivity
x = specimen thickness.

The variable time solution for the temperature difference at distance r from a point source at the origin is also analytical (Carslaw and Jaeger, 1959) and is of the form

$$\Delta T = \frac{Q}{2\pi\, k\, x}\left[1 - \mathrm{erf}\frac{r}{(4\,D\,t)^{1/2}}\right] \tag{7}$$

where erf denotes the error function and t the time.

Both these expressions are dominated by the heat flux, and the calculated temperature rise as a function of probe current density is shown in Figure 2 for a typical ceramic using a small electron probe. The absolute values of the beam current

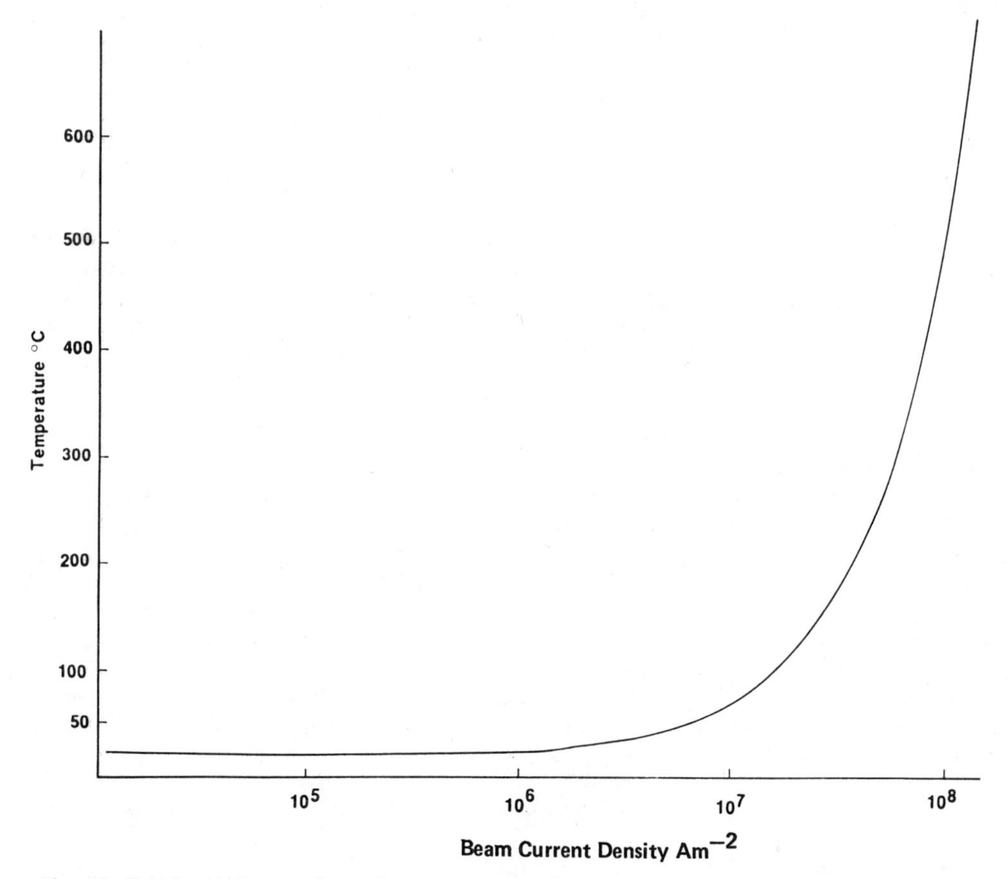

Figure 2 Relationship between the specimen temperature adjacent to an electron probe and the beam current density calculated from Eq. (6).

density predicted in this way will always be inaccurate because neither the thermal conductivity nor the specimen thickness are constant; nevertheless, the effects of beam heating will certainly become overwhelming over a quite narrow range of electron intensity that can be readily measured. This is not a fundamental problem—only a practical inconvenience—because the operation of the electron microscope at electron-beam currents below such intensities will eliminate it.

III. THE COLLECTION OF SPECTRA

An assessment of the significance of radiation damage to the application of analytical electron microscopy might begin by comparing some relevant cross sections for damage with those that give rise to discerptible spectra. For both electron energy-loss and energy-dispersive x-ray spectroscopy (EELS/EDXS), the signal detected (S) is related to the electron dose to the specimen (i t) by an expression of the form

$$S = K \epsilon N_o i t \tag{8}$$

where

K = appropriate ionization cross section
ϵ = an efficiency factor
N_o = number of atoms of the material being detected that can come into contact with the electron beam.

To serve as a general guide, it is sensible to calculate the electron doses that would be necessary to identify some elements present in reasonable concentrations and then to compare these with the electron doses that bring about substantial changes in composition to the regions of the specimen influenced by the electron probe.

In a calculation of the detection efficiency of popular energy-loss spectrometers, the values can vary from close to unity to well below 10^{-2} depending upon whether one needs to confirm the presence of a particular edge or explore the entire spectrum for unknowns. With the advent of parallel recording techniques (Jones *et al.*, 1982), both investigative methods can be highly efficient; this is not true with regard to x-ray detection in TEM/STEM. Here, efficiencies at best approach 0.02 ξ, where the angular efficiency factor (ξ) falls rapidly at energies much below 1.5 keV. In the case of K-shell ionization, therefore, if the effects of absorption and multiple scattering within the specimen are ignored, the signal after an x-ray detector will be given by

$$S_{K\alpha} = 0.02\, \xi\, \sigma_K\, N_o\, w\, i\, t\, \frac{I_{K\alpha}}{I_{K\alpha} + I_{K\beta}}. \tag{9}$$

The signal behind the energy-loss spectrometer can be represented in essentially the same way if a cross section representing the scattering of the incident electron within the acceptance angle is substituted for the total angular cross section, and the fluorescent yield (w), the fraction of emission devoted to Kα, and the factor of 0.02 ξ are omitted.

The cross sections used in quantifying electron-energy-loss spectra are referred to as partial cross sections and are the cross sections for shell scattering by a particular element through angles up to α (the acceptance angle of the spectrometer), including only the energy losses that extend to an energy (ΔE) above the edge. These cross sections can be readily calculated (Egerton, 1979; Leapman *et al.*, 1980). They are measured, for all the atoms in the beam path, as the ratio of the electron intensity in the region of the edge ($E_{th} \rightarrow E_{th} + \Delta E$) to the electron intensity over the same range (ΔE) starting at the zero-loss peak (e.g., Egerton, 1982a). Because the latter intensity is not exactly the same as the incident-electron intensity, this cross section is not strictly defined in quite the right way for our purposes. However, the errors involved in its incorporation will be, at most, a few percent.

The detection of the signal will also depend upon its statistical definition in the presence of the background noise originating from other elements in the specimen and from the collection device itself. Assuming Poisson statistics, the signal-to-noise ratio is given by

$$S/\text{noise} = \frac{S}{\sqrt{S + h\,I_b}} \tag{10}$$

where I_b is the intensity of the background noise, and h is a parameter that is probably ~2 for EDXS (Zaluzec, 1981) and possibly as high as 14 in EELS (Egerton, 1982b). The significance of the background will be to increase the number of counts necessary to identify the element; hence, any number of counts could serve as a reference upon which to base a comparison with radiation damage. The signal-to-noise ratio needed to identify an edge by eye, in the presence of only statistical noise ($\sqrt{S}$), will probably be ~5. This is the value appropriate to the visibility of circular discs of uniform intensity (Rose, 1948). Twenty-five counts would therefore represent an ideal case for visual recognition, and the value of 100 chosen for comparison is close to it.

In Tables II through IV, the electron-beam currents necessary to deliver 100 counts in EDXS and EELS are given for various edges at different electron voltages.

TABLE II

Electron-Beam Current Necessary to Provide 100 Counts (EELS) From a 180-Å-Radius Probe on a Foil 500 Å Thick, in Coulombs/cm^{-2} (α = 100 mrad; Δ = 100 eV).†

Element	Atomic Number (Z)	Electron-Energy-Loss Spectroscopy		
		50 keV	100 keV	200 keV
C	6	1.6	2.5	3.7
Mg	12	57.1	79.9	111.3
Al	13	121.0	164.8	226.5
Si	14	0.3*	0.5*	0.7*
Ti	22	2.6*	4.0*	5.9*
Fe	26	4.7*	7.0*	10.1*
Cu	29	8.16*	11.8*	16.7*

*Denotes an L rather than a K edge.
†Partial cross sections are from Egerton (1983).

TABLE III

Electron-Beam Current Necessary to Provide 100 Counts (EDXS) From a 180-Å-Radius Probe Incident on a Foil 500 Å Thick, in Coulombs/cm^{-2}.*

		Energy-Dispersive X-Ray Spectroscopy									
		50 keV		100 keV		200 keV		500 keV		1 MeV	
	Atomic Number	Edge		Edge		Edge		Edge		Edge	
Element	(Z)	K	L	K	L	K	L	K	L	K	L
Mg	12	2.2		3.8		6.6		14.0		25.4	
Al	13	2.3		3.9		6.7		14.2		25.5	
Si	14	2.3		3.7		6.4		13.4		23.9	
Ti	22	1.3	13.3	2.0	23.1	3.2	41.0	6.2	89.1	10.6	162.1
Fe	26	0.9	5.9	1.4	10.1	2.1	17.8	3.9	38.3	6.5	69.2
Cu	29	1.1	4.1	1.5	6.9	2.3	12.1	4.2	25.8	6.8	46.5
Ge	32	2.5	6.8	3.3	11.5	4.8	19.9	8.5	42.2	13.5	75.6
Zr	40	4.3	1.2	4.9	1.9	6.6	3.3	10.8	6.8	16.5	12.1
Mo	42	3.4	0.8	3.7	1.2	4.8	2.1	7.8	4.3	11.8	7.6
Ag	47	5.7	1.6	5.5	2.6	6.7	4.3	10.4	8.8	15.2	15.6
Sn	50	12.2	1.3	10.7	2.1	12.7	3.4	18.8	6.9	27.1	12.2
W	74		0.9	38.7	1.3	27.3	2.0	30.0	4.0	37.2	6.7
Au	79		0.9	67.7	1.3	38.4	2.1	39.0	4.0	46.6	6.7
U	92		1.3	269.9	1.7	104.5	2.6	81.4	4.8	88.3	8.0

*Ionization cross section and fluorescent yields are from calculation of Mayer *et al.* (1981).

TABLE IV

Electron-Beam Current Necessary to Provide 100 Counts (EELS and EDXS) From 1000 Atoms, in Thousands of Coulombs.†

	Atomic Number	50 keV		100 keV		200 keV		500 keV		1 MeV	
Element	(Z)	EDXS	EELS	EDXS	EELS	EDXS	EELS	EDXS	EELS	EDXS	EELS
C	6		8.85		13.9		20.8				
Mg	12	9.1	233.9	15.5	326.9	26.9	455.1	57.6		103.8	
Al	13	7.0	364.1	11.7	496.0	20.2	681.7	42.7		76.3	
Si	14	5.6	0.8*	9.4	1.3*	16.0	1.8*	33.5		59.8	
Ti	22	3.7	7.5*	5.7	11.6*	9.0	17.1*	17.5		29.9	
Fe	26	4.0	20.2*	5.8	30.1*	8.8	43.4*	16.5		27.5	
Cu	29	4.7	35.1*	6.5	50.7*	9.7	71.8*	17.6		28.7	
Ge	32	5.6		7.4		10.7		19.0		30.4	
Zr	40	2.6*		4.2*		7.3*		15.0*		36.6*	
Mo	42	2.7*		4.0*		6.9*		14.2*		25.1*	
Ag	47	4.8*		7.8*		12.9*		26.4*		44.5	
Sn	50	2.5*		4.0*		6.5*		13.1*		23.2*	
W	74	2.9*		4.2*		6.4*		12.8*		21.4*	
Au	79	2.7*		3.9*		6.3*		12.0*		20.1*	
U	92	3.1*		4.1*		6.2*		11.5*		19.2*	

*Denotes an L rather than a K edge.
†Ionization cross section and fluorescent yields are from calculation of Mayer *et al.* (1981); partial cross sections are from Egerton (1983).

The two cases considered are: a foil, 500-Å-thick, composed of the pure element in its normal crystal structure at room temperature, with the probe extending over an area of 10^{-3} μm^2 (Tables II and III); and 1000 atoms within the definition of the probe (Table IV). The values in these tables can be linearly extrapolated to assess the electron dose required to identify either a volume of different material that is to be analyzed or the concentration of a dispersed element.

IV. THE SIGNIFICANCE OF RADIATION DAMAGE AND QUANTITATIVE ESTIMATES FOR EDXS AND EELS

The electron dose that represents the upper limit to which a radiation-damaged specimen can be exposed will depend upon the techniques of electron microscopy that are used to evaluate it. In the case of microstructural analysis or microdiffraction at a resolution that is small compared to the extent of the beam imposed on the specimen, the doses to induce any type of detectable transformation will be appropriate. In the case of EDXS and EELS in the AEM, however, the probe size and resolution are comparable, so that radiation effects are inconsequential when they are confined to atomic rearrangement within the influence of the probe, e.g., radiation-induced transformations or any form of enhanced vacancy diffusion (Cosslett, 1979; Urban, 1982), unless material is lost from the surfaces (desorption).

In inorganic materials, atoms can leave the vicinity of the electron probe either during the scattering process or subsequently during the migration or desorption of atoms that are produced by displacement and ionization. In the case of an electron-atom interaction, the displaced atoms will be heavily peaked in the forward direction because of the large values of E_d. In aluminum at 500 keV, for example, the displaced atoms are confined to a cone with a semiangle of 15°; the displacement of atoms outside these cones will only occasionally result from electron-atom collisions involving electrons that have already suffered high-angle collisions. Displacements produced by ionization events, such as the focused collisions observed in the alkali halides, will not themselves be peaked in the forward direction but will originate from within the more confined trajectories of the scattered electrons which are forward peaked (Figure 1).

The signal from characteristic x-rays is derived from a volume larger than that of the electron probe, whereas the energy-loss signal is derived from a volume close to it. For a 500-Å-thick specimen, one would expect that the effective probe diameter giving rise to x-rays at 100 keV should be between 10 and 100 Å larger than the probe diameter itself, depending upon the average atomic weight of the specimen.

The signal derived from the elements within the probe can be estimated by using numerical methods to calculate the rates at which the atoms are depleted from these various volumes and then summing the contributions made by those remaining to the x-ray or energy-loss spectra. In the case of self-interstitial diffusion, however, the activation energy for migration is of the order of tenths of an electron volt. Here the atom transport at moderate probe currents is so rapid that the rate of atom depletion from these volumes is determined by the interstitial generation rate. When this is true, and the distinction between the various volumes is ignored, the problem can be solved algebraically in a straightforward manner.

The rate at which atoms are depleted from a probe is given by

$$\frac{dn}{dt} = -n \sigma_T i \tag{11}$$

where

n = number of atoms of the particular species within the probe
σ_T = its displacement cross section (cm^2)
i = beam current (electrons cm^{-2} s^{-1}).

Integrating this expression over time gives the instantaneous number of remaining atoms

$$n = n_o \exp(-\sigma_T i t) \tag{12}$$

in terms of the original number (n_o).

Some examples of this depletion are shown in Figure 3. When the interstitials are generated by ionization processes, the damage cross section (σ_D) is substituted for σ_T, and these cross sections will invariably be larger.

The count rate derived from the atoms in the probe for EELS or EDXS is given by a similar expression

$$\frac{dc}{dt} = n \sigma_c i \tag{13}$$

where σ_c is the cross section for signal generation. (The values of σ_c can be obtained from the electron doses in Table IV by dividing them into 1.6×10^{-20}.) Integrating this expression gives the total number of counts that can be collected, in the presence of radiation damage, as a function of time, i.e.,

$$c = n_o \frac{\sigma_c}{\sigma_T} [1 - \exp(-\sigma_T i t)] . \tag{14}$$

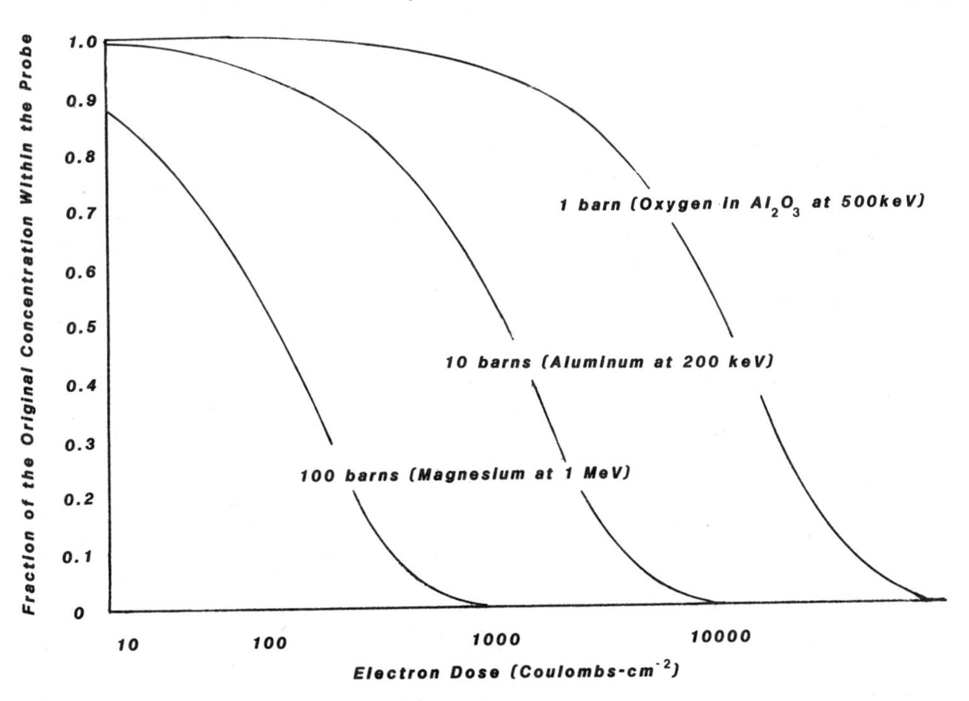

Figure 3 Concentration of unaffected atoms in a specimen, as a function of time, for various displacement cross sections at a beam current of 10 A/cm^{-2}.

Values of the collected counts as a function of electron dose are shown in Figure 4 for 100 aluminum ions with various displacement cross sections and in Figure 5 for various probe sizes in magnesium at 200 keV. In both cases, the curves reflect the reduction of the counts at high dose, which leads to the limits of detection in terms of number of atoms within the probe volume.

The expression can also be rearranged to provide a direct comparison with the electron doses in Table IV, i.e.,

$$\mathrm{i\,t} = -\ell\mathrm{n}\left(1 - \frac{\sigma_{\mathrm{T}}}{10\,\sigma_{\mathrm{c}}}\right)^{1/\sigma_{\mathrm{T}}} \tag{15}$$

or similarly for the other tables.

The example of atom depletion at this rate of defect production will be an exaggeration of the radiation effect in almost all materials. The process will, in reality, be slowed by the time it takes for atoms to diffuse to the boundary of the irradiated volume and their ability to diffuse on into the bulk or to desorb from the surfaces. The diffusion within the probe volume for all the species will be enhanced by the presence of any vacancies that are produced, and electron-stimulated desorption will be confined to the entry and exit surfaces of the electron beam. These restrictions to the depletion of material will not change the general characteristics of the behavior—only the magnitude of the variations to the count rate.

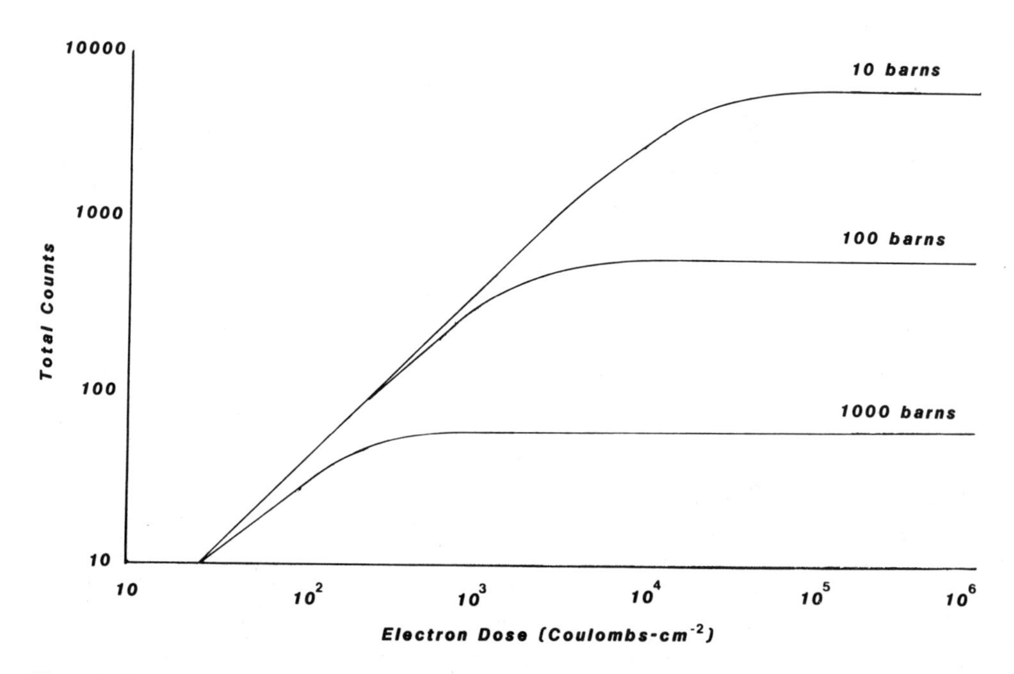

Figure 4 Values of the total counts collected in EDXS at 200 keV as they vary with electron dose for 100 aluminum ions with various displacement cross sections.

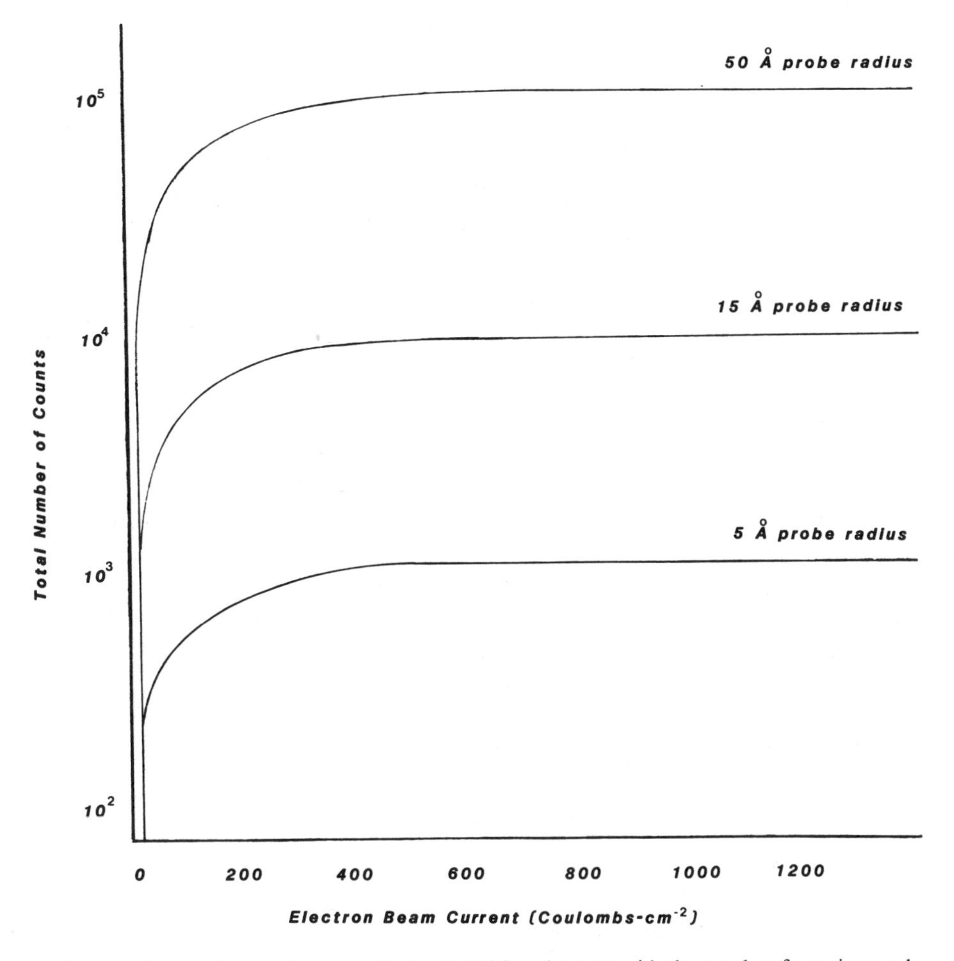

Figure 5 Values of the total counts collected in EELS as they vary with electron dose for various probe sizes in pure magnesium at 200 keV.

V. RADIATION DAMAGE FROM ION-THINNING PROCESSES

The 5- to 6-keV ion beam that impinges onto the surface of a specimen in an ion-milling machine will be composed mostly of argon ions that have maintained all but a few of their 18 electrons. The distribution of valence states and impurity ions will depend strongly upon the source design, but one might expect neutral and singly charged ions in roughly equal proportions to be the major components (>95%) (Radjabov and Kadirov, 1978). Details of the ionization and plasma formation mechanisms have been the subject of some study (Dearnaley *et al.*, 1973). Unfortunately, in small ion guns, the particle wall collisions often dominate over the particle-to-particle collisions, which will give rise to contaminating ions (Freeman and Sidenius, 1972). The lateral spread of the ion beam will be dominated by the collimation from the guns rather than by interaction with the specimens. The current density of the beam will nominally be between 10^{-2} and 10^{-3} $\mu A/cm^2$, extending over an area that can vary from 0.1 to 10^{-3} cm^2, depending upon the type of the ion-milling machine.

There is little doubt that the scattering events that give rise to sputtering are far removed from the initial ion-specimen collision. The experiments of Rol *et al.* (1960) in copper, for example, show that even at shallow incidence, the angular distribution of sputtered atoms is symmetrical. The evaluation of ion sputtering based upon a multiple-scattering model (Sigmund, 1969) predicts this behavior and also that the peak thinning rates should occur at high angles of incidence to the surface normal, indeed close to the angles found from experiment—60° for copper (Fluit, 1962).

The angle of incidence is important in the preparation of thin foils because it determines the thickness of the specimen surface layer that is affected by the ions and hence the volume fraction of the specimen that has credibility. The depth of ion penetration varies roughly as the cosine of the angle of incidence. The variation of the projected range (the average distance an ion travels into the foil), as well as the maximum penetration depth, is shown in Figure 6 for argon ions in a silicon foil.

The significance of the various depths is shown in Figure 7, which is a semi-quantitative plot of the concentration of deposited argon ions, displaced host atoms, and ionization events following the irradiation of a specimen with 6-keV ions. It should be remembered that 6 keV is the energy of a singly ionized atom, and ions with higher multiples of this energy will exist in the beam and penetrate to greater depths (Figure 8).

A 6-keV argon beam will deposit ~85% of its energy to displacement processes (Smith, 1977). At a current of 0.01 $\mu A/cm^2$, this translates into ~10^{22} displacements $cm^{-3}\ s^{-1}$ averaged over the ion range, which corresponds to the relocation of ~10% of the atoms each second. The sputter yield for most materials to 5-keV argon ions is between 0.5 and 3 (Franks, 1978) and represents only the removal of the occasional atom for the hundreds of displacement events that occur in the specimen.

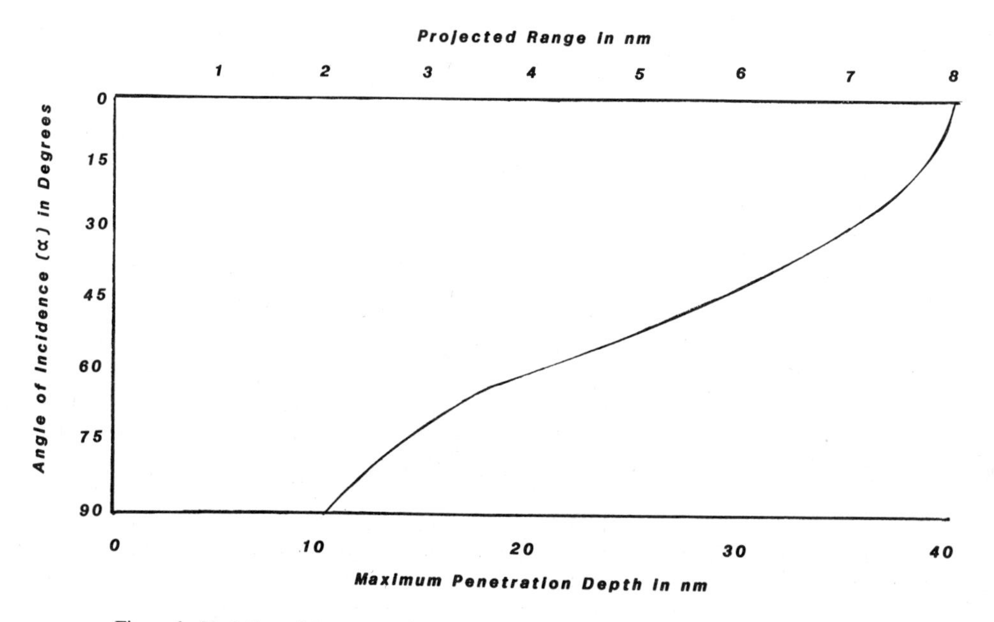

Figure 6 Variation of the penetration depth to the inclination of a 6-keV argon-ion beam.

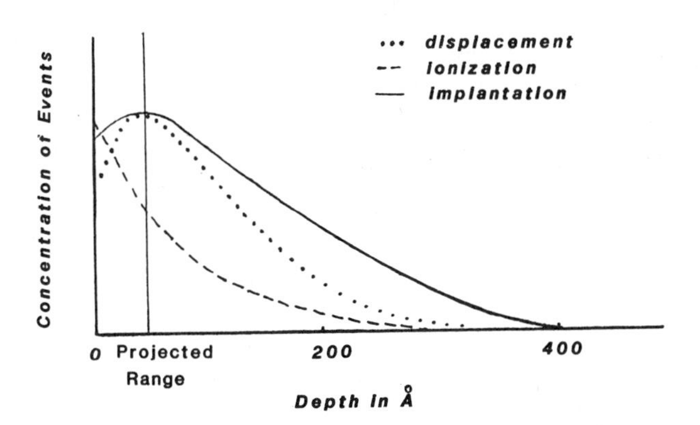

Figure 7 Calculated distribution of radiation-damage processes from the implantation of 6-keV argon ions (Ar^+) into a silica slab.

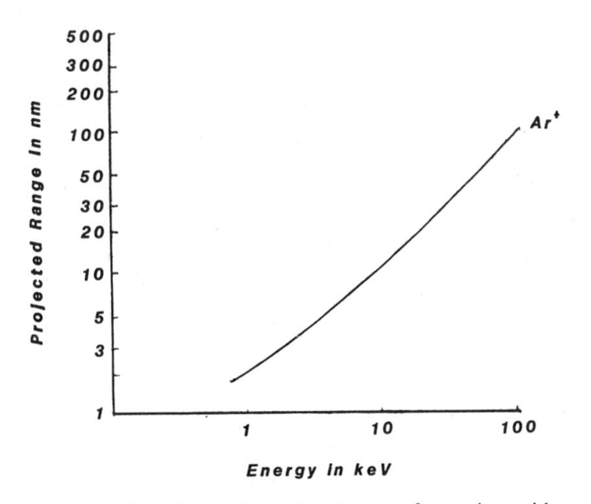

Figure 8 Variation in the projected range of argon ions with energy.

The surface layers of ion-milled specimens should, therefore, be considered to be a viscous mixture of specimen and argon ions, and even after outgassing, they will bear little resemblance to the original specimen. At low thinning angles, the maxima for all these processes can be contained within a few atom layers of the specimen surface. Significant effects could, however, be detectable at depths many times this value, depending upon the distribution of ion species and the ability of atoms in this surface layer to diffuse to the interior of the foil to form aggregated defects.

TABLE OF CHAPTER VARIABLES

c	Velocity of light (2.99×10^{10} cm s^{-1})
C	Number of signal counts collected at a detector
D	Thermal diffusivity
e	Charge on an electron (1.6×10^{-19} Coulombs)
E	Energy that can be transferred by an electron at a point inside the specimen ($E < E_{max}$)

E_d	Energy to displace an atom from its lattice site in electron volts (eV)
E_{max}	Maximum energy that can be transferred from an electron incident on a specimen
E_{th}	Electron energy at which atoms can be displaced in electron volts (eV)
G	Number of ionization damage events that accompany the absorption of 100 eV of energy
h	A magnification factor used to correct the proportion of background signal that is observed to contribute to the noise level in detection systems
i	Electron beam current (1 Ampere = 6.2×10^{18} electrons s^{-1})
I_b	Number of background counts
$I_{K\alpha}$	Intensity of the K_α x-ray
$I_{K\beta}$	Intensity of the K_β x-ray
k	Thermal conductivity
K	Ionization cross section for signal production from an electron in barns
m_o	Rest mass of the electron (9.1×10^{-28} g; 5.1×10^5 eV)
M	Mass of an atom (1 amu = 9.3×10^8 eV)
M(y,E)	Ratio of Mott to Rutherford cross section
n	Number of atoms in the specimen that are in the beam path during electron exposure
n_o	Number of atoms in the specimen that were originally in the beam path
N	Number of atoms in unit volume
N_o	Number of atoms defined in a volume
Q	Heat flux
r	Perpendicular distance from the center of the electron beam
r_o	Half the width of the electron beam
S	Amplitude of a detected signal as counts
t	Time (seconds)
T	Temperature (degrees Kelvin)
V	Electron velocity
w	Fluorescence yield
x	Thickness of a TEM specimen
y	Ratio of displacement energy to electron energy
Z	Atomic number
β	Ratio of the velocity of an electron to that of light
ϵ	Collection efficiency as a fraction
θ	Angle between the direct and off-axis trajectories in the laboratory frame for a colliding particle
ξ	Angular efficiency factor for a solid-state x-ray detector
σ_c	Cross section for signal production measured by the x-ray or electron detector (1 barn = 10^{-24} cm^2)
σ_D	Ionization cross section
σ_K	A K-shell ionization cross section
σ_T	Displacement cross section

ACKNOWLEDGMENTS

The author gratefully acknowledges support from the U. S. Department of Energy (Contract No. DE-AT03-79ER10437) and the provision of unpublished cross sections by Ray Egerton.

REFERENCES

Audouard, A.; Balogh, J.; Dural, J.; and Jousset, J. C. (1982), Radiat. Eff. *62*, 161.

Bethe, H. A. (1933), Handb. Phys. (Berlin) *24*, 273.

Carslaw, H. S., and Jaeger, J. C. (1959), "Conduction of Heat in Solids," Oxford.

Corbett, J. W., and Bourgoin, J. C. (1975), in "Point Defects in Solids" (J. H. Crawford, Jr., and L. M. Slifkin, ed.), Plenum Press, New York, vol. 2, pp. 1-161.

Corbett, J. W., and Watkins, G. D. (ed.) (1971), "Radiation Effects in Semiconductors," Gordon and Breach, New York.

Cosslett, V. E. (1979), Inst. Phys. Conf., London, Ser. 52, Electron Microscopy and Analysis *5*, 277.

Das, G., and Mitchell, T. E. (1974), Radiat. Eff. *23*, 49.

Dearnaley, G.; Freeman, J. H.; Nelson, R. S.; and Stephen, J. (1973), "Ion Implantation," North-Holland, Amsterdam.

Egerton, R. F. (1979), Ultramicroscopy *4*, 169.

——— (1982a), Phil. Trans. R. Soc. Lond. *A305*, 521.

——— (1982b), Proc. 40th EMSA Mtg., Washington, DC, p. 489.

——— (1983), private communication.

Faust, W. E.; O'Neal, T. N.; and Chaplin, R. L. (1969), Phys. Rev. B. *183*, 609.

Fisher, S. B. (1970), Radiat. Eff. *5*, 239.

Fluit, J. M. (1962), in "Ion Bombardment, Theory and Applications," Intl. Symp. of NSRC, Gordon and Breach, New York.

Franks, J. (1978), Adv. Electron. & Electron Phys. *47*, 1.

Freeman, J. H., and Sidenius, G. (1972), Nucl. Instrum. Meth. *10*, 477.

Gale, B., and Halc, K. F. (1961), Brit. J. Appl. Phys. *12*, 115.

Glaeser, R. M. (1979), "Introduction to Analytical Electron Microscopy" (J. J. Hren, J. I. Goldstein, and D. C. Joy, ed.), Plenum Press, New York.

Hobbs, L. W. (1975), "Surface and Defect Properties of Solids" (M. W. Roberts and J. M. Thomas, ed.), The Chemical Society, London, vol. 4, pp. 152-250.

——— (1979), "Introduction to Analytical Electron Microscopy" (J. J. Hren, J. I. Goldstein, and D. C. Joy, ed.), Plenum Press, New York.

Hobbs, L. W.; Howitt, D. G.; and Mitchell, T. E. (1978), in "Electron Diffraction 1927 – 1977" (P. J. Dobson, J. B. Pendry, and C. J. Humphreys, ed.), Inst. Phys. Conf., London, Ser. 41, pp. 402-10.

Hobbs, L. W.; Hughes, A. E.; and Pooley, D. (1973), Proc. Roy. Soc. *A332*, 167.

Howitt, D. G. (1984), J. Electron Micros. Technique *1*, 405.

Howitt, D. G., and DeNatale, J. F. (1983), J. Rad. Phys. & Chem. *21*, 5, 445.

Howitt, D. G., and Thomas, G. (1977), Radiat. Eff. *34*, 209.

Iseler, G. W.; Dawson, H. I.; Mehner, A. S.; and Kauffman, J. W. (1966), Phys. Rev. *146*, 468.

Jones, B. L.; Rossouw, C. J.; and Booker, G. R. (1982), Proc. 10th Intl. Cong. on E. M., Hamburg, vol. 1, p. 587.

Jung, P., and Lucki, G. (1975), Radiat. Eff. *26*, 99.

Kyser, D. F. (1979), "Introduction to Analytical Electron Microscopy" (J. J. Hren, J. I. Goldstein, and D. C. Joy, ed.), Plenum Press, New York.

Leapman, R. D.; Rez, P.; and Mayer, D. F. (1980), J. Chem. Phys. *72*, 1232.

Makin, M. J. (1971), in "Electron Microscopy in Material Science" (U. Valdre, ed.), Acad. Press, London, pp. 388-461.

Maury, F.; Biget, M.; Vajda, P.; Lucasson, A.; and Lucasson, P. (1976), Phys. Rev. B. *14*, 5305.

Maury, F.; Vajda, P.; Lucasson, A.; and Lucasson, P. (1974), Radiat. Eff. *21*, 65.

Mayer, D. M.; Joy, D. C.; Ellinghom, M. B.; Zaluzec, N. J.; and Mochel, P. E. (1981), Proc. AEM MAS Workshop, Vail, Colorado, July 1981.

Mott, N. F. (1929), Proc. Roy. Soc. *A124*, 426.

Mott, N. F., and Massey, H. S. W. (1965), "The Theory of Atomic Collisions," Clarendon Press, Oxford.

Newton, A. S. (1963), "Radiation Effects on Organic Materials" (R. O. Bolt and J. G. Carrol, ed.), Acad. Press, New York.

Norris, D. I. R. (1975), in "Electron Microscopy in Materials Science" (E. Ruedl and U. Valdre, ed.), Commiss. Euro. Communities, Luxembourg, vol. III, pp. 1099-1144.

Oen, O. S. (1973), "Cross Sections for Atomic Displacements in Solids by Fast Electrons," Oak Ridge Natl. Lab. Rpt. ORNL-4897.

Pacansky, J. (1982), private communication.

Peterson, N. L., and Harkness, S. M. (ed.) (1976), "Radiation Damage in Metals," ASM, Metals Park, Ohio.

Pooley, D. (1975), in "Radiation Damage Processes in Materials" (C. H. S. Dupuy, ed.), Noordhof, Leyden, pp. 309-23.

Radjabov, T. D., and Kadirov, A. G. (1978), "Low Energy Ion Beams," Inst. Phys. Conf., London.

Reimer, L. (1975), in "Physical Aspects of Electron Microscopy and Microbeam Analysis" (B. M. Siegel and D. R. Beaman, ed.), Wiley, New York, pp. 231-45.

Rizk, R.; Vajda, P.; Lucasson, A.; and Lucasson, P. (1973), Phys. Stat. Sol. A. *18*, 241.

Rol, P. K.; Fluit, J. M.; and Kistemaker, J. (1960), Physical *26*, 1000.

Rose, A. (1948), Adv. Electron. *1*, 131.

Sigmund, P. (1969), Phys. Rev. *2*, 184.

Smith, B. S. (1977), "Ion Implantation Range Data for Silicon and Germanium Device Technologies," Learned Information, Oxford.

Sonder, E., and Sibley, W. A. (1972), in "Point Defects in Solids" (J. H. Crawford, Jr., and L. M. Slifkin, ed.), Plenum Press, New York, vol. 1, pp. 201-90.

Thornburg, D. D., and Wayman, C. M. (1973), Phys. Stat. Sol. A. *15*, 449.

Urban, K. (1982), Proc. 10th Intl. Cong. on E. M., Hamburg, vol. 2, p. 439.

Yoshiie, T.; Hwanga, H.; Shibata, N.; Ichihara, M.; and Takeuchi, S. (1979), Phil. Mag. *A40 2*, 297.

Zaluzec, N. J. (1981), Proc. AEM MAS Workshop, Vail, Colorado, July 1981.

CHAPTER 12

HIGH-RESOLUTION MICROANALYSIS AND ENERGY-FILTERED IMAGING IN BIOLOGY

H. Shuman

Pennsylvania Muscle Institute
University of Pennsylvania, Philadelphia, Pennsylvania

I. INTRODUCTION

Previous chapters in this book have considered how the composition of such biological entities as cells or subcellular organelles can be determined at electron microscopic spatial resolution using either x-ray (EPMA) or electron energy-loss (EELS) microanalysis. Because the sensitivity and spatial resolution of x-ray microanalysis is restricted by the low collection efficiency of x-ray detectors, its utility for microanalysis or for the generation of elemental distribution maps is limited to a few special cases.

Electron energy-loss spectroscopy has already been shown to offer significantly improved microanalytical sensitivity because (a) the primary ionization event, rather than the subsequent x-ray fluorescence, is detected; (b) magnetic-sector electron spectrometers collect at least 50% of the inelastically scattered electrons, compared to the 1% collection of x-rays; (c) the lower energy (50 eV to 350 eV), higher cross-section $L_{2,3}$ ionization edges of biologically important elements are readily observed with EELS. These advantages can be exploited to yield a technique capable of producing high-resolution spectra, or images of the distribution of elements, from ultrathin biological specimens (Isaacson and Crewe, 1975; Isaacson and Johnson, 1975; Johnson, 1979; Colliex, 1981; Ottensmeyer and Andrew, 1980).

This chapter describes recent advances in high-spatial-resolution microanalysis of biological systems and discusses the use of energy-filtered imaging for studies of elemental distribution.

II. EXPERIMENTAL ARRANGEMENT

The effective use of energy-filtered imaging requires an electron spectrometer optimized for this purpose. The spectrometer used in the work described here is a magnetic-sector with circular pole-face edges corrected for second-order aberration. It is mounted below the projection chamber of a Philips EM400 equipped with a field-emission gun. The sector, detection system, and postsector optics are shown schematically in Figure 1 for two optical configurations. Figure 1(a) shows the sector used as a conventional spectrometer, and Figure 1(b) shows it used for energy-filtered TEM imaging. The sector calculations are based on Enge's focusing coefficients (Enge, 1967) and are carried out by a straightforward FORTRAN program (Shuman, 1980). Given bending angle ϕ, effective gap G between the pole faces, and object distance H_o to the entrance of the sector, the program computes optimum image distance H_i from the exit of the sector to the dispersion plane so that second-order aberrations affecting energy resolution are zero.

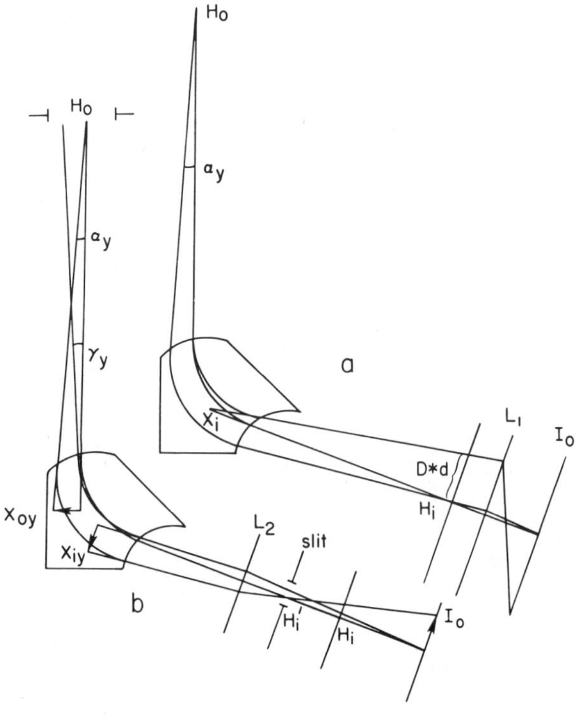

Figure 1 Experimental arrangement for (a) energy-loss spectroscopy and (b) energy-filtered imaging (electron trajectories and spectrometer dimensions are shown in scale).

The outputs from the program, along with H_i, are the tilts and radii of the entrance and exit pole-face edges of the sector. A sample output is shown in Table I for the sector used to obtain the data described in this chapter. The program also computes the equations describing the position of a ray originating at H_o, with angles relative to the optic axis of α_y and α_z as it crosses the H_i plane. The Y-axis is in the dispersion direction (the radial direction), and the Z-axis is in the direction of the magnetic field (the axial direction). The energy-dependent terms are given in units of relative momentum deviation $d = \Delta P/P$. The leading terms in Y/R and Z/R (Eq. (1) and (2)) are zero, indicating that the sector is indeed first-order stigmatic focusing. The center of the energy-dispersion object plane (H_o) coincides with the differential pump aperture of the EM400, which is also the back focal plane of the microscope imaging lenses. In the microscope imaging mode, the specimen diffraction pattern is centered at H_o, with camera length CL = 400 mm/M, where M is the microscope magnification. The sector object size is therefore defined by the objective aperture and the imaging magnification.

TABLE I

Sector Parameters

ϕ = 70° bending angle
R = 16.51 cm bending radius
PA = 11.744° entrance tilt
R1/R = 0.707 entrance radius
PB = 28.785° exit tilt
R2/R = −0.603 exit radius
H_i/R = 2.3840 dispersion image plane
H_o/R = 3.600 dispersion object plane
$GAP*I_2/R$ = 0.023900 effective pole face gap
D/R = 3.7601 dispersion
X_i/R = 0.505 achromatic image plane
X_{oy}/R = −0.611 radial achromatic object plane
X_{oz}/R = −0.962 axial achromatic object plane

Energy Dispersion Aberrations

(1) $Y/R = 0.000\alpha_y + 3.760d + 0.000\alpha_y d + 0.000\alpha_y^2 - 5.858d^2 + 0.000\alpha_z^2$
(2) $Z/R = 0.000\alpha_z - 9.133\alpha_y\alpha_z + 37.443\alpha_z d$

Imaging Aperature Aberrations

(3) $Y/R = 0.000\gamma_y^2 + 0.000d + 1.356\gamma_y d + 0.509\gamma_y^2 + 0.787d^2 - 0.783\gamma^{z2}$
(4) $Z/R = 0.000\gamma_z + 0.193\gamma_y\gamma_z - 0.948\gamma_z d$

Imaging Field Aberrations

(5) $Y = 0.827y + 0.000d + 0.866yd + 0.273y^2 + 0.787d^2 - 0.676z^2$
(6) $Z = 1.069z - 2.100yz - 1.458zd$

Manipulation of the sector object with the imaging system of the microscope has been discussed by Johnson (1980). The unscattered and elastically scattered electrons at H_o are focused to second order by the sector to a point at H_i, on the dispersion plane, as shown by the terms in α_y^2 and α_z^2 in Table I. The energy-loss electrons are focused to a straight line (perpendicular to the dispersion direction) also on the H_i plane. The spectrum can be scanned across a slit placed at H_i by changing the high tension, the sector current, or with a deflective yoke after the sector. Alternatively, a photographic plate or position-sensitive electronic detector can be placed at the dispersion plane, and a large portion of spectrum can be recorded in parallel.

Many electronic recording devices are now commercially available. One- or two-dimensional cameras based on Vidicon, photodiode array, and CCD technology are readily interfaced to the output of an electron spectrometer (Johnson *et al.*, 1981; Shuman, 1981). Direct electron (100-keV) bombardment of these silicon devices leads to increased dark current (Shuman, 1981) and changes in channel-to-channel signal gain. Dark current can be reduced by cooling to 77 K (Egerton, 1981), and gain uniformity can be partially recovered by annealing the solid state devices. The high-energy electron signal can also be converted to light with enough efficiency to detect the photons produced by a single electron with a cooled slow-scan detector or in a light-intensified detector. (For review, see Monson *et al.*, 1982).

Since the energy dispersion of this sector is small (4 μ/eV at 80 keV) compared to the spatial resolution of available position-sensitive detectors, the dispersion plane is magnified 45 times by an EM round lens (the intermediate lens from an RCA EMU-4), at L_1 in Figure 1(a), unto a P20 transmission phosphor (Grant Scientific) at I. Two optical lenses are used to transfer the spectrum on the phosphor to the photocathode of a silicon-intensified-target (SIT) television camera (PAR 1254). Television-rate observation of a line spectrum on an X-Y oscilloscope monitor makes the optical alignment relatively simple. A spectrum of $CaCO_3$ displayed on the monitor is shown in Figure 2(a) along with its digitized profile, Figure 2(b).

Postsector optics have been used to form energy-filtered images with Castaing (Ottensmeyer, 1980) or Omega (Krahl, 1981) filters. In this case, the normal imaging lenses of the electron microscope are used to magnify the energy-filtered image onto the microscope viewing screen. The single sector mounted below the projection chamber can also be used to form energy-filtered TEM images as shown in Figure 1(b). This arrangement has the practical advantage of not requiring modification of the microscope column, an important consideration for high-vacuum instruments, such as the EM400. The point in the center of the microscope image at a distance X_{oy} from the sector (in this case, the object is virtual) is focused by the sector to X_{iy}, coincident with the achromatic point of the sector. This is the point from which (for a small change in electron energy—or magnet current) the incident ray H_oX_{oy} appears to originate at the exit, independent of energy. The achromatic image is a distance X_i from the sector exit and is virtual. The calculated values for these imaging planes are given in Table I.

The virtual image is focused with a round EM lens at L_2 unto the phosphor at I. The lens at L_2 also refocuses the dispersion plane from H_i to H_i'. A slit placed at H_i' allows the passage of rays of only a desired energy and the observation of an energy-filtered TEM image with the same phosphor and TV system used for spectral recording. The achromatic focus is found experimentally by opening the slit, modulating the magnet current with a small ac component, and adjusting L_2 until the elastic image remains stationary at I. The sector is designed to be stigmatic for the pair of points H_o and H_i. For any other pair of points, the sector is astigmatic. Figure

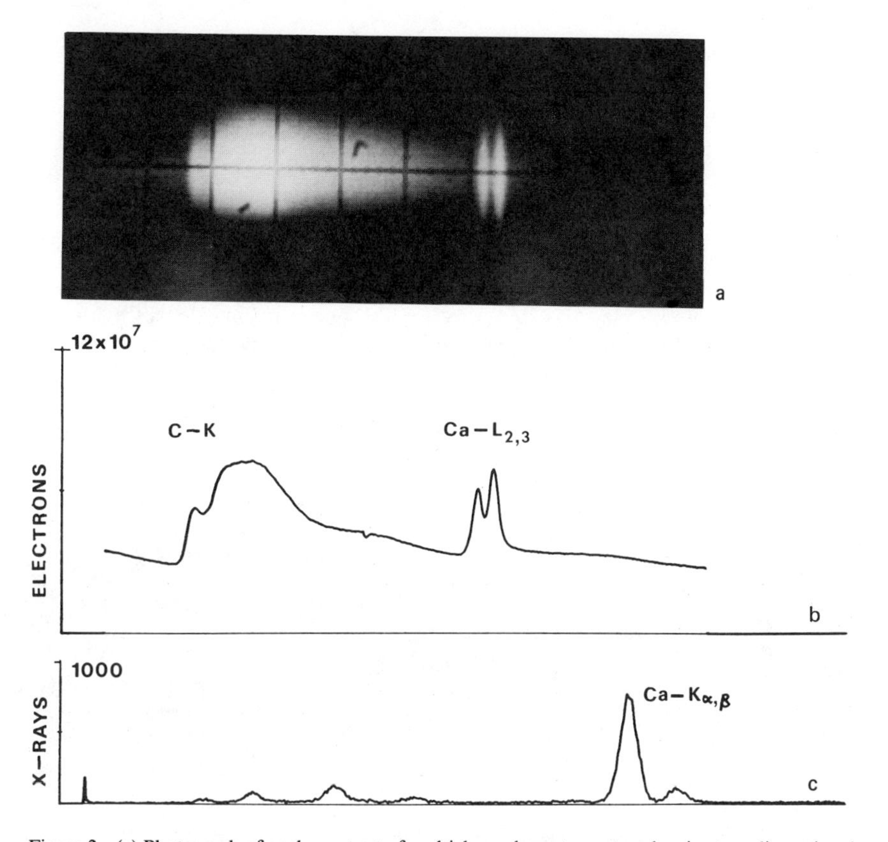

Figure 2 (a) Photograph of analog output of multichannel spectrometer, showing two-dimensional image of an energy-loss line spectrum from a $CaCO_3$ film; (b) digitized signal of integrated Vidicon line scans, showing Ca $L_{2,3}$ edge at 350 eV; (c) x-ray fluorescence spectrum obtained simultaneously (Ca K_α x-ray at 3690 eV).

1 shows the focusing properties of the sector along the Y-axis (radial focusing). The sector focuses another object point X_{oz} to X_i along the Z-axis (axial focusing). The position of the object plane between X_{ox} and X_{oz} is adjusted with the objective lens focus, while the substantial sector astigmatism can be corrected with the objective stigmators.

The imaging quality of the sector is determined by the aperture aberrations (dependent on the angles γ_x, γ_y, defined in Figure 1(b), the field aberrations (distortion and astigmatism), and the energy-dispersion aperture aberration (dependent on the angles α_y, α_z). For a uniform field sector with circular pole faces, the second-order aperture aberrations are zero for only one pair of image and object planes, in this case H_i and H_o. In general, the second-order terms in γ and the third-order terms in α are not zero. The program used here computes the field aberrations and both sets of second-order aperture aberrations. For point objects at X_{oy} and X_{oz}, the image spread at X_i in the two focusing directions is calculated from the image aberration equations (3) and (4) of Table I to be

$$\Delta Y = 165 \text{ mm } (0.509\gamma_y^2 - 0.783\gamma_z^2)$$

$$\Delta Z = 165 \text{ mm } (0.193\gamma_y\gamma_z) \ .$$

The differential pump aperture at H_0 limits γ to be

$$\gamma_y \approx \gamma_z \backsimeq 0.25 \text{ mrad}$$

so that the worst-case second-order point spread at X_i is

$$\Delta Y \backsimeq 13 \text{ nm}$$

$$\Delta Z \backsimeq 2 \text{ nm} .$$

Since the sector is placed after the microscope imaging lenses, this limiting resolution (when referred to the specimen) is decreased by the overall microscope magnification M. For a magnification of M = 100 times, the resolution at the specimen is 0.13 nm. The resolution is limited by the microscope, the detection system, and the scattering process observed. For example, at a microscope magnification of 280,000 times, the lattice fringes of graphite were readily observed in an *elastic* image displayed on the TV monitor and shown in Figure 3.

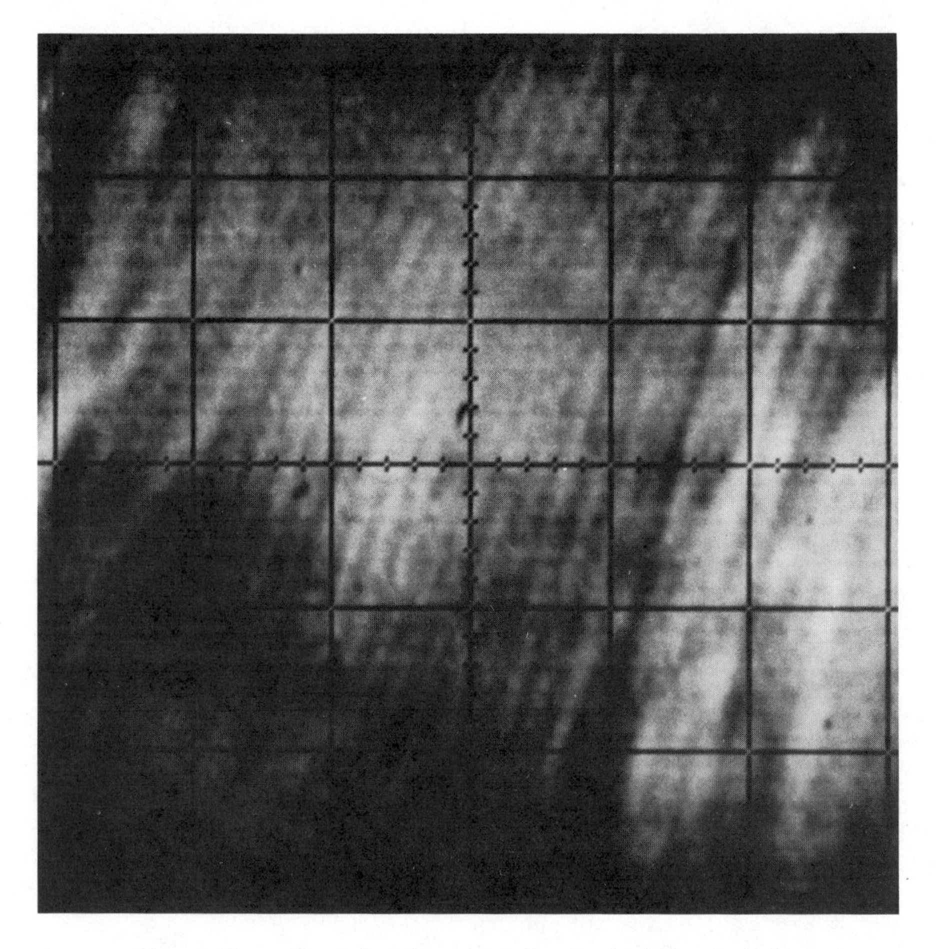

Figure 3 Energy-filtered elastic image of graphite, showing 0.34-nm lattice fringes.

The field distortion of the sector is, however, large and readily observed. The zero-loss image of a carbon replica grating, photographed from the TV monitor, is shown in Figure 4. With an EM magnification of 3700 times, the spacing of replica is 1.7 mm/line at the sector entrance. The graticule spacing on the viewing monitor corresponds to 3 mm at the exit phosphor at I_0. The square grating is distorted into a rectangular grating with axial and radial magnifications of 1.069 and 0.827 times, respectively, by the first-order terms in Table I Eq. (5) and (6). The second-order distortion term yz in Eq. (6) turns the rectangle into a trapezoid, whereas the z^2 term in Eq. (5) converts straight lines in the axial direction into parabolas. The predicted distortion is in good agreement with the image of Figure 4.

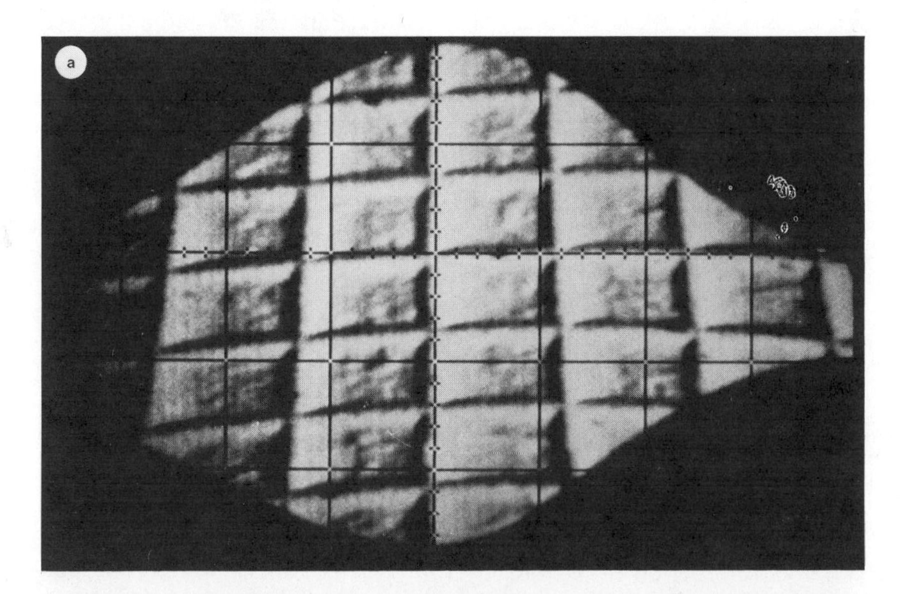

Figure 4 Elastic image of a carbon replica grating with energy-selecting slit widths of (a) 32 eV and (b) 2 eV.

The imaging field of view is limited by the third-order energy-dispersion aberrations and the width ΔE of the energy-selecting slits. If the sector had no aperture aberrations, the slit edges would not appear until $\Delta E = 0$. This is the electron optical equivalent of the Foucault knife-edge test in light optics and can be used to measure the magnitude of the sector aberrations. Other intermediate slit-width images were obtained, and the field of view S at I as a function of slit width along the axial symmetry direction was found to be $S \cong 4.4$ mm $(\Delta E)^{0.3}$ (ΔE in eV), indicating that the dominant energy resolution and field-limiting aberration is third order. The spherical aberration coefficient C_s of the sector is computed from these data to be $C_s = 1.34$ m. For the digital recording of biological EELS images, the central 6-mm square of I was magnified (2 times) onto the TV photocathode. For a slit width of $\Delta E = 9$ eV as used for Ca $L_{2,3}$ edge imaging, $S = 8.5$ mm, and the entire TV image can be used for recording.

III. ELEMENTAL ANALYSIS WITH EELS

Calcium plays a unique and vital role in the regulation of many biological processes. Measurement of the distribution of low concentrations and small absolute amounts of Ca is a major interest in analytical microscopy of biological specimens.

Theoretical predictions and experimental estimates of the sensitivity of EELS indicate that, for ultrathin specimens, it is at least one order of magnitude more sensitive for the detection of low Ca concentrations in organic matrices than EPMA (Shuman *et al.*, 1980). A direct comparison of the signals obtained from evaporated $CaCO_3$ with EELS and EPMA is shown in Figures 2(b) and (c), respectively. The spectra were taken simultaneously with a 30-mm^2 Si(Li) x-ray detector and with the EELS spectrograph described previously. The ratio of the number of electrons counted in the Ca $L_{2,3}$ white lines to the number of Ca $K_{\alpha,\beta}$ x-rays detected is 0.35×10^5. Unfortunately, the background produced by the organic matrix of a typical biological specimen is higher for EELS than for EPMA.

A pair of EPMA and EELS spectra obtained simultaneously from a concentration standard of 30 mM/kg Ca EGTA dissolved in polyvinylpyrrolidone (PVP) is shown in Figures 5(a) and (b). Although the Ca x-ray lines are visible over the background of the EPMA spectrum, the Ca $L_{2,3}$ edges are obscured by the C-K edge background in the EELS spectrum. The ratio of the backgrounds in the two spectra in the region of the Ca signal is $\sim 4 \times 10^6$. The problem of quantitating the Ca concentration with EELS becomes one of extracting the large signal from the even larger background.

Three factors will limit accurate background subtraction: (a) statistical noise in the background; (b) systematic variations in the background caused by specimen thickness and/or modulation of the carbon K edge EXELFS; and (c) experimentally induced variations in the background caused by nonuniform gain in the detection system. One simple and straightforward method of background subtraction is to collect another spectrum from a specimen identical to the unknown, but containing no calcium, and subtract it to obtain the difference spectrum. A spectrum from PVP with no added Ca was obtained, and the difference between it and the 30-mM spectrum is shown in Figure 5(c). The Ca $L_{2,3}$ edges are now clearly visible over the statistical noise of the background. An estimate of the sensitivity of the two spectroscopies for measuring Ca concentration can now be made. In both cases, signal S is obtained by subtracting a background from the raw data, so that

$$S = I_x - I_B$$

where

I_x = number of counts, either x-rays or electrons, in the unknown specimen spectrum in the region of the calcium peaks
I_B = number of counts in an equivalent Ca-free spectrum.

The variance of the signal σ_S^2 is then

$$\sigma_S^2 = \sigma_{I_x}^2 + \sigma_{I_B}^2 = I_x + I_B$$

or since (for low concentrations)

$$I_x = I_B$$

$$\sigma_S^2 = 2I_B \ .$$

The signal and standard error for the two spectra are therefore

$$S_{Ca} = 516 \pm 48 \text{ x-ray counts (EPMA)}$$

$$S_{Ca} = 15 \times 10^6 \pm 8.7 \times 10^4 \text{ electrons (EELS).}$$

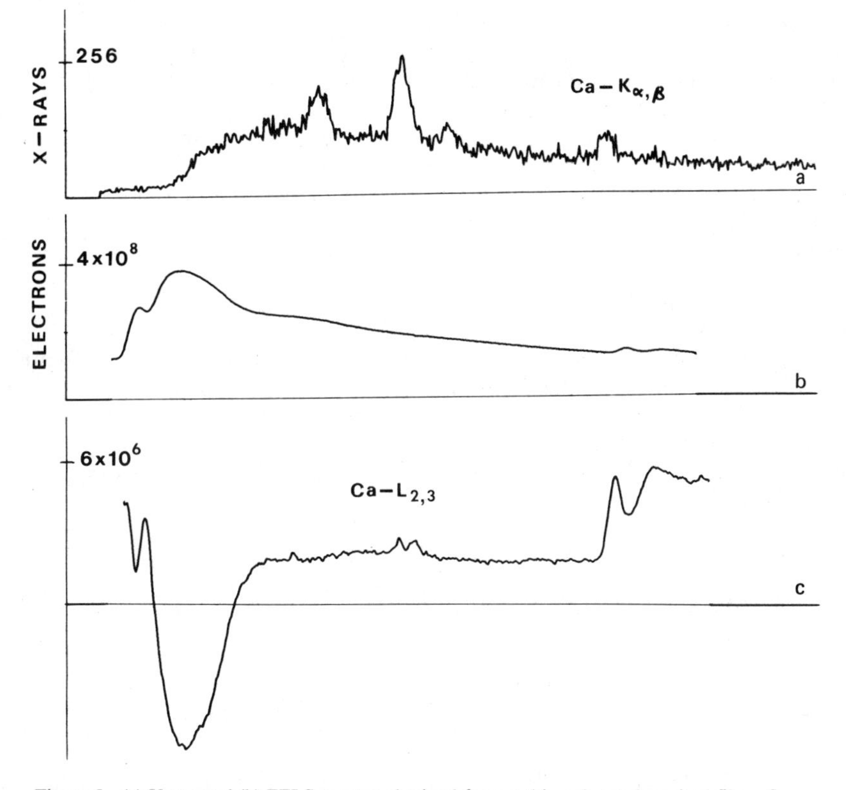

Figure 5 (a) X-ray and (b) EELS spectra obtained from a thin solvent deposited film of polyvinylpyrrolidone doped with 360 ppm calcium; (c) background subtracted from (b) to show Ca $L_{2,3}$ edge.

As observed previously, the EELS signal is 0.30×10^5 higher than the EPMA signal, while the relative error σ_s/S in the presence of an organic matrix is 16 times lower for EELS than for EPMA.

Background subtraction based on empirical data, although reasonable, is rather cumbersome since a new background is needed for every thickness. Therefore, several workers have taken advantage of the fact that the background follows closely the relationship AE^{-r} ($r = 3$ to 4), and the most reasonable value of r (dependent on angle) can be obtained by a multiple least-squares fit of the pre-edge spectrum (Egerton, 1978; Maher, 1979; Colliex and Trebbia, 1979). The thickness-dependent background variation is due to plural inelastic scattering. Electrons that have ionized the carbon K-shell electrons can also excite the organic matrix "plasmon" loss at ~23 eV. Since this effect also causes a loss of signal from the Ca $L_{2,3}$ edge, specimens thinner than the mean free path for plasmon excitation should preferably be used. For thin specimens, the plural scattering in the background will also be small, and the single scattering distribution (SSD) can be found by deconvoluting the raw data with experimentally determined zero loss and plasmon spectra (Leapman and Swyt, 1981). The SSD can then be used for further quantitation of elemental concentrations.

An alternate method of background subtraction is by differentiation, either before or after the spectra are collected. In Auger spectroscopy, the incident electron energy is modulated at some frequency f, and the secondary electron yield is phase-sensitively detected at f or 2f to give, respectively, either the first or second derivative of the original high-background spectrum. In EPMA, the background can be removed after the spectrum is collected by numerically differentiating with a "top-hat" filter.

Both methods can be applied to EELS spectra. In the *a posteriori* approach, a gain normalization curve is first obtained by uniformly illuminating the detector with electrons and dividing this gain curve into the energy-loss spectrum. A numerically computed second derivative of the normalized spectrum can then be obtained. For the *a priori* second derivative, three spectra were acquired with the spectrometer current changed by the equivalent of $\epsilon = 2$ eV. Figure 6(a) shows the differential spectrum $S_2(E)$ obtained numerically:

$$S_2(E) = 2S(E) - S(E + \epsilon) - S(E - \epsilon) .$$

The effect of the nonuniform gain of the parallel detector has been virtually eliminated without gain normalization in the *a priori* second derivative, and Ca $L_{2,3}$ edges for a 30-mM/kg standard are readily observed. The dramatic improvement in the signal-to-background is obtained at the expense of a loss in the signal-to-noise ratio.

Numerical analysis of the signal-to-noise ratio of $S_2(E)$ will require a simple model. The Taylor expansion of $S_2(E)$, assuming that S(E) is a continuous function, is

$$S_2(E) \equiv -(\epsilon)^2 \, d^2S/dE^2 .$$

We can model the Ca $L_{2,3}$ white line and background as a Gaussian peak on a E^{-r} background

$$S_o(E) = A\left(\frac{E_K}{E}\right)^r + C \exp\left(-(E - E_{Ca})^2/\Delta E^2\right)$$

and the peak-to-background ratio at $E = E_{Ca}$ in S_o is

$$(P/B)_o = C/A \times (E_{Ca}/E_K)^r$$

where

A = step height for the carbon K-edge at E_K
C = peak height above the background of a Ca $L_{2,3}$ white line, at E_{Ca}
ΔE = observed width of the white line.

For small ϵ,

$$S_2(E) \text{ at } E = E_{Ca} \text{ is}$$

$$S_2(E_{Ca}) = -\epsilon^2 \left[r(r+1)\, AE_K^r/E_{Ca}^{r+2} - 2C/\Delta E^2\right] .$$

The peak-to-background ratio of S_2 at $E = E_{Ca}$ is

$$P/B_2 = 2CE_{Ca}^{r+2}/r(r+1)\, AE_K^r \Delta E^2$$

$$= 2E_{Ca}^2\,(P/B)_o/r(r+1)\,\Delta E^2 .$$

For $\Delta E = 2$ eV, $E_{Ca} = 350$ eV, and $r = 3$:

$$(P/B)_2 \cong 5000\,(P/B)_o$$

which explains the increased visibility of the Ca $L_{2,3}$ edge.

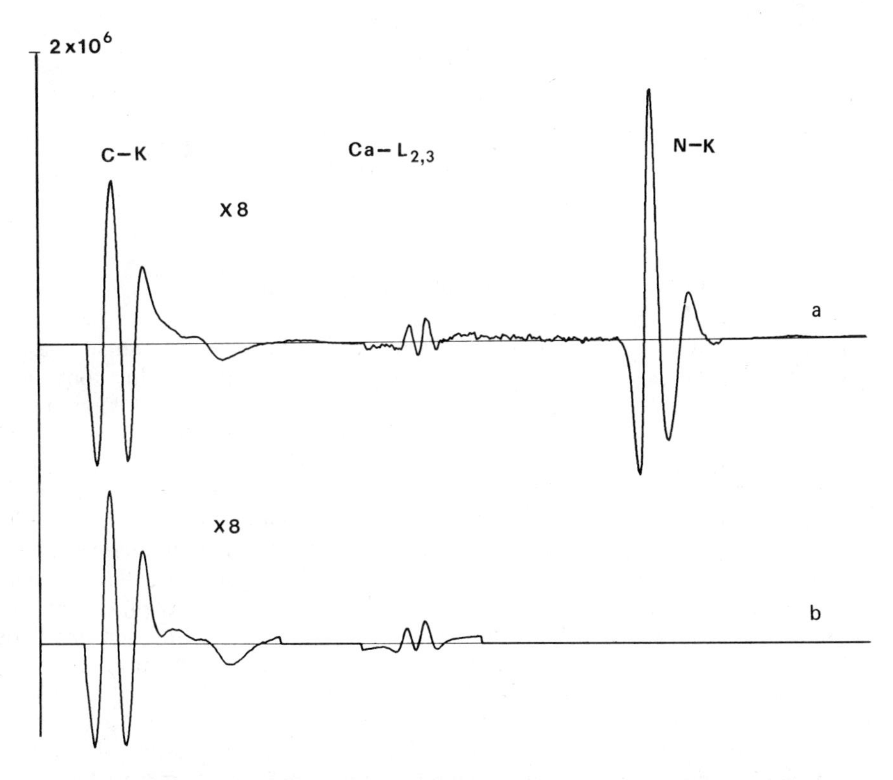

Figure 6 (a) Experimentally obtained second-difference EELS spectrum from 360 ppm Ca in polyvinylpyrrolidone; (b) result of a least-squares fit of a high-concentration standard of Ca and a polynomial to the experimental data.

Since the noise in S_o or S_2 is due to the statistical fluctuations of the background, the respective signal-to-noise ratios $(S/N)_o$ and $(S/N)_2$ are simple to calculate. The variance σ_2^2 of S_2 is

$$\sigma_2^2 = 4\sigma_{S(E)}^2 + \sigma_S^2(E + \epsilon) + \sigma_S^2(E - \epsilon)$$

or for small peaks

$$\sigma_2^2 = 6\sigma_{S(E)}^2 \cong 6S(E)$$

and as we have shown, the peak height for the Gaussian model over background in S_2 is

$$P_2 = 2C, \text{ for } \epsilon = \Delta E \ .$$

If we had simply added three unshifted spectra, the variance σ_o^2 of S_o would be

$$\sigma_o^2 = 3S(E)$$

and the peak height would be

$$P_o = 3C \ .$$

Then

$$(S/N)_o = P_o/\sigma_o = 3C/[3S(E_{Ca})]^{1/2}$$

and

$$(S/N)_2 = P_2/\sigma_2 = 2C/[6S(E_{Ca})]^{1/2} = 0.47(S/N)_o \ .$$

The signal-to-noise ratio of the derivative is a factor of two worse than the original data. The differentiation also reduces the effect of the modulations in the background caused by plural inelastic scattering. Since the modulations have a width ~23 eV, the ratio of the L-edge/plural-scattering modulations will be increased by $(23/2)^2$ ~130, again making the Ca $L_{2,3}$ white lines more visible.

The analysis of low concentrations of calcium in cryosections of normal muscle with EPMA is complicated by the presence of high concentrations of potassium. The K K_β and the Ca K_α x-ray lines are separated by 101 eV. Although numerical methods for separating these closely spaced peaks are available, small systematic errors in fitting the K peaks can introduce relatively large errors in the Ca quantitation. Three second-derivative spectra obtained from a thin cryosection of frog semitendinosus muscle are shown in Figure 7. The spectrum in Figure 7(a) was obtained from a 50-nm region of a terminal cisterna of the sarcoplasmic reticulum, the upper curve of Figure 7(b) from a neighboring region of I-band, and the lower curve of Figure 7(b) from the A-band. The positions of the C, K, Ca, and N absorption edges are indicated, and it is clear that there is no interference between the K and Ca edges. EPMA measurements of Ca concentration in TC and in the I-band are 120 mM/kg and 3 mM/kg, respectively.

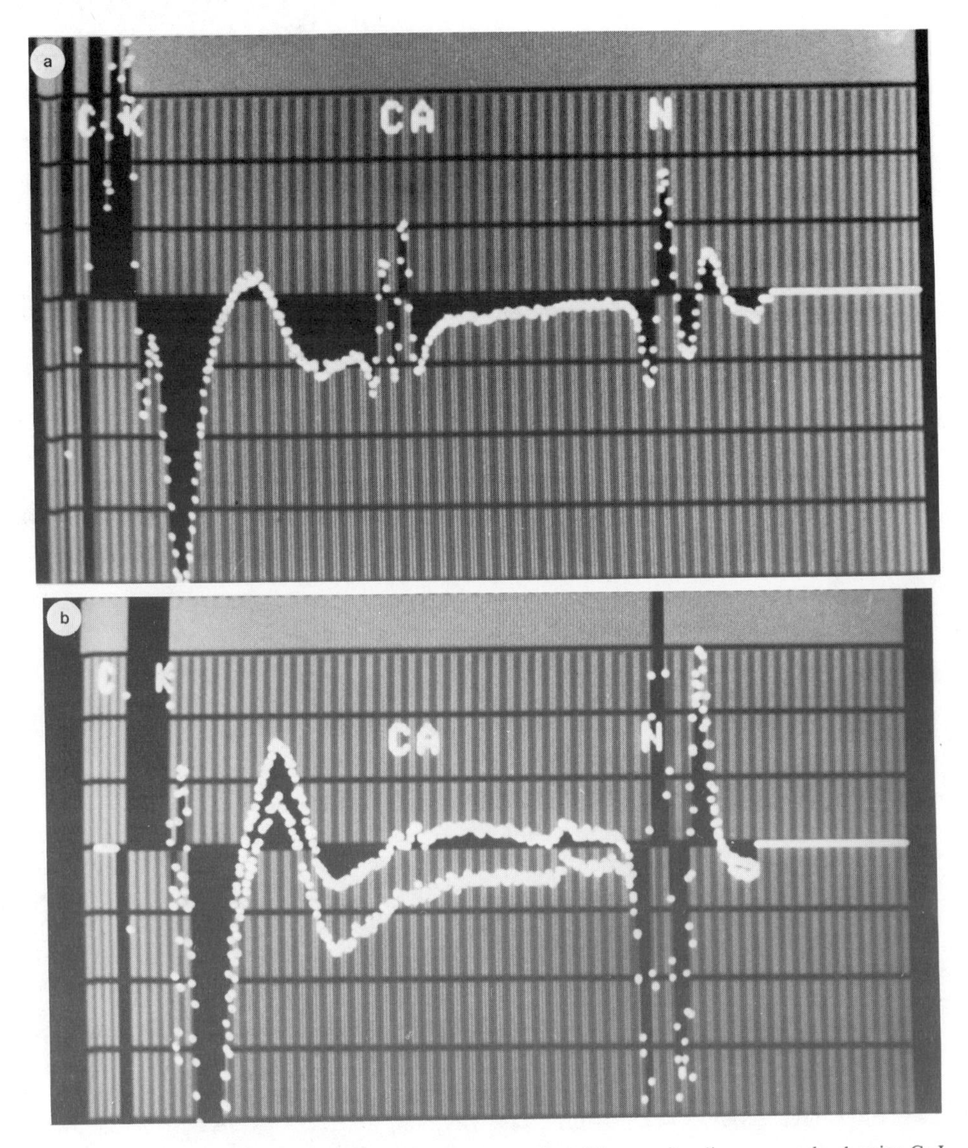

Figure 7 Second-difference EELS spectra from a cryosection of frog semitendinosus muscle, showing Ca $L_{2,3}$ edge for the range of concentrations present in: (a) terminal cisternae of the sarcoplasmic reticulum; (b) I-band (upper curve) and A-band (lower curve).

Several methods for the quantitation of elemental concentrations with EELS are discussed elsewhere in this volume. In the method of Joy (1982), the ratio of the partial integrated-scattering cross sections (integrated over a finite energy window and scattering angle) of two elemental absorption edges is shown to be proportional to the ratio of atomic fractions of the two elements. If the energy windows are identical for the two edges and the collection angles are sufficiently large (Egerton, 1981), the effect of plural plasmon scattering will be the same for both edges, and the ratio should be independent of thickness.

The highest signal-to-noise ratio for the Ca $L_{2,3}$ edges is in an energy window of ~7 eV containing the two white lines. The ratio of Ca edges and carbon K edges integrated over a 7-eV window should therefore give the highest sensitivity measure of calcium concentration in biological specimens. The cross sections in this energy region are, however, most sensitive to the chemical state of the atom. Biochemically bound calcium is usually coordinated with six oxygen atoms so that its near-edge fine structure is almost constant. The near-edge fine structure of the carbon K edge shows significant variation among compounds. How much this will affect accurate quantitation can at present only be determined experimentally. The constancy of the Ca $L_{2,3}$ white lines, although disappointing for determining its chemical binding, and possibly representing only a "final state" due to radiation damage, allows the use of a linear fit of the unknown S_2 spectrum (for example, Figure 6(a)) with a known high-concentration S_2 spectrum (from Figure 2(b)) of $CaCO_3$ plus a polynominal background:

$$S_2\ (\text{unknown}) = S_2\ (\text{CaCO}_3) + A + B\,E + CE^2\ .$$

The result of the fit is shown in Figure 6(b). For the C K edge and Ca $L_{2,3}$ edges separately, the scale has been reduced on the carbon edge by a factor of 8.

IV. ELEMENTAL IMAGING

The anticipated improvement of mass and concentration sensitivity of EELS over EPMA has been demonstrated for calcium in model protein systems. This enhanced sensitivity can be expected only for ultrathin specimens. The reduced sensitivity of EELS measurements from relatively thick sections is due to the effects of plural scattering on the background and the removal of characteristic loss electrons from the edge region. The inelastic mean free path λ_p for 80-keV electrons is ~60 nm in an amorphous carbon film. For specimen thickness equal to λ_p, 63% of the characteristic absorption edge signal will be convoluted by plural inelastic plasmon scattering, making interpretation and quantitation of the spectra difficult. Dehydrated cryosections originally cut 350-nm thick are equivalent to ~60-nm carbon films. Spectra and elemental images obtained from 150-nm thick cryosections (t ~26-nm carbon) significantly reduce the plural scattering artifacts.

The large improvement in mass sensitivity of EELS has been exploited by forming high-resolution elemental images (Ottensmeyer and Andrew, 1980; Costa *et al.*, 1978; Shuman and Somlyo, 1982). Energy-loss images can be obtained either in a scanning transmission microscope (Costa *et al.*, 1978; Hainfeld and Isaacson, 1978) or by using the optical properties of a magnetic sector (Shuman and Somlyo, 1982) or combination of sectors (Castaing *et al.*, 1967; Krahl *et al.*, 1981) to directly form an energy-filtered TEM image. In either case, it is necessary to obtain at least two images (Johnson, 1979), one formed with background inelastic electrons (pre-edge image) and a second with a combination of background and characteristic energy-loss electrons (edge image). The difference between the two images gives the spatial distribution of the appropriate element. STEM imaging has a distinct advantage over TEM since several characteristic and background images, as well as bright- and dark-field images, can be collected simultaneously, thereby minimizing specimen damage and the effects of stage drift. High-resolution multiple-image STEM requires a field-emission gun and high-speed parallel recording and spectral processing EELS electronics. Energy-filtered

TEM imaging has the temporary advantage of being far simpler to do. The TV system described previously has the nominally conventional format of 256 lines and 512 picture points (pixels) per line.

The edge and pre-edge images for iron in ferritin, phosphorus in ribosomes and calcium in muscle are shown in Figures 8, 9, and 10. The ferritin molecule images (Figure 8) are used to test the spatial resolution and sensitivity of EELS microanalysis. The molecule has up to 4000 iron atoms in the 7.5-nm-diameter core, and the protein shell is 12 nm in diameter. Figure 8(a) is the image taken with electrons that had lost 55 to 60 eV of energy, corresponding to the maximum of the iron $M_{2,3}$ absorption edge, while Figure 8(b) was taken with 45- to 50-eV-loss electrons. The two images were scaled so that the carbon support film appears with the same intensity in both. The difference image was computed numerically and is shown in Figure 8(c). The iron core is clearly resolved, and some substructure may be observed. The carbon K-edge image taken with 290- to 295-eV-loss electrons is shown in Figure 8(d). An estimate of the minimum detectable mass (MDM) determined by the counting statistics indicates

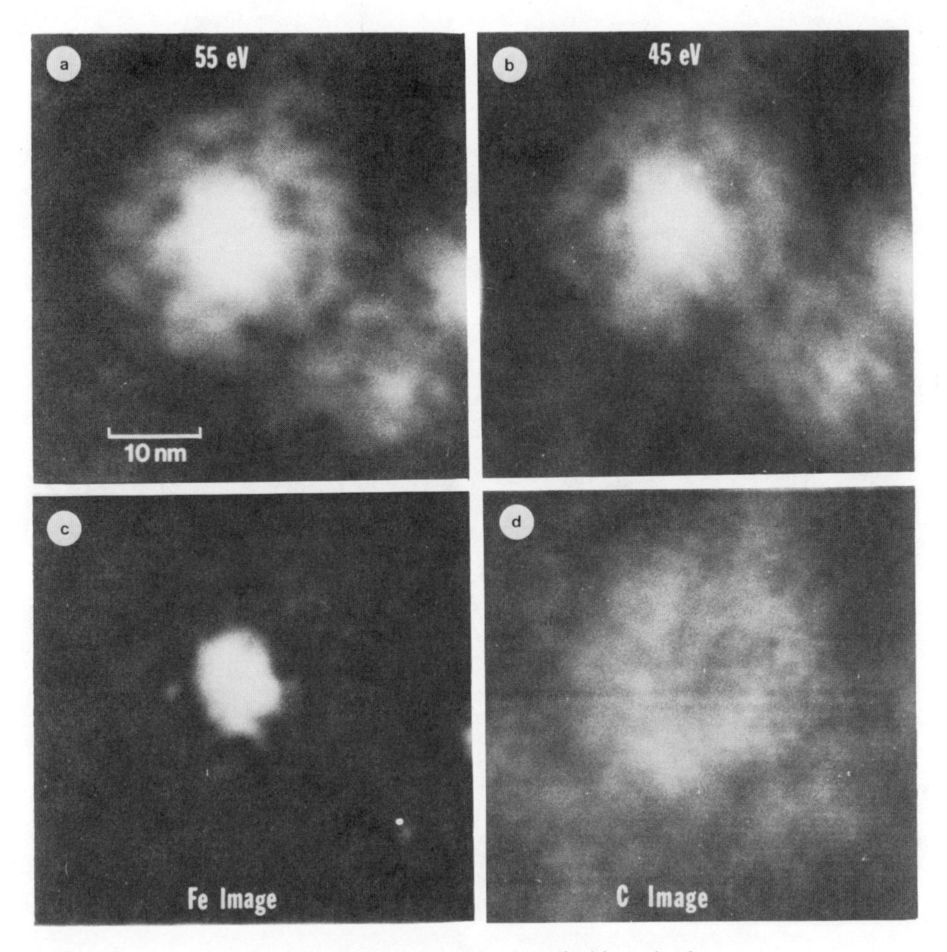

Figure 8 Filtered images of a single ferritin molecule.

an MDM of 50 iron atoms, with a total dose of 40 C/cm^2 (Shuman and Somlyo, 1982). In Figure 8(d), the contrast is due partly to the distribution of carbon in the protein shell and partly to the loss of electrons by large-angle elastic scattering from the iron core.

The ribosome is the cellular structure responsible for protein synthesis. The 80s eukaryotic ribosome has a molecular weight of 4.5 $\times$ 10^6, of which ~55% is RNA. The remaining 45% consists of ~70 different proteins: 40 on the large (60s) subunit and 30 on the small (40s) subunit. About 10% of the mass of RNA is phosphorus, and EELS images with the P $L_{2,3}$ absorption edge may help determine the RNA distribution within ribosomes. The phosphorus edge image in Figure 9(a) was taken with 150- to 164-eV-loss electrons, and the pre-edge image of Figure 9(b) with 122- to 136-eV-loss electrons. The contrast in Figure 9(b) is primarily mass contrast from nonspecific inelastic scattering and demonstrates the need for taking two images. The two pictures are subtracted to give the image due only to the phosphorus in the

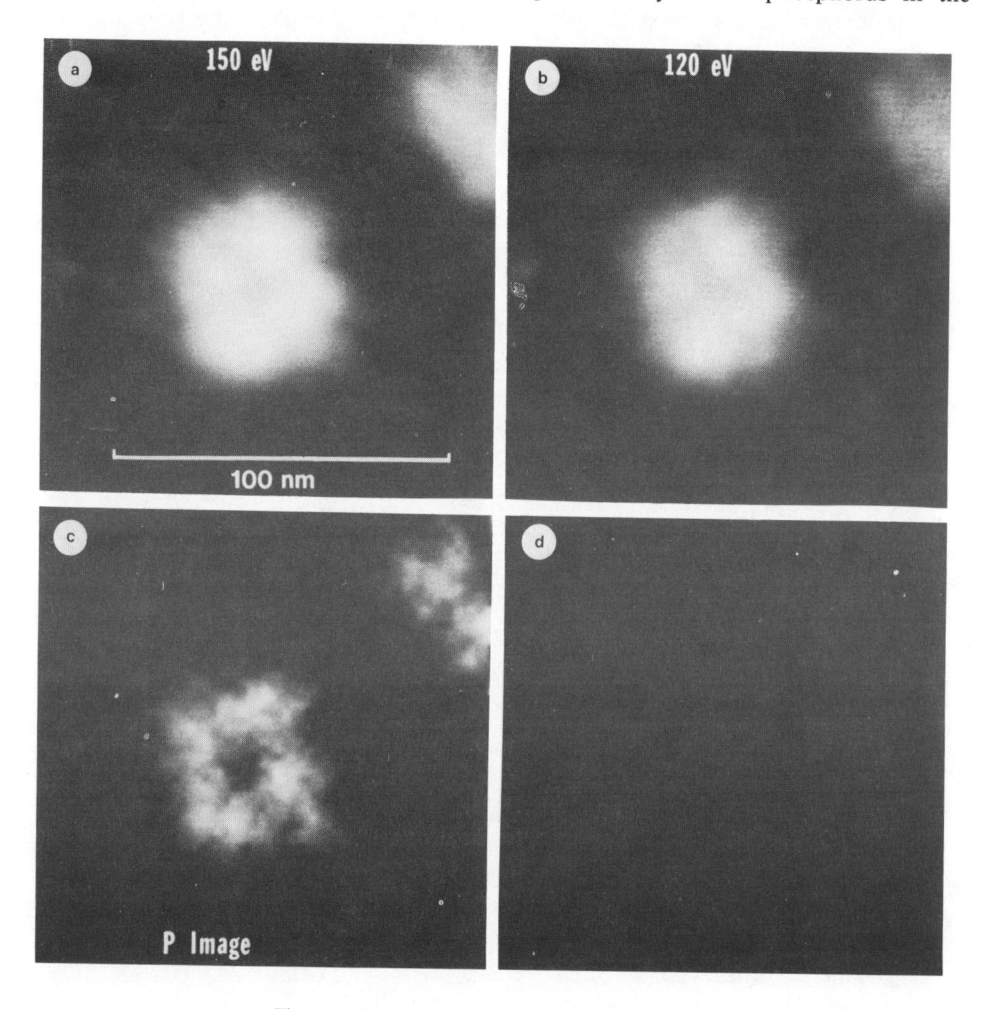

Figure 9 Filtered images of a ribosome tetramer.

ribosome shown in Figure 9(c). There is a noticeable amount of stage drift in Figure 9(a), accounting for most of the noise in Figure 9(c). The total dose required for both pictures is 100 C/cm^2, so that the ultimate spatial resolution will be limited by radiation damage.

The elemental distributions of Ca and P in a 150-nm-thick cryosection of frog semitendinosus muscle are shown in Figure 10. The Ca edge and pre-edge image are shown in Figures 10(a) and (b), respectively. A triadic junction, with a swollen T-tubule in the center of the field and two terminal cisterna on either side, is visible. The Ca difference image of Figure 10(c) shows the calcium localization in the cisterna. Quantitative electron-probe analysis of similar areas indicates a calcium concentration of ~120-mM/kg dry weight (Somlyo *et al.*, 1981), so that each of the bright regions in Figure 10(c) corresponds to ~30,000 Ca atoms. A phosphorus difference image obtained from the same area is shown in Figure 10(d) and illustrates that most of the phosphorus is associated not with the Ca in the lumen of the terminal cisternae, but with the surrounding membrane of the sarcoplasmic reticulum.

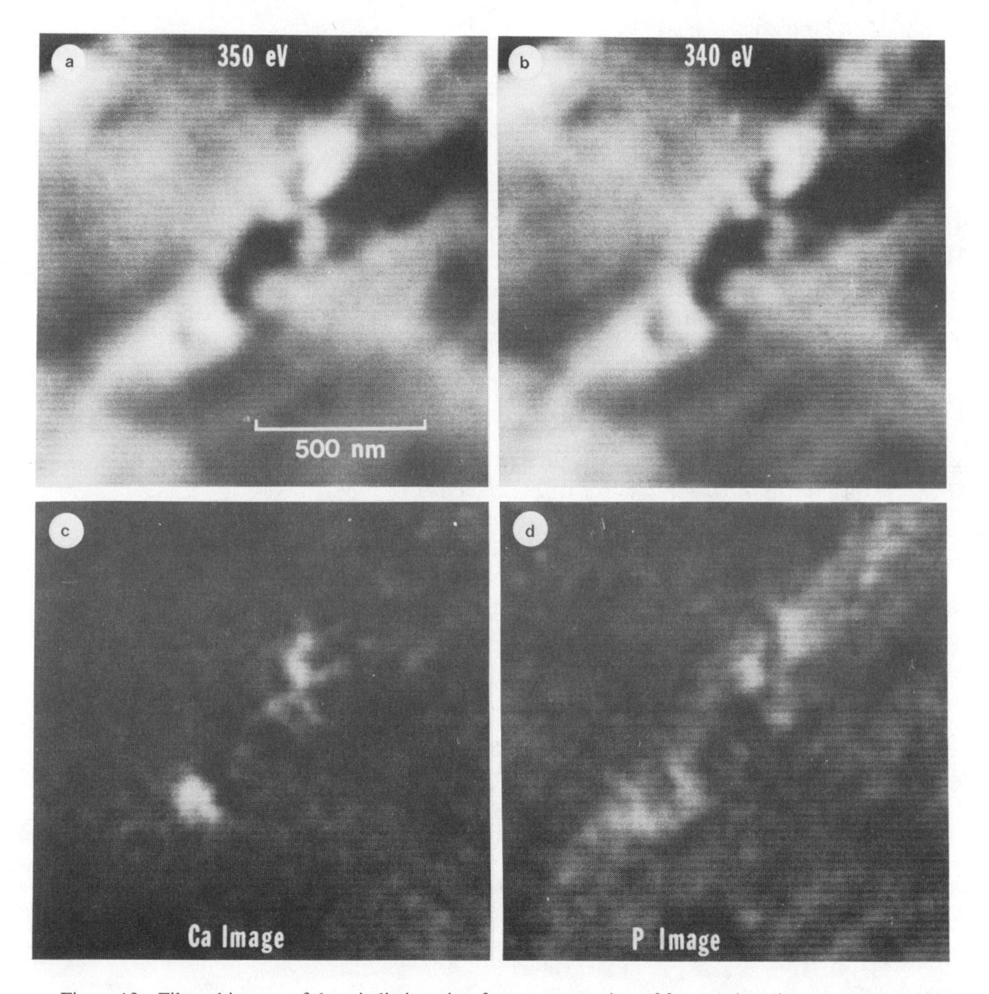

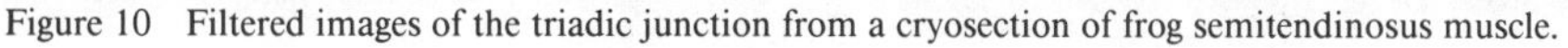

Figure 10 Filtered images of the triadic junction from a cryosection of frog semitendinosus muscle.

The major technical problems in realizing the potential of EELS compositional imaging are: (a) preparing thin specimens so that background subtraction is not complicated by plural inelastic scattering, and (b) collecting the images with high spatial resolution and large dynamic range. Both of these problems appear to be surmountable, and the advantages expected from the use of this electron optical technique are sufficiently important to warrant further improvements in data collection and specimen preparation.

TABLE OF CHAPTER VARIABLES

CL	Camera length of microscope
C_s	Spherical aberration of sector
d	Relative momentum change of loss electron
E	Energy loss of electron
E_o	Incident electron energy
G	Sector polepiece gap
H_i	Distance of dispersion plane to sector exit
H_o	Distance of dispersion object to sector entrance
I_b	Background signal
I_x	Total signal
M	Magnification of microscope
P	Momentum of incident electron
R	Radius of electron trajectory in sector
S	Characteristic signal, x-ray photons or electrons
S_2	Second difference EEL spectrum
X_i	Distance of achromatic image to sector exit
X_o	Distance of achromatic object to sector entrance
α	Acceptance angle of sector
γ	Angular extent of dispersion object
ΔE	Range of electron energies in filtered image
ΔP	Absolute momentum change of loss electron
λ_p	Mean free path for total inelastic scattering
σ^2	Variance of signal
ϕ	Bending angle of sector

ACKNOWLEDGMENT

This work was supported by NIH Grant HL15835 to the Pennsylvania Muscle Institute, University of Pennsylvania, Philadelphia, Pennsylvania.

REFERENCES

Castaing, R.; Hennequin, J. F.; Henry, L.; and Slodzian, G. (1967), in "Focusing of Charged Particles" (A. Septier, ed.), Acad. Press, New York, pp. 265-93.

Colliex, C.; Jeanguillaume, C.; and Trebbia, P. (1981), "Microprobe Analysis of Biological Systems" (T. Hutchinson and A. P. Somlyo, ed.), Acad. Press, New York.

Colliex, C., and Trebbia, P. (1979), in "Microbeam Analysis in Biology" (C. Lechene and R. Warner, ed.), Acad. Press, New York, pp. 65-87.

Costa, J. L.; Joy, D. C.; Maher, D. M.; Kirk, K. L.; and Hui, S. W. (1978), Science *200*, 537-39.

Enge, H. A. (1967), in "Focusing of Charged Particles" (A. Septier, ed.), Acad. Press, New York, vol. II, ch. 4.

Egerton, R. F. (1978), Ultramicroscopy *3*, 243-51.

——— (1981), Ultramicroscopy *7*, 207-10.

Egerton, R. F., and Cheng, S. C. (1982), J. Micros. *127*, RP3.

Hainfeld, J., and Isaacson, M. S. (1978), Ultramicroscopy *3*, 87-95.

Hall, T. A., and Gupta, B. J. (1979), in "Introduction to Analytical Electron Microscopy" (J. J. Hren, J. I. Goldstein, and D. C. Joy, ed.), Plenum Press, New York, pp. 169-97.

Hutchinson, T. E. (1979), Int. Rev. Cytol. *58*, 115-58.

Hutchinson, T. E., and Somlyo, A. P. (ed.) (1981), "Microprobe Analysis of Biological Systems," Acad. Press, New York.

Isaacson, M. S., and Crewe, A. V. (1975), Annl. Rev. Biophys. and Bioengr. *4*, 16.

Isaacson, M. S., and Johnson, D. E. (1975), Ultramicroscopy *1*, 33-52.

Joy, D. C. (1982), SEM II, pp. 505-15.

Johnson, D. E. (1979), in "Introduction to Analytical Electron Microscopy" (J. J. Hren, J. I. Goldstein, and D. C. Joy, ed.), Plenum Press, New York, pp. 245-57.

——— (1980), Ultramicroscopy *5*, 163-74.

Johnson, D. E.; Csillag, S.; Monson, K. L.; and Stern, E. A. (1981), Proc. 39th Annl. Mtg. EMSA (G. W. Bailey, ed.), Claitor's Pub. Div., Baton Rouge, pp. 368-69.

Krahl, D.; Herrmann, K.-H.; and Zeitler, E. (1981), Proc. 39th Annl. Mtg. EMSA (G. W. Bailey, ed.), Claitor's Pub. Div., Baton Rouge, pp. 366-67.

Leapman, R. D., and Swyt, C. R. (1981), Proc. 39th Annl. Mtg. EMSA (G. W. Bailey, ed.), Claitor's Pub. Div., Baton Rouge, p. 636.

Lechene, C. P. (1977), Am. J. Physiol. *232*, F391-96.

Maher, D. M. (1979), in "Introduction to Analytical Electron Microscopy" (J. J. Hren, J. I. Goldstein, and D. C. Joy, ed.), Plenum Press, New York, pp. 259-94.

Monson, K. L.; Johnson, D. E.; and Csillag, S. (1982), SEM IV, pp. 1411-19.

Ottensmeyer, F. P., and Andrew, J. W. (1980), J. Ultrastructure Res. *72*, 336-48.

Shuman, H. (1980), Ultramicroscopy *5*, 45-53.

——— (1981), Ultramicroscopy *6*, 163-68.

Shuman, H., and Somlyo, A. P. (1982), Proc. Natl. Acad. Sci. *79*, 106-107.

Shuman, H.; Somlyo, A. V.; and Somlyo, A. P. (1976), Ultramicroscopy *1*, 317-39.

——— (1980), in "Microbeam Analysis" (D. B. Wittry, ed.), San Francisco Press, San Francisco, pp. 275-79.

Somlyo, A. P., and Shuman, H. (1982), Ultramicroscopy *8*, 219-34.

Somlyo, A. V.; Gonzalez-Serratos, H.; Shuman, H.; McClellan, G.; and Somlyo, A. P. (1981), J. Cell Biol. *90*, 577-94.

CHAPTER 13

A CRITIQUE OF THE CONTINUUM NORMALIZATION METHOD USED FOR BIOLOGICAL X-RAY MICROANALYSIS

C. E. Fiori, C. R. Swyt, and J. R. Ellis

Biomedical Engineering and Instrumentation Branch
National Institutes of Health, Bethesda, Maryland

I. INTRODUCTION

X-rays observed by the energy-dispersive detector in an analytical electron column instrument arise from two types of inelastic, or energy-loss, interactions between fast beam electrons and target atoms. In one case a beam electron interacts strongly with a core electron and imparts enough energy to remove it from the atom. The ejected core electron can have any energy up to the beam energy less the characteristic shell energy. The beam electron is depreciated in energy by whatever kinetic energy the ejected electron has acquired plus the characteristic energy required to remove it. A characteristic x-ray is occasionally emitted when the ionized atom relaxes to a lower energy state by the transition of an outer-shell electron to the vacancy in the core shell.

The x-ray is called "characteristic" because its energy equals the energy difference between the two levels involved in the transition, and this difference is characteristic of the element. The relaxation may take place in one step or by transitions to progressively lower levels. Since the energies of these levels are well known, the photon energy distribution that results from a statistically meaningful quantity of x-rays, from the same transition and from an ensemble of atoms of the same atomic number, generally suffices to identify the atom species. This x-ray "line" is characterized by a "natural width," normally specified by the full width at half the peak maximum (FWHM). Because there is usually instrumental broadening when an x-ray line is measured, we must make the distinction between the measured distribution and the line. Consequently, we call the measured distribution a "peak" and specify its width, also, as FWHM. The difference between these distributions is very large when a solid state energy-dispersive detector is used. For example, the natural width of the MnK_α line is 1.48 eV (Krause and Oliver, 1979). This line is broadened to a peak of typically 155 eV by a 30-mm^2 lithium-drifted silicon detector and its associated electronics.

The second type of inelastic interaction we must consider occurs between a fast beam electron and the nucleus of a target atom. A beam electron can decelerate in the Coulomb field of an atom, which consists of the net field due to the nucleus and core electrons. A photon is emitted that can have an energy ranging from near zero up to the energy of the beam electron, depending on the deceleration. X-rays that emanate because of this interaction process are referred to in this chapter as continuum x-rays. They are called alternatively "background," "white," or "bremsstrahlung" x-rays.

By counting characteristic x-rays, we obtain a measure of the number of analyte atoms present in the volume of target excited by the electron beam. By counting continuum x-rays, we obtain a measure approximately proportional to the product of the average atomic number and the mass thickness of the excited volume. If the average atomic number of the target is known, the continuum signal gives a measure of target mass thickness. In the ratio of a characteristic signal to a continuum signal, the factors that affect both signals equally cancel. Since the two signals are recorded simultaneously in the usual energy-dispersive x-ray analysis system, the effects of the recording time and incident probe current cancel. Marshall and Hall (1966) developed a highly successful analytical procedure based on the above, which is widely used in biological applications for thin targets in the scanning electron microscope and microprobe and the analytical electron microscope.

The purpose of this tutorial chapter is to comment on certain difficulties encountered in forming a meaningful characteristic-to-continuum ratio for the analytical electron microscope operating between 70 and 200 keV, and on certain properties of the ratio itself. We restrict our discussion to K radiation from elements of atomic number 11 to 37. We will derive, in a tutorial context, the Marshall-Hall equation from first principles. We will comment on certain shortcomings of the procedure when it is applied in the relativistic energy regime of the analytical electron microscope and will recommend some "variants of the theme." None of our comments should be construed as a criticism of this elegant and exceedingly powerful analytical procedure.

Parts of this chapter were presented at the 1982 joint meeting of the Microbeam Analysis Society and the Electron Microscope Society of America in Washington, DC, and appear in the MAS proceedings (Fiori *et al.*, 1982).

II. THE THIN TARGET

Because relatively few inelastic collisions suffered by the incident-beam electrons in any target result in inner-shell ionization, most inelastic collisions result in the loss of only very small amounts of energy through mechanisms such as ionization of an atom by the ejection of an outer-shell electron or collective excitation of outer-shell electrons. Also, only a few electrons are multiply scattered in a thin target. Therefore, most scattered beam electrons lose very little energy in total in traversing a thin target, and almost none are absorbed. That this is the case, in fact defines a target as thin. We may therefore conveniently assume that the energy of the electrons involved in collisions producing x-rays from a thin target is the beam energy.

Let us first consider in some detail the equation describing the generation of characteristic x-rays.

A. The Characteristic Signal

We can predict the number of characteristic x-rays generated into 4π steradians from the following relation:

$$I_{ch} = (N_o\, \rho C_A/A_A)Q_A\omega_A F_A N_e dz \qquad (1)$$

where

N_o = Avogadro's number
ρ = density of the target in the analyzed volume
C_A = weight fraction of the analyte in the volume
A_A = atomic weight of element A
Q_A = ionization cross section for the shell of interest; it has dimensions of area
ω_A = fluorescent yield; it is the probability that an x-ray will be emitted because of the ionization of a given shell
F_A = probability of emission of the x-ray line of interest relative to all the lines that can be emitted because of ionization of the same shell
N_e = number of electrons that have irradiated the target during measurement time
dz = target thickness in the same units of length as used in Q and ρ.

Terms in Eq. (1), other than the cross section, that are subject to uncertainty are ω_A and F_A. Values for both of these quantities have been obtained primarily by measurement. The use of values determined from mathematical fits to various experimental determinations masks the underlying uncertainty due to measurement. To demonstrate this point, in this chapter we use some of the primitive data to which fits have been made (Bambynek *et al.*, 1972; Heinrich *et al.*, 1979; Slivinsky and Ebert, 1972). Plots of the theoretical characteristic-to-continuum ratio as a function of Z, to be shown later, manifest the experimental scatter. For analytical procedures, use of fitted values is perfectly reasonable and probably desirable. Results can be found, for example, in Krause (1979).

The quantity in parentheses in Eq. (1) is the number of atoms of analyte per cm^3 of target and is obtained by the following rationale. By definition, the weight fraction of an element, A, is the mass of A divided by the total mass. The mass of A per cm^3

is the weight fraction of A times the density, ρ, in the analytical volume given in g/cm^3. We emphasize that density is a measured quantity and refers to the mass of all the atoms per unit volume. To convert mass to number of atoms, we use Avogadro's number, $N_o = 6.02 \times 10^{23}$, the number of atoms in a mole of an element. Therefore, the number of atoms of element A per cm^3 is $(C_A \times \rho) \times (N_o/A_A)$, where A_A is the gram atomic weight of a mole of element A. The total number of atoms in a volume of 1 cm^3, then, is the sum over all the elements "i" in the volume

$$N = N_o \rho \sum_i (C_i/A_i) \ . \qquad (2)$$

In the following paragraphs we present a number of characteristic cross sections that have appeared in the recent literature. We will not attempt to show that one has merit over any other since there is simply insufficient experimental data, obtained at AEM energies, in the literature to make such a comparison. However, we are not in agreement with certain aspects of several of the cross sections and so state. Our purpose in presenting the following choices is to demonstrate the variety of mathematical form and substantial numerical disagreement that exists among them. Unfortunately, this is the current state of affairs. We will show later that for the purposes of continuum normalization in biological x-ray microanalysis, the choice of characteristic cross section is not the most critical consideration, and good results can be obtained from several of the various options.

The basic functional form of the characteristic cross section was derived by Bethe (1930)

$$Q = 6.51 \times 10^{-20} \, [n_s \, b_s/(UV_c^2)] \ln(C_s \, U) \qquad (3)$$

where

the product, πe^4 = 6.51×10^{-20} (in keV^2-cm^2; e = charge of an electron)
n_s = number of electrons that populate the sth shell or subshell of interest ($n_K = 2$)
V_c = energy required to remove an electron from a given shell or subshell (the critical excitation energy)
U = overvoltage ratio V_o/V_c (where V_o = energy of the impinging beam electron)
b_s and C_s = constants for the sth shell or subshell.

For the remainder of this chapter, we use the symbol, V, to denote electron kinetic energy and the symbol, E, to denote photon energy. The "area" in cm^2 of Q is essentially the size of the K-, L-, or M-shell "target" that a beam electron must "hit" to produce an ionization of that shell. In general, this area is ~100 times smaller than the cross-sectional area of the entire atom.

Powell (1976a,b) reviewed a number of semiempirical cross sections for the beam energy range commonly found in the electron microprobe (<40 keV). For the K-shell, he recommended that $b_K = 0.9$ and $C_K = 0.65$ when the energies are expressed in keV. Powell favorably notes a modification to the Bethe formula by Fabre

$$Q_F = 6.51 \times 10^{-20} \, \{n_K \ln(U)/[V_c^2 \times a(U+b)]\} \qquad (4)$$

where a = 1.18 and b = 1.32 are constants for the K-shell. The range of the overvoltage ratio is recommended to be $1.5 < U < 25$, and the subscript, F, on Q denotes "Fabre." The other terms are as above. It was not intended that these cross sections should be applied above ~40 keV, and it should cause no great surprise if they fail in the range of beam energies found in the analytical electron microscope.

Several forms have been suggested for the ionization cross section for the energy range 70 to 200 keV. This is an energy region where the effects of relativity should not be ignored and a generally useful cross section should accommodate the effect explicitly, or at least provide empirical adjustment. Zaluzec (1978, 1979), for example, has recommended the formulation given in Mott and Massey (1965) of the relativistic cross section derived by Bethe and Fermi (1932) and Williams (1933)

$$Q = [6.51 \times 10^{-20} n_K a_K/(V_c V_o)] [\ln(b_K U) - \ln(1 - \beta^2) - \beta^2] \quad (5)$$

where

all energies are given in keV

$a_K = 0.35$ (K-shell)
$b_K = 0.8U/[(1 - e^{-\gamma})(1 - e^{-\delta})$
$\gamma = 1250/(V_c U^2)$
$\delta = 1/(2V_c)$
$\beta = v/c = \sqrt{1-[1+(V_o/511)]^{-2}}$
= the relativistic correction factor.

The coefficients were determined by Zaluzec by fitting to values reported in the literature. For a more detailed discussion on the fitting of the coefficients in the Bethe-Fermi-Williams equation, the reader is referred to Williams *et al.* (1984) and Newbury *et al.* (1984).

It should be noted that Powell (1976a,b) has concluded from an extensive survey of the extant data that no significant variation of the Bethe parameters occurs as a function of atomic number. A similar conclusion was reached independently by Quarles (1976), again for K-shell ionization.

Schreiber and Wims (1981a,b) have proposed a highly empirical variant of the Bethe equation with coefficients determined indirectly from their relative measurements of characteristic lines at two beam energies, 100 and 200 keV, using an uncalibrated x-ray detector and specimens of unknown mass-thickness. Their formulation is unique in that the coefficients are functions of atomic number. They give the cross section as

$$Q = \pi e^4 m_s b \ln U/(V_c^2 U^d) \quad (6)$$

where

$b = 8.874 - 8.158(\ln Z) + 2.9055(\ln Z)^2 - 0.35778(\ln Z)^3$ for $Z \leq 30$
$b = 0.661$ for $Z > 30$
$d = 1.0667 - 0.0046Z$.

In their paper, these authors include coefficient expressions for the L- and M-shells, again, determined indirectly. It should be noted that the cross section coefficients that they propose are machine and procedure dependent and are not generally applicable.

Kolbenstvedt (1967, 1975) has given a relatively simple approximate formula for K ionization by relativistic electrons much higher in energy than the ionization energy. He has obtained good agreement with experimental data for the elements silver and tin for energies from <100 keV to 2 MeV. His equation is written as

$$\sigma_K = \frac{0.275}{I}\frac{(V_o'+1)^2}{V_o'(V_o'+2)}\left[\ln\frac{1.19(V_o'+2)}{I} - \frac{V_o'(V_o'+2)}{(V_o'+1)^2}\right]$$
$$+\frac{0.99}{I}\frac{(V_o'+1)^2}{V_o'(V_o'+2)}\left[1-\frac{I}{V_o'}\left\{1-\frac{V_o'^2}{2(V_o'+1)^2}+\frac{2V_o'+1}{(V_o'+1)^2}\ln\frac{V_o'}{I}\right\}\right] \text{(barns)} \quad (7)$$

where V_o is the kinetic energy of the incident electron and I is the ionization energy, both in units of the electron rest energy. For our purposes we require the following conversions

$$Q_K = 10^{-24}\ \sigma_K\ (\text{cm}^2)$$

$$I = V_c/511$$

$$V_o' = V_o/511$$

where V_c and V_o are the critical excitation energy and beam energy, respectively, in keV.

We have presented these currently used cross sections not to make a critical evaluation, because there simply are insufficient data in the literature to permit such an evaluation, but only to demonstrate their functional form and significant differences between the various possibilities. They are plotted as a function of Z for the range $11 < Z < 37$ for the beam energies of 100 and 200 keV in Figures 1 and 2, respectively. It is usual practice to plot characteristic cross sections as a function of overvoltage. However, since most analysis is performed at one beam voltage, we are more concerned here with the behavior of the cross section as a function of atomic number.

It is apparent from the plots that the first difficulty in formulating a theoretical characteristic-to-continuum ratio is choosing a cross section to use in calculating Eq. (1), the numerator. We have considered here only K radiation; sufficient data are not available for L and M radiation. Clearly, further work is required. Unfortunately, because of the confined specimen region, the analytical electron microscope is not an ideal instrument in which to make the required measurements. It is difficult, at best, to remove the contribution of scattered electrons and radiation from the measured characteristic intensities. On the other hand, the accurate determination of mass-thickness is one of the more difficult aspects of making good cross-section measurements. If the AEM is equipped with an energy-loss spectrometer, it may be possible to determine the specimen mass-thickness simultaneously with the characteristic x-ray yield from the electron-beam-specimen interaction volume. However, the use of the AEM to measure x-ray cross sections is not a trivial matter and must be done with extreme care.

B. The Continuum Signal

The interaction of a large number of beam electrons with a thin foil of a given element will produce an emitted continuum spectrum having a distribution

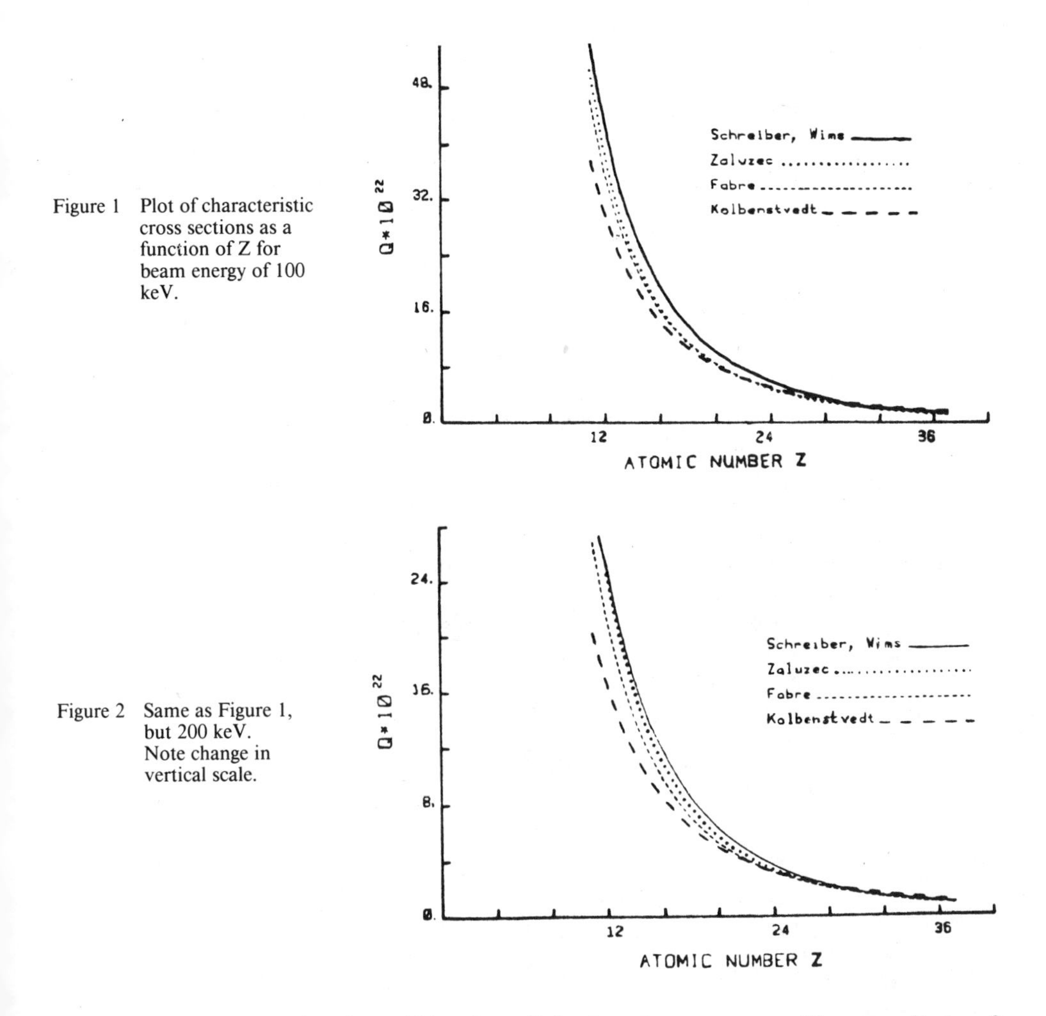

Figure 1 Plot of characteristic cross sections as a function of Z for beam energy of 100 keV.

Figure 2 Same as Figure 1, but 200 keV. Note change in vertical scale.

approximately proportional to 1/E, where E is the photon energy. The magnitude of this distribution is proportional to the number of atoms comprising the foil. Other factors, such as beam current and measurement time, scale the magnitude linearly. The beam energy, V_o, and observation angle, θ, affect the overall shape of the distribution in a complex manner. We define the observation angle as the angle between the incident-beam direction and the x-ray detector. We can predict the number of continuum x-rays generated into a unit steradian from the following relation

$$I_{co} = \sum_i (N_i Q_i) N_e \tag{8}$$

where

N_i = number of atoms of element i
Q_i = continuum cross section for that element; it has dimensions of area
N_e = number of electrons that have irradiated the N_i target atoms during measurement time.

Q is usually assumed to be differential in photon energy and observation angle. Consequently, it is not necessary to include dE and dθ terms in Eq. (8). Using the fact that the total number of atoms in 1 cm^3 is

$$N_o \rho \sum_i (C_i/A_i) \tag{9}$$

Eq. (8) can be rewritten for a thin film of thickness dz as

$$I_{co} = N_o \rho \sum_i (C_i Q_i/A_i) N_e dz \tag{10}$$

where dz has the same units of length as in Q_i and ρ.

One should note the following differences between the generation of characteristic and continuum radiation. For our purposes, the ejection of a core electron by a fast electron and the occasional subsequent emission of a characteristic x-ray photon are independent events. A characteristic photon has equal probability of being emitted in any direction (isotropy) after the ionization. Relativistic considerations apply only to the probability of the ionization. The probability of continuum emission, on the other hand, is intimately related to the probability that a fast beam electron will decelerate in the Coulomb field of an atom. Indeed, the emission is a direct consequence of the deceleration. Furthermore, the probability of continuum emission is directionally dependent and peaked toward the forward direction as defined by the direction of travel of the beam electrons (Figure 3). The "angle" and magnitude of the maximum of each of the lobes shown in the figure depend on the photon energy being observed, beam energy, and the average atomic number of the target. Target tilt has no effect other than to change the apparent thickness of the target. Because of this anisotropy, continuum radiation is usually expressed as emission into a unit steradian at a specified angle, rather than uniformly into 4π steradians, as is the characteristic cross section.

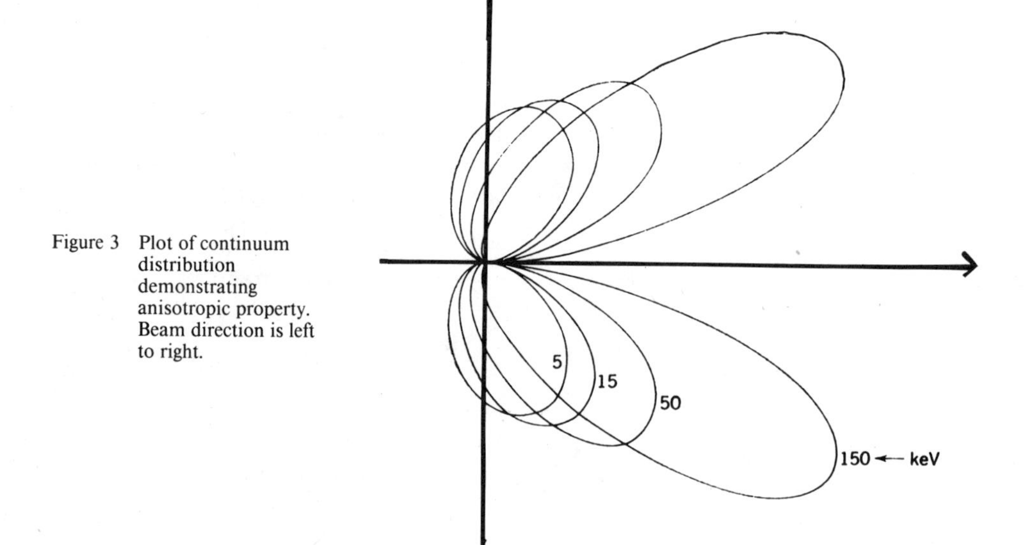

Figure 3 Plot of continuum distribution demonstrating anisotropic property. Beam direction is left to right.

Dyson (1959) has observed that the angle of maximum intensity, θ, for photons having an energy near the beam energy approximately satisfies the equation

$$\beta\,(1 + \sin^2\theta) = \cos\theta \tag{11}$$

where β is defined above.

We continue our discussion of the continuum cross section with the purpose of deriving the equation used in the Marshall-Hall procedure for biological microanalysis. We first note two important properties of continuum radiation.

(a) The maximum energy that can be given up by an incident-beam electron is its kinetic energy, which is numerically equal to the beam voltage, V_o. Consequently, there is a highest photon energy, E_o, in a continuum spectrum due to beam electrons of energy V_o. This highest energy is the so-called high-energy limit, or the Duane-Hunt limit.

(b) In a thin target, both theory and experiment indicate that the amount of emitted energy in an energy interval, dE, is approximately uniformly distributed from near zero up to the high-energy limit (Compton and Allison, 1935) (Figure 4).

From (b), we can define the fraction of the total emitted continuum energy, E_t, in the interval from E to E + dE, as $dE/(E_o - 0)$. The efficiency of the generation of continuum in a thin film is defined to be the total continuum energy (from near zero to E_o) generated by electrons that lose an amount, dV, of their energy divided by the energy lost. Kirkpatrick and Wiedmann (1945) have determined from the theory of Sommerfeld that this efficiency of production is $2.8 \times 10^{-9}\ ZV_o$, where V_o is the beam voltage in kV and Z is the atomic number. We can thus express the total radiated continuum energy as

$$E_t = 2.8 \times 10^{-9}\ ZV_o\ dV\ . \tag{12}$$

The fraction of this quantity in the energy interval, dE, provides us with the number of photons, I_{co}, of energy E. The number of photons in the vanishingly small interval, dE, is obtained by dividing the amount of energy in that interval by the photon energy, E. Consequently, the number of photons in dE is given by the quantity $(E_t \times dE/E_o)/E$ or

$$I_{co} = 2.8 \times 10^{-9}\ (Z/E)\ dE\ dV\ . \tag{13}$$

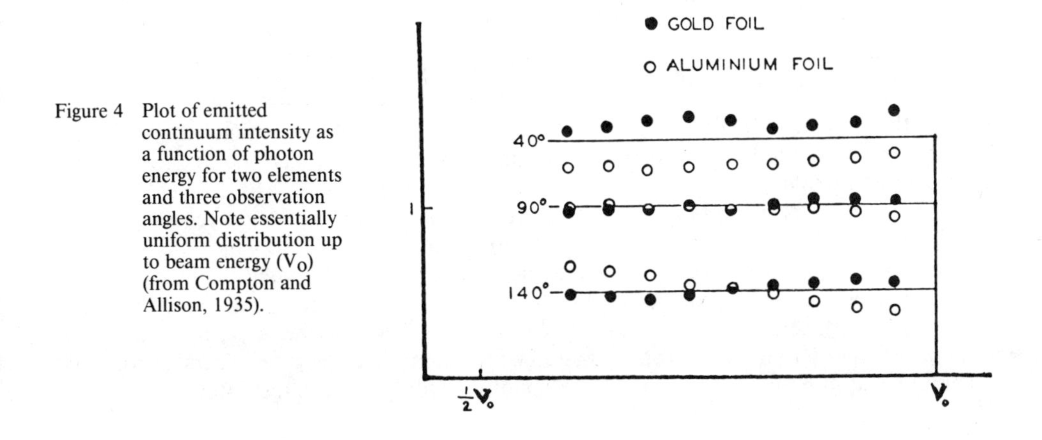

Figure 4 Plot of emitted continuum intensity as a function of photon energy for two elements and three observation angles. Note essentially uniform distribution up to beam energy (V_O) (from Compton and Allison, 1935).

The energy loss, dV, of electrons traversing a thin film of a single element of thickness dz can be given by the Bethe equation for the deceleration of an electron (Bethe, 1933)

$$-dV = 2\pi e^4 N_o Z\rho/(V_o A) \ln(1.166 V_o/J)\, dz \quad (14)$$

where

J = mean excitation energy of an atom (Bethe, 1933; Heinrich, 1981)
1.166 = square root of half the base of the natural logarithms
dz = target thickness in cm if ρ is given in g/cm^3.

The other terms are as defined earlier. The derivation of the equation is based on the assumption that the energy transferred to the relatively massive nucleus when an electron interacts with an atom is negligible, so that the interaction can be assumed to involve only the atomic electrons. (For a more detailed discussion of the assumptions used by Bethe in deriving the equation and of potential errors from its application, see Heinrich, 1981, pp. 226-32).

Although the continuum results from interactions of beam electrons and atomic nuclei, and Eq. (14) applies to energy assumed lost only in interactions between the beam electrons and atomic electrons, we note that the energy loss in Eq. (14) can be used in Eq. (12). This is because Eq. (12) describes only the "efficiency of production"; no connection between energy loss and continuum production is implied. Combining these two equations, we obtain the number of continuum photons, I_{co}, with energy E, generated in interval dE by a beam electron with energy V_o, traversing a thin film of thickness dz of one element

$$I_{co} = 220 Z^2 \rho/(A V_o E) \ln(1.166 V_o/J)\, dz\, dE\ . \quad (15)$$

The energy terms are in eV. Equation (15) can be expressed in units of a cross section, differential in photon energy, by multiplying by the quantity $A/(N_o \rho dz)$, giving

$$Q_{iu} = 3.65 \times 10^{22}\, Z^2/(V_o E) \ln(1.166 V_o/J)\, dE\ . \quad (16)$$

Subscripts iu on Q distinguish this cross section by its principal characteristics: isotropy, and uniformity in energy distribution from continuum radiation property (b), above. The resulting cross section now has the required dimensions of area in cm^2 per beam electron. Several versions of the J factor are available in the literature. We present as an example the Sternheimer formulation (Heinrich, 1981)

$$J = Z(9.76 + 58.82 Z^{-1.19}) \qquad (eV)\ . \quad (17)$$

Use of the J factor in Eq. (14) permits the application of the equation, originally derived from a quantum mechanical treatment of the hydrogen atom, to higher-atomic-number elements.

Equation (16) is an apparently simple expression of the continuum cross section. The simplicity is, however, a result of the approximations and assumptions used in the derivation. The cross section takes no account of the strong relativistic effects for beam energies above several keV. Furthermore, the equation poorly predicts the continuum distribution as a function of photon energy and, as mentioned, it takes no account of the strong anisotropy of the continuum. The degree of anisotropy is a complicated function of beam energy, photon energy, and target atomic number.

However, despite these considerable shortcomings, the cross section, as used by Marshall and Hall, gives remarkably good results. Care must be taken to apply the equation only under the experimental conditions they carefully specified. We will subsequently discuss cross sections that take more exact account of the physical processes involved in continuum generation. However, since we have derived the continuum cross section (Eq. (16)) used by Marshall-Hall, we continue our discussion of their procedure after a short digression concerning continuum cross sections in general.

The generalized formulation for generation of continuum radiation cannot be solved in closed form. As a result, all the currently used cross sections are derived from approximations of varying accuracy. However, they have a common form from which it is possible to factor out the square of the atomic number. Generally, the factored cross sections retain additional terms involving Z. In the more rigorously derived cross sections discussed below, the residual Z terms include a photon energy dependence. In Eq. (10), which predicts the number of continuum photons in a given material for a given electron flux, we note that a square of the atomic number term factored from Q is divided by the atomic weight. The quantity, Z^2/A, is approximately equal to $Z/2$ throughout the periodic table (i.e., $A \approx 2Z$). We stress the word "approximately." As can be seen in Figure 5, the function does not plot as a straight line. Indeed, it is not even monotonic; that is, the same value of Z^2/A occurs at more than one Z. Consequently, according to Eq. (10), normalized continuum radiation from pure element foils, as a function of atomic number, has the undesirable shape as plotted in Figure 5.

The argument can be put forth that by writing the continuum intensity in terms of atomic concentration, such as in Eq. (8), one obtains a smoothly varying continuum

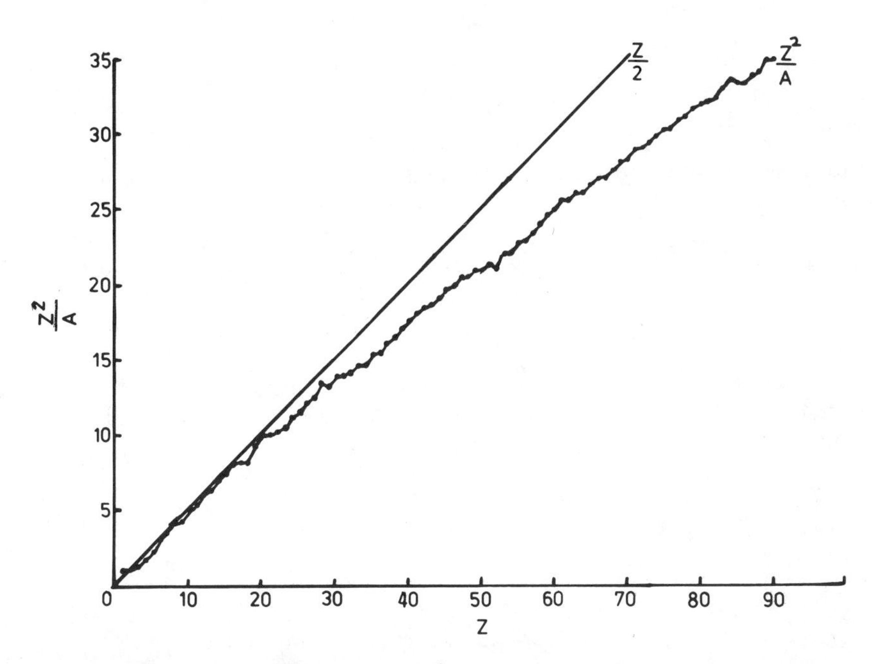

Figure 5 Plot of Z^2/A and $Z/2$ versus Z.

distribution, approximately proportional to the square of Z, as a function of single atomic number. However, specimens do not come with a known number of atoms in any given volume. Consequently, we must use Eq. (10) rather than Eq. (8) for most practical situations.

III. DERIVATION OF THE HALL PROCEDURE IN TERMS OF X-RAY CROSS SECTIONS

The density of the target within the analytical volume of the electron probe is generally not known unless the material being irradiated is a standard reference material. We also note that the thickness of a target at the site of impact of the electron beam is rarely a known quantity. However, it is obvious that a change in either local density or thickness causes a proportionate change in the number of atoms with which a beam electron is likely to interact in passing through a thin target. Characteristic and continuum generation are equally affected. To treat these changes more easily, density and thickness are frequently combined into a single quantity, "mass-thickness," denoted by ρdz, which has the somewhat confusing dimensions of g/cm^2 (i.e., $g/cm^3 \times cm = g/cm^2$).

We stress that the concept of mass-thickness is useful only if the beam energy does not change by more than a very small amount in traversing the target thickness, dz. A typical biological target should be less than several thousand angstroms in thickness for this assumption to be sufficiently accurate. With this caveat, we form the ratio of Eq. (1) and (10)

$$\frac{I_{ch}}{I_{co}} = \frac{(N_o\, C_A/A_A)\, Q_A\, \omega_A\, F_A\, N_e\, (\rho dz)}{N_o \sum_i (C_i\, Q_i/A_i)\, N_e\, (\rho dz)} \,. \tag{18}$$

We note that ρdz, Avogadro's number, and the number of electrons incident on the film during the measurement period all cancel in the ratio. To further simplify the formula, we take advantage of the fact that for the analysis of a given element, the fluorescent yield and associated relative transition probability are constants that can be gathered into one grand constant, k. Thus we have

$$\frac{I_{ch}}{I_{co}} = k\, \frac{(C_A\, Q_A/A_A\,)}{\sum_i (C_i\, Q_i/A_i)} \tag{19}$$

where Q_A is the characteristic cross section for a particular element, A, and Q_i is the continuum cross section for the ith element. Subscript i must span all the elements present in the electron-beam-target interaction volume.

Equation (19) is the most general formula for the method proposed by Hall and co-workers. We emphasize the flexibility of the Hall method when expressed in this form: both characteristic and continuum cross sections more appropriate to the relativistic regime of the AEM can be used. However, we will proceed with the derivation, with the cross sections originally used by Marshall and Hall for their lower energy application.

We now put into Eq. (19) the cross sections given by Eq. (3) and (16) and collect all numerical constants into k to obtain

$$\frac{I_{ch}}{I_{co}} = k \frac{C_A/(A_A\, U V_c^2) \ln (C_s U)}{\sum_i [(C_i\, Z_i^2/A_i)/(V_o\, E) \ln (1.166 V_o/J_i)]} dE \; . \tag{20}$$

We can further simplify this formula by taking advantage of the fact that the operating voltage, V_o, is held constant during analysis and that the energy band, dE, of continuum measurement (centered at energy E) is usually not changed during analysis. Consequently, dE, E, and V_o (and hence U) can be absorbed into k, resulting in the simpler form

$$\frac{I_{ch}}{I_{co}} = k \frac{C_A/A_A}{\sum_i [(C_i\, Z_i^2/A_i) \ln (1.166 V_o/J_i)]} \tag{21}$$

where J_i is a function of atomic number such as given in Eq. (17). This is the usual representation of the 1966 Marshall-Hall equation, which is a more rigorous expression of the continuum correction concept proposed earlier by Hall *et al.* (1966) and often referred to as the "Hall" method or correction. Usually, the equation is used in the following manner: A measured characteristic-to-continuum ratio from a characterized material is set equal to the right side of Eq. (21). Since the target being irradiated is a reference standard, we presumably know the atomic numbers (Z_i), atomic weights (A_i), and weight fractions (C_i). Since all other terms except the constant, k, are known, k can be calculated for the given set of experimental conditions. Next, we hold these conditions constant and measure the characteristic and continuum intensities from the specimen. To calculate a weight fraction of analyte, C_A, from the specimen, we must know the weight fraction of each and every one of the elements that comprise that part of the specimen irradiated by the electron beam. Crucial to the Hall procedure is the following assumption: The quantity

$$\sum_i (C_i Z_i^2/A_i)$$

for the biological specimen to be analyzed is dominated by the matrix so that the unknown contribution of the analyte, C_A, to the sum may be neglected. Furthermore, the value of the sum is known or can be estimated from other information about the detail being analyzed. A table of values for a number of typical biological materials is presented in Table I (Hall, 1977).

Equation (16) is an apparently simple expression of the continuum cross section. The simplicity is, however, a result of using approximations and assumptions in the derivation. We next consider three continuum cross sections that take more exact account of the physical processes involved in continuum generation. It must be noted, however, that these more "exact" cross sections also result from certain simplifications since the formulation of the cross section from which these are derived cannot be solved in closed form. There are two fundamentally different types of simplifications of particular relevance to applications in the AEM. We shall discuss representative cross sections using each.

TABLE I

$\sum(C_i Z_i^2/A_i)$ Values for a Number of Typical Biological Materials.

	Number of Atoms						
Material	H	C	N	O	P	S	Z^2/A
Water	2.0			1.0			3.67
Fatty Acid (oleic)	34.0	18.0		2.0			2.87
Triglyceride	107.0	60.0		12.0			2.98
Glucose	12.0	6.0		6.0			3.40
Deoxyribose	10.0	5.0		4.0			3.33
Nucleic Acid Cytidine Monophosphate	14.0	9.0	3.0	8.0	1.0		3.78
Same without P							3.41
Protein (with S)	112.0	66.7	18.3	25.0		1.0	3.28
Protein (no S)	112.0	66.7	18.3	25.0			3.20
Araldite	8.4	5.8	0.02	1.19		0.08	3.15
Nylon	11.0	6.0	1.0	1.0			3.01
Polycarbonate	14.0	16.0		3.0			3.08

IV. THE COULOMB APPROXIMATION: THE THEORY OF SOMMERFELD

Sommerfeld (1931, 1950) developed a theory of continuum generation that neglects the presence of atomic electrons (screening) and assumes a pure Coulomb field about a point nucleus. Relativity, electron spin, and retardation of potential are neglected, and the incoming electron and the scattered electron are represented by plane waves that occupy all space. Other assumptions that affect the spectral distribution at both the extreme high- and low-energy limits have been made but have little effect on the distribution in the range of interest to us (Stephenson, 1967).

The Sommerfeld theory is troublesome to evaluate because of its mathematical form. Numerical estimates for selected atomic numbers and values of photon and electron energies have been made by Kirkpatrick and Wiedmann (1945) following the method of Weinstock (1942, 1944). These estimates are accurate to within 2% of the fully calculated theory. The Sommerfeld theory is valid only for incident-beam energies less than several keV. It is possible, however, to apply a relativistic correction factor and extend the range of utility to encompass the AEM range. Neglecting the screening effects of core electrons, especially in heavy atoms, can be expected to cause some error.

The algebraic fit, $Q_{k\text{-}w}$, of Kirkpatrick and Wiedmann to the Sommerfeld theory is given by

$$Q_{k\text{-}w} = [I_x (1 - \cos^2\theta) + I_y (1 + \cos^2\theta)]/E \qquad (22)$$

where

$$I_x = (300Z^2/V_o)[0.252 + a(E/V_o - 0.135) - b(E/V_o - 0.135)^2]\ 1.51 \times 10^{-28}$$

$$I_y = (300Z^2/V_o)\{-j + k/[(E/V_o) + h]\}\ 1.51 \times 10^{-28}$$

and

$$a = 1.47B - 0.507A - 0.833 \qquad A = \exp(-0.223V_o/300Z^2) - \exp(-57V_o/300Z^2)$$

$$b = 1.70B - 1.09A - 0.627 \qquad B = \exp(-0.0828V_o/300Z^2) - \exp(-84.9V_o/300Z^2)$$

$$h = \frac{-0.214y_1 + 1.21y_2 - y_3}{1.43y_1 - 2.43y_2 + y_3} \qquad j = (1 + 2h)\, y_2 - 2(1 + h)\, y_3$$

$$k = (1 + h)\,(y_3 + j) \qquad y_1 = 0.22[1 - 0.39 \exp(-26.9V_o/300Z^2)]$$

$$y_2 = 0.067 + 0.023/[(V_o/300Z^2) + 0.75] \qquad y_3 = -0.00259 + 0.00776/[V_o/300Z^2 + 0.116]$$

and E and V are expressed in eV.

$Q_{k\text{-}w}$ can be made differential in energy and angle by multiplying by $d\Omega$ and dE, the differential of solid angle subtended by the detector at angle θ and the differential of photon energy, respectively.

Zaluzec (1978, 1979) has recommended the relativistically corrected form derived by Sommerfeld

$$Q_{k\text{-}w} = \left\{ I_x \left[\frac{1 - \cos^2\theta}{(1 - \beta\cos\theta)^4} \right] + I_y \left[1 + \frac{\cos^2\theta}{(1 - \beta\cos\theta)^4} \right] \right\} / E \tag{23}$$

where β, the relativistic correction factor, is as defined previously. Kulenkampff *et al.* (1959) have observed that the angular intensity distribution predicted by this equation is incorrect. A detailed derivation of the error in the equation is given in Robertson (1979, p. 133).

Motz and Placious (1958) have recommended another formulation of relativistic correction that is markedly different in form and results

$$Q_{k\text{-}w} = \left\{ I_x \left[\frac{1 - \cos^2\theta}{(1 - \beta\cos\theta)^2} \right] + I_y \left[1 + \frac{1 + \cos^2\theta}{(1 - \beta\cos\theta)^2} \right] \right\} / E\ . \tag{24}$$

(Note that in both forms the relativistic correction made to the continuum radiation is zero for a 90° observation angle.) They derived this formulation by comparing the relativistic and nonrelativistic cross sections given in Heitler (1954, Eq. (13) and (17), p. 242). As pointed out by Motz and Placious, Heitler's cross sections are derived with free particle-wave functions. Consequently, the corrected cross section can be considered only a rough estimate of the exact cross section that would be obtained with relativistic Coulomb wave functions. The Motz-Placious correction has been used by Statham (1976) for a lower beam energy ($\approx$20 keV) Monte Carlo study in bulk specimens (photon energies $\approx 1-10$ keV) in which continuum anisotropy is examined. A more involved relativistic correction to the Kirkpatrick-Wiedmann equation that includes approximations for screening and retardation effects is given by Robertson (1979, p. 41).

V. THE BORN APPROXIMATION

The Born theory also assumes free particle-wave functions. An extensive review and bibliography of continuum cross sections obtained by a number of workers using the Born approximation is given in Koch and Motz (1959). As pointed out by these authors, such cross sections are available in a relatively simple analytical form for relativistic energies, with or without screening correction. For the Born approximation to be valid, the initial and final electron kinetic energies in an electron-nuclear interaction must be large enough to satisfy the two conditions

$$(2\pi Z/137\beta_o),\ (2\pi Z/137\beta) \ll 1$$

where β is as defined in Eq. (5). Despite the fact that the beam and photon energies in the AEM often do not satisfy these conditions, the theory is useful in predicting the shape of the continuum distribution. Koch and Motz report that the absolute error for total emission is within a factor of two in the worst case and can be considerably less for lower-atomic-number specimens, 90° observation angle, and higher AEM beam energies (>150 keV).

A representative Born cross section, derived by Bethe and Heitler and reported in Koch and Motz (1959), differential in photon energy and angle is given by

$$Q_{E,\theta} = \frac{dk}{k}\frac{p}{p_o} SZ^2\, d\Omega \left\{ \frac{8\sin^2\theta(2E_i^2+1)}{p_o^2\Delta^4} - \frac{2(5E_i^2+2E_fE_i+3)}{p_o^2\Delta^4} - \frac{2(p_o^2-k^2)}{G^2\Delta^2} + \frac{4E_f}{p_o^2\Delta} \right.$$
$$+ \frac{L}{pp_o}\left[\frac{4E_i\sin^2\theta(3k-p_o^2E_f)}{p_o^2\Delta^4} + \frac{4E_i^2(E_i^2+E_f^2)}{p_o^2\Delta^2} + \frac{2-2(7E_i^2-3E_fE_i+E_f^2)}{p_o^2\Delta^2} \right.$$
$$\left.\left. + \frac{2k(E_i^2+E_fE_i-1)}{p_o^2\Delta} \right] - \left(\frac{4\epsilon}{p\Delta}\right) + \left(\frac{\epsilon^G}{p^G}\right)\left[\frac{4}{\Delta^2} - \frac{6k}{\Delta} - \frac{2k(p_o^2-k^2)}{G^2\Delta}\right] \right\} \tag{25}$$

where

$k = E/511$
$dk = dE/511$
$E_i = (V_o/511) + 1$
$E_f = E_i - k$
$p_o = (E_i^2 - 1)^{1/2}$
$p = (E_f^2 - 1)^{1/2}$
$S = 2.31 \times 10^{-29}\ cm^2$
$G^2 = p_o^2 + k^2 - 2p_o \times \cos\theta$

$$L = \ln\left(\frac{E_fE_i - 1 + pp_o}{E_fE_i - 1 - pp_o}\right)$$

$\Delta = E_i - p_o\cos\theta$

$$\epsilon = \ln \left(\frac{E_f + p}{E_f - p} \right)$$

$$\epsilon^G = \ln \left(\frac{G + p}{G - p} \right)$$

$d\Omega$ = solid angle subtended by the detector at angle θ
θ = observation angle of the detector relative to the electron-beam axis as described previously
E = photon energy (keV)
V_o = energy of the beam electrons (keV).

This equation is essentially Eq. (2BN) of Koch and Motz (1959), which is given in units of mc^2. We have included the conversions of beam and photon energies to keV.

Robertson (1979), Ferrier *et al.* (1976), and Chapman *et al.* (1983) have reported a modification to the Bethe-Heitler equation. Their formulation includes a screening correction derived by Gluckstern and Hull (1953) and a correction for the Coulomb effect derived by Elwert (1939). These authors report good agreement between their experimental measurements and the modified Bethe-Heitler equation, given as

$$Q_{E,\theta} = \frac{dk}{k} \frac{p}{p_o} SZ^2 \, d\Omega \frac{\beta_o}{\beta} \frac{[1 - \exp(-2\pi Z/137\beta_o)]}{[1 - \exp(-2\pi Z/137\beta)]} \left\{ \frac{8\sin^2\theta(2E_i^2 + 1)}{p_o^2\Delta^4} \right.$$

$$- \frac{2(5E_i^2 + 2E_fE_i + 3)}{p_o^2\Delta^2} - \frac{2(p_o^2 - k^2)}{G^2\Delta^2} + \frac{4E_f}{p_o^2\Delta} + \frac{L}{pp_o}$$

$$\times \left[\frac{4E_i \sin^2\theta \, (3k - p_o^2 E_f)}{p_o^2\Delta^4} + \frac{4E_i^2(E_i^2 + E_f^2)}{p_o^2\Delta^2} + \frac{2 - 2\,(7E_i^2 - 3E_iE_f + E_f^2)}{p_o^2\Delta^2} \right.$$

$$\left. \left. + \frac{2k(E_i^2 + E_fE_i - 1)}{p_o^2\Delta} \right] - \left[\frac{4\epsilon}{p\Delta} \right] + \left[\frac{\epsilon^G}{p^G} \right] \left[\frac{4}{\Delta^2} - \frac{6k}{\Delta} - \frac{2k(p_o^2 - k^2)}{G^2\Delta} \right] \right\} \quad (26)$$

where

$\beta_o = p_o/E_i$
$\beta = p/E_f$
$k = E/511$
$dk = dE/511$
$E_i = V_o/511$
$E_f = E_i - k$
$p_o = (E_i^2 - 1)^{1/2}$
$p = (E_f^2 - 1)^{1/2}$
$S = 2.31 \times 10^{-29}\ \text{cm}^2$

$$L = \ln\left[\frac{(E_f E_i - 1 + pp_o)^2}{k^2\, p_o^2\, \alpha^{-2}\, \Delta^{-2}}\right]$$

$$\Delta = E_i - p_o \cos\theta$$

$$E = \ln\left[\frac{E_f + p}{E_f - p}\right]$$

$$G^2 = p_o^2 + k^2 - 2p_o k \cos\theta$$

$$\alpha = 108/Z^{1/3}$$

$$\epsilon^G = \frac{1}{2}\ln\left[\frac{(G + p)^4}{4(k^2\Delta^2 + p_o^2\,\alpha^{-2})}\right] .$$

We have included the conversions of beam and photon energies to keV.

Again, we have presented these currently used cross sections not to make a critical evaluation but merely to demonstrate the functional form and numeric variation for the energy range of interest to us. Clearly, more work needs to be done in this area. The work of Chapman *et al.* (1983) is an important contribution. However, continuum measurements are extremely difficult to perform in the AEM for the same reasons as are characteristic radiation measurements. Additionally, it is a nontrivial limitation that the x-ray detector is confined to one observation angle in the usual AEM. Since continuum emission is a complicated anisotropic function of both beam and photon energy, as well as target atomic number, it is desirable to determine the emission at more than one observation angle.

The continuum cross sections can be normalized by dividing each cross section by Z^2 and multiplying by the photon energy. Plots then reveal deviations from nominal dependence on these two quantities. Figures 6, 7, and 8 show the forms of the three cross sections discussed in this chapter plotted for one observation angle (158°), one atomic number (11), and three beam energies: 50, 100, and 200 keV. Two relativistically corrected forms of the Kirkpatrick-Wiedmann equation are plotted. Notation MP denotes the Motz-Placious form, and notation Z denotes the Zaluzec form. Notation MH denotes the cross section used by Marshall-Hall, BH denotes the Bethe-Heitler cross section, and MBH denotes the modified Bethe-Heitler cross section. We observe the strong deviation from a $1/E$ form and the change in spectral distribution as a function of beam energy for all the cross sections except the Marshall-Hall. Note the change in vertical scale among the figures.

Figures 9 and 10 plot MP, MH, BH, and MBH against photon energy for a fixed beam energy (100 keV), two observation angles (90° and 158°), and three atomic numbers (11, 19, and 26). We use only the Motz-Placious form of the Kirkpatrick-Wiedmann equation. These plots demonstrate (a) the deviation of the MH, MP, and MBH cross sections from nominal Z^2 dependence, and (b) the change in the cross section distribution with change in observation angle generated from the Sommerfeld and Born theories. The BH cross section does not exhibit a deviation from a Z^2 dependence since this is one of the assumptions made in the approximation. The plot of the BH cross section is most useful here to examine the dependence on photon and beam energy and to point out the magnitudes of the two corrections applied in the MBH equation. A discussion of the significance of these distributions is the purpose of the next section.

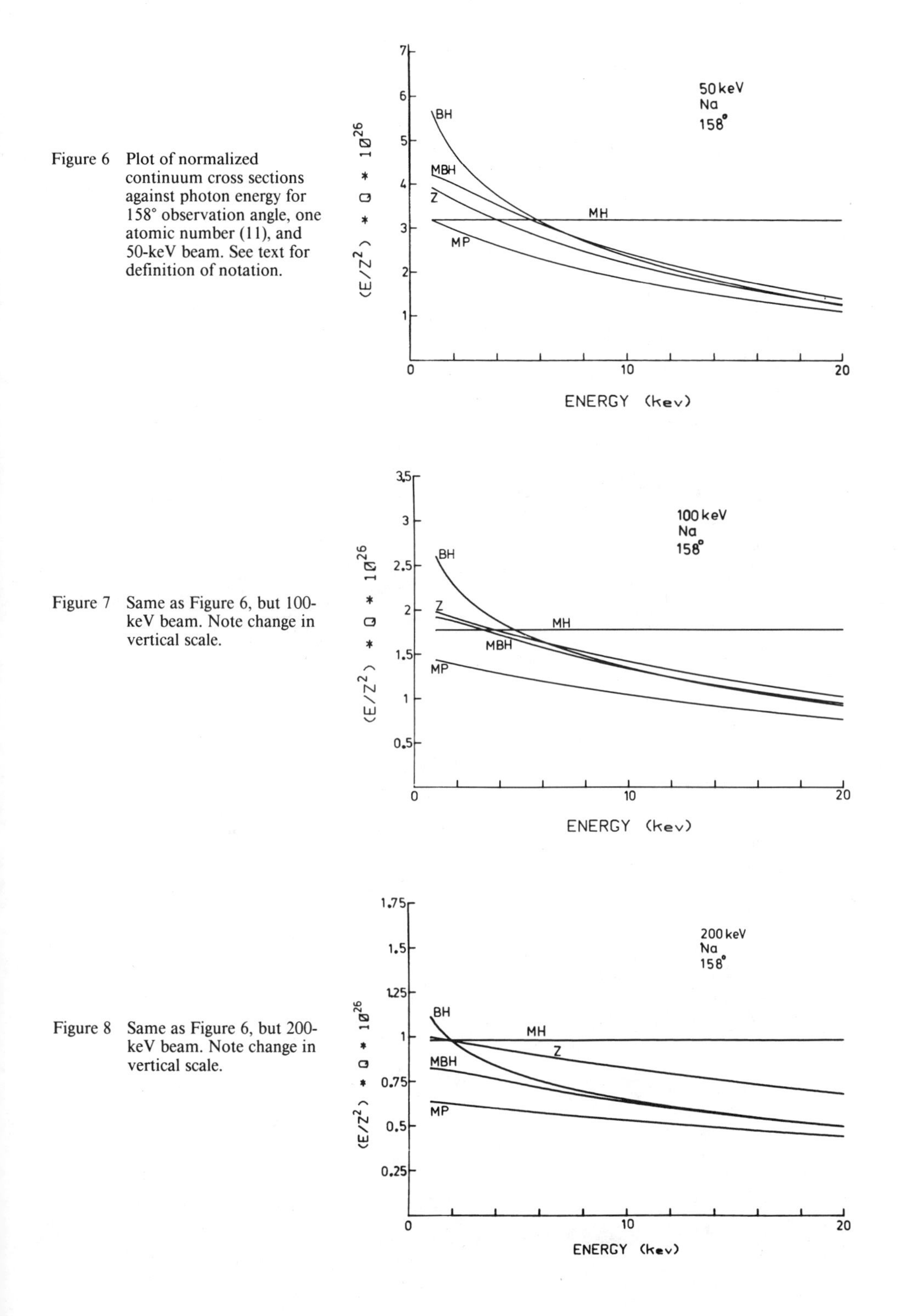

Figure 6 Plot of normalized continuum cross sections against photon energy for 158° observation angle, one atomic number (11), and 50-keV beam. See text for definition of notation.

Figure 7 Same as Figure 6, but 100-keV beam. Note change in vertical scale.

Figure 8 Same as Figure 6, but 200-keV beam. Note change in vertical scale.

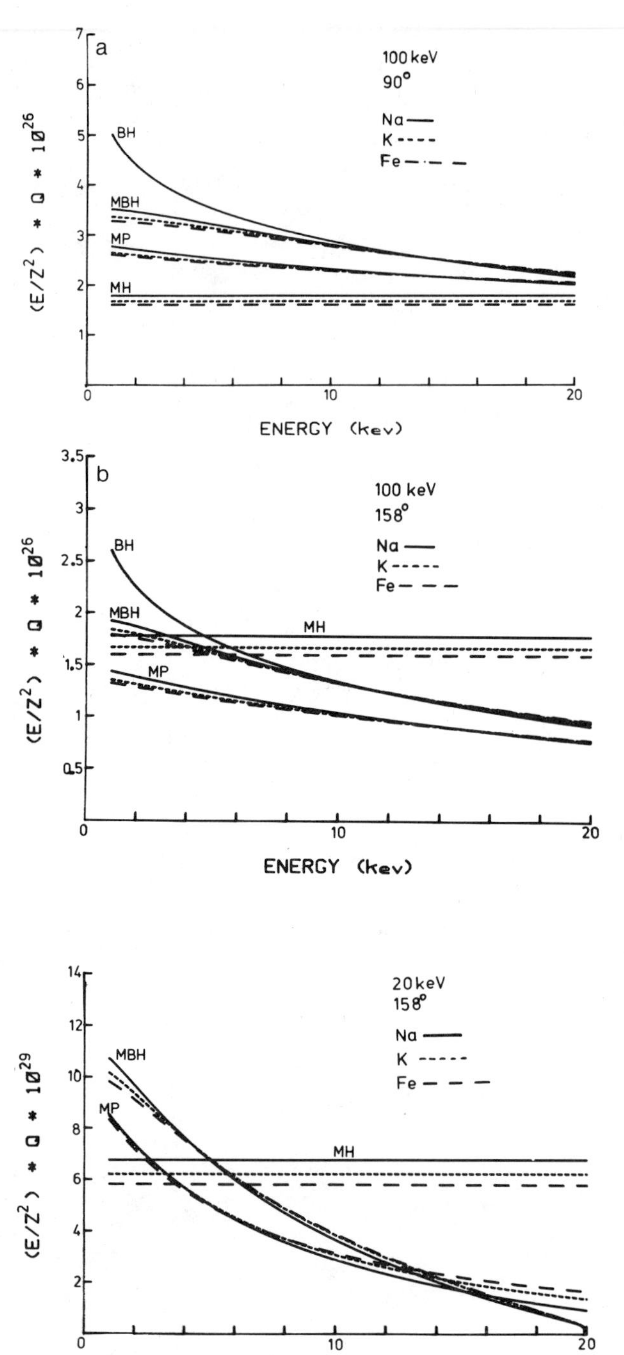

Figure 9 Plot of normalized continuum cross sections against photon energy for fixed beam energy (100 keV) and three atomic numbers (11, 19, 26). Note change in vertical scale and change in distribution shape. (a) 90° observation angle; (b) 158° observation angle.

Figure 10 Same as Figure 9(b), but 20-keV beam energy. The Born approximation is not valid in this energy region, and the MBH curve is included merely to demonstrate the "crossing" property.

VI. THE NOMINAL Z^2 AND 1/E DEPENDENCE OF CONTINUUM CROSS SECTIONS: THE "BEST" ENERGY AT WHICH TO MEASURE THE CONTINUUM

The accuracy of the Marshall-Hall method, when a standard of average atomic number different from that of the specimen is used, relies among other things on the ability of the continuum cross section to accommodate the change. While the major dependence of the continuum emission on average atomic number is predicted by the quantity

$$\sum_i C_i Z_i^2 / A_i$$

there is a nontrivial residual Z dependence. The Marshall-Hall cross section approximately accounts for the residual effect by the J factor in the logarithmic term in Eq. (13). The dependence, however, is uniform with photon energy, as can be seen from Figures 9(a) and 9(b). The Sommerfeld and modified Bethe-Heitler theories, on the other hand, predict residuals that are not uniform with photon energy and, indeed, reverse the Z dependence at a photon energy $\sim$14 keV for a 100-keV beam. (This occurs at $\sim$4 keV for a beam energy of 20 keV, Figure 10.) Figures 9(a) and 9(b) reveal that the two predictions of the residual Z dependence best agree for photon energies $<\sim$4 keV.

There has been considerable debate about the best energy at which to measure the continuum signal for the Marshall-Hall procedure. The choice is difficult and involves a consideration of statistics (there is more continuum signal at low photon energies); instrumental effects (where is there a minimum of spectral artifacts, overlaps, effects of scattered electrons, etc?); and physics (how accurate are the models?). Considering only the physics involved and assuming that the Sommerfeld and modified Bethe-Heitler predictions of the residual Z dependence are correct, it would seem that the optimal region to measure the continuum for the original Marshall-Hall procedure would be in the energy region below $\sim$4 to 5 keV. It should be noted that there are implementations of the Marshall-Hall procedure that neglect the logarithmic term of Eq. (13). For those implementations, Figures 9(a) and 9(b) indicate that the optimal region would be near a photon energy of 14 keV, for a 100-keV beam.

In principle, it is possible to extend the idea of the Hall procedure to the "one-standard" concept advocated by Russ (1974), at least over a restricted range of photon energy. In the one-standard method, the ratio of Eq. (1) and (10) are set equal to a measured characteristic and continuum signal obtained from a standard. By appropriate substitution of the quantities in the equations for another nearby element, it is possible to predict the ratio of intensities that would be emitted by a standard containing the new element. To do this, however, one should not use the cross sections in the procedure as originally formulated for lower beam energy applications, for two reasons: the complicated Z dependence of the continuum radiation and the error in the assumption of a 1/E dependence evidenced by an examination of Figures 6 through 10. Unless the continuum cross section can accommodate a change in measured photon energy, the measured interval must remain fixed at some energy during the course of an analysis. This is an unnecessary restriction, which will be discussed below. Note again that the shape of the generated continuum distribution, as predicted by both the Born and Sommerfeld theories, changes with observation angle. This is an important point to consider if one is to formulate a method generally useful for instruments of different manufacture with different observation angles. A relativistically corrected Kirkpatrick-Wiedmann equation or the modified Bethe-Heitler equation, for example, could be used to replace the original cross section of Marshall and Hall.

The exact choice of characteristic cross section for a one-standard extension of the Marshall-Hall procedure as applied to the AEM is not so critical since several of the various options have the same form for the restricted photon energy range over which they would be used, even though they differ significantly in absolute value.

VII. THE CHARACTERISTIC-TO-LOCAL-CONTINUUM RATIO: A WAY AROUND THE ABSORPTION CORRECTION

It is useful to define a ratio of the characteristic intensity to the continuum intensity in an energy interval, dE, centered at the energy of the characteristic line. Comparison of this ratio experimentally determined to a theoretical predication is extremely useful as an indicator of analytical suitability of a particular electron column-detector interface (Zaluzec, 1978, 1979; Nicholson, 1982). Any contribution to the measured ratio from beam-induced contamination, stray x-rays, etc., is apparent.

Furthermore, such a ratio can be useful as an analytical signal. For example, in the Hall procedure, "thicker" specimens may be analyzed since the distribution in depth of the characteristic and the local continuum radiation tends to be the same, and specimen self-absorption effects will cancel in the ratio. This is an important point and, alone, is sufficient justification for the use of an appropriate continuum cross section.

A standardized method of forming the ratio independent of spectrometer resolving power would permit the comparison of data obtained from different analytical microscopes. To make the ratio independent of the spectrometer resolution, an observed x-ray peak must be integrated over its full width, so as to provide a measure of the intensity of the originating x-ray line.

There are two ways to specify the denominator. One method is to use a continuum energy interval that would be as wide as the "natural" x-ray line. However, the width of the natural lines varies with the energy of the line and, furthermore, the reported widths are not in complete agreement, especially for the L and M x-ray lines. A second way is to settle on a fixed energy interval, such as 10 eV. This particular interval has two attributes: Most multichannel analyzers used with x-ray detectors on an AEM are calibrated for 10 eV per channel; and 10 eV is a value not much larger than the natural x-ray line widths. Consequently, a ratio formed with this value will be in the same order as the ratio obtained if the natural line width had been used. For the remaining discussion, we will use an energy interval of 10 eV.

VIII. THE THEORETICAL CHARACTERISTIC-TO-LOCAL-CONTINUUM RATIO

It is obvious from our examination of both the characteristic and continuum x-ray cross sections that the value of the ratio formed by dividing Eq. (1) by Eq. (10) will depend significantly on the choice of cross section used. Figures 11 and 12 display plots of the ratios obtained by using each of the four characteristic cross sections discussed, compared to both the MBH and Motz-Placious corrected Sommerfeld continuum cross sections. The ratios are plotted against atomic numbers from 11 to 37. The characteristic cross sections used are designated by the length of the dash, and the continuum cross sections are denoted with vertical braces. The beam energy used is 100 keV for 90° and 158° observation angles. The letter, P, is used to denote "characteristic," and the letter, B, is used to denote "continuum." Figures 13 and 14 are the same as 11 and 12, but a 200-keV beam is assumed. As mentioned, the use of measured values for the fluorescent yields and relative x-ray transition probabilities in the numerator causes the plots to exhibit irregularity. These plots are useful for comparison to measured characteristic-to-continuum ratios obtained from pure-element foils for the stated conditions.

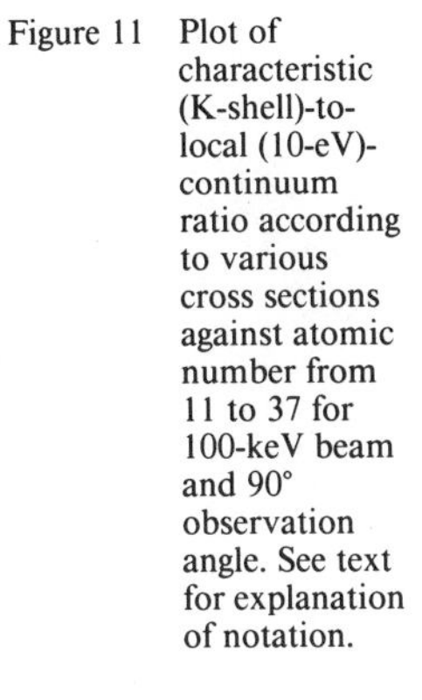
Figure 11 Plot of characteristic (K-shell)-to-local (10-eV)-continuum ratio according to various cross sections against atomic number from 11 to 37 for 100-keV beam and 90° observation angle. See text for explanation of notation.

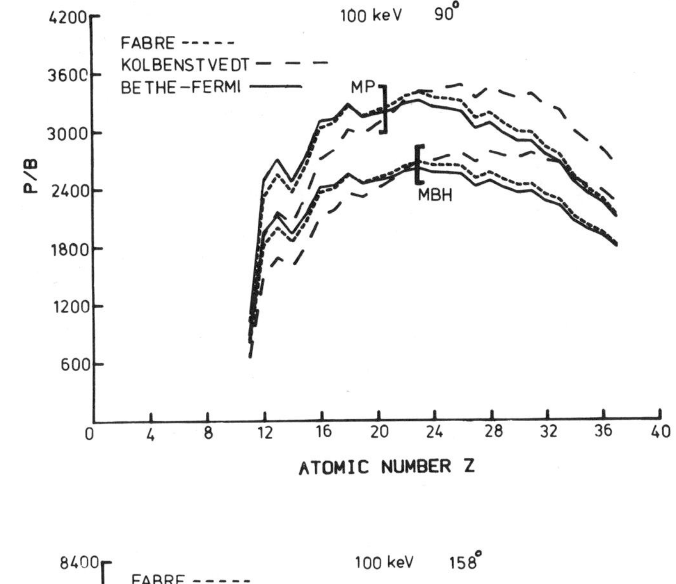

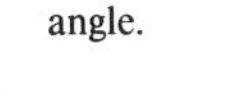
Figure 12 Same as Figure 11, but 158° observation angle.

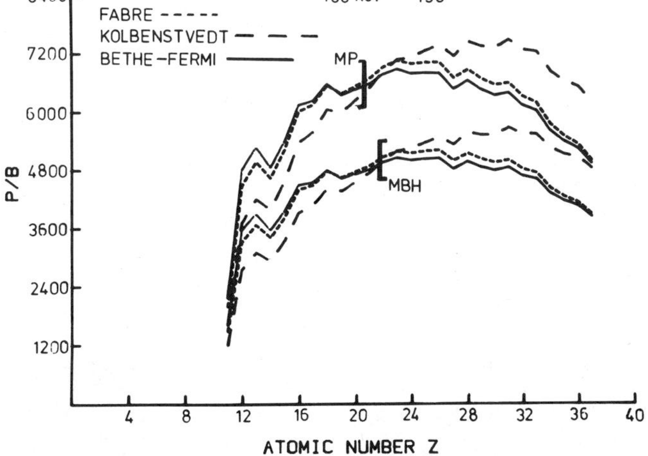

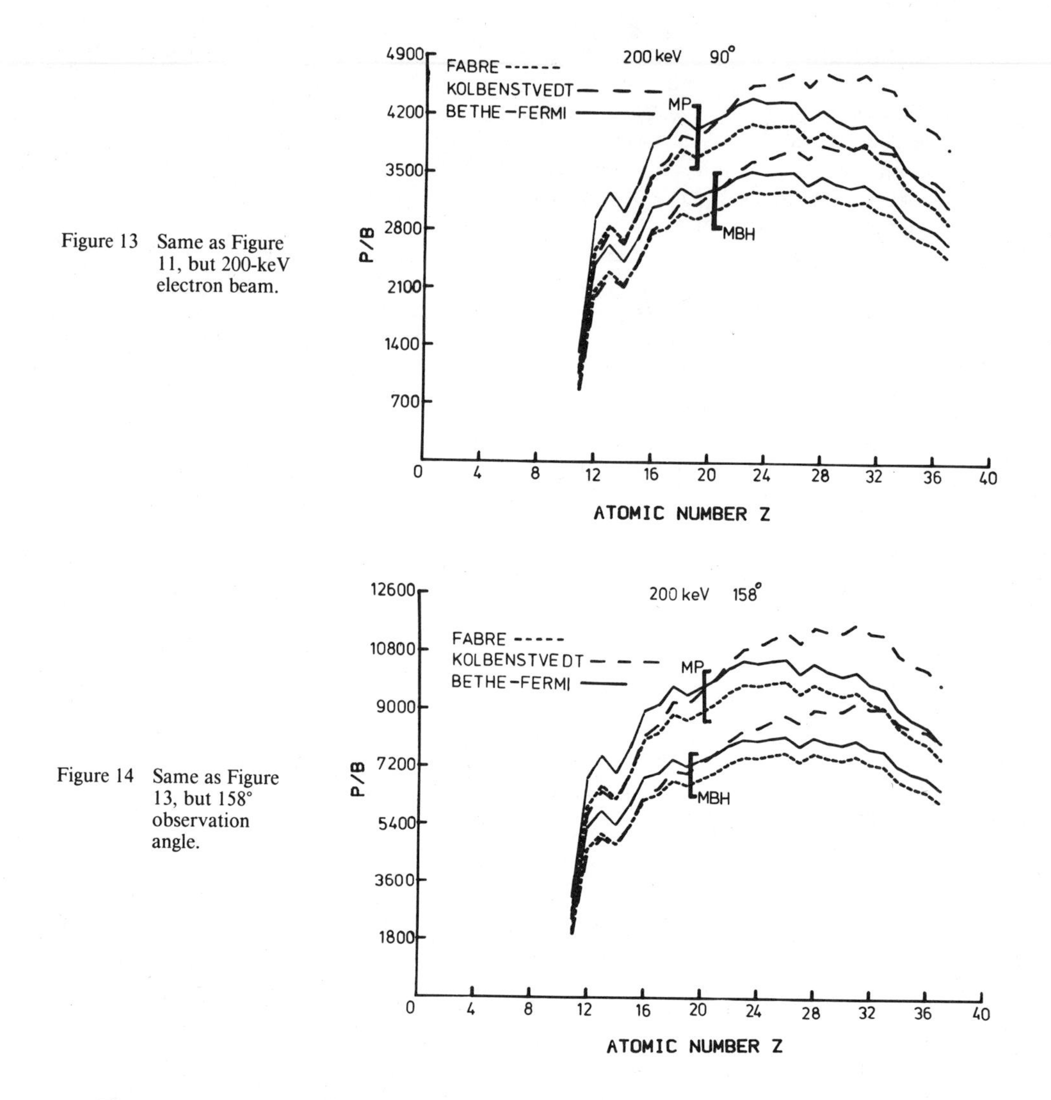

Figure 13 Same as Figure 11, but 200-keV electron beam.

Figure 14 Same as Figure 13, but 158° observation angle.

Figures 15 and 16 plot the characteristic-to-local-continuum ratio as a function of observation angle for two elements (Na and Fe) at a beam voltage of 100 keV for several of the cross sections discussed in this chapter. These plots reveal that the theoretical ratio increases with increasing observation angle. However, it is the experience of the authors that of the two commercial configurations, "side-looking" at either 90° or 120°, or "top-looking," where the x-rays are extracted through the objective lens polepiece at an angle of 158°, the side-looking configuration is the one most likely to approach the theoretical predictions. This is because the effects of scattered radiation are less. Furthermore, the specimen-detector distance is smaller in the side-looking case (~15 mm vs ~35 mm for a top-looking detector). Statistical considerations imply that any advantage for a higher characteristic-to-continuum ratio obtained with the higher observation angle is more than offset by the reduced collection efficiency. The plots also reveal the highly anisotropic nature of the

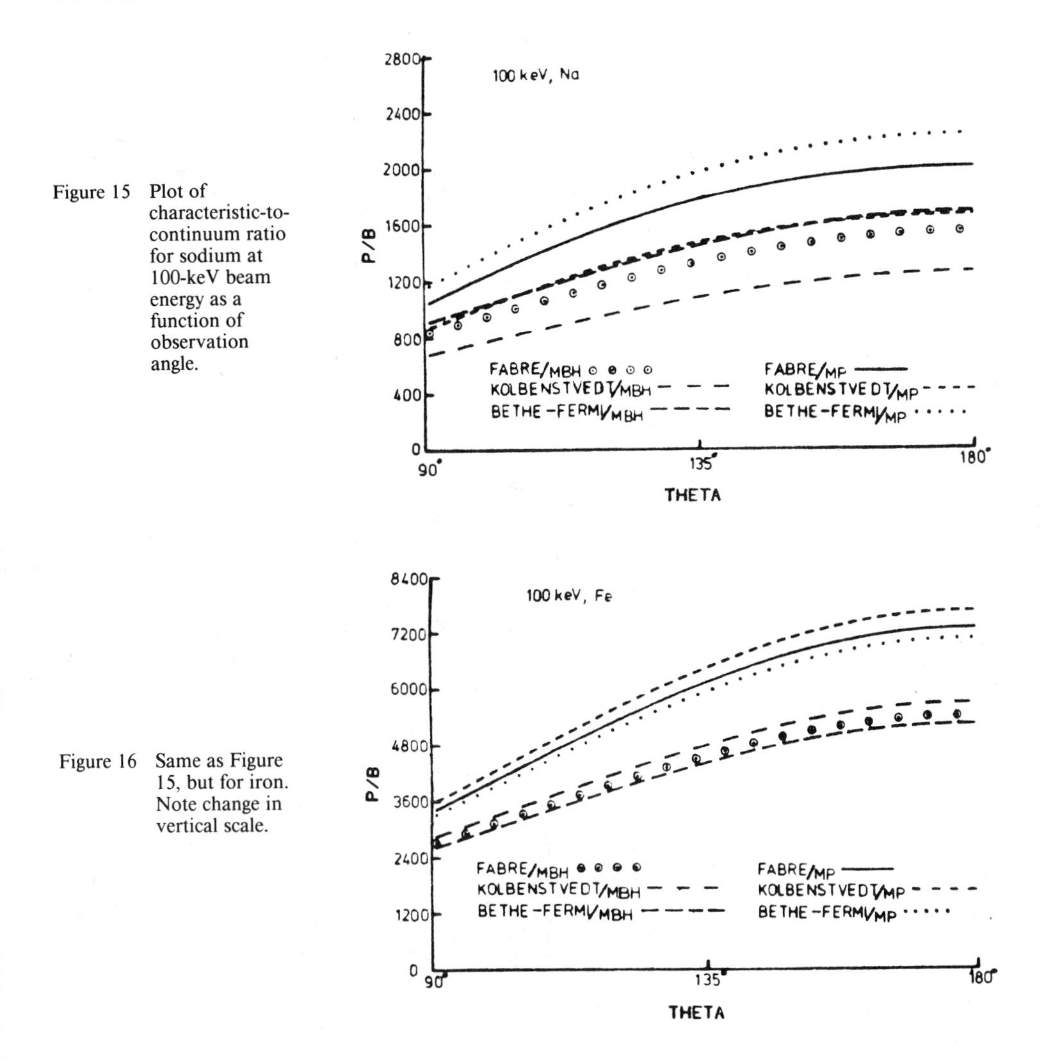

Figure 15 Plot of characteristic-to-continuum ratio for sodium at 100-keV beam energy as a function of observation angle.

Figure 16 Same as Figure 15, but for iron. Note change in vertical scale.

continuum radiation as a function of target atomic number. This is indicated by the change in the ratio of a given curve in the two plots as a function of angle. The degree of anisotropy depends strongly on the cross section.

IX. STATISTICAL PROPERTIES OF CHARACTERISTIC-TO-CONTINUUM RATIOS

Because the quantities desirable for analysis are not necessarily the same as those that are measurable, certain transformations are necessary to extract the former from the latter. Specifically, in a given energy band, an x-ray detector cannot determine whether an event was generated by a characteristic or a continuum process. However, there are standard methods whereby components P (peak or characteristic) and B (background or continuum) can be estimated from T (the total measured). The various ratios that can be made from these signals have attractive natural normalization properties. The purpose of this section is to derive certain statistical properties of these ratios.

We will assume in this discussion that our variables satisfy certain criteria and relationships forming a framework that is independent of particular values encountered. Variables used in our analysis are not the same as measurements made in the experimental context. Measurements are estimates to be related to the variables used. They need not be perfect to be useful, valid quantities, nor to establish estimates of uncertainty.

Suppose Total and Continuum signals are estimated from different spectral portions of a single acquisition by one of the standard procedures used for this purpose. The functions P/B = (T − B)/B, [T(S) − B(S)]/[T(1) − B(1)], and [P(S)/B(S)]/[P(1)/B(1)], are slightly different algebraic forms that all represent estimates of characteristic intensity with continuum correction. P( ) represents characteristic; B( ), continuum; and T( ), total signal; with P( ) = T( ) − B( ). S denotes specimen quantity, and 1 denotes standard quantity. If these functions are to be used with confidence, limits on their possible errors need to be established. It is accepted practice to represent an estimate as a mean value plus or minus a standard deviation. Other specifications, such as the confidence level of an obtained hypothesis, can be obtained by somewhat different treatment of the same data. The statistical principles and functional framework are not changed.

Assume that relationships among the mean, variance, and other moments of the variables composing these ratio functions are known. Statistical properties of functions of these variables can be derived from the following propagation-of-errors framework. Consider a general function of variables x, y, ... to avoid inferences that this framework is restricted to quotients of characteristic and continuum. Let f = f(x,y,...), and let subscript o indicate evaluation of a variable or a function at the mean estimate. Then the total differential of this function is

$$\Delta f = \left(\frac{\partial f}{\partial x}\right)_o \Delta x + \left(\frac{\partial f}{\partial y}\right)_o \Delta y + \left(\frac{\partial^2 f}{\partial x^2}\right)_o \frac{\Delta x^2}{2} + \left(\frac{\partial^2 f}{\partial x \partial y}\right)_o \Delta x\, \Delta y + \left(\frac{\partial^2 f}{\partial y^2}\right)_o \frac{\Delta y^2}{2} + \ldots \tag{27}$$

where $\Delta x = x - x_o$, $\Delta y = y - y_o$, etc.

In most cases of interest here, second- and higher-order terms should be negligible. They are, of course, well defined and easily calculated. The variance of this function over an ensemble is derivable in a straightforward manner.

$$\begin{aligned} \text{var}\,[f(x,y,\ldots)] &= \overline{(f - f_o)^2} \\ &= \overline{\left[\left(\frac{\partial f}{\partial x}\right)_o \Delta x + \left(\frac{\partial f}{\partial y}\right)_o \Delta y + \ldots\right]^2} \\ &= \overline{\left[\left(\frac{\partial f}{\partial x}\right)_o^2 (\Delta x)^2 + \left(\frac{\partial f}{\partial y}\right)_o^2 (\Delta y)^2 + 2\left(\frac{\partial f}{\partial x}\right)_o \left(\frac{\partial f}{\partial y}\right)_o (\Delta x\, \Delta y) + \ldots\right]} \\ &= \left[\left(\frac{\partial f}{\partial x}\right)_o^2 \text{var}\,(x) + \left(\frac{\partial f}{\partial y}\right)_o^2 \text{var}(y) + 2\left(\frac{\partial f}{\partial x}\right)_o \left(\frac{\partial f}{\partial y}\right)_o \text{cov}(x,y) + \ldots\right] \end{aligned} \tag{28}$$

The overbar indicates an appropriate ensemble average of a variable or function.

A normalized variance is often more meaningful than a raw one. The ratio of standard deviation to mean of a random variable is known as its coefficient of variation. By appropriate manipulation, squared coefficients of variation can be obtained

$$\frac{\mathrm{var}(f)}{f_o^2} = \frac{\left(\frac{\partial f}{\partial x}\right)_o^2}{\left(\frac{f}{x}\right)_o^2} \frac{\mathrm{var}(x)}{x_o^2} + \frac{\left(\frac{\partial f}{\partial y}\right)_o^2}{\left(\frac{f}{y}\right)_o^2} \frac{\mathrm{var}(y)}{y_o^2}$$

$$+ \frac{2\left(\frac{\partial f}{\partial x}\right)_o \left(\frac{\partial f}{\partial y}\right)_o}{\left(\frac{f}{x}\right)_o \left(\frac{f}{y}\right)_o} \frac{\mathrm{cov}(x,y)}{x_o\, y_o} + \ldots \tag{29}$$

For example, when $f(x,y) = x/y$, then

$$\mathrm{var}(x/y) = (1/y)_o^2\,\mathrm{var}(x) + (-x/y^2)_o^2\,\mathrm{var}(y) + 2\,(1/y)_o\,(-x/y^2)_o\,\mathrm{cov}(x,y) + \ldots$$

This may be written in the symmetric form

$$\frac{\mathrm{var}(x/y)}{(x/y)_o^2} = \frac{\mathrm{var}(x)}{x_o^2} + \frac{\mathrm{var}(y)}{y_o^2} - \frac{2\,\mathrm{cov}(x,y)}{x_o y_o} + \ldots \tag{30}$$

Note that, to first order, the normalized variances add, while the normalized covariances subtract (covariances will exist, for example, if the average effects of either the characteristic or continuum signals are incorrectly subtracted). Therefore, if the factors of a quotient are statistically independent, the variation of a quotient is the sum of the variations of its components. If the factors are completely dependent, e.g., y is a multiple of x, then the variation of the quotient is zero.

Letting $x = P$ and $y = B$ gives

$$\frac{\mathrm{var}(P/B)}{(P/B)_o^2} = \frac{\mathrm{var}(P)}{P_o^2} + \frac{\mathrm{var}(B)}{B_o^2} - \frac{2\,\mathrm{cov}(P,B)}{P_o B_o} + \ldots \tag{31}$$

For a pure Poisson process, which is a good approximation for both characteristic and continuum count generation in electron-beam microanalysis, if P and B are estimated separately,

$$\mathrm{var}(P) = P_o\,, \text{ and } \mathrm{var}(B) = B_o\,. \tag{32}$$

If the samples are independent, then

$$\mathrm{cov}(P,B) = 0\,. \tag{33}$$

This is also a good approximation for correctly estimated characteristic and continuum since each is the sum of variables that are often assumed to be mutually independent. Nonzero covariances can be introduced into this formulation by miscalculation of either component. However, in many cases, P is calculated as the difference between T and B. Then, one must be careful about variances of differences. From the basic variance equation, one obtains

$$\mathrm{var}(x - y) = \mathrm{var}(x) + \mathrm{var}(y) - 2\,\mathrm{cov}(x,y)\ . \tag{34}$$

Thus the variance of the ratio, P/B, where P = T − B, with T and B Poisson, is

$$\frac{\mathrm{var}[(T-B)/B]}{[(T-B)/B]_o^2} = \frac{\mathrm{var}(T-B)}{(T-B)_o^2} + \frac{\mathrm{var}(B)}{B_o^2} - \frac{2\mathrm{cov}[(T-B),B]}{(T-B)_o B_o}$$

$$= \frac{T_o^2}{(T-B)_o^2}\left[\frac{\mathrm{var}(T)}{T_o^2} + \frac{\mathrm{var}(B)}{B_o^2} - \frac{2\mathrm{cov}(T,B)}{T_o B_o}\right]. \tag{35}$$

In this case, the covariances are necessary to keep track of variances introduced by algebraic operations.

For the ratio [T(S) − B(S)]/[T(1) − B(1)], we obtain the following squared coefficient of variation.

$$\frac{\mathrm{var}\{[T(S)-B(S)]/[T(1)-B(1)]\}}{\{[T(S)-B(S)]/[T(1)-B(1)]\}_o^2} = \frac{\mathrm{var}[T(S)-B(S)]}{[T(S)-B(S)]_o^2} + \frac{\mathrm{var}[T(1)-B(1)]}{[T(1)-B(1)]_o^2} - \frac{2\mathrm{cov}\{[T(S)-B(S)]\,,[T(1)-B(1)]\}}{[T(S)-B(S)]_o\,[T(1)-B(1)]_o}$$

$$= \frac{\mathrm{var}[T(S)] + \mathrm{var}[B(S)]}{[T(S)-B(S)]_o^2} + \frac{\mathrm{var}[T(1)] + \mathrm{var}[B(1)]}{[T(1)-B(1)]_o^2} - \frac{2\mathrm{cov}[T(S),B(S)]}{[T(S)-B(S)]_o^2} - \frac{2\mathrm{cov}[T(1),B(1)]}{[T(1)-B(1)]_o^2} - \frac{2\mathrm{cov}\{[T(S)-B(S)]\,,[T(1)-B(1)]\}}{[T(S)-B(S)]_o\,[T(1)-B(1)]_o} \tag{36}$$

A similar, slightly more complicated result can be written for the ratio

$$\{[T(S) - B(S)]/B(S)\}/\{[T(1) - B(1)]/B(1)\}\ .$$

The first portion of this result, containing variances obtained by propagation of errors, is shown in Heinrich (1981). A different approach, oriented toward determining counting times necessary for appropriate levels of confidence, is reported by Ancey *et al.* (1977). An expanded development is given in Ellis and Pun (1986).

X. SUMMARY

The characteristic-to-continuum ratio has several desirable qualities for applications in electron-beam microanalysis. Meaningful use of this ratio requires understanding its properties. Discussion of the characteristic/continuum ratio leads naturally to separate consideration of each component.

Each component depends critically on the appropriate cross section. Several models for cross sections have been proposed for the range of energies used in the AEM. We have displayed their substantially different functional forms. Regions of validity of the models and extrapolation beyond these regions have been discussed, with an emphasis on the implications for biological x-ray microanalysis at AEM energies.

Major sources of error for each cross-section model and variable region were examined. A general, usable, statistical framework for determining the estimation uncertainty of the ratio from the data was presented.

TABLE OF CHAPTER VARIABLES

A_A	Atomic weight of element A
B	Background or continuum signal in an x-ray spectrum
B( )	Continuum (background) signal, where parameter S denotes specimen quantity and 1 denotes standard quantity
c	Speed of light, in cm/s
C_A	Weight fraction of analyte in the analyzed volume
$d\Omega$	Differential of solid angle subtended by the detector at angle θ
dE	Differential of photon energy
dV	Energy loss of electrons traversing a thin film of matter
dz	Target thickness in the same units of length as used in Q and ρ (cm in this chapter)
e	Charge of an electron
E_t	Total radiated continuum energy
F_A	Probability of emission of the x-ray line of interest relative to all the lines that can be emitted because of ionization of the same shell
I_{ch}	Number of characteristic x-rays generated into 4π steradians
I_{co}	Number of continuum x-rays generated into a specified solid angle at a specified observation angle
I	Ionization energy in units of the electron rest energy
J	Mean excitation energy of an atom
k	Energy of the emitted photon, in m_0c^2 units
n_s	Number of electrons that populate the sth shell or subshell of interest ($n_K = 2$)
N_e	Number of electrons that have irradiated the target during the measurement time
N_o	Avogadro's number $= 6.02 \cdot 10^{23}$
p_o, p	Initial and final momentum of the electron in a collision, in m_0c units
P	Peak or characteristic signal in an x-ray spectrum
P( )	Characteristic signal, where parameter S denotes specimen quantity and 1 denotes standard quantity
Q_A	X-ray cross section, either characteristic or continuum, depending on context; it has dimensions of area
r_o	Classical electron radius; $2.82 \cdot 10^{-13}$ cm
T	Total measured characteristic-plus-continuum signal in a spectrum
T( )	Total signal, characteristic plus continuum, with P( ) = T( )−B( ), where parameter S denotes specimen quantity and 1 denotes standard quantity
U	Overvoltage ratio V_o/V_c
v	Velocity of an electron, in cm/s
V_c	Energy required to remove an electron from a given shell or subshell (critical excitation energy)
V_o, V	Initial and final total energy of the electron in a collision, in m_0c^2 units
V_o'	Kinetic energy of the incident electron, in units of the electron rest energy
Z	Atomic number
Z_i	Atomic number of element i in a multi-element specimen

β Relativistic correction factor = v/c = $\sqrt{1-[1+(V_o/511)]^{-2}}$
θ Observation angle of the detector relative to the electron beam axis
ρ Density of the specimen in the analyzed volume
ω_A Fluorescent yield; the probability that an x-ray will be emitted because of the ionization of a given shell

REFERENCES

Ancey, M.; Bastenaire, F.; and Tixier, R. (1977), Proc. 8th Intl. Cong. X-Ray Optics and Microanalysis, Boston, pp. 49-56.

Bambynek, W., *et al.* (1972), Rev. Mod. Phys. *44*, 716.

Bethe, H. (1930), Annl. Phys. (Leipzig) *5*, 325.

—— (1933), Handb. Phys. *24*, 273, Springer-Verlag, Berlin.

Bethe, H., and Fermi, E. (1932), Zeit. f. Physik *77*, 296.

Chapman, J. N.; Gray, C. C.; Robertson, B. W.; and Nicholson, W. A. P. (1983), "X-Ray Production in Thin Films by Electrons With Energies Between 40 and 100 keV (I) Bremsstrahlung Cross Sections," X-Ray Spectrometry *12*(4), 153-62.

Compton, A. H., and Allison, S. K. (1935), "X-Rays in Theory and Experiment," 2nd ed., Van Nostrand, New York.

Dyson, N. A. (1959), Proc. Phys. Soc. *LXXII*(6), 924-36.

Ellis, J. R., and Pun, T. (1986), J. Micros. Accepted for publication.

Elwert, G. (1939), Annl. Physik *34*, 178.

Ferrier, R. P.; Chapman, J. M.; and Robertson, B. W. (1976), in "Electron Microscopy 1976" (D. G. Brandon, ed.), Tal Intl. Pub. Co., Jerusalem, vol. 1, pp. 411-13.

Fiori, C. E.; Swyt, C. R.; and Ellis, J. R. (1982), in "Microbeam Analysis - 1982" (K. F. J. Heinrich, ed.), San Francisco Press, San Francisco.

Gluckstern, R. L., and Hull, M. H. (1953), Phys. Rev. *90*, 1030.

Hall, T. A. (1977), in "Microbeam Analysis in Biology" (C. Lechenne and R. Warner, ed.), Acad. Press, New York.

Hall, T. A., *et al.* (1966), in "The Electron Microprobe" (McKinley, Heinrich, and Witry, ed.), John Wiley and Sons, New York, p. 805.

Heinrich, K. F. J. (1981), "Electron Beam X-Ray Microanalysis," Van Nostrand Reinhold, New York.

Heinrich, K. F. J.; Fiori, C. E.; and Myklebust, R. L. (1979), J. Appl. Phys. *50*(9), 5589.

Heitler, W. (1954), "The Quantum Theory of Radiation," 3rd ed., Oxford Univ. Press, New York.

Kirkpatrick, P., and Wiedmann, L. (1945), Phys. Rev. *67*(11 and 12), 321-39.

Koch, H. W., and Motz, J. W. (1959), Rev. Mod. Phys. *31*(4), 920-55.

Kolbenstvedt, H. (1967), J. Appl. Phys. *38*(12), 4785-87.

—— (1975), J. Appl. Phys. *46*, 2771.

Krause, M. O. (1979), J. Phys. Chem. Ref. Data *8*(2).

Krause, M. O., and Oliver, J. H. (1979), J. Phys. Chem. Ref. Data *8*(2), 329-38.

Kulenkampff, H.; Sheer, M.; and Zeitler, E. (1959), Zeit. f. Physik *157*, 275.

Marshall, D. J., and Hall, T. (1966), in "X-Ray Optics and Microanalysis" (Castaing, Descamps, and Philibert, ed.), Hermann, Paris, p. 374.

Mott, N. F., and Massey, H. S. W. (1965), "The Theory of Atomic Collisions," 3rd ed., Oxford Univ. Press, Oxford.

Motz, J. W., and Placious, R. C. (1958), Phys. Rev. *109*(2), 235-42.

Newbury, D. E.; Williams, D. B.; Goldstein, J. I.; and Fiori, C. E. (1984), in "Analytical Electron Microscopy - 1984" (D. B. Williams and D. C. Joy, ed.), San Francisco Press, San Francisco, pp. 276-78.

Nicholson, W. A. P., *et al.* (1982), J. Micros. *125*, pt. 1, pp. 25-40.

Powell, C. J. (1976a), Rev. Mod. Phys. *48*, 33.

——— (1976b), in "Use of Monte Carlo Calculations in Electron Probe Microanalysis and Scanning Electron Microscopy" (Heinrich, Newbury, and Yakowitz, ed.), NBS Spec. Pub. 460, US Govt. Printing Office, Washington, DC, pp. 97-104.

Quarles, C. A. (1976), Phys. Rev. *A13*, 1278.

Robertson, B. W. (1979), PhD. Thesis, Univ. of Glasgow, UK.

Russ, J. C. (1974), Proc. 9th MAS Conf., paper 22.

Schreiber, T. P., and Wims, A. M. (1981a), Ultramicroscopy *6*, 323-34.

——— (1981b), "Microbeam Analysis - 1981" (R. H. Geiss, ed.), San Francisco Press, San Francisco, pp. 313-16.

Slivinsky, V. W., and Ebert, P. J. (1972), Phys. Rev. *A5*, 1581.

Sommerfeld, A. (1931), Annl. Physik *11*(5), 257.

——— (1950), "Wellenmechanik," Frederick Ungar, New York, ch. 7.

Statham, P. J. (1976), X-Ray Spect. *5*, 154-68.

Stephenson, S. T. (1967), in "Encyclopedia of Physics," Springer-Verlag, vol. XXX, pp. 337-69.

Weinstock, R. (1942), Phys. Rev. *61*, 585.

——— (1944), Phys. Rev. *65*, 1.

Williams, D. B.; Newbury, D. E.; Goldstein, J. I.; and Fiori, C. E. (1984), J. Micros. *136*, 209.

Williams, E. J. (1933), Proc. Roy. Soc. *139*, 163.

Zaluzec, N. J. (1978), "An Analytical Electron Microscope Study of the Omega Phase Transformation in a Zirconium-Niobium Alloy," PhD. Thesis, Univ. of Illinois, Urbana-Champaign.

——— (1979), in "Introduction to Analytical Electron Microscopy" (Hren, Goldstein, and Joy, ed.), Plenum, New York, p. 121.

INDEX

*Some authors prefer EDS; others, EDXS. They are the same thing.